AN ATLAS OF HISTOLOGY

Springer
New York
Berlin
Heidelberg
Barcelona
Hong Kong
London
Milan
Paris
Singapore
Tokyo

AN ATLAS OF HISTOLOGY

SHU-XIN ZHANG

With 309 Color Plates

Springer

Shu-xin Zhang, M.D., Ph.D.
Department of Anatomy and Neurobiology
University of Kentucky
Lexington, KY 40536, USA

Library of Congress Cataloging-in-Publication Data
Zhang, Shu-xin.
An atlas of histology/Shu-xin Zhang.
p. cm.
Includes index.
ISBN 0-387-94954-2 (pbk. : alk. paper)
1. Histology—Atlases.
QM557.Z44 1998
611' .018—dc21 97-47335

Printed on acid-free paper.

Production managed by Lesley Poliner; manufacturing supervised by Rhea Talbert.
Typeset by A Good Thing, Inc., New York, NY.
Printed and bound by R.R. Donnelley and Sons, VA.
Printed in the United States of America.

9 8 7 6 5 4 3 2 1

ISBN 0-387-94954-2 Springer-Verlag New York Berlin Heidelberg SPIN 10565638

Dedicated to
My wife, Fengfa,
My daughter, Wei, and
My son, Si (Jack)

Acknowledgments

An Atlas of Histology contains over 300 hand-drawn illustrations, that represent a huge amount of work, including designing, drawing, modifying, labeling, computer editing, writing and typing, which took me almost all the weekends and holidays in the last six years. Fortunately, I received support and assistance from many people during the preparation of this book, and I am indebted to all these individuals.

I am most grateful to Dr. Harumichi Seguchi of the Department of Anatomy and Cell Biology, Kochi Medical School, Japan, who kindly allowed me to use the microscope and histological preparations in his department as a reference.

I am deeply indebted to Dr. Natalie A. Connors of the Department of Anatomy and Neurobiology, St. Louis University School of Medicine, who enthusiastically reviewed my manuscript and provided many valuable suggestions.

I would particularly like to thank Dr. George A. Vogler of the Department of Comparative Medicine, St. Louis University School of Medicine, for his continuing support, suggestions, help and encouragement.

I wish to express my warmest thanks to Dr. David N. Menton of the Department of Anatomy and Neurobiology, Washington University School of Medicine, who reviewed my manuscript.

Special thanks are due to Mr. Wei Yun-shao of the Department of Histology and Embryology, Nanfang Medical College, China, who provided me with numerous valuable histological preparations; to Dr. M. Waheed Rana of the Department of Anatomy and Neurobiology, St. Louis University School of Medicine, for his encouragement and suggestions; also to Miss Christine Tanaka of St. Louis University School of Medicine, who conscientiously checked the whole manuscript for grammar.

I wish to express my gratitude for their help in the preparation of this manuscript to Dr. John M. Robinson, Dr. Thomas G. Hayes, Dr. Richard W. Burry and Ms. Ann Osterfeld of the Department of Cell Biology, Neurobiology and Anatomy, the Ohio State University College of Medicine; to Dr. Teruhiko Okada, Dr. Toshihiro Kobayashi and Ms. Rumi Yamanaka of the Department of Anatomy and Cell Biology, Kochi Medical School, Japan; to Dr. Vernon W. Fischer, Dr. Paul A. Young, Dr. Yunxi Tan and Mr. Norman Bamber of the Department of Anatomy and Neurobiology, St. Louis University School of Medicine; also to Dr. James W. Geddes, Dr. Don M. Gash, and Dr. Kurt F. Hauser of the Department of Anatomy and Neurobiology, University of Kentucky College of Medicine.

I express gratitude to my publishers, Dr. Robert C. Garber and Dr. William F. Curtis, and the staff at Springer-Verlag New York, Inc., for their support, cooperation, and editing. It was a pleasure to work with them.

Finally, I am extremely grateful to my wife, Fengfa, my daughter, Wei, and my son, Si (Jack) for their all-out support and forbearance. I must say I could not have completed this book without their support.

Lexington, Kentucky

Shu-xin Zhang
January 1999

Contents

List of Color Plates

Introduction

The beginning student of histology is frequently confronted by a paradox: diagrams in many books that illustrate human microanatomy in a simplified, cartoon-like manner are easy to understand, but are difficult to relate to actual tissue specimens or photographs. In turn, photographs often fail to show some important features of a given tissue, because no individual specimen can show all of the tissue's salient features equally well. This atlas, filled with photo-realistic drawings, was prepared to help bridge the gap between the simplicity of diagrams and the more complex reality of microstructure.

All of the figures in this atlas were drawn from histological preparations used by students in my histology classes, at the level of light microscopy. Each drawing is not simply a depiction of an individual histological section, but is also a synthesis of the key structures and features seen in many preparations of similar tissues or organs. The illustrations are representative of the typical features of each tissue and organ. The atlas serves as a compendium of the basic morphological characteristics of human tissue which students should be able to recognize.

Instead of using abbreviations and symbols, each drawing has been clearly labeled to aid students in recognizing the relevant structures, and is accompanied by a legend to provide a concise summary of the structural, functional and physiological features of the tissue or organ depicted. Features that are labeled in figures are highlighted as bold terms when they appear in the legend. The stain used in preparing each specimen is indicated, as is an approximate magnification ("low"= 20× to 100×, "medium"=100× to 400×, "high"=400× to 1200×). In nearly all cases the organ from which the tissue specimen was derived is identified, as is the organism. Most figures are drawn from human tissues, but in a number of cases the features of certain tissues are best revealed in another mammal. The tissue sources used in this atlas are the same as those commonly used in histology laboratory courses.

This atlas will be useful to students of a wide range of biological and medical disciplines including human and veterinary medicine, dentistry, mammalian biology, pharmacology, and nursing. I have tried my best to assure the accuracy of the illustrations and text in this book, however if any errors or omissions are noted, I sincerely welcome any criticisms and suggestion to help improve future editions of this book.

1. Epithelial Tissue

Epithelial tissues are cellular sheets, covering the surfaces and lining the tracts or cavities of the body. The functions of the epithelial tissues are associated with protection, secretion, and absorption. There are several common features.

1. The epithelial cells fit together tightly, with little intercellular substance.
2. They have a basement membrane, which separates the epithelial cells and the underlying connective tissue.
3. They have a free, or apical, surface and a basal surface. The free surface is adjacent to the lumen; the basal surface rests on the basement membrane.
4. They lack a vascular supply, excepting the stria vascularis in the cochlear duct.
5. They are rich in nerve endings.

The epithelial tissues are classified as:

Simple squamous epithelium
Simple cuboidal epithelium
Simple columnar epithelium
Ciliated pseudostratified columnar epithelium
Stratified squamous epithelium
Stratified columnar epithelium
Transitional epithelium

Fig. 1-1. Epithelium: Isolated Cells

Figure 1-1 shows isolated cells from various epithelial tissues that are different in size and shape. Some have specializations at the cell surface, such as the striated border of simple columnar epithelium and the cilia of ciliated pseudostratified columnar epithelial cells.

Pigment cells (Fig. 1-1A), which comprise the pigmented epithelium of the retina in the bovine eyeball, are hexagonal in shape and fit together. They are characterized by the presence of numerous brown, round or oval **pigment granules** in the cytoplasm. The portion of the cytoplasm where the **nucleus** is located shows less numerous pigment granules (see **Fig. 16-6**).

Melanocytes (Fig. 1-1B) are present in the choroid of the bovine eyeball (not those in the epidermis of the skin, as shown in **Fig. 15-3**). They are not epithelial cells, but it is convenient to study them with the pigmented cells from the eyeball. Melanocytes also contain numerous **pigment granules** and have a round **nucleus** located in the center of the cell, but their cell bodies are star-shaped, which differentiates them from the pigmented cells. The pigment granules both in pigmented cells and in melanocytes in the eyeball are capable of the absorption of light rays that pass through the retina (see **Fig. 16-6**).

Figure 1-1C shows cells from ciliated pseudostratified columnar epithelium of the rat trachea. The **ciliated cell** is columnar in shape with an oval nucleus. On the free surface of the cell there are numerous **cilia.** Electron microscopic studies have demonstrated that the cilia are elongated motile structures; in the center is one pair of microtubules surrounded by nine pairs of microtubules. In living cells, the back-and-forward movement of the cilia help remove materials such as dust granules and mucus from the cell surface. The **intermediate cell** is fusiform with an oval nucleus in the middle portion of the cell. The **basal cell** is triangular-shaped, and its nucleus is round or oval (see **Fig. 1-8**).

The **columnar cell,** comprising the bulk of the simple columnar epithelium in the small intestine (**Fig. 1-1D**), is tall or elongated and has an oval **nucleus** in the center of or toward the base of the cell. It shows on its free surface a **striated border** formed by closely packed, parallel microvilli, which can be recognized with electron microscopy, that greatly increase the absorptive area of the small intestine. The **goblet cell** is cuplike in shape with a small nucleus at the bottom of the "container". The goblet cells, which also occur in the ciliated pseudostratified columnar epithelium (see **Fig. 1-8**), secrete mucus, lubricating the epithelia.

The transitional epithelium (**Fig. 1-1E**) consists of basal cells, intermediate cells, and superficial cells. The **basal cells** are cuboidal in outline, with an oval or round nucleus located centrally. The **intermediate cells** are polygonal, with a round nucleus. The **superficial cells** are rounded, larger than other cells, and may contain two nuclei (not shown in this figure). The basal surface of the superficial cell shows several **indentations** that accommodate the underlying intermediate cells (see **Fig. 1-12** and **Fig. 1-13**).

Figure 1-1. Epithelium: Isolated Cells
Unstained • High magnification

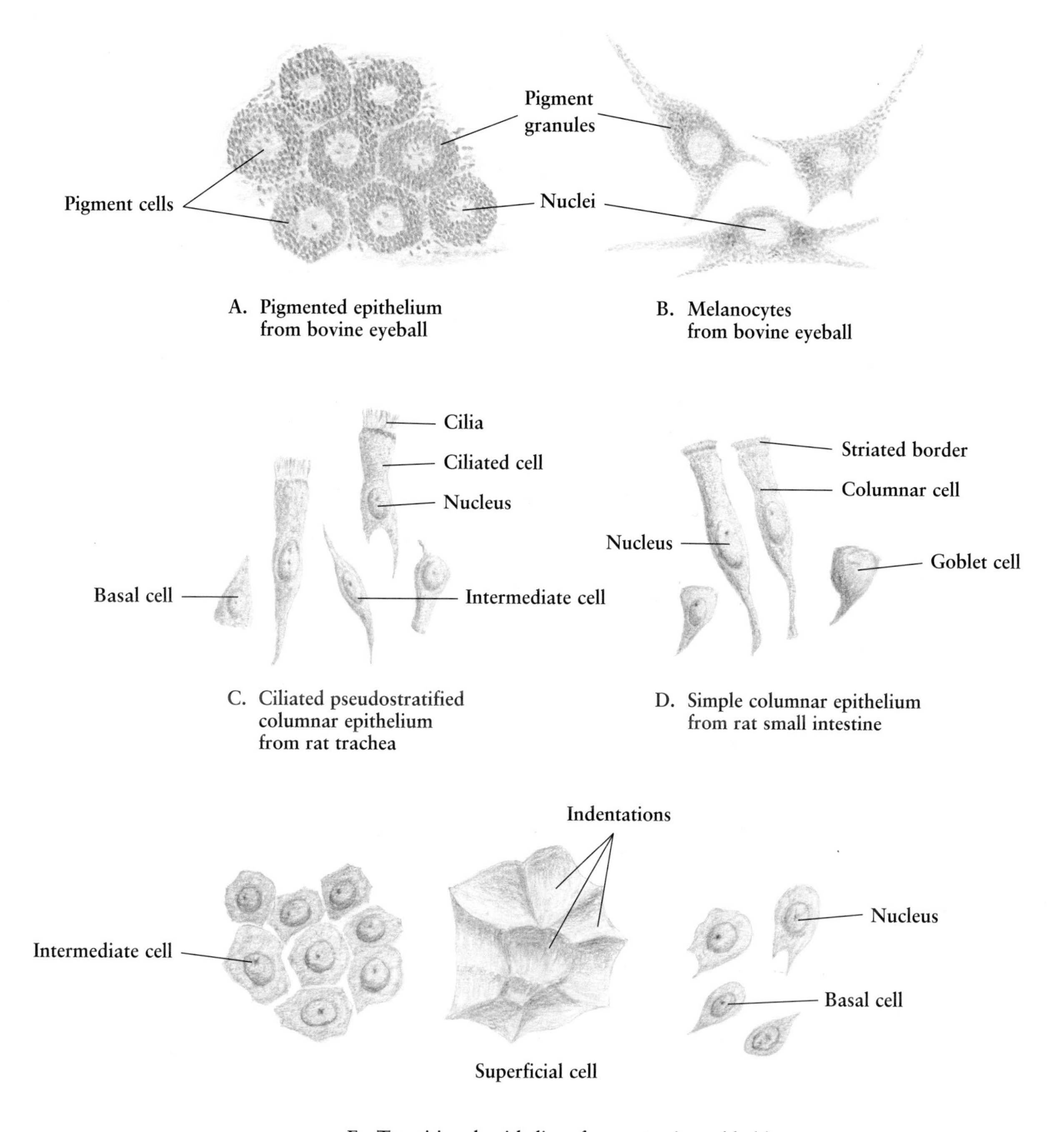

A. Pigmented epithelium from bovine eyeball

B. Melanocytes from bovine eyeball

C. Ciliated pseudostratified columnar epithelium from rat trachea

D. Simple columnar epithelium from rat small intestine

E. Transitional epithelium from rat urinary bladder

Figs. 1-2 and 1-3.

Simple Squamous Epithelium

The **simple squamous epithelium** consists of a thin layer of flattened epithelial cells and a basement membrane. It comprises the *endothelium,* the lining of blood and lymph vessels, and the *mesothelium,* the serous lining of pericardium, pleura, and peritoneum.

Figure 1-2 shows a surface view of **simple squamous epithelium** from the mesothelium of mesentery, demonstrating the **border** between polygonal **epithelial cells** outlined by silver impregnation. Note that the nuclei of epithelial cells are not visible by this method.

Figure 1-3 is a vertical section from epicardium in which the **mesothelial cells** are flattened. The portion with the nucleus is generally thicker than other parts of the cytoplasm. The **basement membrane** cannot be recognized easily in this epithelium. Beneath the epithelium and basement membrane is the **connective tissue,** containing **capillaries, fibroblasts,** and **pericytes.**

Figure 1-3 also demonstrates the **endothelial cells** lining a capillary. These cells have an oval nucleus and very thin cytoplasm, except for the portion containing the nucleus. The endothelial cells of the capillary are also supported by a thin layer of basement membrane. In addition, **pericytes,** which are considered to be undifferentiated cells, are often found surrounding the capillary. Under certain conditions, such as inflammation or injury, the pericytes are believed to develop into fibroblasts, participating in the wound healing process.

The simple squamous epithelium is involved in passive transport of gases, such as in the lung alveoli, and active transport of fluid by pinocytosis, such as in the endothelium and mesothelium. Mesothelial cells also secrete a small amount of fluid, lubricating the surface of the serosa.

Figure 1-2. Simple Squamous Epithelium
Surface of mesothelium • Mesentery spread • Rat • Silver impregnation • High magnification

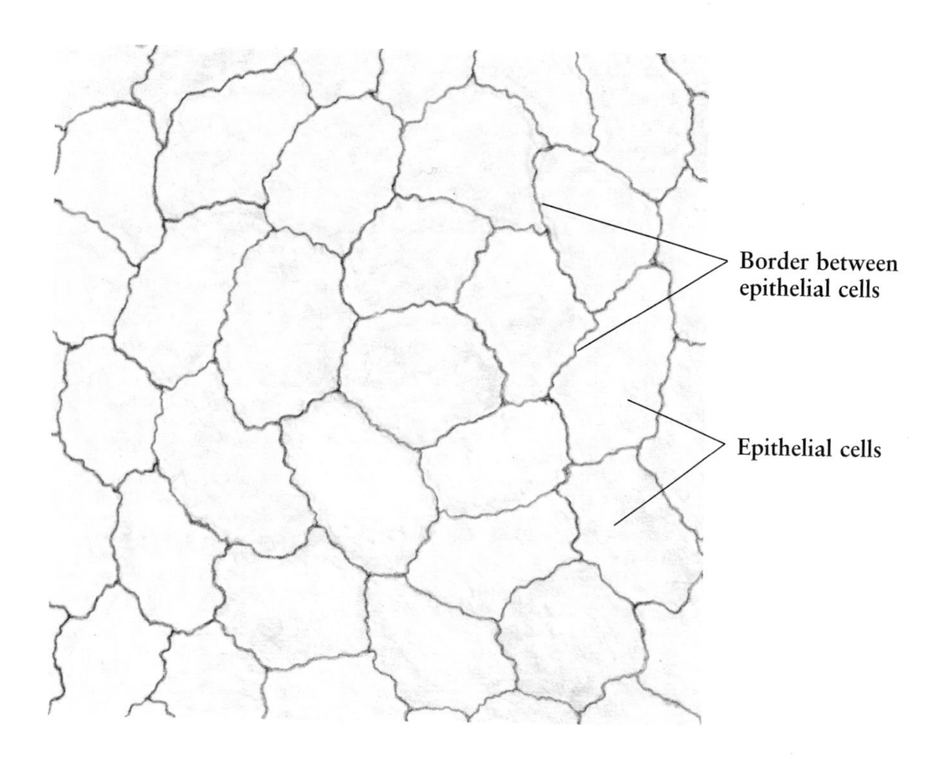

Figure 1-3. Simple Squamous Epithelium
Mesothelium and endothelium of epicardium • Human • H.E. stain • High magnification

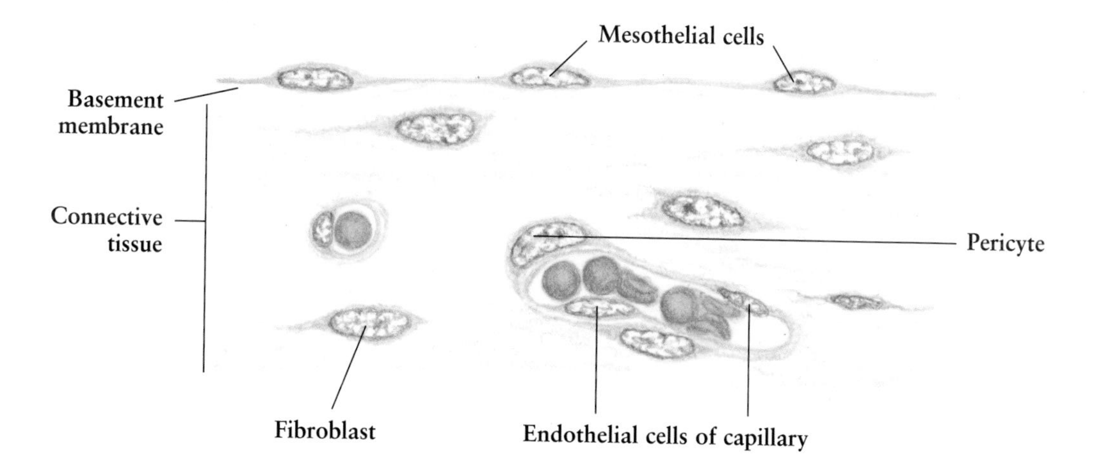

Figs. 1-4 and 1-5.

Simple Cuboidal Epithelium

Simple cuboidal epithelial cells are hexagonal when seen from the surface of the tissue they cover. In vertical sections, the cuboidal epithelial cell appears square with a round nucleus in the center. The cell boundaries are distinct. The **basement membrane** is often recognized. Simple cuboidal epithelium is often found in the proximal, distal, and **collecting tubules (Fig. 1-4)** of the kidney, follicles of the thyroid gland, and pigmented epithelium of the retina (see **Fig. 1-1**).

Figure. 1-5 demonstrates a **follicle** of the thyroid gland, which is lined with simple cuboidal epithelial cells. In the **lumen** of the follicle there is **colloid,** which is darkly stained by the eosin. Around the follicle is a thin **connective tissue** that contains **fibroblasts** and **capillaries.**

The function of the simple cuboidal epithelium is associated with secretion, such as in the follicles of the thyroid gland, and absorption, such as in the proximal, distal, and collecting tubules of the kidney and in the follicles of the thyroid gland.

Figure 1-4. Simple Cuboidal Epithelium
Collecting tubule of kidney • Human • M.G. stain • High magnification

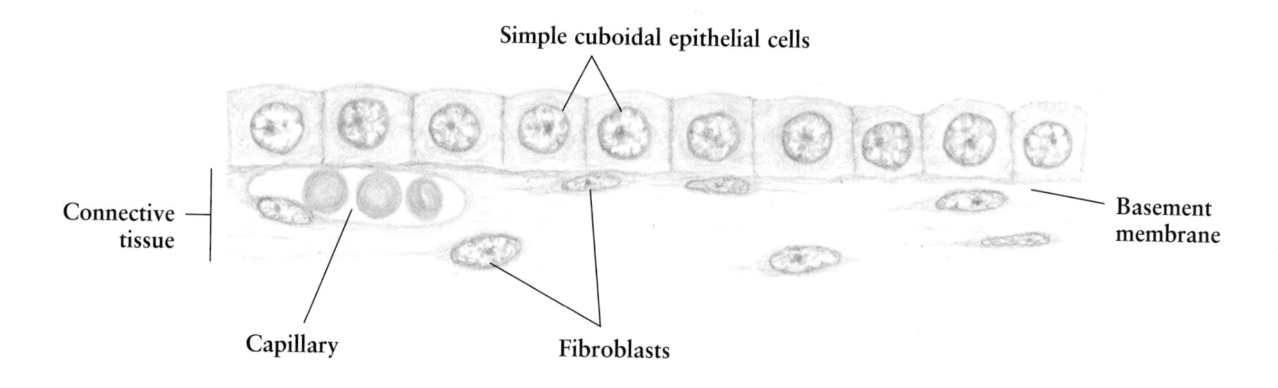

Figure 1-5. Simple Cuboidal Epithelium
Follicle of thyroid gland • Human • H.E. stain • High magnification

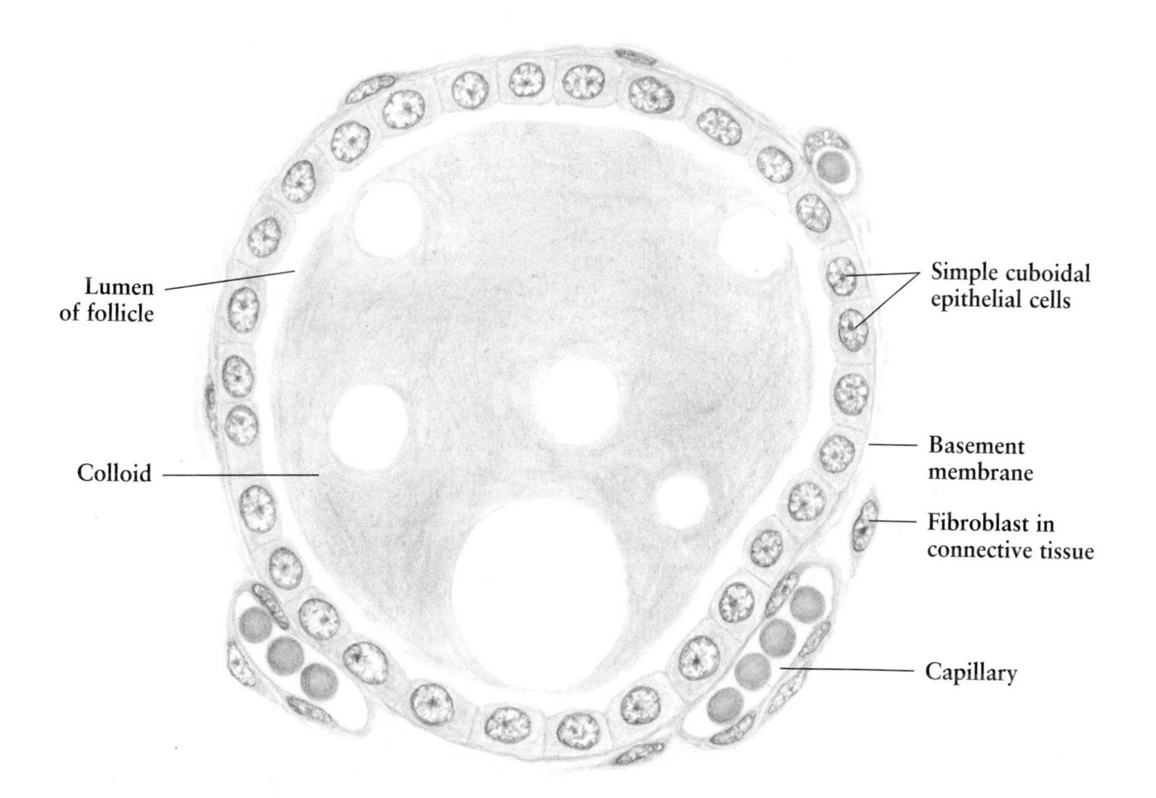

Figs. 1-6 and 1-7.

Simple Columnar Epithelium

Simple columnar epithelium most often lines highly absorptive surfaces such as in the intestine, and is responsible for absorption and secretion. It is composed of a single layer of slender prismatic cells, the columnar epithelial cells (see **Fig. 1-1D**).

In the vertical section shown in **Fig. 1-6,** the **columnar epithelial cell** appears taller than its width. The elongated nucleus is often located near the base of the cell. These cells have clear cytoplasm and visible boundaries. Touching the **basement membrane,** the epithelial cells are connected by junctional complexes, which can be recognized under the electron microscope . In the mucosa of the digestive tract, from duodenum to rectum, there is a **striated border** on the free surface of the epithelial cells. Electron microscopy has shown that the striated border is composed of numerous **microvilli,** which greatly increase in the absorptive area of the small intestine.

Scattered among the columnar cells are the **goblet cells (Fig. 1-6),** which are cup- or goblet-shaped in outline. The cytoplasm of goblet cells is filled with mucous secretion that lubricates and protects the epithelium. The **nucleus** is flattened or cuplike and located in the basal part of the cell. The goblet cells are found in the epithelium of the small and large intestines, nasopharynx, larynx, trachea, bronchi, and conjunctiva.

Figure 1-7 shows the simple columnar epithelium of the gallbladder. The **epithelial cells** are provided on the free surface with **short microvilli,** which do not form the striated border seen in the columnar cells of the intestine. The columnar epithelial cells of the gallbladder are involved in the absorption of water from the bile, which thus becomes concentrated. Note that no goblet cells can be observed among the epithelial cells of the gallbladder.

Figure 1-6. Simple Columnar Epithelium
Human duodenum • H.E. stain • High magnification

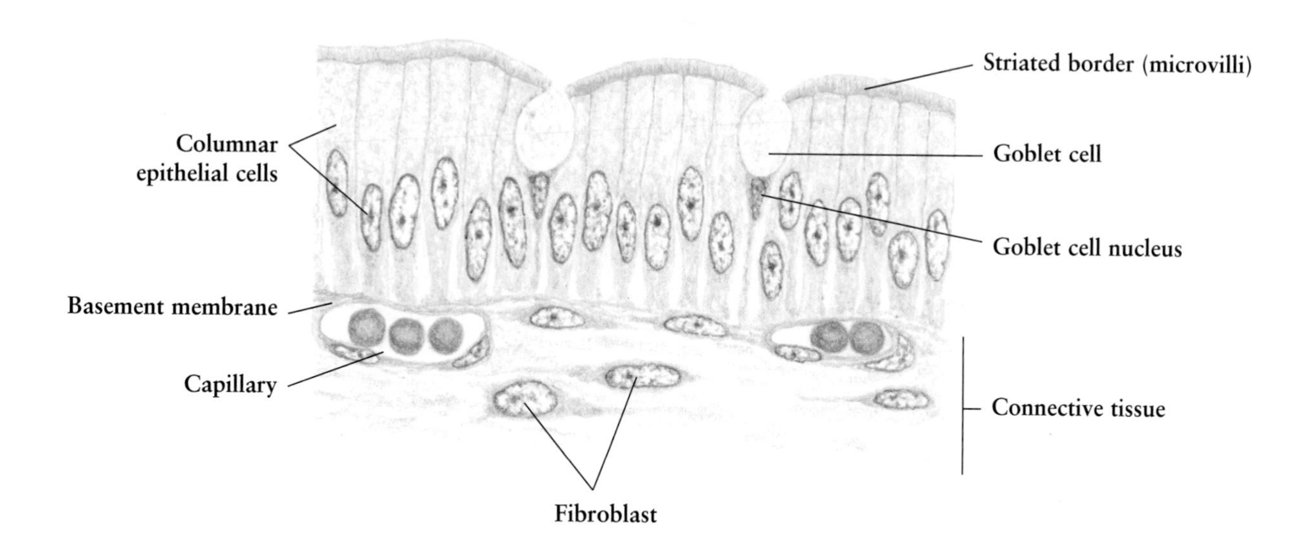

Figure 1-7. Simple Columnar Epithelium
Human gallbladder • H.E. stain • High magnification

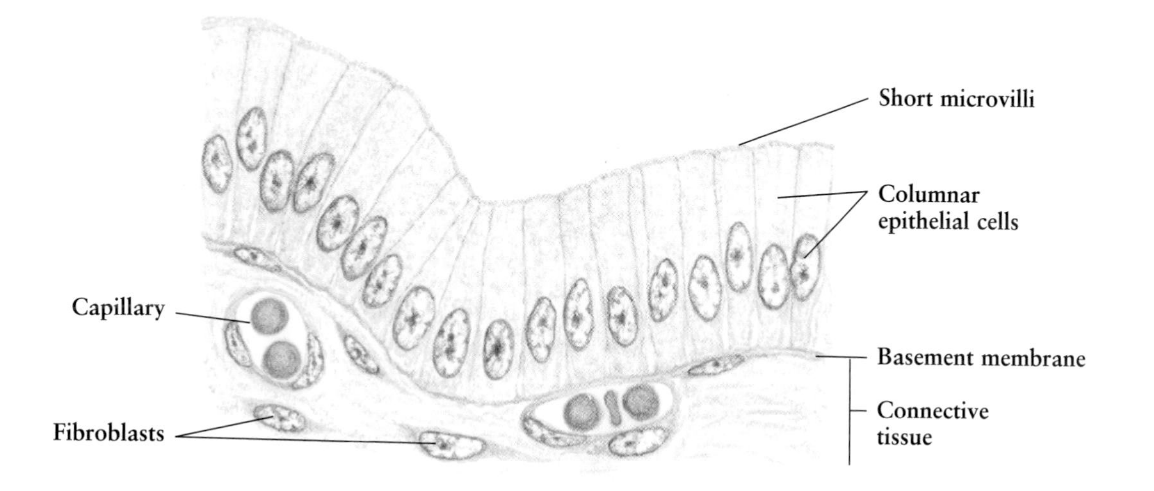

Fig. 1-8. Ciliated Pseudostratified Columnar Epithelium

Ciliated pseudostratified columnar epithelium, distributed in the respiratory tract, consists of columnar ciliated cells, conical intercalated (intermediate) cells, and basal cells (see **Fig. 1-1C**).

The **columnar ciliated cells** are tall and are provided on the free surface with **cilia.** The cilia undergo a rapid forward beat with a slower recovery stroke, thus moving material such as mucus in one direction along the surface of the epithelium. The **conical intercalated (intermediate) cells** are fusiform, without reaching the free surface of the epithelium. They are considered to be in course of differentiation into the ciliated cells. The **basal cells** are triangular in shape, capable of differentiating into other cell types of the epithelium. All the cells rest on the **basement membrane,** which is more prominent and thicker than that in other epithelia. The nuclei of the epithelial cells are usually round or oval, varying in shape according to the shapes of the cells to which they belong. They lie at different levels, which therefore gives the tissue the appearance of a multiple cell layers, hence "pseudostratified".

The **goblet cells,** responsible for the secretion of mucus, are usually scattered among the columnar ciliated cells. In the underlying **connective tissue** adjacent to the epithelium, a few **plasma cells** or **macrophages** can be found in addition to **fibroblasts.**

Fig. 1-9. Stratified Columnar Epithelium

Stratified columnar epithelium is composed of truly stratified cells: superficial columnar cells, intermediate cells, and basal cells. The **columnar cells** are tall or low columnar, forming the outmost layer among which **goblet cells** can be found; **intermediate cells** are fusiform, and **basal cells** are triangular in outline. The shapes of nuclei vary with the shapes of cells in which they reside. Only basal cells touch the **basement membrane.** The stratified columnar epithelium lines the conjunctival fornices, spongy part of the male urethra, ducts of sweat glands, and the stria vascularis. Its function is associated with protection or secretion.

In addition to **fibroblasts,** some **plasma cells** and **lymphocytes** may be seen in the **connective tissue** under the basement membrane.

Figure 1-8. Ciliated Pseudostratified Columnar Epithelium
Human trachea • H.E. stain • High magnification

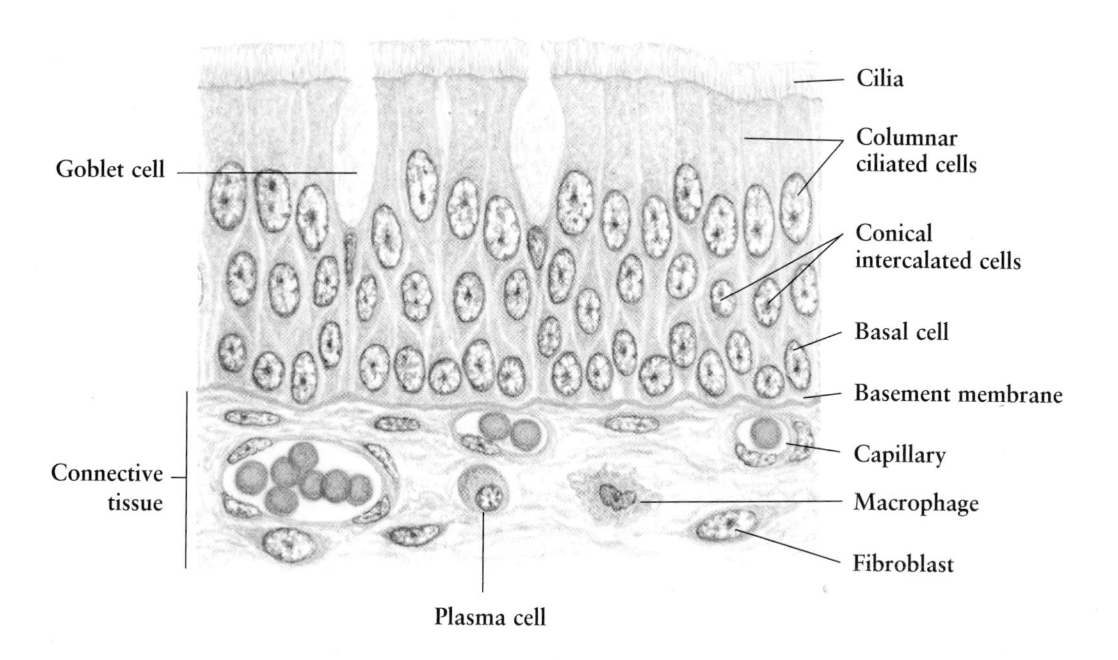

Figure 1-9. Stratified Columnar Epithelium
Conjunctiva of human eyelid • H.E. stain • High magnification

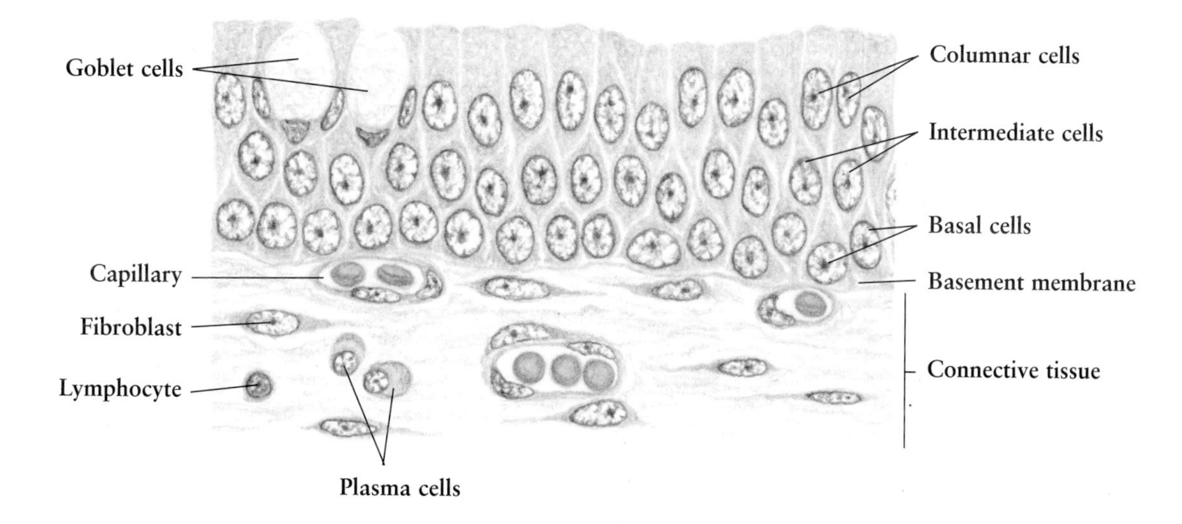

Figs. 1-10 and 1-11.

Stratified Squamous Epithelium

The cells of **stratified squamous epithelium** are arranged in several layers, in which only the deepest cells contact the basement membrane. In vertical sections, the **basal cells** in the deepest layer are small and columnar or cuboidal in shape, the **intermediate cells** in the middle layer become larger and **polyhedral,** and the **superficial cells** in the surface layers gradually become flattened. The shapes of their nuclei also change, from oval in the deepest layers to round in intermediate layers to flattened in surface layers. The **basement membrane** appears uneven when observed under a low-power microscope. In addition, some other cell types, such as **melanocytes,** Merkel's cells, and Langerhans' cells, may be found in stratified squamous epithelium (see **Fig. 15-3**).

The stratified squamous epithelium functionally forms the most protective type of lining in the body and is usually subdivided into two types: keratinized and nonkeratinized. The **keratinized epithelium (Fig. 1-11)** is characterized by the fact that the superficial cells become filled with keratin and die to form a **stratum corneum.** The keratinized epithelium comprises the epidermis covering the whole outer surface of the body, and also occurs in the epithelial lining of the gingiva, filiform papillae of the tongue, and part of the hard palate of the oral cavity. In contrast, the **nonkeratinized epithelium (Fig. 1-10)** contains no keratinized cells, and all cells are viable. They are distributed chiefly in the linings of the majority of the oral cavity, pharynx, esophagus, vagina, and cornea. Actually, all stratified squamous epithelium contains filamentous proteins of the keratin type, but the term "keratinized epithelium" has long been used to described only stratified squamous epithelium forming a layer of dead keratin-filled keratinocytes.

Beneath the epithelium is the **connective tissue,** containing **capillaries, fibroblasts,** and collagen fibers. In the oral cavity, **lymphocyte** and **plasma cells** are often found in the connective tissue. In thick skin, the connective tissue projects to the epithelium (epidermis), forming the dermal papillae.

Figure 1-10. Stratified Squamous Epithelium: Nonkeratinized
Epithelium of human oral cavity • H.E. stain • High magnification

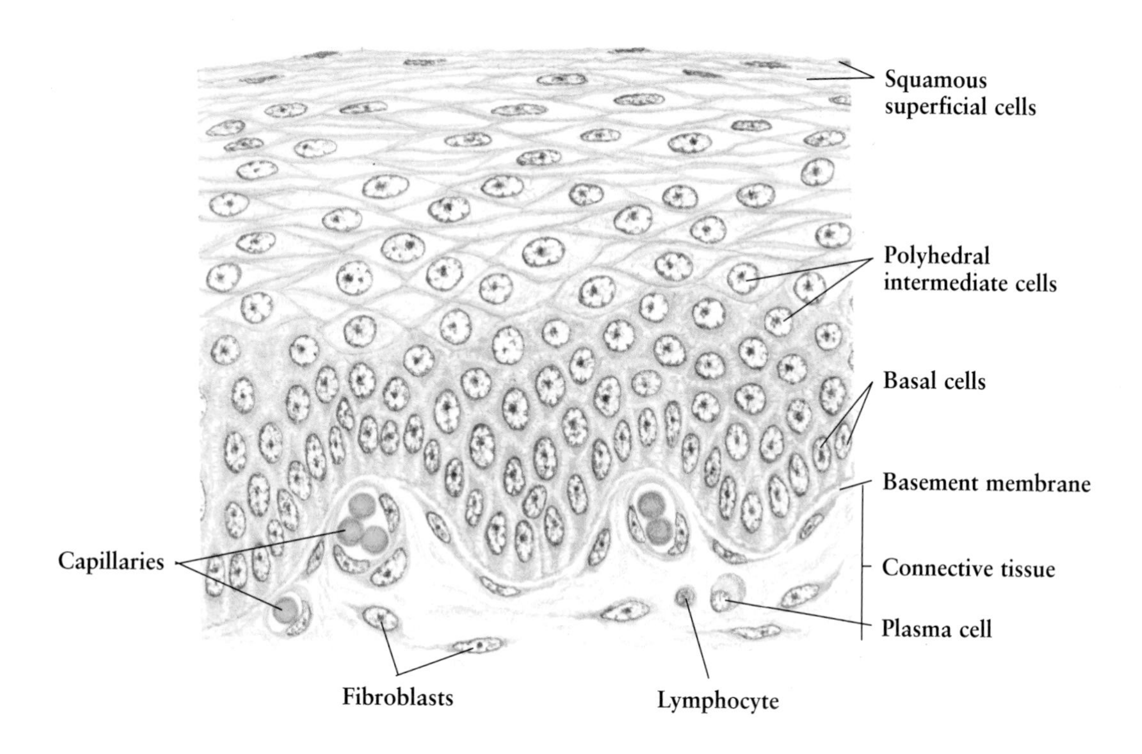

Figure 1-11. Stratified Squamous Epithelium: Keratinized
Skin of human eyelid • H.E. stain • High magnification

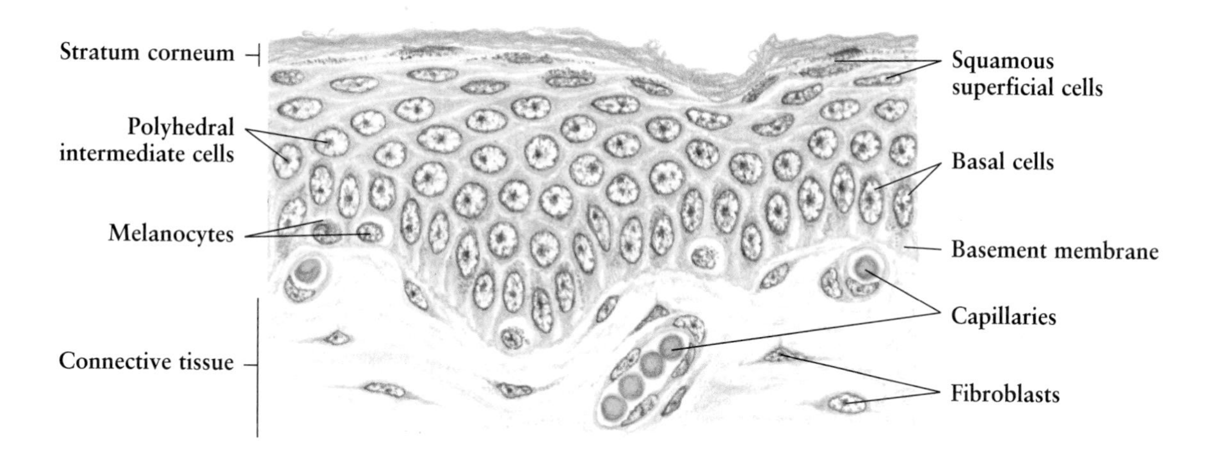

Figs. 1-12 and 1-13.

Transitional Epithelium

Transitional epithelium is a stratified epithelium composed of a layer of superficial cells, a few layers of intermediate cells, and a layer of basal cells. It is common to consider that only basal cells rest on the **basement membrane.**

The **basal cells** in the deepest layer are small, cuboidal in shape, with pale-stained cytoplasm and an oval or round nucleus in the center of the cell. The **intermediate cells** are polygonal. The cytoplasm is pale-stained, containing a round nucleus. The **superficial cells** are larger than other epithelial cells and present a scalloped outline on the luminal surface. They often cover two or three underlying intermediate cells, with indentations to accommodate the latter (see **Fig. 1-1E**). Many superficial cells have two nuclei. The cytoplasm close to the luminal surface appears to be more condensed and darkly stained with eosin in H.E.-stained (hematoxylin and eosin) preparations. This is due to the presence of numerous discoidal vesicles which, as demonstrated by electron microscopy, are made up of asymmetrical unit membranes that comprise the luminal cytoplasmic membrane of the superficial cells. The superficial cells protect the epithelium from the toxicity of urine.

In the **connective tissue** beneath the basement membrane, there is a dense population of **fibroblasts** and a network of **capillaries.**

The transitional epithelium is distributed in the pelvis of the kidney, ureters, urinary bladder, and the first part of the urethra. When the urinary bladder contracts, the epithelium increases in thickness with more layers of cells (**Fig. 1-12**). When the urinary bladder distends, the epithelium becomes thinner and consists of only three layers of cells. The large superficial cells also become flattened in this state (**Fig. 1-13**).

Figure 1-12. Transitional Epithelium
Contracted urinary bladder • Human • H.E. stain • High magnification

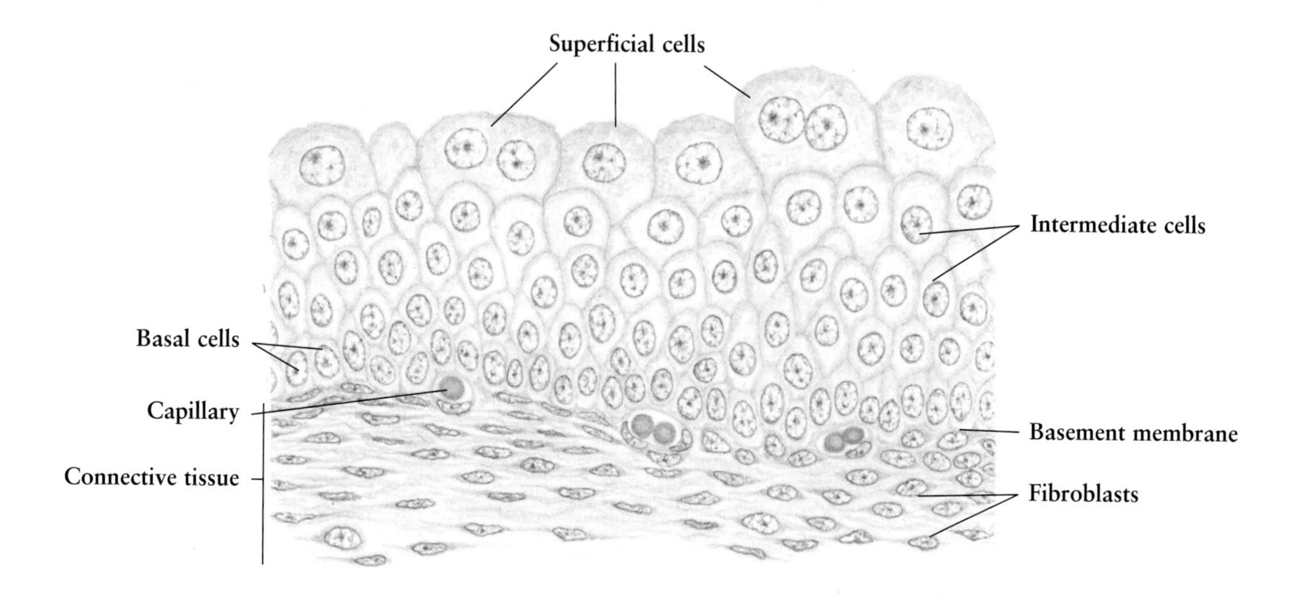

Figure 1-13. Transitional Epithelium
Distended urinary bladder • Human • H.E. stain • High magnification

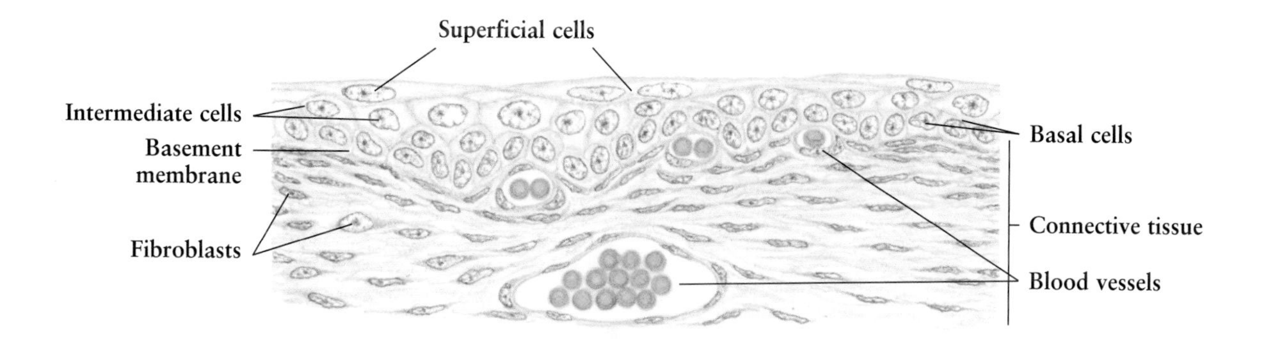

Fig. 1-14.

Glandular Epithelium

Epithelium whose major function is involved in secretion is called the **glandular epithelium.** Glandular epithelium forms secretory organs, the *glands.* Glands are classified as two types: *exocrine* and *endocrine.* An exocrine gland transports its secretion via a duct system to a body surface; an endocrine gland directly transports its secretion, the *hormone,* into the blood (see Chapter 14). The **acinus** is the secretory unit of the exocrine gland, composed of a group of glandular cells surrounding a small central **lumen.** Each acinus is supported by a layer of **basement membrane** and is connected to a narrow **intercalated duct. Figure 1-14** illustrates the glandular epithelium from three kinds of exocrine glands: the *serous gland,* the *mucous gland,* and a *mixed gland.*

In the **serous gland (Fig. 1-14A),** the acinus consists of pyramidal glandular cells that form a small lumen. The **acinar cell** has a round nucleus, which lies near the base of the cell, and at the apical cytoplasm the secretory product, **zymogen granules,** are heavily stained by the H.E. method. **Myoepithelial cells** are located between acinar cells and the **basement membrane.** The **intercalated duct** consists of squamous epithelial cells, connecting to the acinus. The cells of the **striated duct** are cuboidal or low columnar with striations, which are formed by the basal infoldings of cytoplasmic membrane and mitochondria as identified by electron microscopy. The parotid gland (see **Fig. 10-19**), pancreas (see **Figs. 10-55** and **10-56**), lacrimal gland (see **Fig. 16-9**), and serous glands of von Ebner (in the tongue, see **Fig. 10-17**) are pure serous glands. The secretion of the serous gland is a watery, clear fluid, usually containing enzymes.

The acinus of **mucous glands (Fig. 1-14B),** is composed of cuboidal cells with a small and irregular **lumen.** The **acinar cells** are clear and vacuolated, slightly stained in H.E. preparations; the nuclei are small, dark, and flattened, displaced toward the base of the cell. **Myoepithelial cells** are interposed between the acinar cells and the **basement membrane.** In addition, the **excretory duct** is composed of simple cuboidal cells. The sublingual gland (see **Fig. 10-21**), cardiac glands (see **Fig. 10-24**), pyloric glands (see **Fig. 10-28**), and duodenal glands (see **Figs. 10-31** and **10-32**) belong to the mucous glands. The secretion of the mucous glands is mucus, containing a mixture of proteins and glycoproteins.

Mixed glands (Fig. 1-14C), contain both **serous** and **mucous acini** or both serous and mucous cells in a single acinus. A small group of serous cells are often arranged in a crescent fashion at the end of the acinus, and this is called the **demilune.** The mixed gland also contains **intercalated ducts** and **striated ducts.** Like serous and mucous glands, the mixed gland also has **myoepithelial cells** surrounding its acini. The submandibular gland (see **Fig. 10-20**), glands in the tracheal wall (see **Fig. 9-8**), and some small glands in the tongue (see **Fig. 10-14**) and in the epiglottis (see **Fig. 9-4**) are classified as mixed glands. The secretion of the mixed glands is composed of a mixture of serous and mucous fluids.

Figure 1-14. Glandular Epithelium
Human • H.E. stain • High magnification

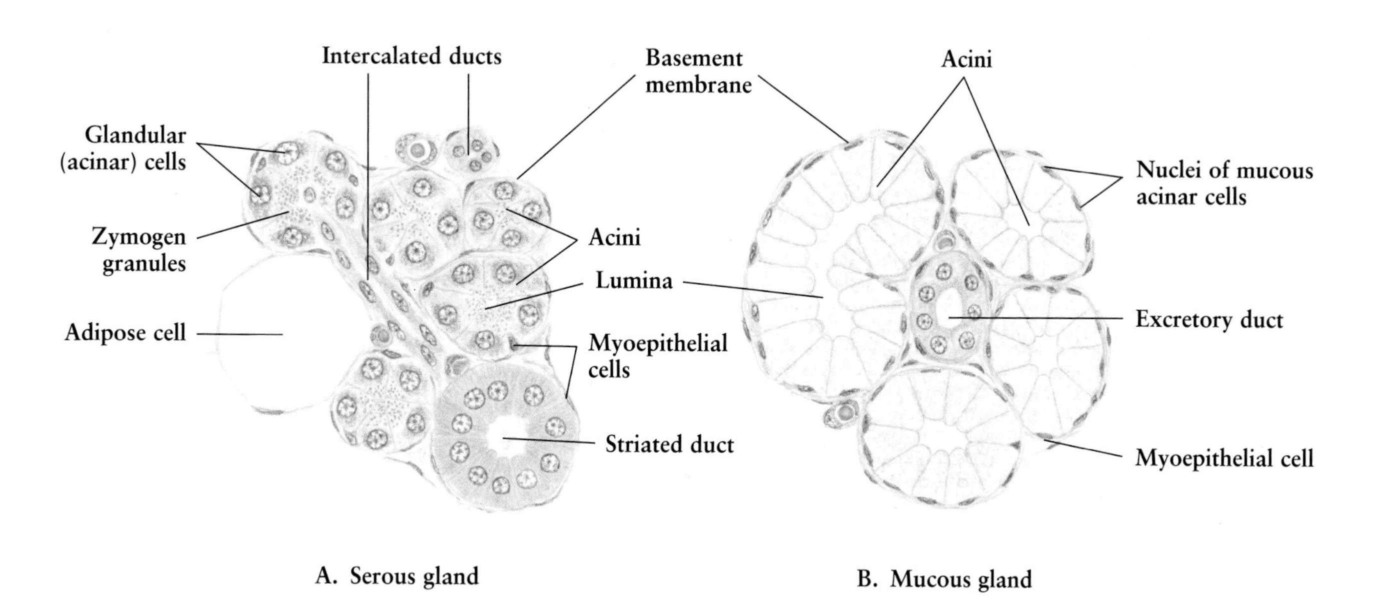

A. Serous gland

B. Mucous gland

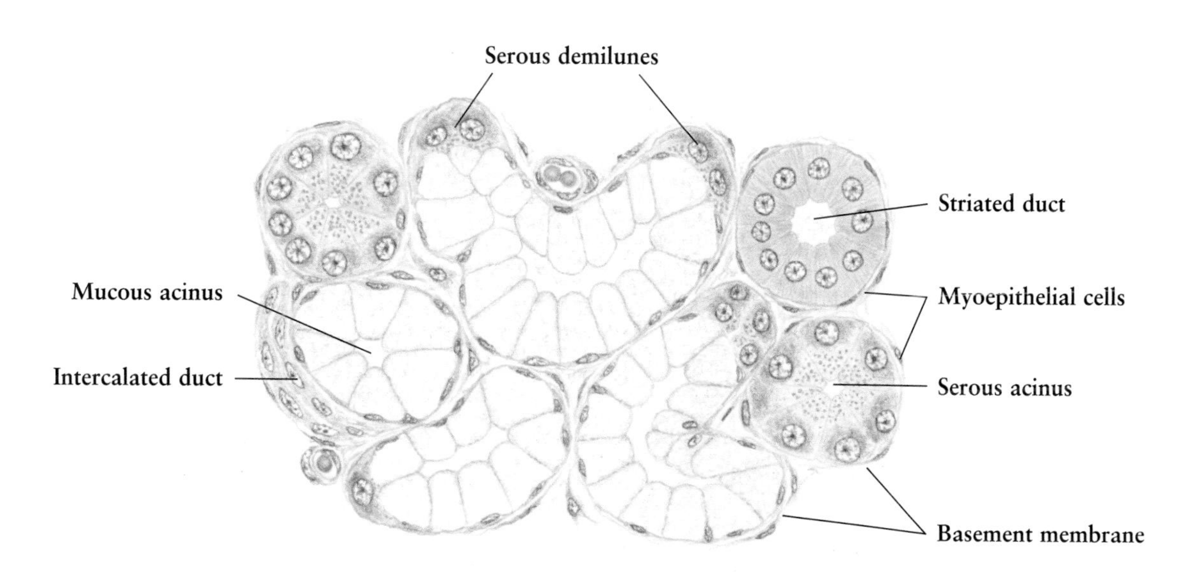

C. Mixed gland

Fig. 1-15.

Glandular Epithelium

Figure 1-15 is a second collection of **glandular epithelium,** which lines the *eccrine sweat gland, apocrine sweat gland, sebaceous gland,* and *intraepithelial gland.*

The acinus of the **eccrine** (ordinary) **sweat gland (Fig. 1-15A)** is lined by a single layer of cuboidal or columnar secretory cells, supported by a distinct **basement membrane.** Two different cell types are present within the acinus: clear cells and dark cells. The **clear cells** are the principal cell type, with a spherical nucleus in the central part of the cytoplasm. **Intercellular canaliculi** (secretory capillaries) occur between the clear cells. The **dark cells** are scattered between the clear cells or lie on the clear cells. Between the secretory cells and the basement membrane, there is a layer of **myoepithelial cells.** The secretion of the clear cells is a watery fluid containing sodium chloride, potassium chloride, ammonia, urea, and uric acid. The dark cells secrete small amounts of glycoprotein (see **Fig. 15-10**).

The acinus of the **apocrine** (odoriferous) **sweat gland (Fig. 1-15B)** is composed of a layer of cuboidal or columnar **secretory cells,** according to the functional situation of the gland. The nucleus is round or elliptical, found toward the base of the cell. The **lumen** of the acinus is larger than that of the eccrine sweat gland, and often filled with **secretory product.** Numerous **myoepithelial cells** are present between secretory cells and **basement membrane.** The secretion of the apocrine sweat gland is thick and become odoriferous after decomposition by skin bacteria (see **Fig. 15-11**).

The secreting portion, the acinus, of the **sebaceous gland (Fig. 1-15C)** is composed of a mass of cells surrounded by a **basement membrane.** The mass of cells contains two types of cells: basal cells and secretory cells. The **basal cell** is cuboidal with a round nucleus, and is present at the periphery of the secreting portion. The **secretory cells** are polygonal in shape, and occur toward the center of the secreting portion. They have a small shrunken nucleus and cytoplasm filled with **lipid droplets** so that the appearance of a secretory cell is somewhat similar to that of an adipose cell. The secretory cells finally die and break down, forming a fatty secretion, the sebum, which lubricates skin and hair (see **Fig. 15-9**).

The **intraepithelial gland (Fig. 1-15D)** is a group of glandular cells that are present in respiratory (see **Fig. 9-2**) and urethral epithelia (see **Fig. 11-12**). This drawing shows the **intraepithelial gland** situated within the **ciliated pseudostratified columnar epithelium.** The **goblet cell** is considered as a *unicellular intraepithelial gland.* The function of the intraepithelial gland is associated with the secretion of mucus.

Figure 1-15. Glandular Epithelium
Human • H.E. stain • High magnification

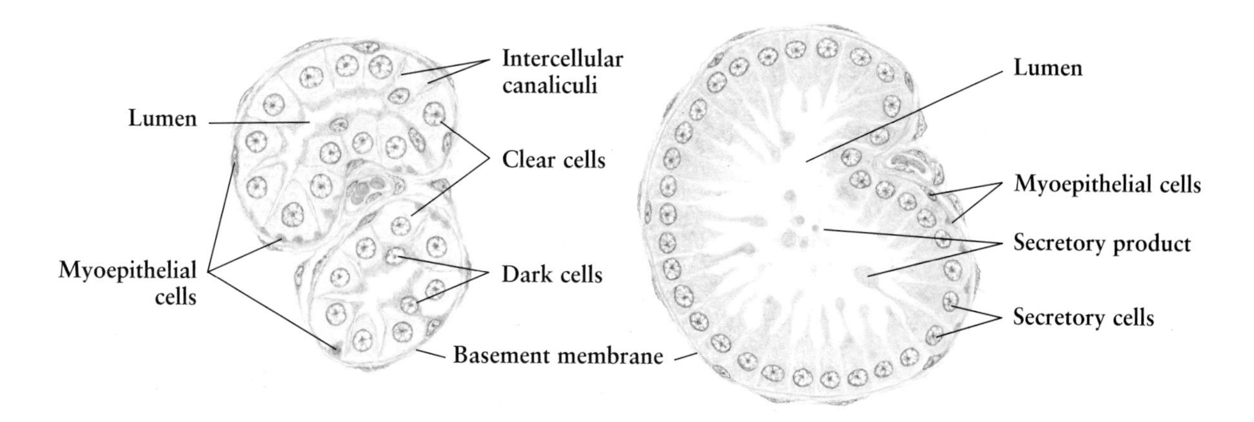

A. Eccrine sweat gland

B. Apocrine sweat gland

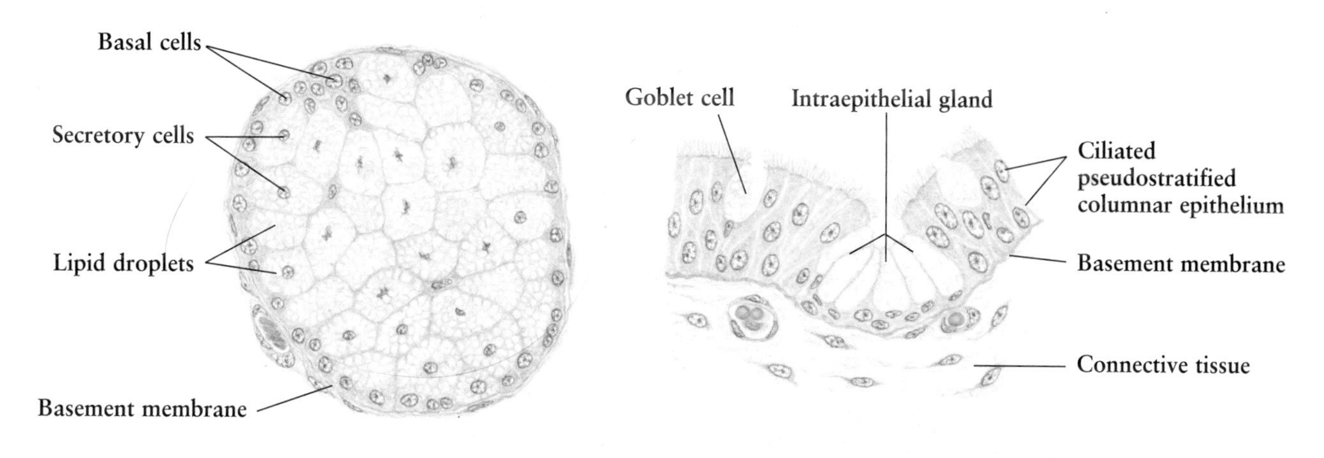

C. Sebaceous gland

D. Intraepithelial gland

2. Connective Tissue

Connective tissue is one of the four basic tissues of the body, forming a diverse group of structures including circulating blood, soft tissue, and hard bone. All connective tissues have a common origin from mesenchyme and are composed of separated *cells* and abundant *intercellular substance,* or *matrix*. The latter is composed of *fibers,* amorphous *ground substance,* and *tissue fluid*. Connective tissues are found throughout the body and in many different forms with diverse functions, such as connection, defense, support, storage, transport, metabolism, and repair following injury.

In general, connective tissues are classified as:

1. **Connective tissue proper**
   - **Loose connective tissue**
   - **Dense connective tissue**
   - **Elastic connective tissue**
   - **Mucous connective tissue**
   - **Reticular connective tissue**
   - **Adipose (connective) tissue**
2. **Blood**
3. **Cartilage and bone**

In this chapter, the connective tissue proper that contains loose connective tissue, dense connective tissue, elastic connective tissue, mucous connective tissue, and adipose tissue is shown. The blood, cartilage, and bone are demonstrated in separate chapters. Reticular connective tissue is presented in the chapter on lymphatic organs.

Fig. 2-1.

Loose Connective Tissue

Loose connective tissue is important for nutrient and gas exchange and is always found in close association with epithelial tissues. Loose connective tissue is distributed in the papillary layer of the dermis, hypodermis, lamina propria, and telae submucosa. It also fills the gaps between organs and between tissues. Loose connective tissue is composed of several cell types and three kinds of fibers, in addition to the **intercellular substance.** These cell types include fibroblast, macrophage, adipose cell, plasma cell, mast cell, undifferentiated cell, **lymphocyte, monocyte,** and **neutrophil.** The fibers are the collagen fiber, elastic fiber, and reticular fiber. This figure was drawn from a section of lamina propria of the human trachea. The following cell types and structures are shown:

Fibroblasts are the most prominent cell type in loose connective tissue. They are spindle-shaped, with a few processes. The cytoplasm of the fibroblast is somewhat basophilic. It has an ovoid nucleus with a clear nucleolus. Fibroblasts produce intercellular matrix including collagen fibers, reticular fibers, elastic fibers, and ground substance. When they are at rest, both their cell body and nucleus become smaller. Fibroblasts play an important role in wound healing.

Macrophages have irregular shapes and various sizes. The cytoplasm of macrophages tends to be acidophilic, containing numerous lysosomes and phagocytic vacuoles. The nucleus is ovoid, sometimes indented, and smaller, with more heterochromatin than that of the fibroblast. The macrophages are transformed from **monocytes** that migrate into the connective tissue from the blood. They are able to phagocytize bacteria, foreign particulate matter, dead cells, and cell debris. Additionally, macrophages transfer immunological signals to B lymphocytes.

Plasma cells are usually spherical or ovoid with rounded nuclei located eccentrically. The cytoplasm is strongly basophilic because of the abundance of rough-surfaced endoplasmic reticulum. The chromatin is mostly condensed and characteristically distributed like a so-called cartwheel nucleus. The plasma cell is responsible for the synthesis of antibodies.

Undifferentiated mesenchymal cells localize along with a capillary, so they are also called **pericytes.** They usually surround the capillary with long cytoplasmic processes. The nucleus is ovoid or somewhat flattened. Pericytes are undeveloped cells with a great capacity to transform into other cell types, such as fibroblasts in some situations, especially in inflammation and regeneration.

Fig. 2-2.

Loose Connective Tissue

The picture in **Fig. 2-2** was drawn from a subcutaneous connective tissue spread of rat, stained with the AFLGOG (aldehyde fuchsin, light green, and orange G) method. The **collagen fibers** are stained blue, and tend to be thick and unbranched. The **elastic fibers** stain purple and are much thinner than the collagen fibers. They branch and connect with each other to form a network. In this specimen, the **nuclei of the fibroblasts** and the mesothelial cells cannot be seen clearly by this staining method, but the **mast cells** show very definite outlines. The mast cells are large and ovoid, with numerous basophilic secretory granules and a pale nuclear area. In general, they are situated along **blood vessels.** The mast cells are responsible for releasing heparin, histamine, ECFA (eosinophil chemotactic factor of anaphylaxis), and some other secretory products, such as serotonin.

Figure 2-1. Loose Connective Tissue
Lamina propria of trachea • Human • H.E. stain • High magnification

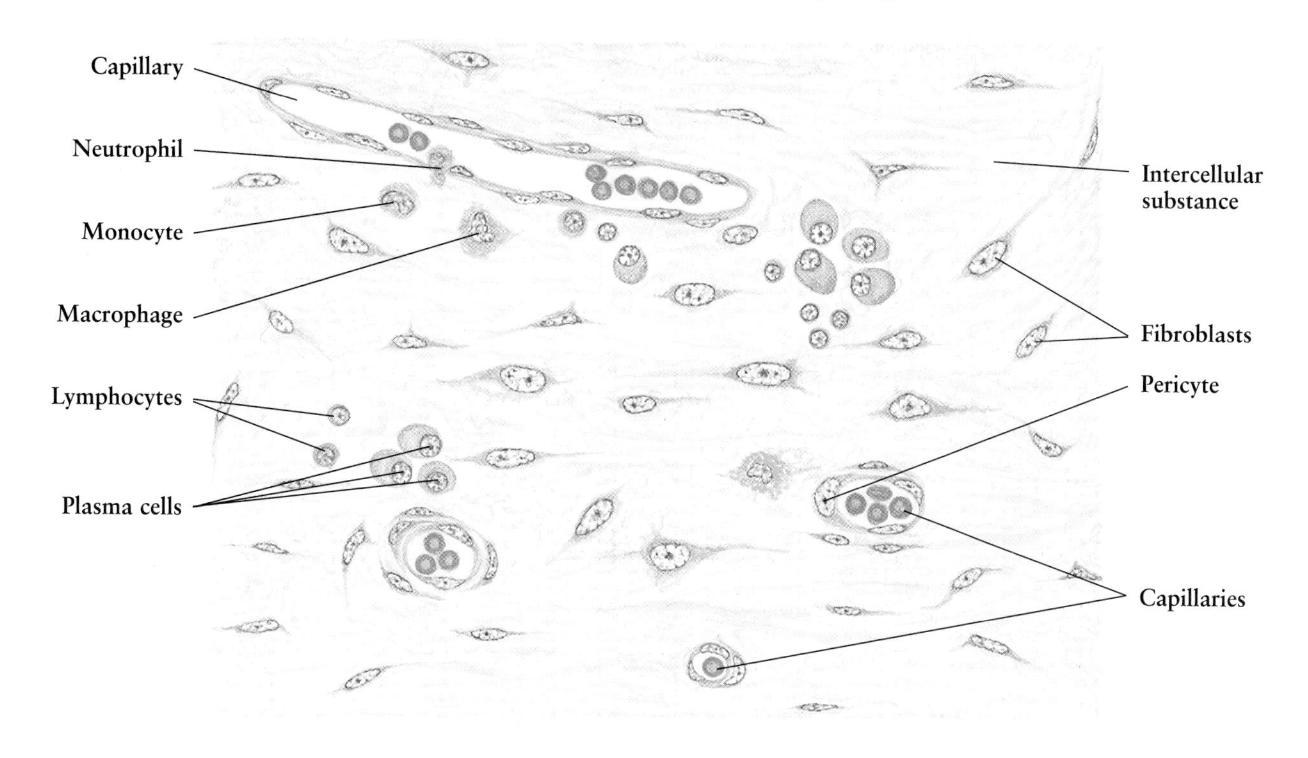

Figure 2.2 Loose Connective Tissue
Subcutaneous connective tissue spread • Rat • AFLGOG stain • High magnification

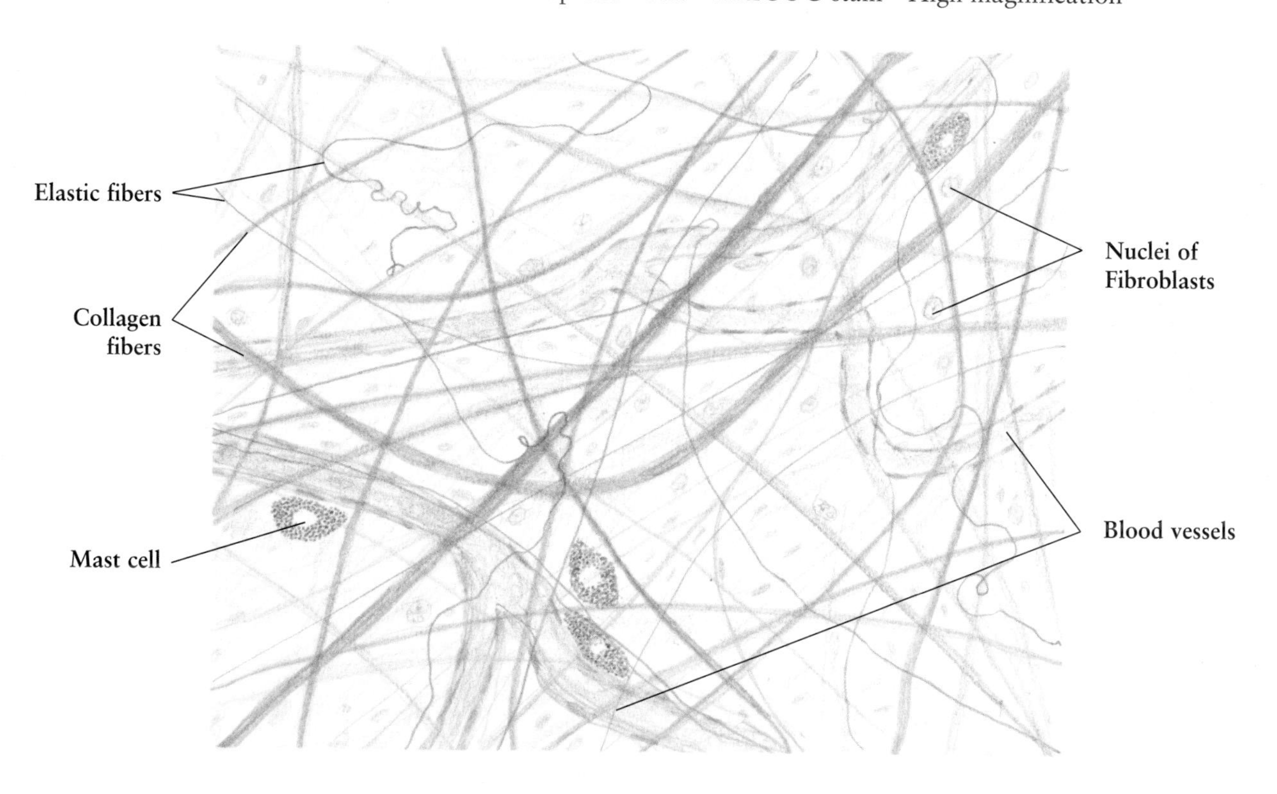

Fig. 2-3.

Dense Connective Tissue: Irregular

Irregular dense connective tissue consists mainly of a large number of **collagen fibers** and a few **fibroblasts,** occurring in the reticular dermis of the skin, perichondrium, periosteum, and capsules and trabeculae of many organs such as lymph nodes and the spleen. In the dermis of the skin, the collagen fibers are arranged in thick bundles in various directions to form a three-dimensional fibrous network capable of resisting stress from all orientations. Between collagen fiber bundles, there are a few **blood vessels, fibroblasts,** and **nerve fibers** surrounded by a small amount of loose connective tissue.

Fig. 2-4.

Dense Connective Tissue: Regular

Regular dense connective tissue is encountered in ligaments, tendons, and aponeuroses. Like irregular dense connective tissue, it is also composed of dense collagen fiber bundles and fibroblasts. In the tendon, which can be regarded as representative of regular dense connective tissue, the **collagen fibers (tendon fibers)** are densely packed in thick unbranching bundles, and run in the same direction, forming structures of great tensile strength. The fibroblasts, or **fibroblast-like cells,** also called **tendon cells,** are flattened and are arranged in rows parallel to the fibers. In longitudinal sections, as shown in this figure, the nuclei of tendon cells are elongated, and the cytoplasm is sparse and basophilic. In cross sections, the tendon cells appear stellate in shape with cytoplasmic processes extending between the collagen fiber bundles.

Figure 2-3. Dense Connective Tissue: Irregular
Dermis of human palm • H.E. stain • High magnification

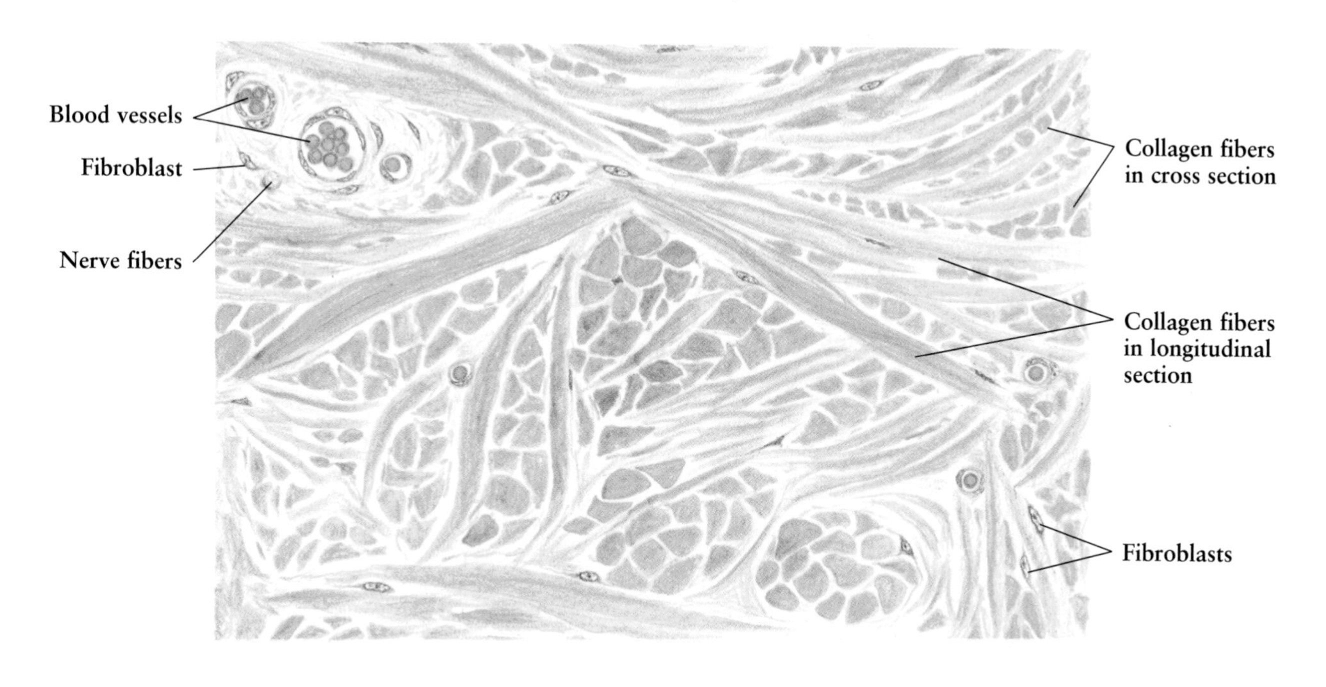

Figure 2-4. Dense Connective Tissue: Regular
Monkey tendon • Longitudinal section • H.E. stain • High magnification

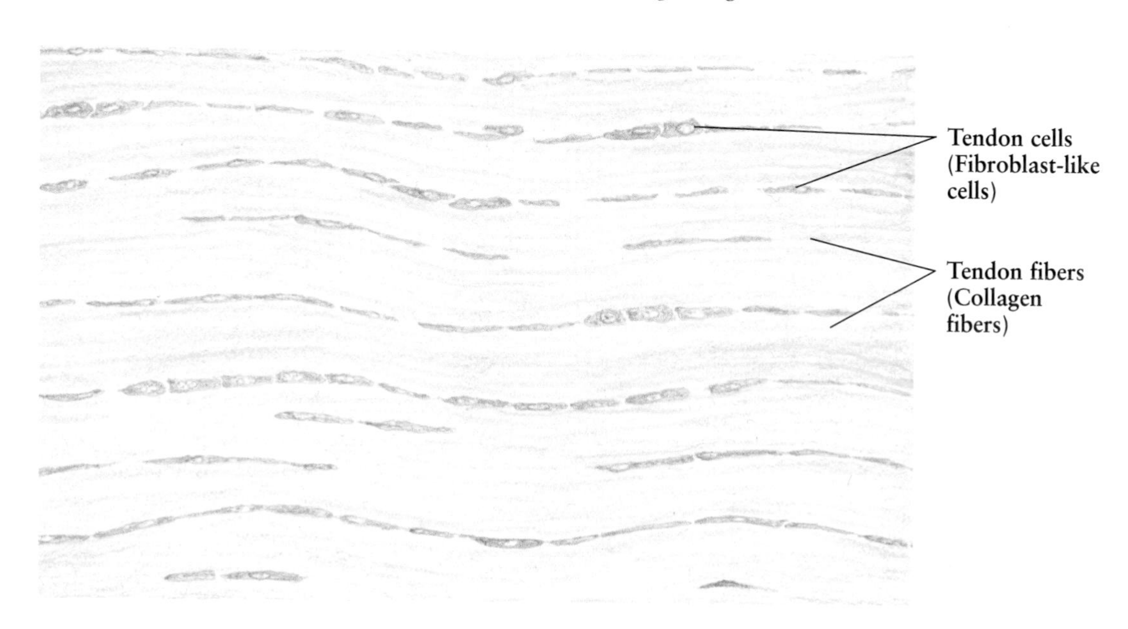

Figs. 2-5 and 2-6.

Elastic Connective Tissue

Elastic connective tissue is a kind of dense connective tissue, composed predominantly of strong elastic fibers. These elastic fibers are branched and are arranged parallel to each other to form a strong bundle with an extremely small number of fine collagen fibrils. Around the **elastic fiber bundle,** there is a small amount of loose connective tissue that contains scattered **fibroblasts** and **blood vessels.** Elastic connective tissue is found in ligamentum nuchae, ligamentum flava, ligamentum stylohyoideum, vocal ligaments, and ligamentum suspensorium penis, as well as the wall of elastic arteries. The elastic fibers enable organs or tissues to recover their shapes and sizes following normal physiological deformation or stretching. **Figures 2-5** and **2-6** are longitudinal and cross sections of bovine ligamentum nuchae, respectively, displaying the arrangement of elastic fiber bundles.

Figure 2-5. Elastic Connective Tissue
Bovine ligamentum nuchae • Longitudinal section •
Resorcin fuchsin stain • High magnification

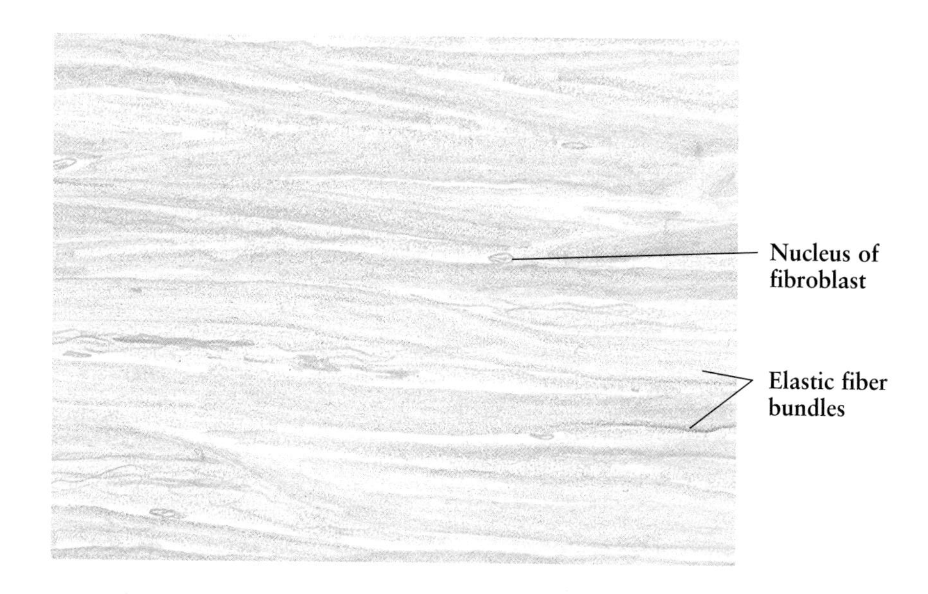

Figure 2-6. Elastic Connective Tissue
Bovine ligamentum nuchae • Cross section •
Resorcin fuchsin stain • High magnification

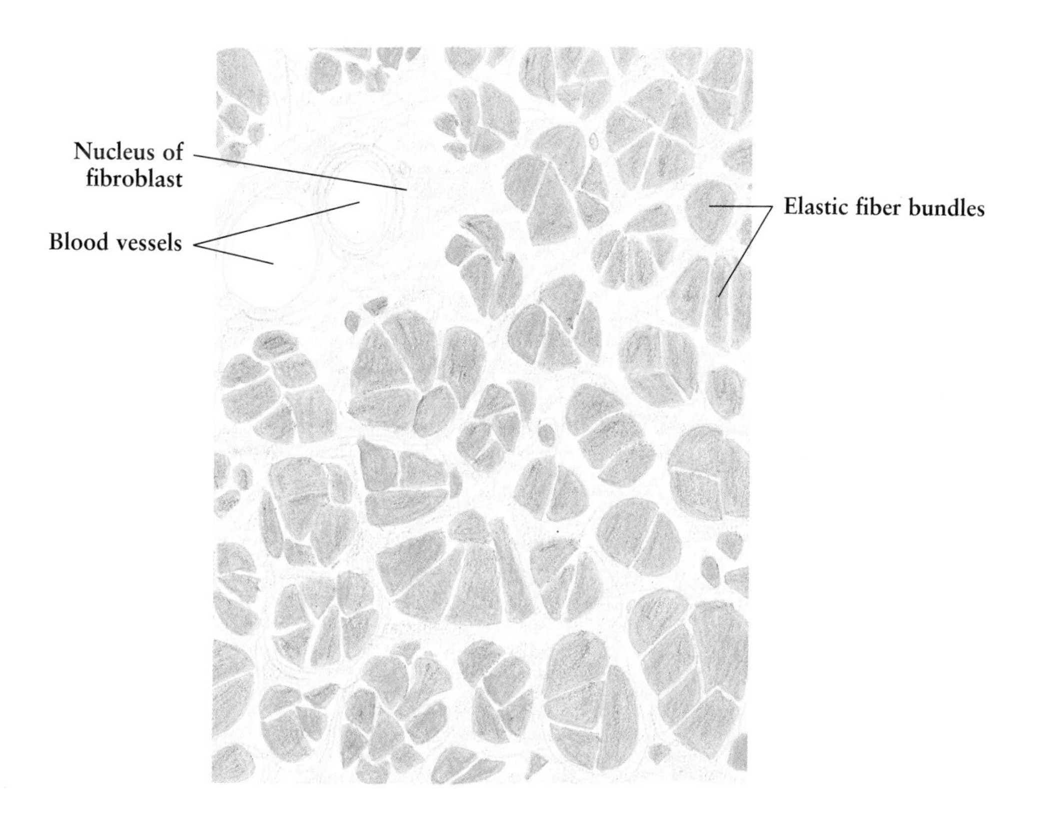

Fig. 2-7. Elastic Connective Tissue

This drawing shows a kind of **elastic connective tissue** present in the tunica media of the human aorta. In this tissue type, the elastin forms the fenestrated **elastic lamellae.** Between lamellae is **matrix** containing smooth muscle cells (see **Fig. 7-9**), which produce elastin and other intercellular substances.

Fig. 2-8. Mucous Connective Tissue

Mucous connective tissue is the embryological tissue that occurs in the normal development and differentiation of connective tissue. It is found in the embryo and is the principal component of the umbilical cord, which is called Wharton's jelly. Mucous connective tissue also appears in a few places in the adult, such as the pulp of young teeth and the vitreous humor of the eye.

This figure is drawn from a section of human **umbilical cord.** The structure of the mucous connective tissue consists of scattered cells and a dominant **intercellular substance** that contains a delicate meshwork of fine **collagen fibers.** The cells are large, stellate **fibroblasts.** The nucleus of the fibroblast is big, round, or ovoid with a prominent nucleolus. The cytoplasm is quite thin, but the processes are often seen connecting with those of neighboring cells. The amorphous ground substance is soft and jelly-like. The mucous connective tissue in the umbilical cord, unlike that in other locations in the embryo, does not have the potential for differentiation.

Figure 2-7. Elastic Connective Tissue
Human aorta • Cross section • Resorcin fuchsin stain • High magnification

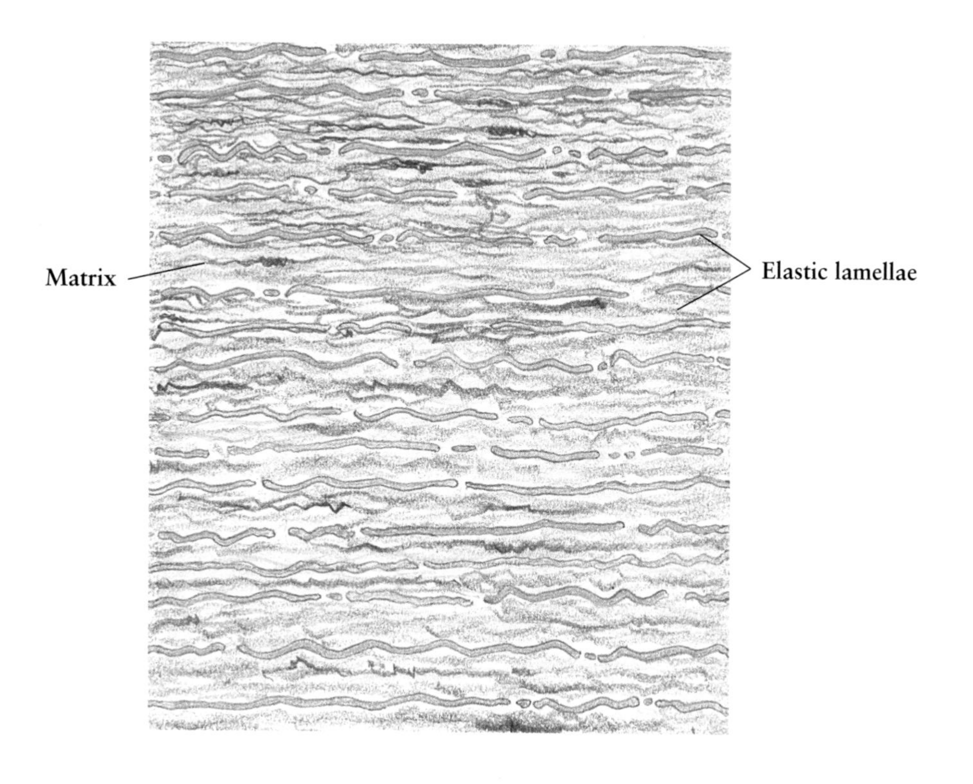

Figure 2-8. Mucous Connective Tissue
Human umbilical cord • H.E. stain • High magnification

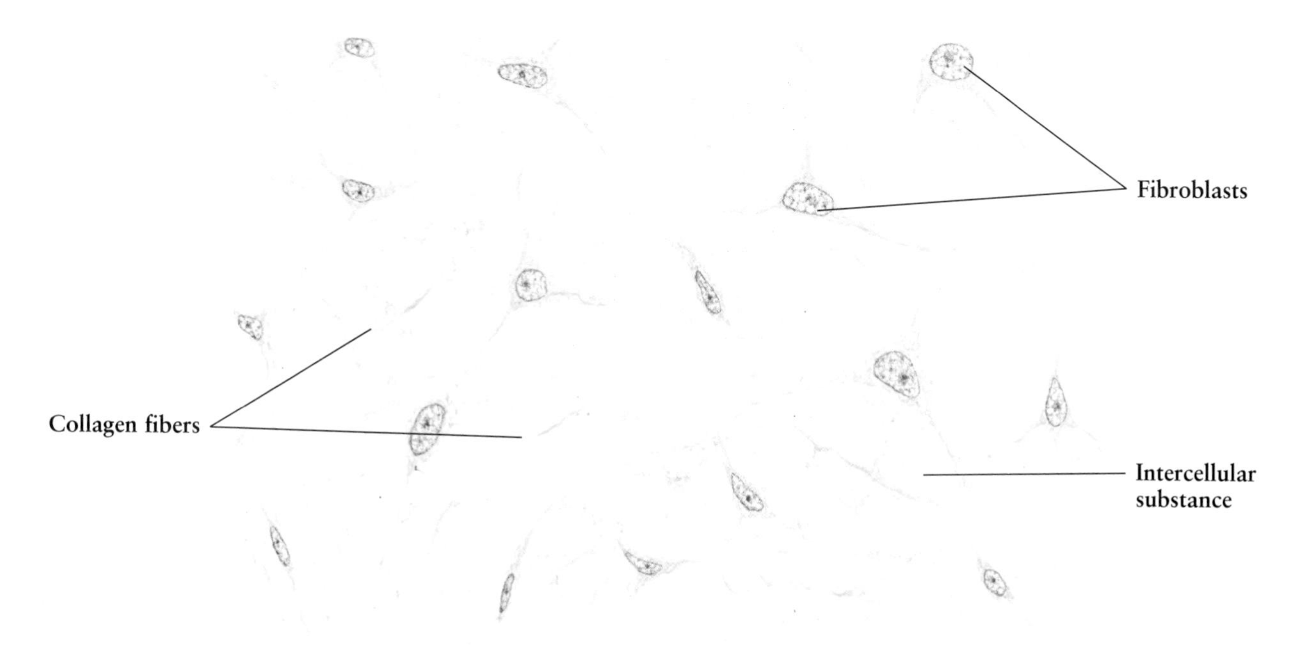

Figs. 2-9 and 2-10.

Reticular Fibers

Reticular fibers are composed of a special type of collagen, and are similar to other collagen fibers in structure. They are extremely fine, and cannot be demonstrated with hematoxylin and eosin, but they appear as a black network by silver impregnation. Reticular fibers are present in all connective tissue, and are particularly abundant in lymph nodes, spleen, red bone marrow, and liver. **Figure 2-9** shows **reticular fibers,** which connect with each other to form a network, in the **medullary cord** and **sinuses** of the lymph node. In **Fig. 2-10,** which was drawn from a liver preparation with silver impregnation, the **reticular fiber** network lies in the **sinusoids** and surrounds the **central vein.**

Figure 2-9. Reticular Fibers
Human lymph node • Silver impregnation • High magnification

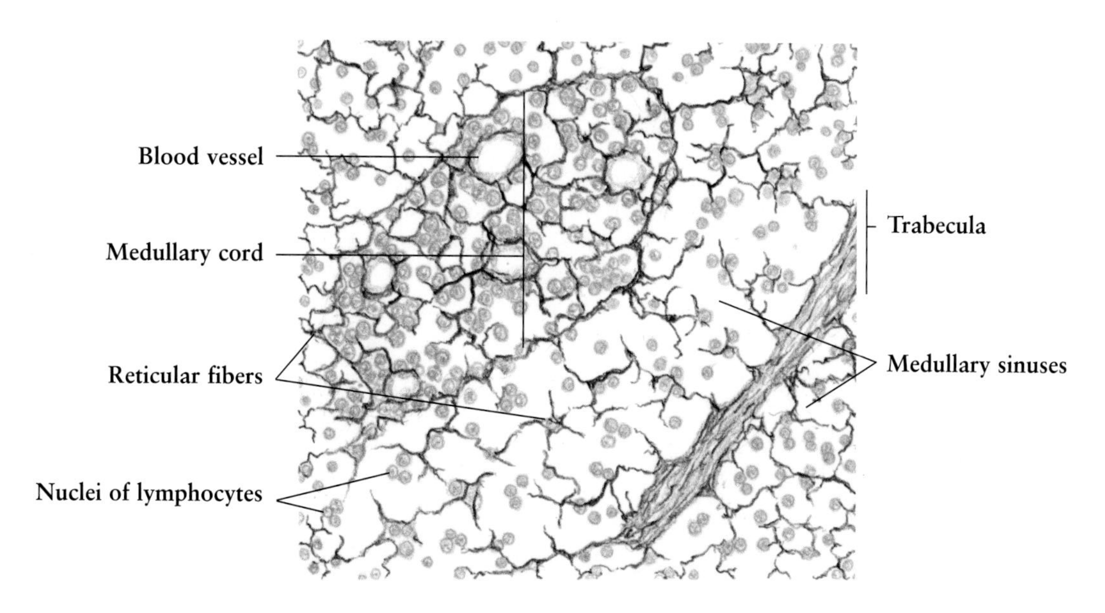

Figure 2-10. Reticular Fibers
Human liver • Silver impregnation • High magnification

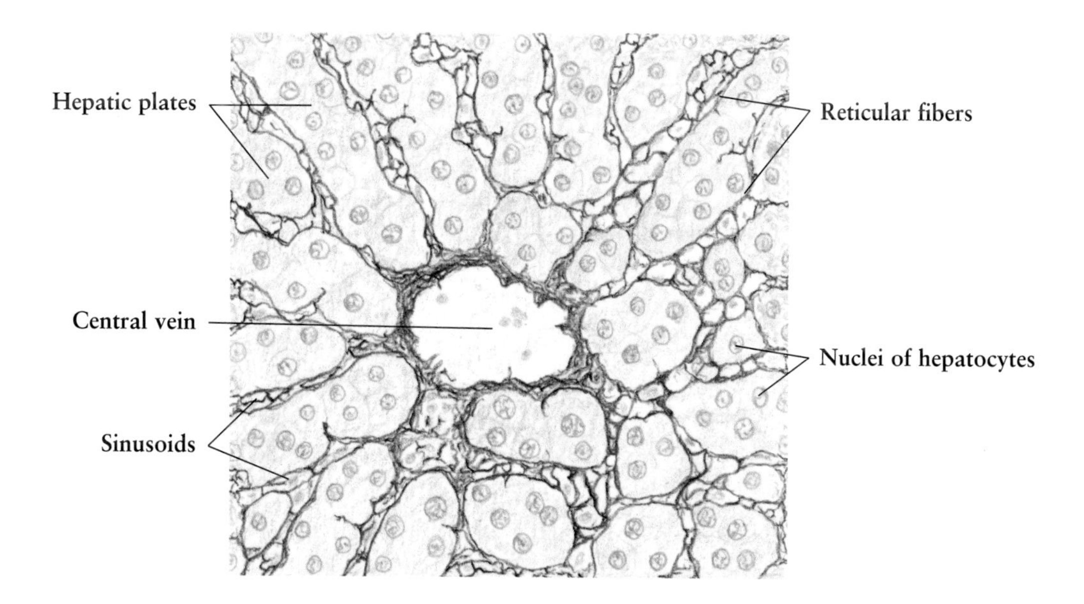

Fig. 2-11.

White Adipose Tissue

Adipose tissue is the storage center for fat. **White (unilocular) adipose tissue** consists of densely packed **adipose cells.** Each adipose cell contains a single large drop of fat (hence the "unilocular" name). The fat is dissolved out by alcohol during the preparation procedure, leaving a large vacuole. The adipose cell has a flattened **nucleus** and sheetlike cytoplasm, which has been pushed into a rim surrounding the fat drop. Between adipose cells there are **capillaries** and a very small amount of connective tissue with reticular fibers, which can be demonstrated by silver impregnation. In this tissue type, adipose cells are usually separated into **lobules** by a **connective tissue septum** containing **blood vessels** and **nerve fibers.**

Under the influence of the hormone *insulin,* the adipose cells can synthesize fat from fatty acids or carbohydrate. The vital balance between deposits and withdrawals of the fat is affected by hormones and by the autonomic nervous system. Except for the eyelid, penis, and scrotum, adipose tissue is present throughout the body, and is especially abundant in the hypodermis, mesenteries, omenta, and around the kidneys. The white adipose tissue, which constitutes as much as 20% of total body weight in a normal male adult and up to 25% in female, acts as a storage site and metabolic source of neutral fat, a cushion against mechanical shock, and a insulator to prevent excessive heat loss or gain through the skin.

Fig. 2-12.

Brown Adipose Tissue

Brown (multilocular) adipose tissue differs from white adipose tissue by both the organization of its lipid and its characteristic color. Brown adipose tissue is abundant in hibernating mammals. In human beings, it is common in the fetus, but almost absent in the adult. Brown adipose tissue is distributed mainly in the shoulder girdle and some other areas. It is composed of densely packed **brown adipose cells.** The cytoplasm contains a number of **lipid droplets** of various sizes; thus, this tissue type is also called multilocular adipose tissue. In addition, abundant in the cytoplasm are the mitochondria and lysosomes, which give the tissue its brown color. The **nucleus** of the multilocular adipose cell is spherical, located centrally or eccentrically. Between the adipose cells is a rich supply of **capillaries.** In hibernating animals and the human fetus, the brown adipose tissue generates heat to keep the body warm during cold temperatures.

Figure 2-11. White Adipose Tissue
Human • H.E. stain • Medium magnification

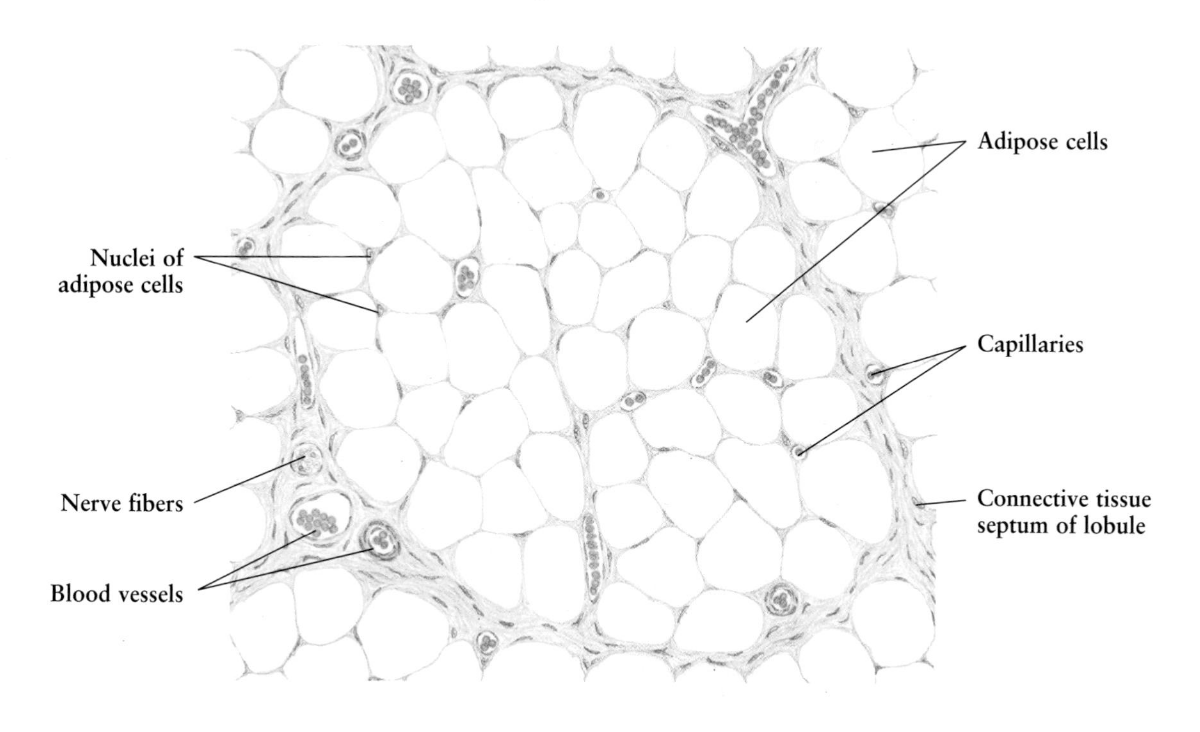

Figure 2-12. Brown Adipose Tissue
Human • H.E. stain • High magnification

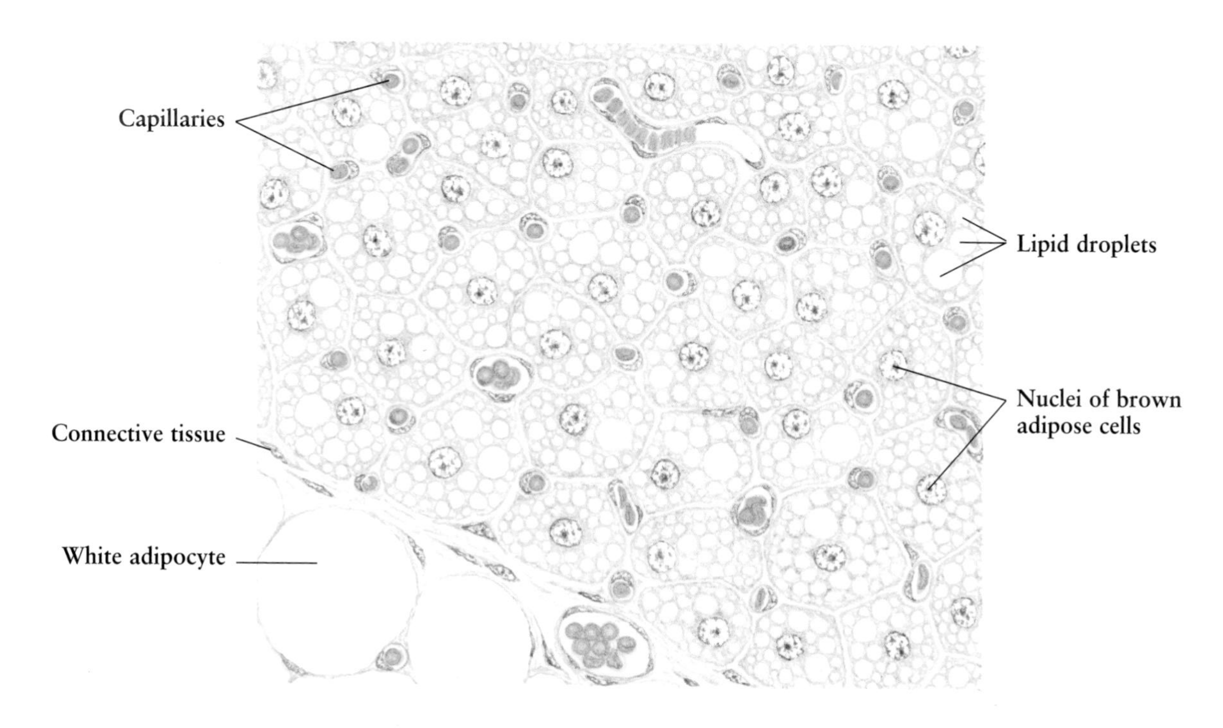

3. Cartilage and Bone

Cartilage and **bone** are the supporting, or skeletal tissues, formed of specialized connective tissues. They provide a rigid protective and supporting framework for most of the soft tissues of the body (such as brain, spinal cord, and trachea). Like general connective tissue, the skeletal tissues are composed of cells (chondrocytes or osteocytes), fibers (collagen or elastic fibers), and ground substance. It is the ground substance that gives rise to the various features or functions of skeletal tissues.

In **cartilage** the cells are called the *chondrocytes,* which are embedded in small cavities known as *lacunae.* The ground substance is composed principally of proteoglycans rich in chondroitin sulfates, which give rise to the solid yet flexible property of cartilage. The intercellular substance contains *collagen* and *elastic fibers.* According to the type and arrangement of the fibers, cartilage is classified as *hyaline, fibrous,* or *elastic.* The cartilage is devoid of blood vessels, lymphatics, and nerves. As a result, the chondrocytes receive all their nutrients by diffusion.

Bone (osseous tissue) is a specialized connective tissue with the property of marked rigidity and strength. Bone forms most of the skeleton of the body. It is composed of cells, *osteocytes,* and intercellular matrix, which contains an organic component, mainly *collagen fibers,* and an inorganic component including calcium phosphate and calcium carbonate. *Lamella,* the basic structure of bone, is composed of osteocytes and collagen fibers. Lamellae form both *compact* and *spongy bone.* Unlike cartilage, bone is supplied with blood vessels and nerves.

This chapter also presents **bone formation,** including intramembranous ossification and endochondral ossification.

Fig. 3-1. Hyaline Cartilage

Hyaline cartilage is composed of a solid glasslike **cartilage matrix** and **chondrocytes** embedded in **cartilage cavities,** the **lacunae.** It forms the cartilages found in the respiratory tract, the costal cartilages, and the articular surfaces of bones within joints.

Except for the articular cartilage, the hyaline cartilage is enclosed by a layer of dense connective tissue, the **perichondrium.** The outer zone of perichondrium contains **fibroblasts** and elastic and collagen fibers. The inner zone is rich in **immature chondrocytes,** also called **chondroblasts,** which develop into typical chondrocytes.

Chondrocytes are spindle-shaped at the periphery and ovoid or spherical in the deeper portion of the cartilage. The nucleus is round with one or two nucleoli; the cytoplasm is basophilic because of the abundant rough-surfaced endoplasmic reticulum. In living cartilage, chondrocytes fill the entire lacunae, but they appear shrunken in paraffin section as a result of fixation and dehydration. Young chondrocytes undergo mitosis as they migrate to the deep portion of the cartilage from the periphery; thus, chondrocytes clustered in a single lacuna are known as an **isogenous group.**

The **cartilage matrix,** containing amorphous ground substance and fibers, is basophilic because of the presence of glycosaminoglycans (hyaluronic acid and proteoglycans). The region surrounding the lacuna exhibits intense basophilia and is referred to as the **cartilage capsule.** Collagen fibers in the cartilage matrix are invisible in routine histological preparations because they have approximately the same refractive index as that of the surrounding intercellular substance.

Fig. 3-2. Fibrous Cartilage

Under the microscope, **fibrous cartilage,** also called fibrocartilage, resembles regular dense connective tissue, except that fibrocartilage has typical chondrocytes in capsules. The intercellular matrix contains thick bundles of **collagen fibers** running parallel with one another. Between bundles of collagen fibers are small regions of hyaline **cartilage matrix** containing round or oval lacunae with enclosed **chondrocytes.** The cells, surrounded by basophilic **capsular matrix,** are arranged singly or as a small group in short rows. Fibrocartilage is totally avascular and lacks a perichondrium. It occurs in the intervertebral discs, in the pubic symphysis, in the attachment of some tendons, and between the dense connective tissue of the ligament and joint capsules.

Figure 3-1. **Hyaline Cartilage**
Human trachea • H.E. stain • High magnification

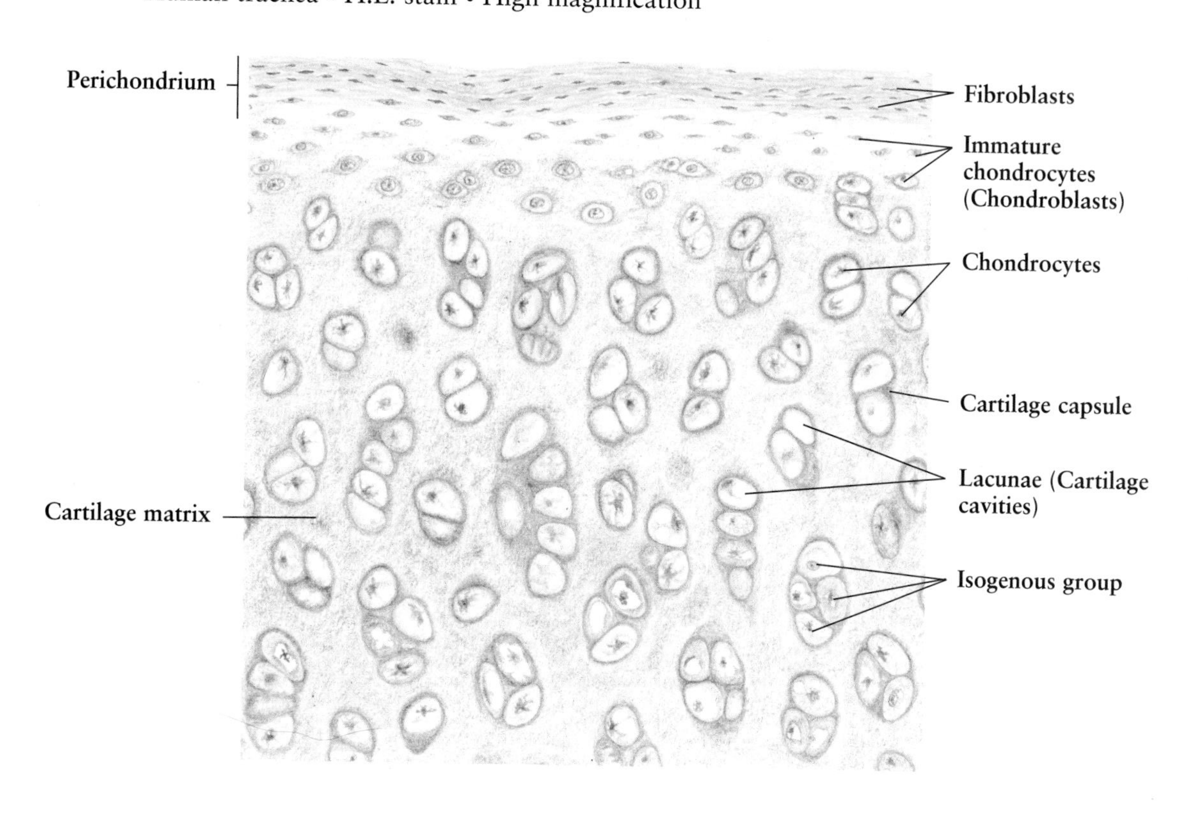

Figure 3-2. **Fibrous Cartilage**
Human intervertebral disc • H.E. stain • High magnification

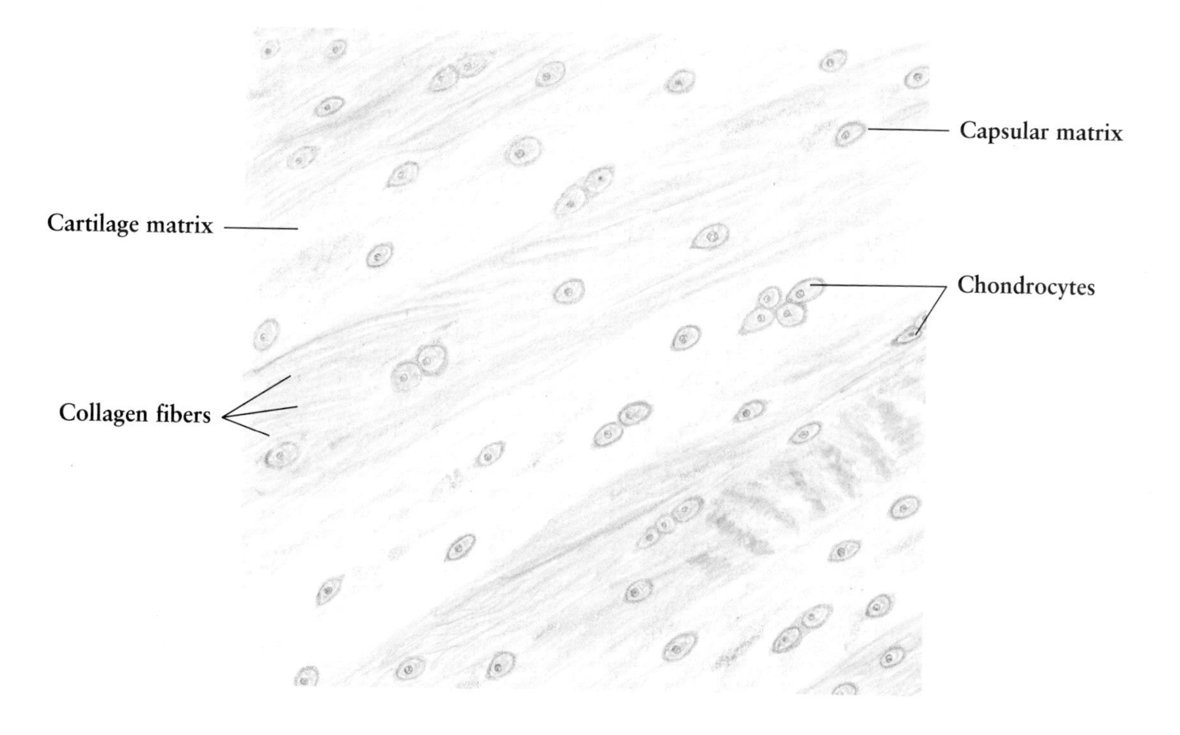

Figs. 3-3 and 3-4.

Elastic Cartilage

Elastic cartilage occurs in locations where support with flexibility is required, as in the epiglottis, auricle of the ear, and external auditory meatus. It is similar to hyaline cartilage in the arrangement of **chondrocytes, cartilage capsule** (**lacunae**), **cartilage matrix,** and **perichondrium.** The difference is that elastic cartilage also contains a network of **elastic fibers** in addition to the masked collagen fibers in the cartilage matrix. These elastic fibers account for the flexibility of structures containing them. The elastic fibers are invisible in hematoxylin-eosin preparations (**Fig. 3-3**), but they may be readily demonstrated by resorcin fuchsin stain (**Fig. 3-4**).

Figure 3-3. Elastic Cartilage
Human epiglottis • H.E. stain • High magnification

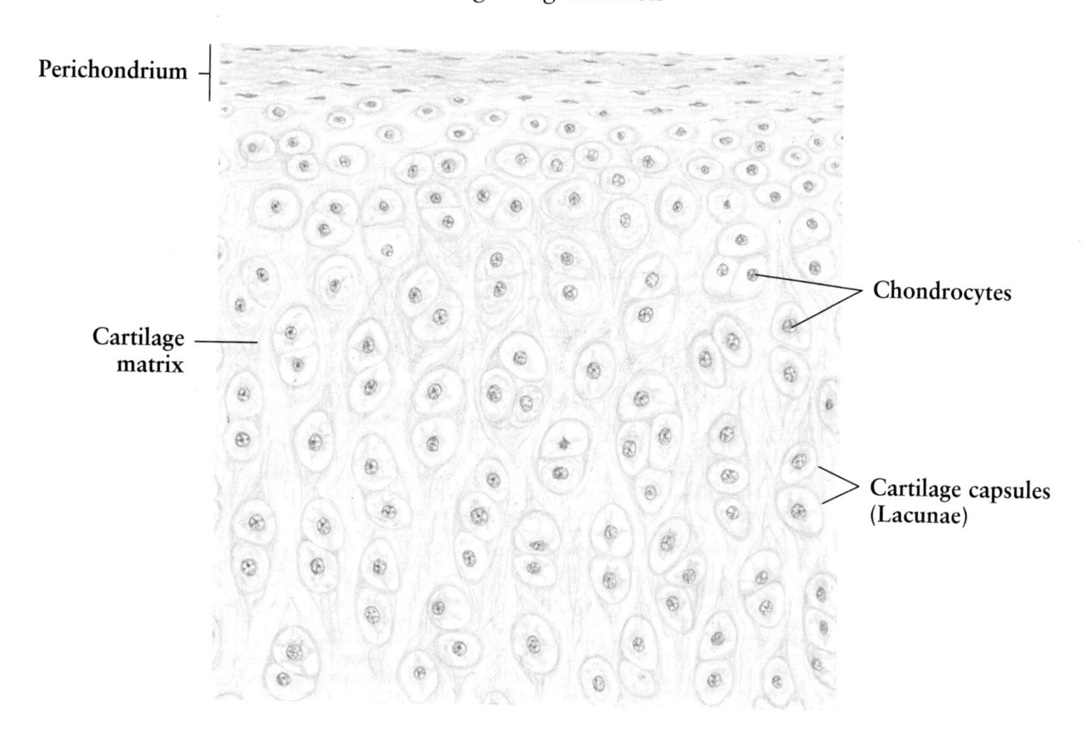

Figure 3-4. Elastic Cartilage
Human auricle • Resorcin fuchsin stain • High magnification

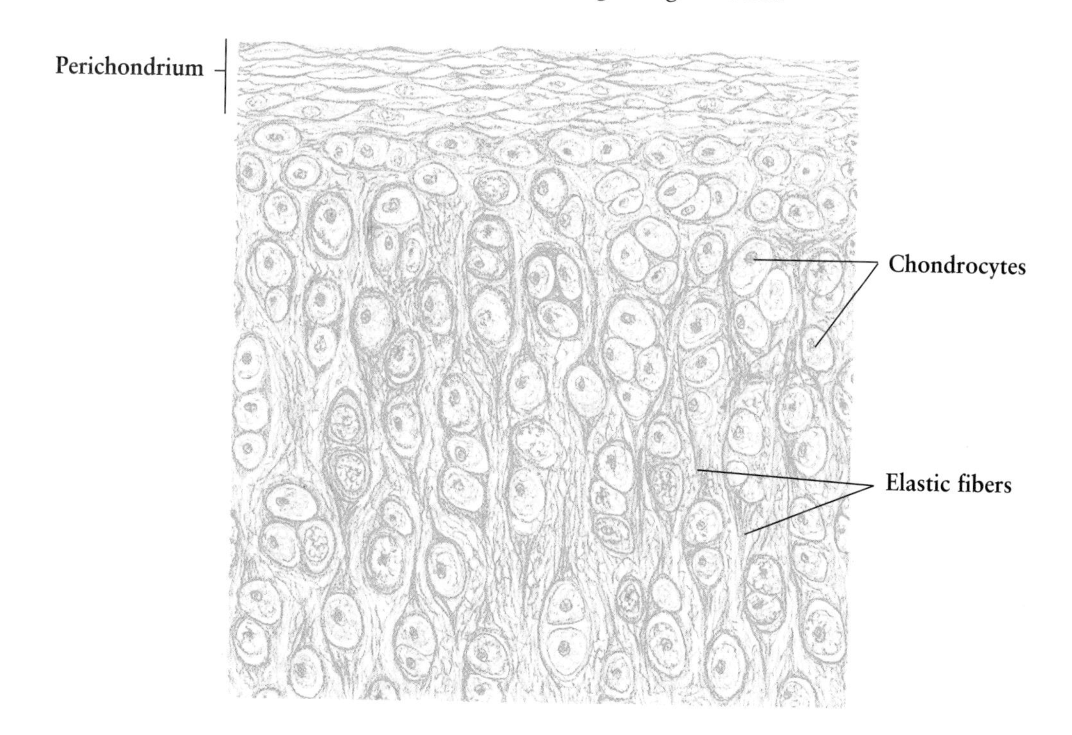

Figs. 3-5 and 3-6.

Bone: Compact

Compact bone, found in the diaphysis in long bone, is organized from outer circumferential lamellae, inner circumferential lamellae, the Haversian systems, and interstitial lamellae. The bone is enveloped by a specialized dense connective tissue, **periosteum** (see also **Fig. 3-9**). The **outer circumferential lamellae** are composed of several layers of lamellae that are parallel with the surface of the bone, located beneath the periosteum. The **inner circumferential lamellae** are collectively thinner than the outer circumferential lamellae and run alongside the marrow cavity, which is lined by **endosteum.** Between the outer and inner circumferential lamellae are the Haversian systems and interstitial lamellae. **Haversian systems** (see also **Figs. 3-7** and **3-8**) consist of concentric lamellae with a **Haversian canal** in the center. **Interstitial lamellae** are a kind of irregular lamellae left by Haversian systems destroyed during growth and remodeling of bone.

Spongy (trabecular) bone is located within the marrow cavity, and is composed of lamellae and lined by endosteum. It forms a meshwork or trabeculae within the marrow cavity.

Figure 3-5 is a cross section, showing the arrangement of lamellae. **Figure 3-6** is a longitudinal section demonstrating that Haversian systems are columnar structures, parallel to one another and to the axis of the long bone. Haversian canals connect with **Volkmann's canals,** which in turn are oblique or transverse to the axis of the long bone.

Figure 3-5. Bone: Compact
Monkey femur • Decalcified • Cross section • Schmorl stain • Low magnification

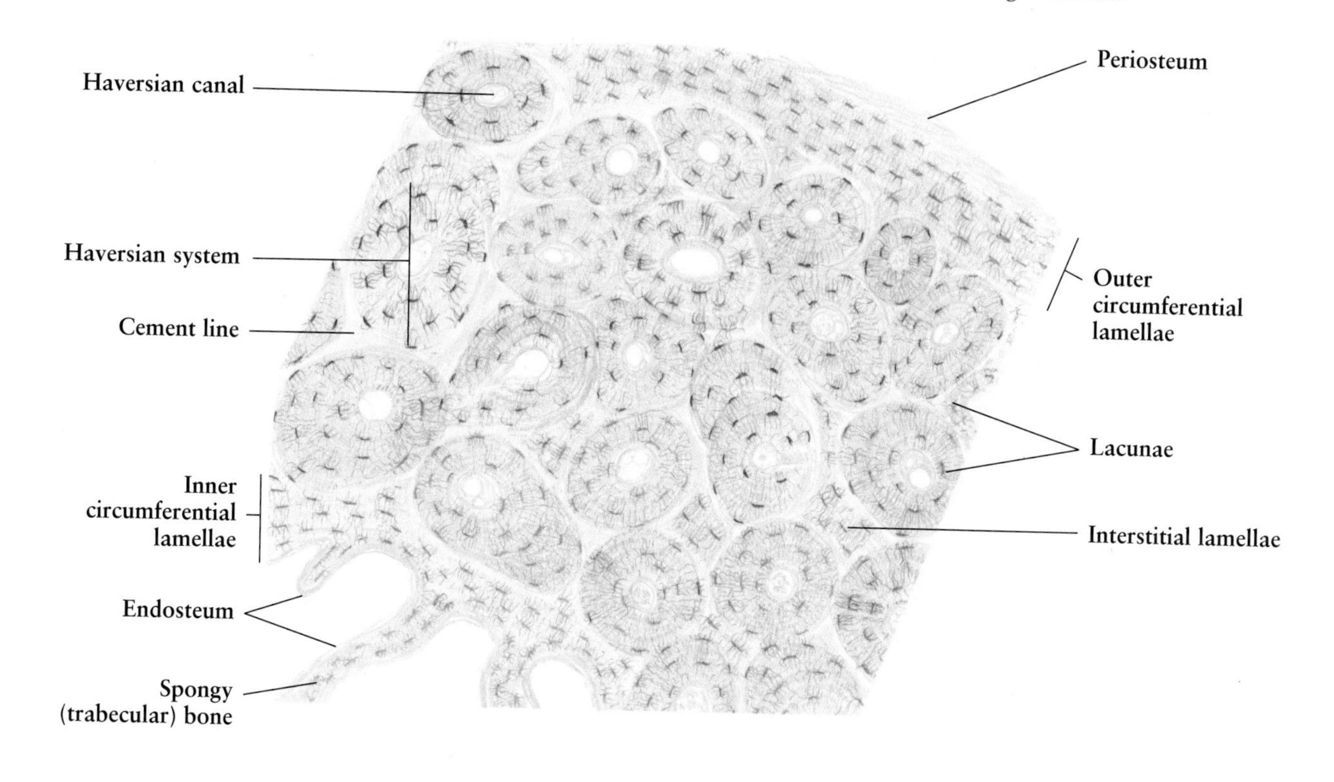

Figure 3-6. Bone: Compact
Monkey femur • Decalcified • Longitudinal section • Schmorl stain • Low magnification

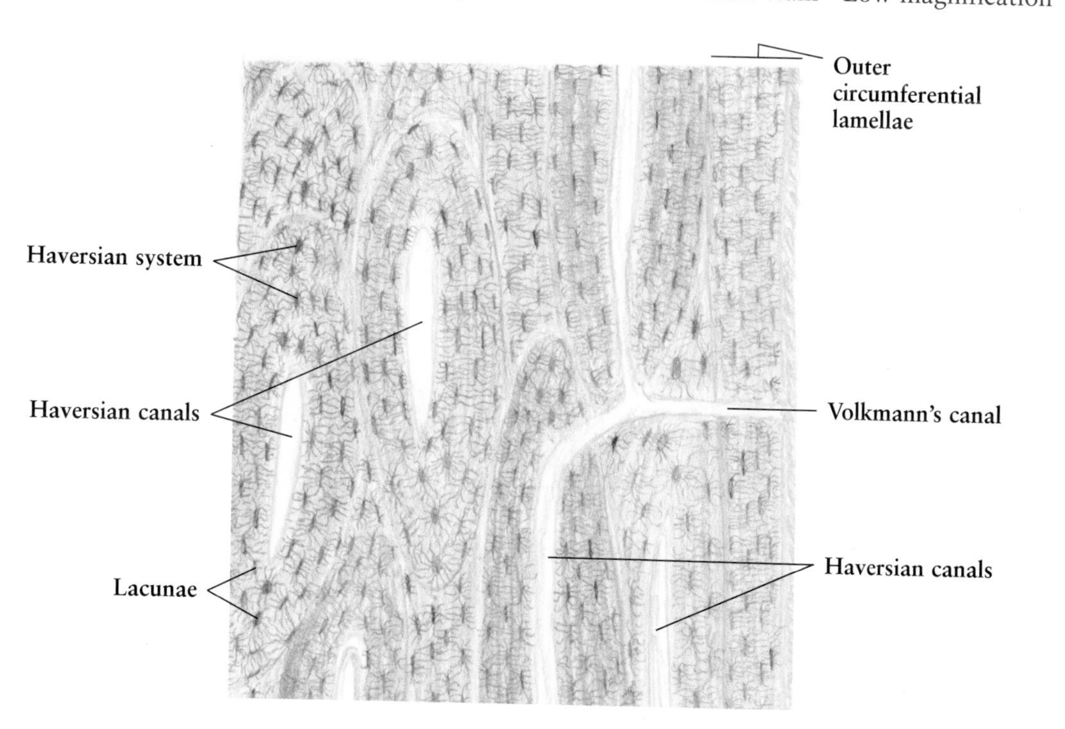

Figs. 3-7 and 3-8.

Bone: Haversian System

A **Haversian system,** or osteon, measuring 100–500 µm in diameter and several centimeters in length, consists of 4 to 20 layers of **Haversian lamellae** arranged concentrically and a **Haversian canal,** which is located in the center of the system. Between lamellae are irregular oval **lacunae** occupied by **osteocytes.** They communicate via fine **canaliculi.** A Haversian canal is filled with loose connective tissue containing **blood vessels, preosteoblasts,** and **macrophages.** Haversian systems are demarcated by a clear line, the **cement line,** from each other or from **interstitial lamellae. Figure** 3-7 is a Schmorl-stained preparation, which clearly shows the lacunae and canaliculi. In **Fig. 3-8,** an H.E.-stained preparation, the osteocytes inside the lacunae can be recognized markedly.

Figure 3-7. Bone: Haversian System

Monkey femur • Decalcified • Cross section • Schmorl stain • High magnification

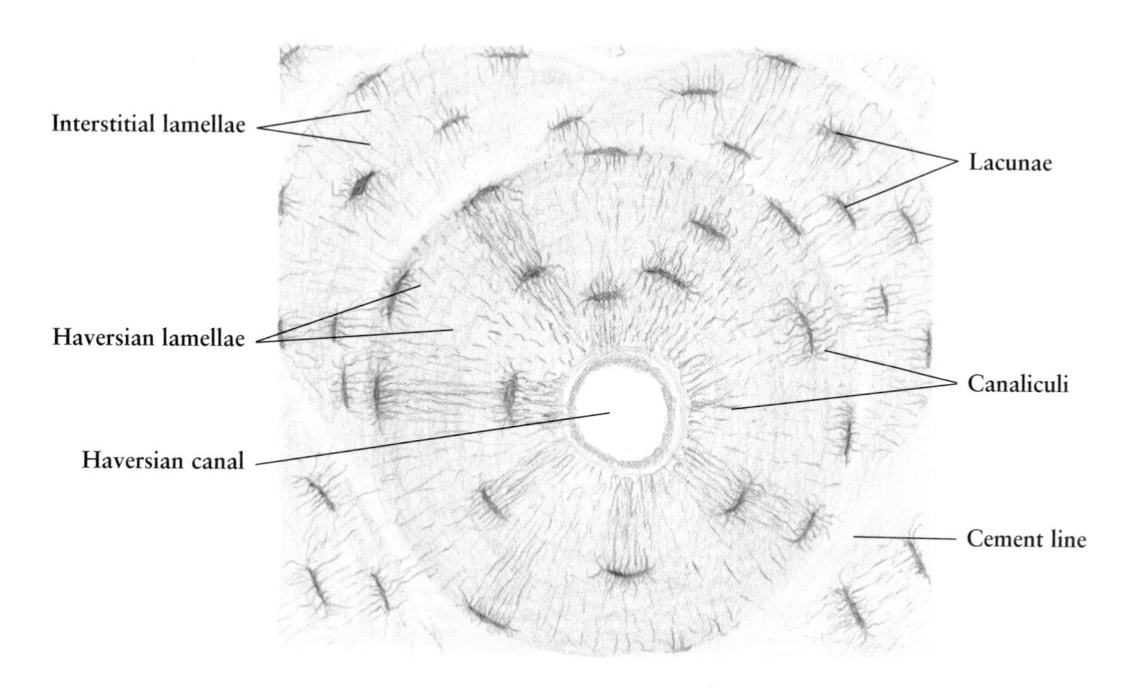

Figure 3-8. Bone: Haversian System

Monkey femur • Decalcified • Cross section • H.E. stain • High magnification

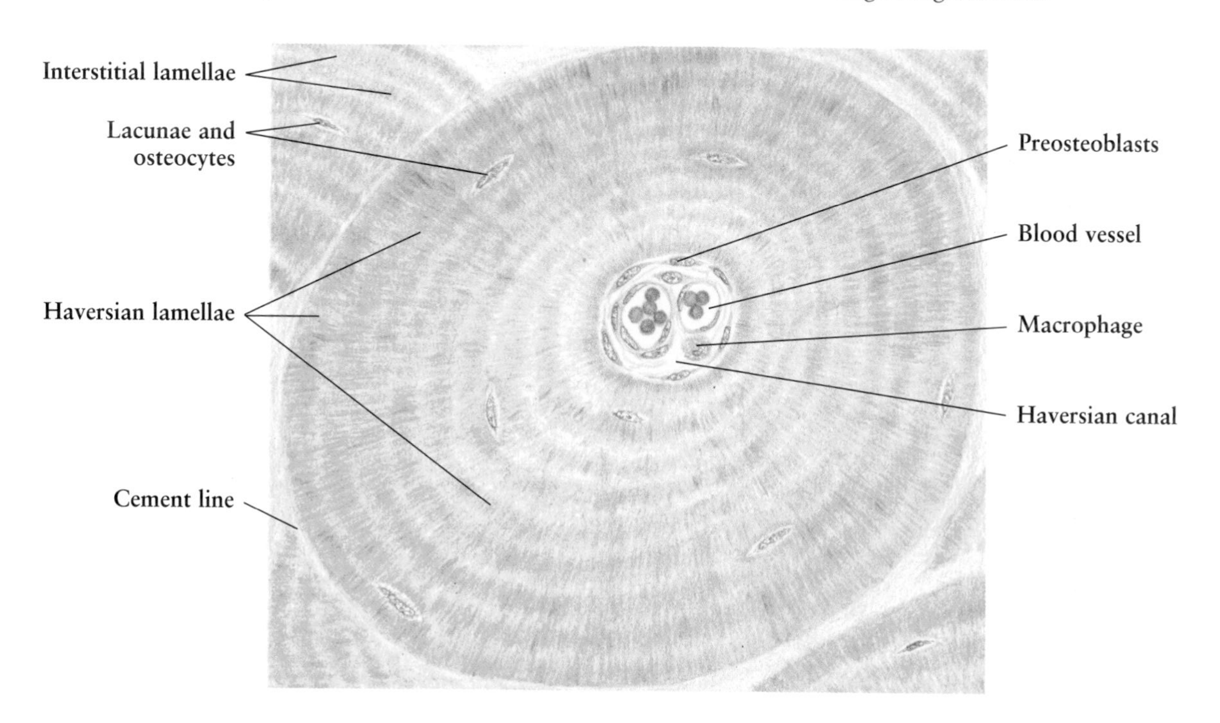

Fig. 3-9.

Bone: Periosteum and Lamellae

Periosteum is a layer of dense connective tissue investing all bones except on the articular surfaces. It consists of outer and inner layers. The outer or **fibrous layer** is very dense fibrous connective tissue. The inner or **osteogenic layer** is more cellular and vascular, and is a kind of loose connective tissue containing blood vessels and fusiform **osteogenic cells** or **osteoblasts** near the surface of the bone. **Sharpey's fibers** are coarse collagen fibers that extend from the fibrous layer into the outer circumferential lamellae and serve to attach the periosteum to the bone. In addition to the periosteum, this picture also shows **outer circumferential lamellae, Haversian systems, interstitial lamellae,** and **cement lines.**

Fig. 3-10.

Bone Formation: Intramembranous Ossification

Intramembranous ossification is one of the ways bone forms that is associated with the periosteum. It occurs not only in the formation of flat bones but also in the thickening of long bone as appositional bone growth. This drawing demonstrates a later stage of intramembranous ossification.

Cells derived from the **inner layer** of active **periosteum,** which is a kind of cellular and vascular loose connective tissue, are transformed into **osteoblasts** and attach to the surface of newly formed **bone.** Osteoblasts are large with a big, round, or ovoid nucleus exhibiting a prominent nucleolus, and have basophilic cytoplasm rich in rough-surfaced endoplasmic reticulum. They are responsible for the formation of the specialized interstitial matrix, the *osteoid,* and for the calcification of the osteoid into bone matrix. As a result of the formation and calcification of osteoids, osteoblasts are surrounded by intercellular matrix and become lacunar **osteocytes.**

On the other hand, **osteoclasts** occur where bone needs to be reconstructed. Osteoclasts are large multinucleate (50–60 nuclei) giant cells found in the concavity (**Howship's lacunae**) of the bone. Their cytoplasm is rich with lysosomes and phagocytic vacuoles. The surface of the osteoclast facing the bone matrix has numerous microvilli forming a **ruffled border,** which serves as an active site of bone resorption. Osteoclasts arise by fusion of monocytes derived from the hematopoietic bone marrow. The main function of the osteoclast is to degrade and absorb old bone matrix during bone remodeling, thus increasing the concentration of calcium in the blood. This activity is stimulated by the *parathyroid hormone* (see **Fig. 14-11**) and is inhibited by the hormone *calcitonin,* secreted by the clear cells in the thyroid gland (see **Fig. 14-9**).

Figure 3-9. Bone: Periosteum and Lamellae
Monkey femur • Decalcified • Cross section • H.E. stain • Medium magnification

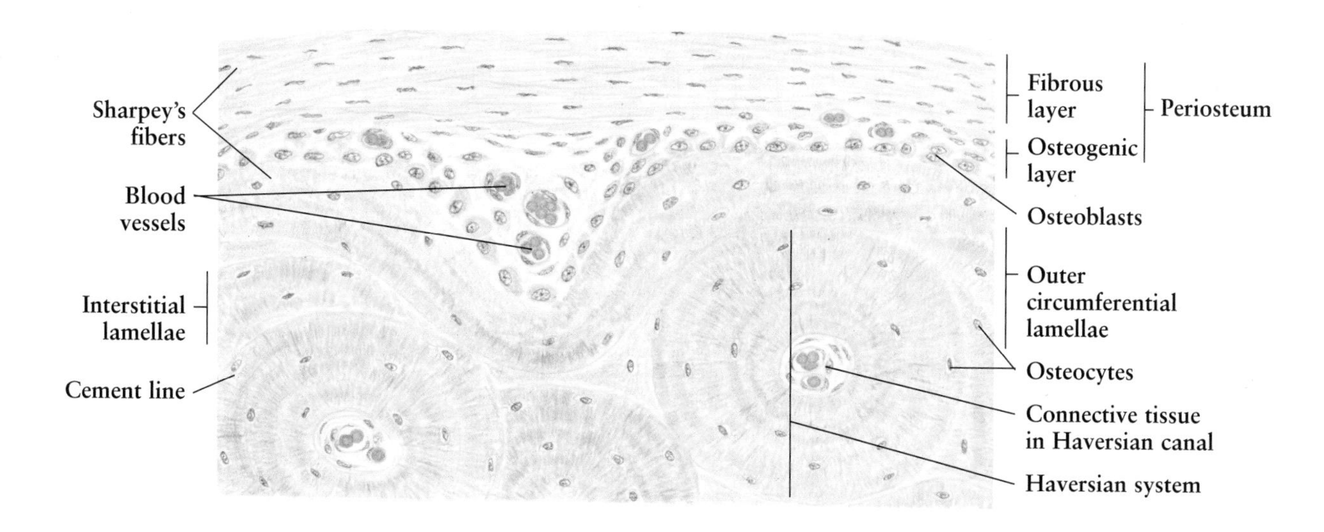

Figure 3-10. Bone Formation: Intramembranous Ossification
Monkey femur • Decalcified • Cross section • H.E. stain • High magnification

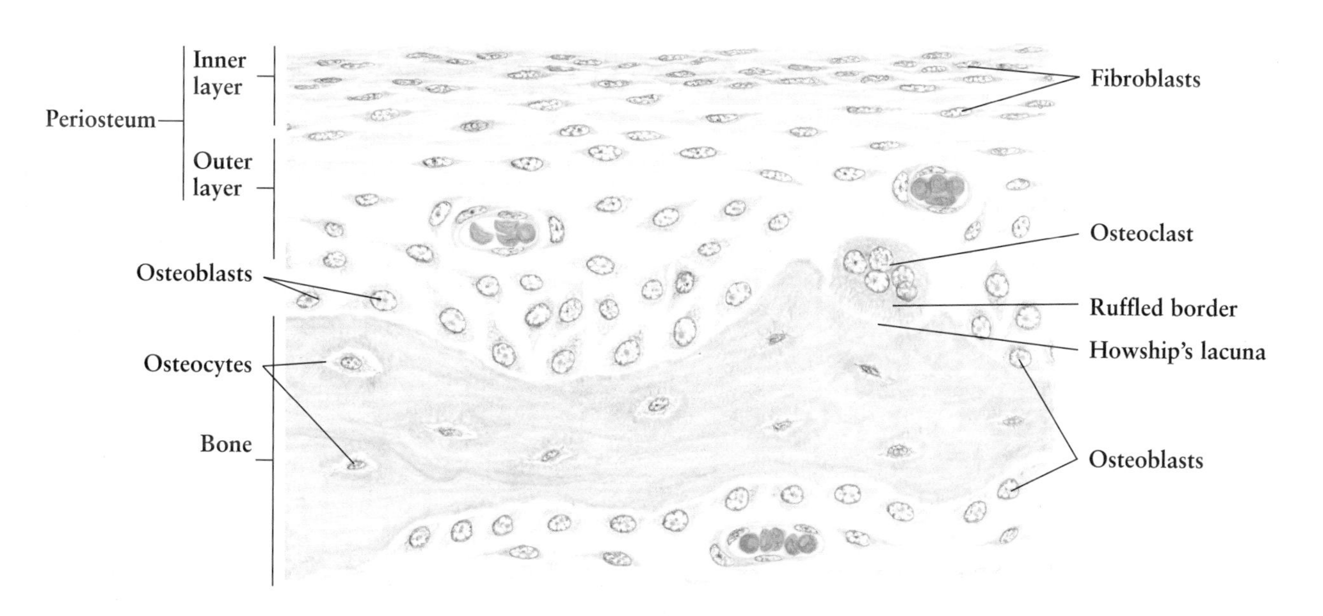

Fig. 3-11.

Bone Formation: Endochondral Ossification

Formation of a long bone takes place by a process of **endochondral ossification** occurring within a hyaline cartilage model, which degrades to be replaced completely by newly formed bone. This includes the appearance of a *primary ossification center* in the middle of the diaphysis and a *secondary ossification center* in each epiphysis. Briefly, the process of formation of the primary ossification center is as follows. Initially, the chondrocytes in the middle of the shaft of the cartilage model hypertrophy and degrade, and the intercellular matrix also degrades and calcifies. Meanwhile, a *bony collar,* a ring of bone surrounding the cartilage, appears beneath the perichondrium (periosteum) by intramembranous ossification. Then blood vessels from the periosteum, along with osteoblasts, osteoclasts, and bone marrow stem cells, penetrate through the bony collar into the calcified matrix where the *primary bone marrow cavity* forms. Subsequently, the formation of *bone matrix* begins. **Figure 3-11** is a longitudinal section of an epiphyseal plate showing a characteristic appearance from epiphysis to diaphysis. The primary ossification center includes five zones.

(1) **Zone of resting cells.** This zone consists of ordinary hyaline cartilage in which **chondrocytes** are not active.

(2) **Zone of cell proliferation.** Chondrocytes undergo proliferation and are arranged in columns parallel with one another and with the long axis of the bone.

(3) **Zone of cell maturation.** The cell division has ceased, and hypertrophic chondrocytes enlarge in size owing to accumulating glycogen and lipid in their cytoplasm. They can no longer divide.

(4) **Zone of cell hypertrophy and calcification.** Chondrocytes become greatly enlarged and vacuolated, then begin to die, and the cytoplasm and nuclei shrink. At the same time, the septa of the cartilage matrix undergo calcification.

(5) **Zone of ossification.** Along with the invasion of **blood vessels** into the **primary bone marrow cavity, osteoclasts** begin to clean the lacunae left by dead chondrocytes and to absorb the **calcified cartilage matrix** to make room for the formation of new bone matrix. Occasionally, **macrophages** can be found cleaning the lacunae. Osteoblasts attach to the remnants of the cartilage matrix to make the bone matrix, in which they become embedded as **osteocytes.** As a result of resorption of bone in the center of the diaphysis by osteoclasts, the primary bone marrow cavity increases in size and becomes the **secondary bone marrow cavity.** At the bottom of this picture, **megakaryocytes** can be observed among the bone marrow stem cells.

At about the time of birth, the *secondary ossification centers* appear in each end of long bones. The process of the bone formation is identical with that which occurs in the primary ossification center, except that the ossification takes place in all directions.

Figure 3-11. Bone Formation: Endochondral Ossification
Epiphysis of monkey femur • Decalcified • H.E. stain • Medium magnification

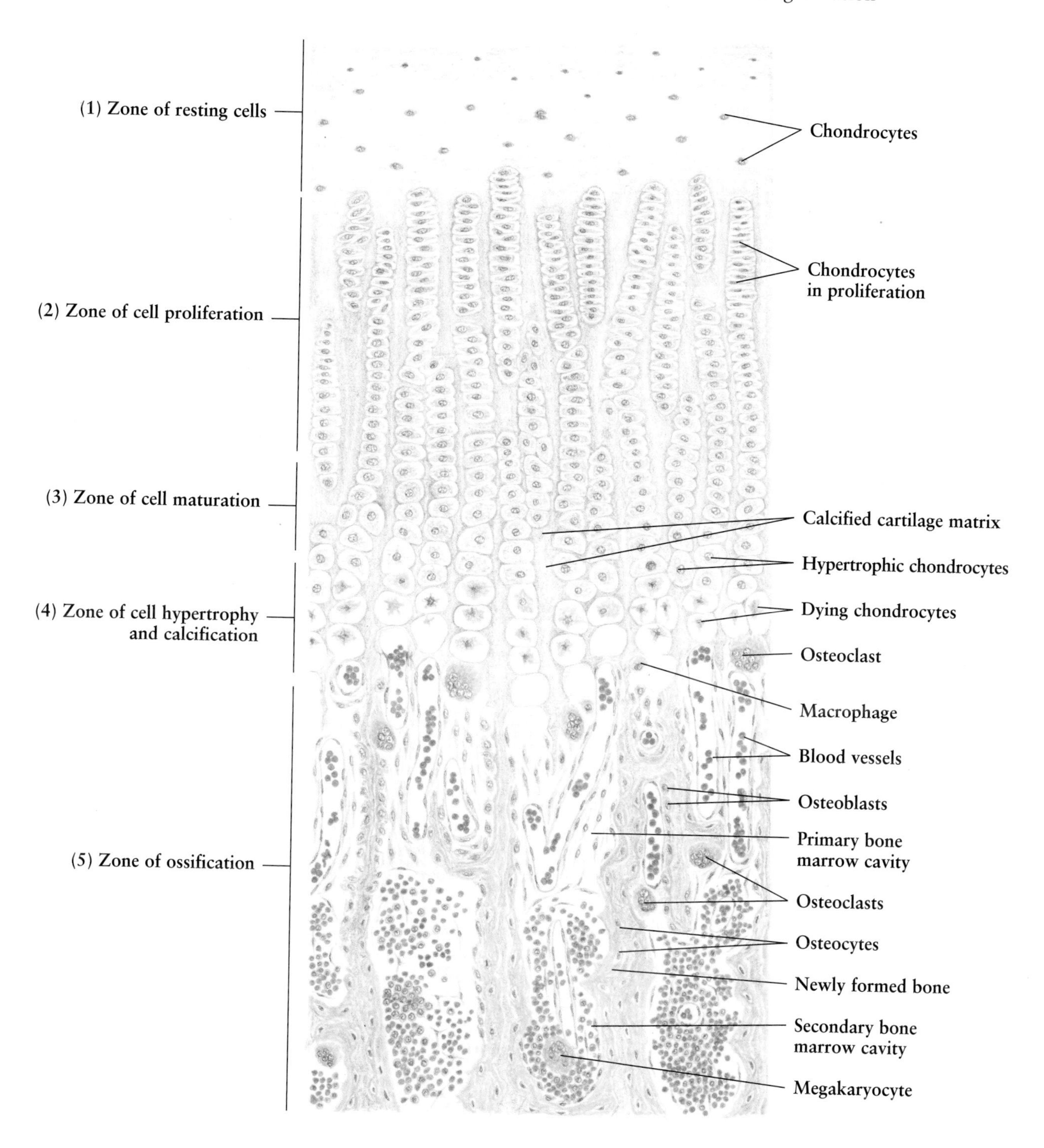

4. Blood Cells and Hemopoietic Cells

Blood is a specialized form of connective tissue, composed of circulating formed elements, the *blood cells,* and a fluid intercellular substance, the blood *plasma.* The total quantity of blood is about 5 to 6 liters, accounting for 6% to 8% of the body weight. Blood serves as a vehicle transporting oxygen, nutrients, metabolic waste products, carbon dioxide, cells, and hormones throughout the body.

Plasma, which constitutes 55% of the total blood volume, is the fluid that contains various inorganic salts, carbohydrates, lipids, and proteins. The plasma proteins, which exert a colloidal osmotic pressure, include albumins, globulins, and fibrinogen. **Serum,** a yellowish fluid, is identical with plasma without fibrinogen, a soluble protein that polymerizes to form the insoluble fibrin during blood clotting.

The formed elements, the **blood cells,** which account for 45% of the total blood volume, consist of *erythrocytes* (red blood cells), *leucocytes* (white blood cells), and *platelets.* The function of the erythrocytes is associated mainly with the transport of oxygen and carbon dioxide. The white blood cells are classified as *neutrophils, eosinophils, basophils, monocytes,* and *lymphocytes,* and play important roles in the defense and immune systems of the body. The *platelets* are cell fragments detached from the *megakaryocytes* in the bone marrow and are a vital component of the clotting mechanism of blood.

All these blood cell types are developed from the **hemopoietic cells** in the bone marrow, by a process known as *hemopoiesis.*

Fig. 4-1.

Erythrocytes (Red Blood Cells)

The **erythrocyte (red blood cell)** in mammals is a cell without a nucleus or organelles and has a life of about 120 days. It is shaped like a biconcave disk, 7.5 µm in diameter and 1.9 µm in its greatest thickness. Erythrocytes contain *hemoglobin,* which gives rise to the red color of blood. The role of erythrocytes is the transport of oxygen and carbon dioxide. In the circulating blood, erythrocytes occur at concentrations of 4.5–5 million/mm^3 in women and 5–5.5 million/mm^3 in men. **Figure 4-1** was drawn from fresh blood samples. Students in the histology laboratory may make similar preparations to examine the characteristic features of the fresh human erythrocytes at the level of the light microscope.

Fig. 4-1A. In a drop of blood just taken from the human ear lobe , erythrocytes tend to adhere to each other along their concave surfaces, a phenomenon known as *rouleaux formation.*

Fig. 4-1B. When a 0.85% (isotonic) saline solution is added to the blood drop, erythrocytes in rouleaux form are separated, and single cells can be seen.

Fig. 4-1C. When a 3% (hypertonic) saline solution is added to the blood drop, erythrocytes shrink in a process called *crenation.*

Fig. 4-1D. When a 1% acetic acid solution is added to the blood drop, erythrocytes are damaged, hemoglobin escapes, and hemolysis occurs. The empty erythrocytes are referred to as *blood shadows* or *blood ghosts.*

Figure 4-1. Erythrocytes (Red Blood Cells)
Fresh human blood • Unstained • High magnification

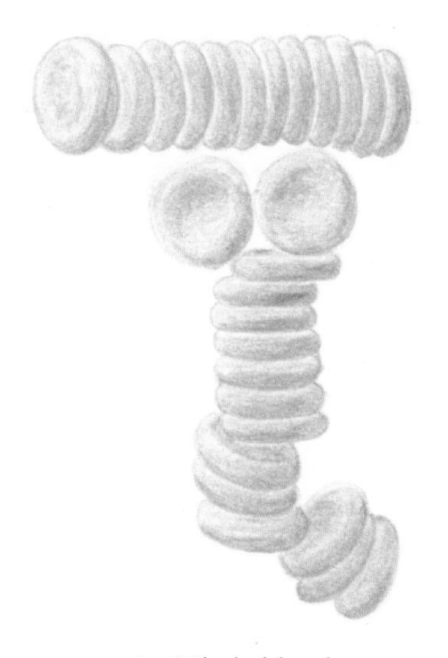

A. Whole blood

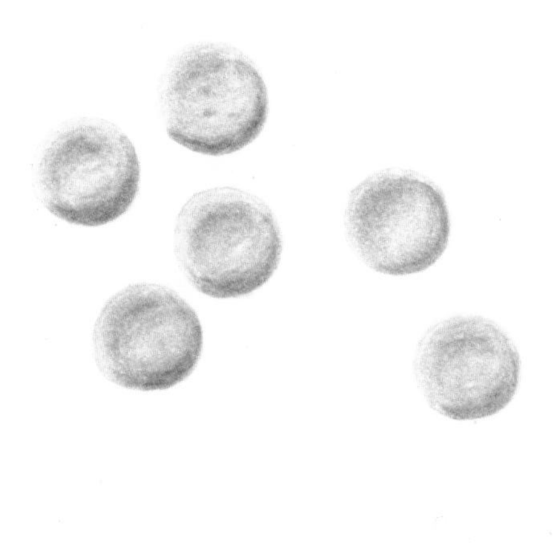

B. 0.85% saline added to blood

C. 3% saline added to blood

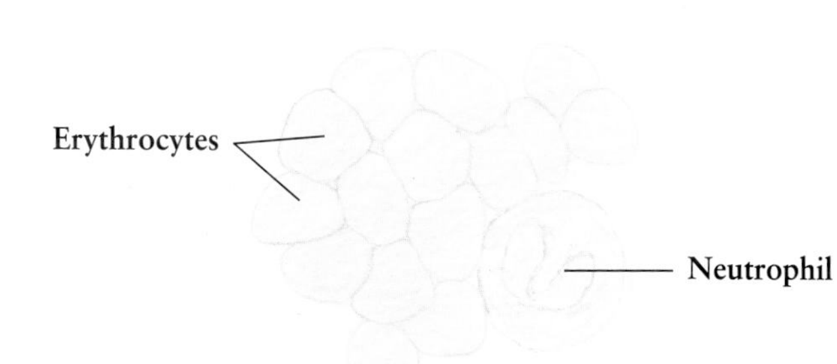

D. 1% acetic acid added to blood

Fig. 4-2. Blood Cells

Figure 4-2 shows a variety of human circulating **blood cells** and **platelets** seen under the light microscope after fixation and staining. A table (**Table 4-1**) summarizes the characteristic features of these cells and their relative size and number in normal blood. These figures can aid in identification of blood cells.

Feature	**Erythrocytes (red blood cells) 4.5–5.5 million/mm³**	**Leukocytes (white blood cells) 6,000–10,000/mm³**					**Platelets, 200,000–300,000/mm³**
		Granulocytes			**Agranulocytes**		
		Neutrophil	**Eosinophil**	**Basophil**	**Monocyte**	**Lymphocyte**[a]	
Percentage in WBC		60%–70%	2%–4%	0%–1%	3%–8%	20%–30%	
Life span	120 days	8 days	8 days	12–15 days	60–90 days	A few days to years	7–9 days
Shape	Biconcave	Spherical	Spherical	Spherical	Spherical	Spherical	Oval, irregular biconvex
Diameter	7.5 μm	12–15 μm	12–15 μm	12–15 μm	12–20 μm	6–18 μm	2–4 μm
Nucleus	None	U-shaped, S-shaped, 2–5 segmented	Segmented, bilobed	Poorly shown, S-shaped	Kidney shaped-	Round	None
Chromatin	None	Condensed	Not clear	Not clear	Delicate network	Relatively condensed	None
Cytoplasm	Pink	Light pink	Not clear	Not clear	Basophilic bluish-gray	Scanty, light blue	Pink
Granules	None	Azurophilic granules; specific granules	Eosinophilic granules	Basophilic granules of different sizes	Small, round azurophilic granules	Few small azurophilic granules	Alpha, delta, gamma granules, basophilic
Function	Carry O_2 and CO_2	Ingest bacteria, release lysozyme, form pus after death	Release histamine	Release histamine, heparin, leukotrienes	See footnote[b]	Involved in immu-nological defense system	Involved in agglutination

[a] **Lymphocytes** may be classified as large, medium, and small lymphocytes, according to their sizes.

[b] The **monocytes** enter the connective tissue and transform into *macrophages,* which engulf bacteria and other foreign bodies, secrete *complement,* and synthesize *interferon.*

Figure 4-2. Blood Cells
Human blood smears • Giemsa's stain • Higher magnification

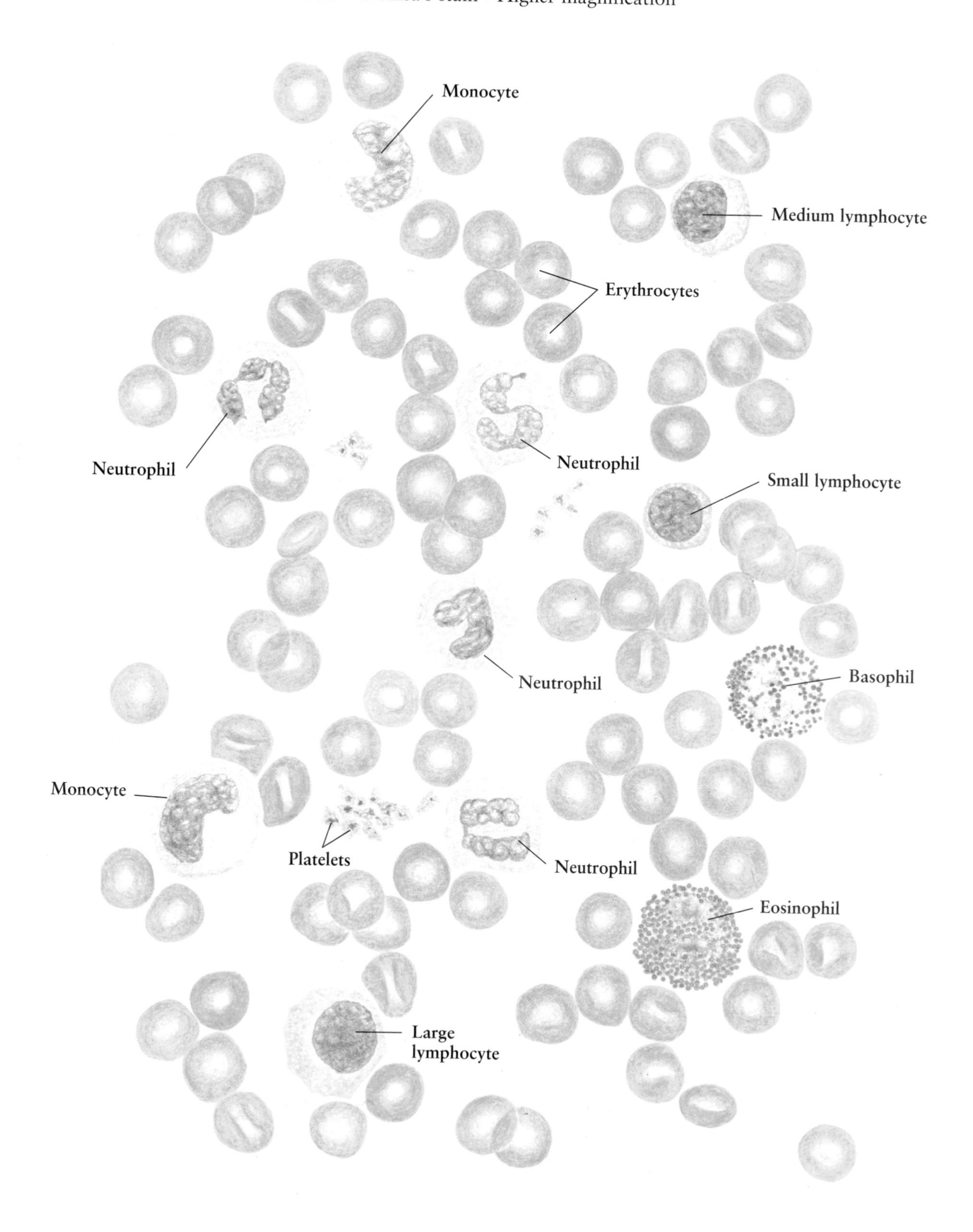

Fig. 4-3.

Hemopoietic Tissue

In the adult body, there are two kinds of **hemopoietic tissues:** *red bone marrow* and *lymphatic tissue.* The former produces erythrocytes, leukocytes, and platelets, and the latter is the source of lymphocytes. The red bone marrow is composed chiefly of sinusoids and three cell types: reticular cells, adipose cells, and hemopoietic cells (blood-forming cells). Along with reticular fibers (not shown in this picture), the **reticular cells,** which have ovoid nuclei and several processes, form a framework of reticular tissue in which the hemopoietic cells are contained. **Adipose cells** are the largest cells in the bone marrow, occupying more and more of the bone marrow cavity with age. **Hemopoietic cells** differentiate into erythrocytes and leukocytes. They are in different developmental stages, and some of them are seen undergoing mitotic division (**in mitosis**). In a marrow section, stages of hemopoietic differentiation are difficult to see, but they can be recognized in the bone marrow smear (see **Fig. 4-5**). Additionally, this figure also shows two giant cells, the platelet-forming **megakaryocytes,** with huge lobulated nuclei (see **Fig. 4-4**). Megakaryocytes are often located near the **sinusoids.**

Fig. 4-4.

Megakaryocyte

Megakaryocytes are the largest among the blood-forming cells in the bone marrow. They may measure 80–100 μm in diameter, with a large volume of acidophilic cytoplasm rich in dense **azurophil granules** (many of them are lysosomes) and microtubules, which may be detected by immunohistochemical methods. The periphery of the megakaryocyte is irregular, with *pseudopods* of different sizes. The **nucleus** is large and lobulated with condensed and coarse heterochromatin. No nucleoli can be recognized. It is from the cytoplasm of megakaryocytes that numerous platelets are formed, which contain a network of microtubules as their skeletons.

Figure 4-3. Hemopoietic Tissue
Human bone marrow section • H.E. stain • High magnification

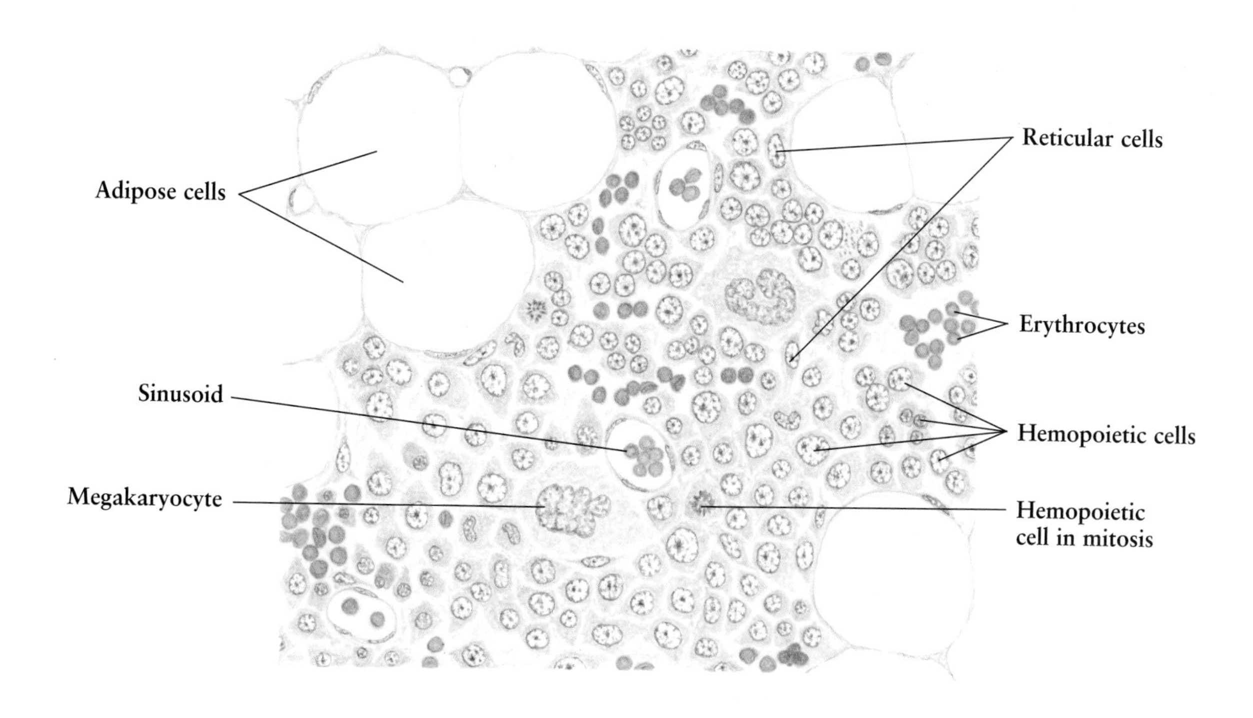

Figure 4-4. Megakaryocyte
Smear of human bone marrow • Grunwald-Giemsa's stain • Higher magnification

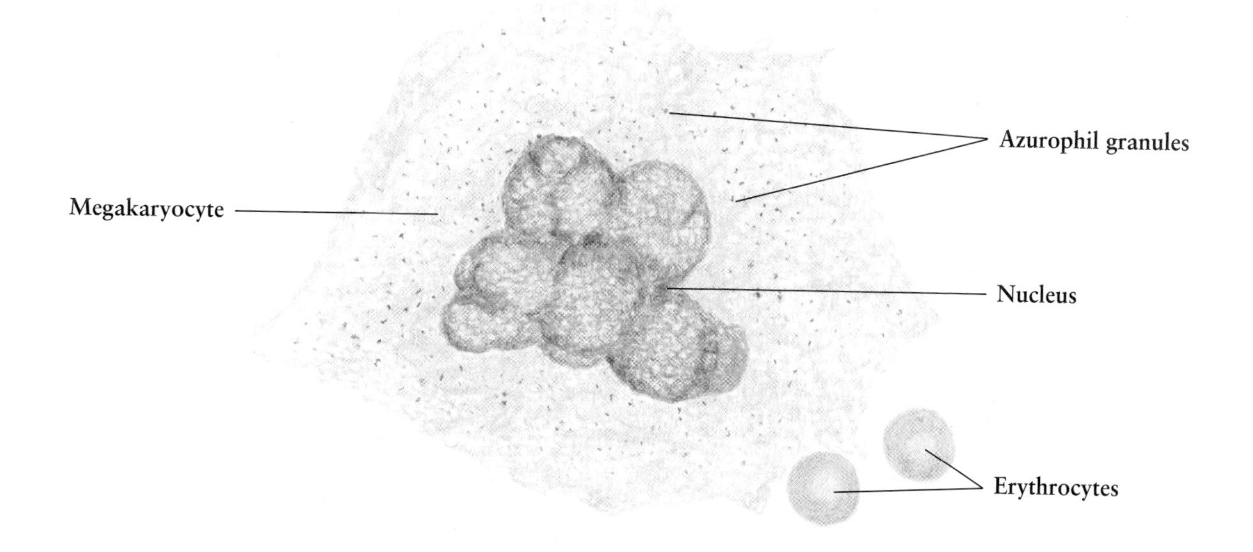

Fig. 4-5.

Formation of Blood Cells (Hemopoiesis)

This figure is a collection of the blood-forming cells of bone marrow in different stages of development from both the erythrocytic series and the granulocytic series. The **erythrocytic series** contains the proerythroblasts, basophilic erythroblast, polychromatophilic erythroblast, orthochromatophilic erythroblast, and reticulocyte; the **granulocytic series** contains the myeloblast, promyelocyte, myelocytes, and metamyelocytes. These cells are drawn from a human bone marrow smear stained by the Wright method. The characteristic features of these cell types are briefly described.

Erythrocytic Series:

Proerythroblast is the earliest cell recognizable in the development of the erythrocyte. It measures up to 20 μm in diameter and has an intense basophilic cytoplasm with an acidophilic rim around the nucleus, indicating the beginning of hemoglobin synthesis. The nucleus is large and spherical, occupying most of the space of the cell, and the chromatin appears as a coarse network. There are two or three nucleoli. After several mitotic divisions, the proerythroblast gives rise to the basophilic erythroblast.

Basophilic erythroblast shows a reduction in size (15–17 μm in diameter). Its nucleus is slightly smaller than that of the proerythroblast. The chromatin becomes condensed, but the nucleolus usually is obscured. The sparse cytoplasm is still strikingly basophilic, and the hemoglobin (an acidophilic cytoplasmic rim around the nucleus) continues to increase in size.

Polychromatophilic erythroblast is formed by mitotic division of a basophilic erythroblast. It becomes smaller (12–15 μm in diameter) than the basophilic erythroblast, and the chromatin of the nucleus becomes more condensed. The cytoplasm has a characteristic grayish-pink color due to the mixture of the acidophilia of the hemoglobin and the basophilia of the ribosomes. The polychromatophilic erythroblast undergoes a number of mitotic divisions to form the normoblast.

Orthochromatophilic erythroblast (**normoblast**) continues to decrease in size, measuring 8–10 μm in diameter. The cytoplasm is usually basophilic, and the nucleus contains coarse condensed chromatin. After three mitotic divisions, the dark nucleus becomes located eccentrically and is finally extruded from the cytoplasm. Thus, the normoblast enters the stage of reticulocyte.

Reticulocyte, an immature erythrocyte, is an anuclear round cell with a diameter of 9 μm. The cytoplasm is filled with hemoglobin and appears pink. The polyribosomes in the cytoplasm can be precipitated by brilliant cresyl blue to form a weblike basophilic reticulum, from which the cell's name has been derived. Reticulocytes account for about 1% of circulating erythrocytes.

Granulocytic Series:

Myeloblast is the earliest cell recognizable in the granulocytic series. The cell body is almost round, with a diameter of about 15 μm. The large spherical or ovoid nucleus shows a delicate chromatin pattern with 3–5 nucleoli. The cytoplasm is basophilic and has a sky-blue color.

Promyelocyte, the largest cell in the granulocytic series, measures 18–24 μm in diameter. The nucleus is ovoid with an eccentric location. It has dense peripheral chromatin and two to three recognizable nucleoli. This cell is characterized by the presence of azurophil granules in the cytoplasm. These granules look like diamonds inlaid on a gauze kerchief. The promyelocyte undergoes mitotic division and differentiates into a myelocyte.

Myelocytes are smaller than promyelocytes, 16–20 μm in diameter. The nucleus is kidney-shaped with condensed chromatin, and sometimes nucleoli are present. The most characteristic feature in the myelocyte stage is the appearance of specific granules accompanied by the reduction of azurophil granules, so that these cells can be recognized as an **eosinophilic myelocyte,** a **neutrophilic myelocyte,** and a **basophilic myelocyte**. After repeated mitotic divisions, myelocytes become metamyelocytes.

Figure 4-5. Formation of Blood Cells (Hemopoiesis)
Human bone marrow • Wright's stain • Higher magnification

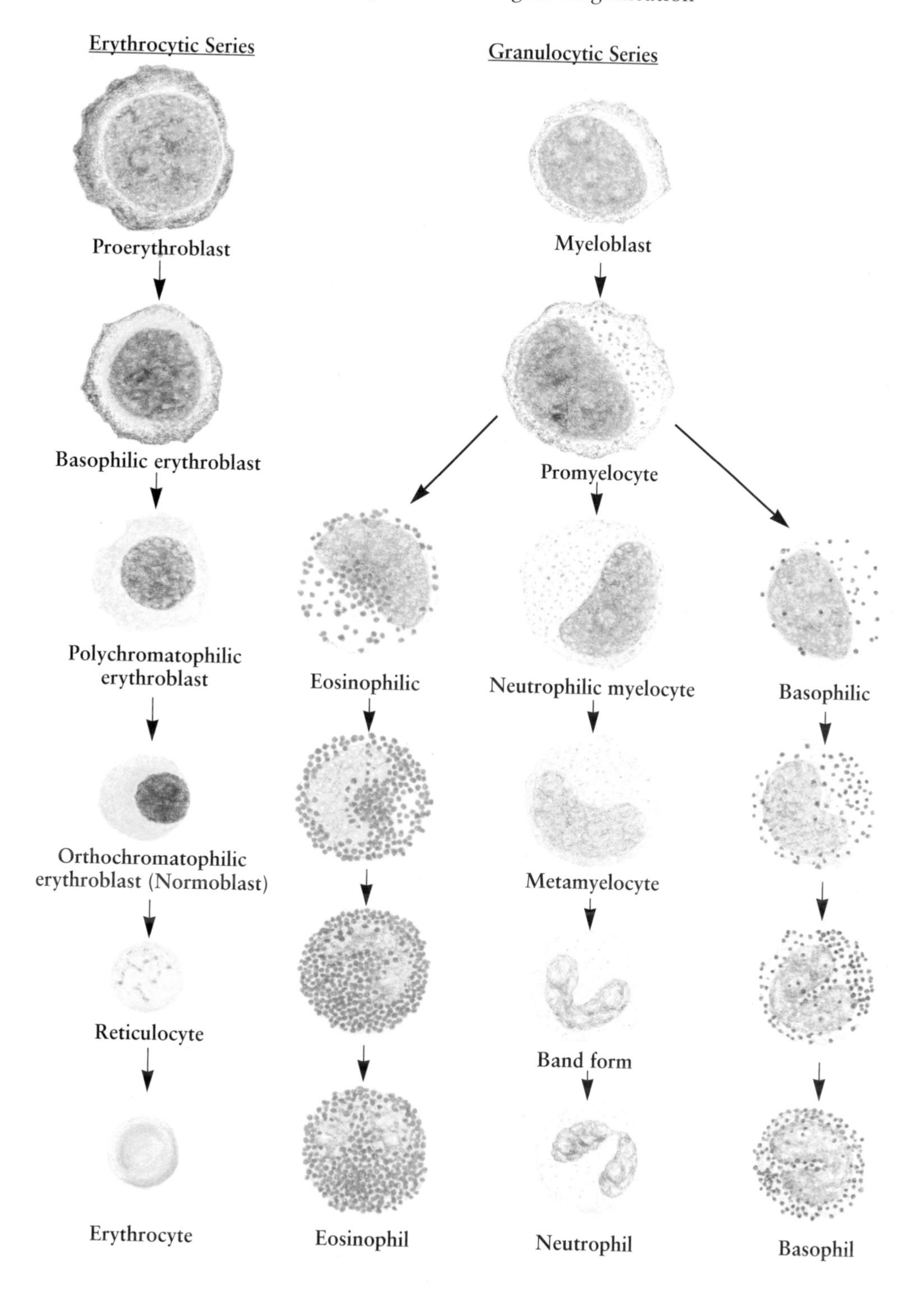

Metamyelocytes continue to decrease in size (about 8–12 μm in diameter). The nucleus has the shape of a horseshoe and gradually indents further, with condensed chromatin and without visible nucleoli. The cytoplasm has a pink color, with an increasing number of specific granules. Metamyelocytes cease division but become the **band form** of a granulocyte in which the nucleus appears as a curved rod. The band form cells make up 3%–5% of circulating granular leukocytes.

5. Muscular Tissue

Muscular tissue is specialized for contractility. It is composed of elongated, contractile muscle cells, which are often referred to as **muscle fibers**. Each muscle fiber contains numerous elongated, threadlike structures, **myofibrils**, which are composed of contractile proteins and are the smallest units of contraction. In muscular tissue, the muscle fibers usually are arranged in bundles, with their long axes parallel to one another. The muscular tissue is provided with a network of capillaries and nerve endings, which conduct the impulse from or toward the central nervous system.

There are three types of muscular tissue distinguished on both a structural and functional basis: smooth muscle, skeletal muscle, and cardiac muscle.

1. **Smooth muscle** is composed of bundles of fusiform muscle cells that do not show any striations under the light microscope. It forms the muscular component in the walls of viscera. The smooth muscle fibers may also function as single contractile units, such as the myoepithelial cells surrounding the acini of exocrine glands. Their contraction is slow and involuntary.

2. **Skeletal muscle** consists of bundles of elongated, cylindrical multinucleated muscle cells that show prominent cross striations. The skeletal muscle, attached to bones or fascia, is responsible for the movement of the skeleton and organs such as the tongue and eyeball. Their contraction is controlled by the will, and therefore is quick, forceful, and voluntary.

3. **Cardiac muscle** is formed of cylindrical and branching muscle cells, which also show cross striations. Cardiac muscle occurs in the wall of the heart and extends into the major veins opening into the heart. The contraction of the cardiac muscle is rhythmic, automatic, and involuntary.

Figs. 5-1 and 5-2.

Smooth Muscle

Smooth muscle is composed of long spindle-shaped cells, the **smooth muscle fibers.** These fibers are enveloped in fine fibroelastic tissue, forming the contractile portion of many organs. A single smooth muscle fiber is 20–500 µm long and 5–10 µm wide, with an elongated ovoid **nucleus** in the middle of the cell. **Figure 5-1** shows individual smooth muscle fibers isolated from guinea pig intestine with 33% KOH. In longitudinal section (**Fig.** 5-2), the smooth muscle fibers are seen arranged with the nuclear (widest) region of one cell aligned with the tapering ends of its adjacent fibers. The **cell processes** of smooth muscle fibers can be clearly recognized in good specimens. These processes are short, narrow fingerlike projections, which make contact with adjacent smooth muscle cells. In cross section, the nucleus is round and is seen only in a few cell profiles.

Smooth muscle, also called visceral muscle, occurs in the digestive tract, blood vessels, lymphatic vessels, respiratory system, urinary system, reproductive system, dermis, and iris and ciliary body of the eye. Smooth muscle is innervated by autonomic (sympathetic and parasympathetic) nerve fibers, and exhibits a slow and continuing contraction. In the digestive tract, **ganglion cells** surrounded by **satellite cells,** a cell type of neuroglia in the peripheral nervous system, occur between two layers of smooth muscle (**Fig.** 5-2).

Smooth muscle enables the viscera to contract in a rhythmic or wavelike fashion for long periods without fatigue. For example, the continuous rhythmic contractions of the smooth muscle in the gastrointestinal tract propel the luminal contents distantly.

Figure 5-1. Isolated Smooth Muscle Fibers
Guinea pig intestine • Unstained • High magnification

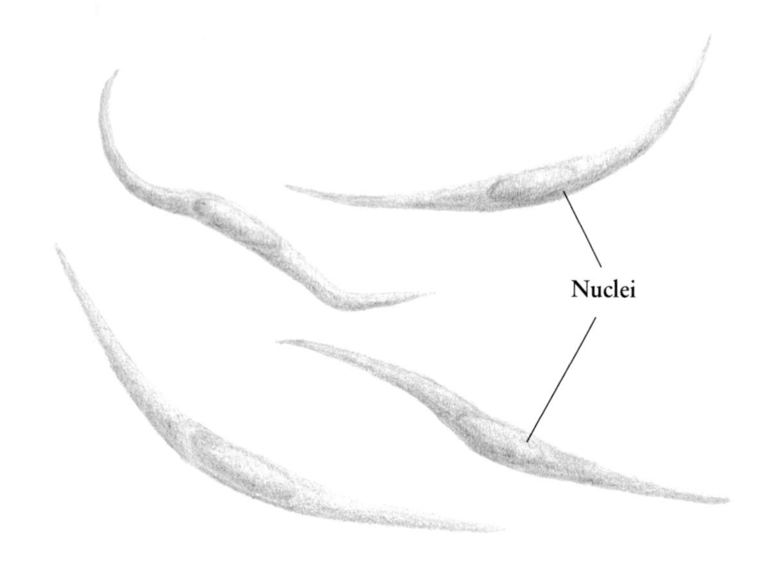

Figure 5-2. Smooth Muscle
Human appendix • H.E. stain • High magnification

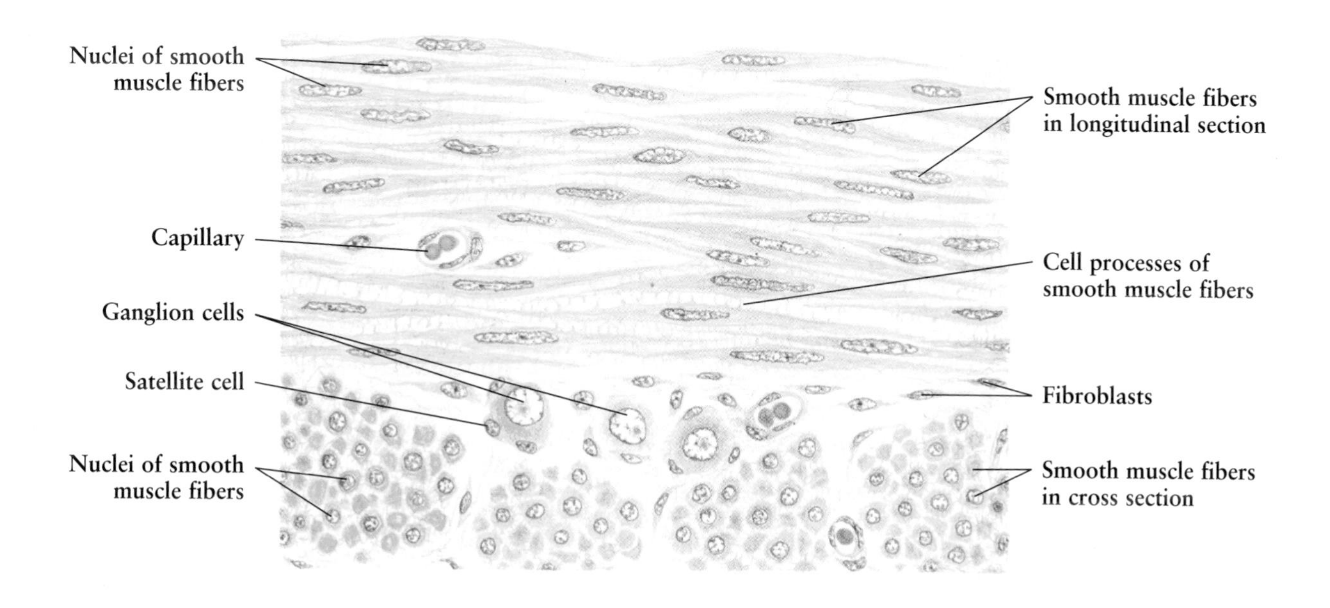

Figs. 5-3, 5-4 and 5-5.

Skeletal Muscle

Skeletal muscle consists of bundles of long cylindrical multinucleated **muscle fibers** (cells) with cross striations. The cells are 1–40 mm long and 10–100 μm wide. A single fiber may contain several hundred elongated ovoid **nuclei** at the periphery of the cell. At high magnification under the light microscope, the striations can be seen composed of alternate light and dark bands. The dark bands are termed **A bands** (anisotropic); the light bands are termed **I bands** (isotropic). In the middle of each I band is a dark line called the Z band or **Z line** parallel to the A and I bands. The distance between two Z lines is called a *sarcomere,* which is the structural and functional unit of the muscle contractile apparatus. Individual muscle fibers are covered by a thin layer of delicate connective tissue called *endomysium,* composed of a mixture of collagen fibers, elastic fibers, and **fibroblasts.** Between muscle fibers there is an abundance of **capillaries** and some **nerve fibers.**

Skeletal muscle, also called voluntary muscle, is innervated by large motor nerves through neuromuscular junctions, the *motor end plate* (see **Fig. 6-23**). The contraction of skeletal muscle is controlled voluntarily and is abrupt and strong yet transient (fatiguable).

Figure 5-3 shows a piece of fresh skeletal muscle fiber separated with 33% KOH from guinea pig, showing the striations and nuclei. **Figure 5-4** is a paraffin preparation and **Fig.** 5-5 is an epon (plastic) preparation. Both **Fig.** 5-4 and **Fig.** 5-5 show **muscle fibers in cross section** (note peripheral nuclei) and **muscle fibers in longitudinal section** (note striations).

Figure 5-3. Isolated Skeletal Muscle Fiber
Guinea pig • Unstained • High magnification

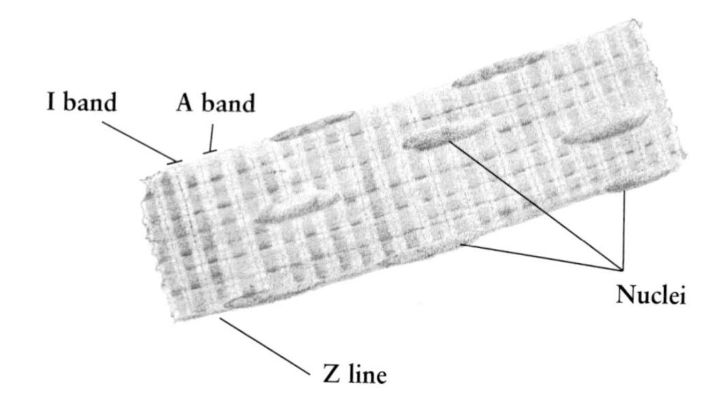

Figure 5-4. Skeletal Muscle
Human tongue • H.E. stain • High magnification

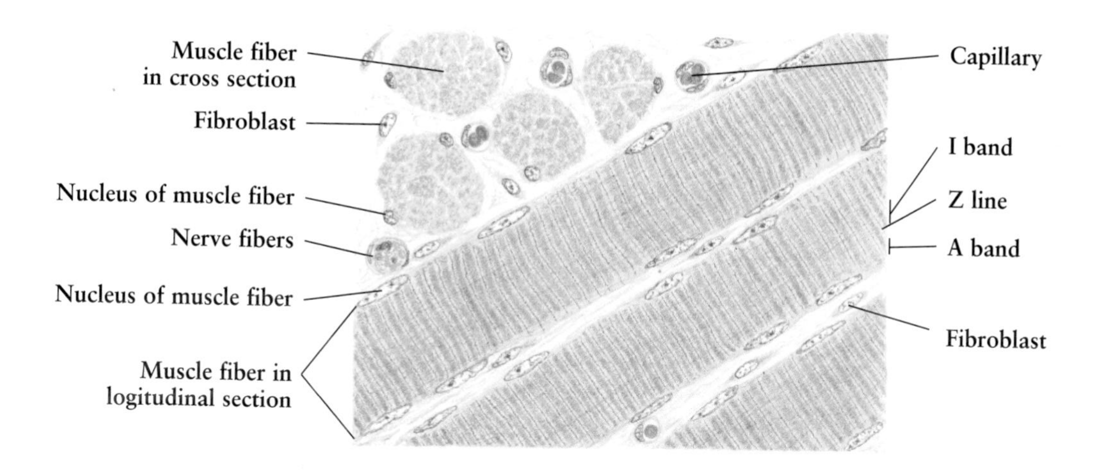

Figure 5-5. Skeletal Muscle
Rat skin • Epon section • Toluidine blue stain • High magnification

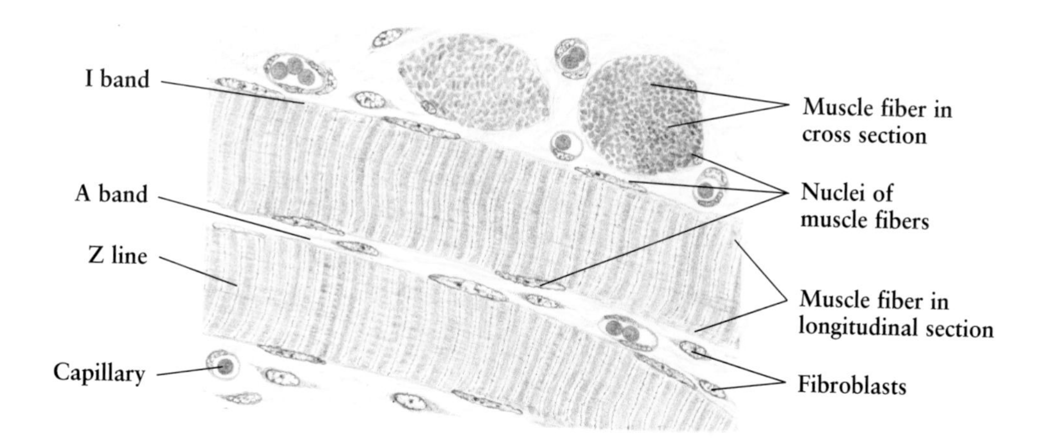

Figs. 5-6, 5-7 and 5-8.

Cardiac Muscle

Cardiac muscle is composed of striated muscle cells that contract involuntarily, rhythmically, and automatically. It is found only in the muscle layer of the heart *(myocardium)* and in the major veins joining the heart.

The **muscle fibers** are cylindrical, having an average length of 100–150 μm and width of 15–20 μm. A single muscle fiber has one or two elongated or ovoid **nuclei,** located centrally. Around the nuclei are fusiform areas of *sarcoplasm* containing many organelles. The **cross striations** formed with the contractile apparatus are less distinct than those in skeletal muscle fibers. Cardiac muscle fibers branch, bifurcate, and anastomose to form a complex three-dimensional network constituting the myocardium. Muscle fibers including their branches join end-to-end, with distinctive junctional zones marked by steplike **intercalated discs.** Between muscle fibers is a layer of connective tissue containing collagen fibers, reticular fibers, and **capillaries** as well as **fibroblasts. Figure 5-6** shows two isolated cardiac muscle fibers with **cross striations** and **nuclei** clearly seen; one of them is branched. In **Fig.** 5-7, muscle fibers in **longitudinal, cross,** and **oblique sections** are seen.

In **Fig. 5-8, Purkinje fibers** are shown just beneath the **subendothelial layer** of the **endocardium.** These are specialized cardiac muscle fibers belonging to the impulse-conducting system of the heart. As cardiac muscle fibers, Purkinje fibers contain cross striations and are marked with **intercalated discs** at the interface of two fibers. However, they are larger and thicker (about 50 μm in diameter) than the usual cardiac muscle fibers and have very few myofibrils and much more central sarcoplasm.

Figure 5-6. Isolated Cardiac Muscle Fibers
Guinea pig heart • Unstained • High magnification

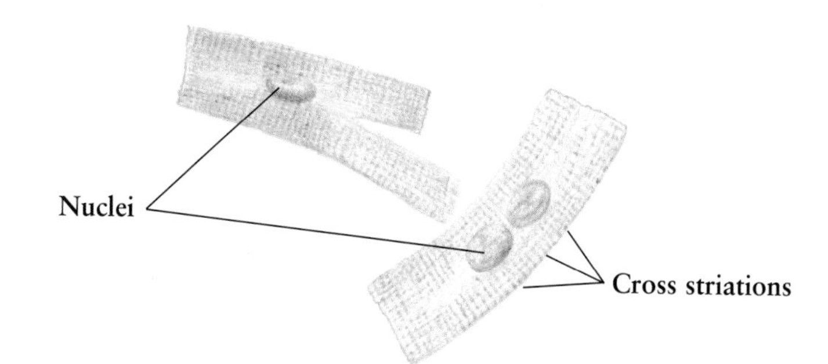

Figure 5-7. Cardiac Muscle
Human heart • H.E. stain • High magnification

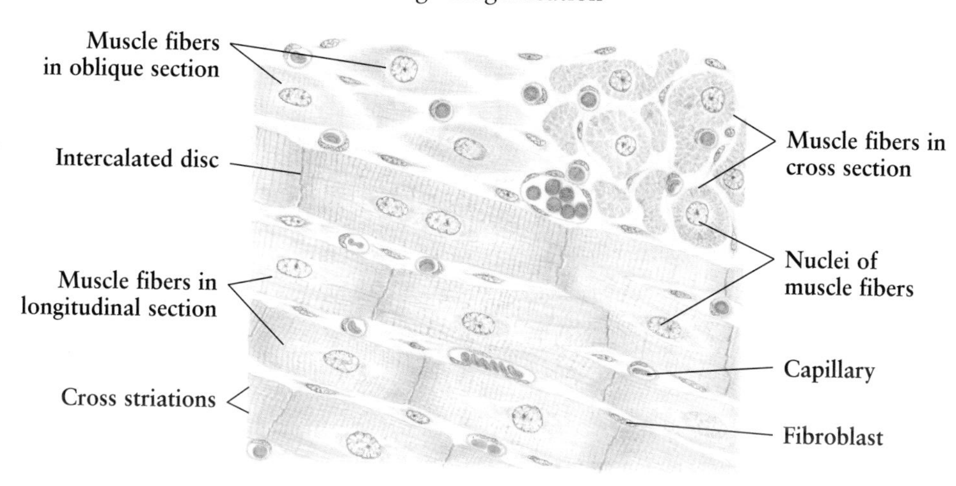

Figure 5-8. Cardiac Muscle: Purkinje Fibers
Human heart • M.G. stain • High magnification

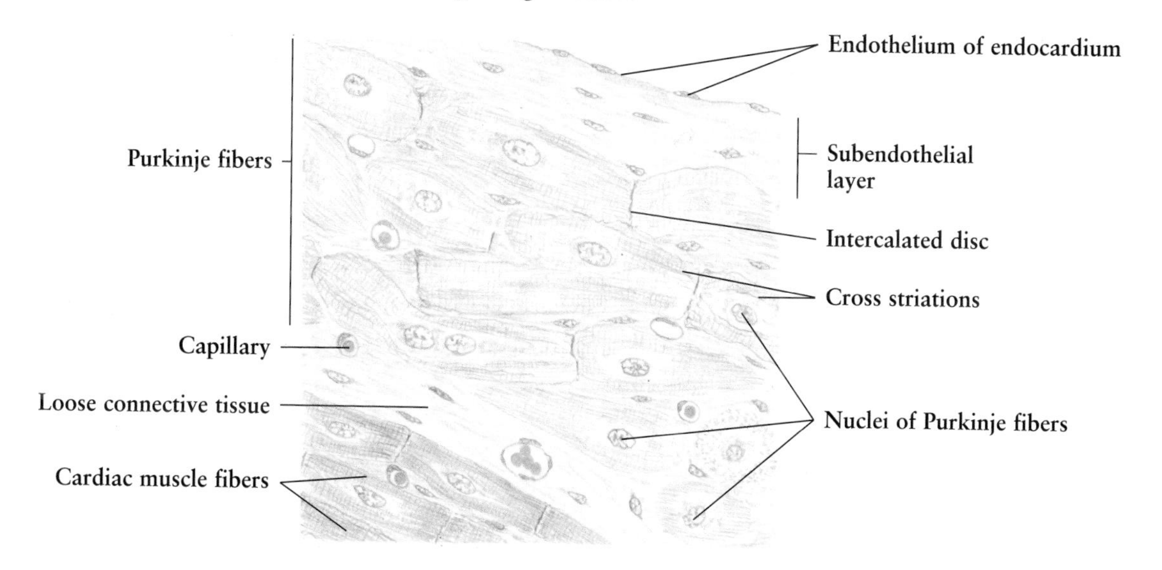

6. Nervous Tissue and Nervous System

Nervous tissue, the most complex of the four basic tissues, is composed of nerve cells, the **neurons,** and supporting cells, the **neuroglia.** Unlike other basic tissues, the nervous tissue almost lacks intercellular substance. Anatomically, the nervous system can be divided into the **central nervous system** (CNS) and the **peripheral nervous system** (PNS). The **CNS** consists of the nervous tissue of the brain and spinal cord; the **PNS** is composed of all other nervous tissue outside the CNS. The PNS collects information from the body surface, viscera, and special sense organs such as eyes and ears, and transmits signals to the CNS where this information is integrated and analyzed to produce appropriate responses in various effector tissues or organs. All these activities are carried out by the neurons, which form an integrated communication network by way of specialized contacts, the *synapses.*

The **neuron** is the structural and functional unit of nervous tissue. It is composed of a cell body and its processes, dendrites and axons. The nerve fibers, collections of dendrites or axons, are classified as either myelinated or unmyelinated fibers. Myelin sheaths of **myelinated fibers** are formed by oligodendrocytes in the CNS and by Schwann cells in the PNS. The **unmyelinated fibers** are nude in the CNS, but are invested by a layer of cytoplasm of Schwann cell in the PNS. In addition to the functions of receiving stimulus and conducting nervous impulse, some neurons of the brain stem secrete hormones.

The **neuroglia** are 10 times more numerous than neurons. They have an important role in supporting, protecting, and nourishing the neurons. They surround the cell bodies of the neurons and their processes, filling the intercellular spaces. According to their structures and functions, the neuroglia are classified as **astrocytes, oligodendrocytes, microglia, ependymal cells,** and **Schwann cells.**

This chapter shows the structure of neurons including their cell bodies and axons at the level of light microscopy. It also demonstrates the fine structure of the **cerebral** and **cerebellar cortices, hippocampal formation,** and **spinal cord.** The **neuroglia, peripheral nerve, ganglion,** and various **nerve endings** are also included.

Fig. 6-1.

Types of Neurons

Structurally, a neuron consists of the **cell body** or *perikaryon* and all its processes, **dendrites** and an **axon.** The cell body of a neuron is round, polygonal, or pyramidal and ranges in diameter from 10 to 120 µm. A neuron has only one axon, but may have more than one dendrite, which can branch, taper, or form a treelike arbor.

According to their processes, most neurons can be morphologically classified as **pseudounipolar, bipolar,** or **multipolar.** The **spinal ganglion cells** are pseudounipolar neurons. These neurons show a single process close to the cell body, which branches into two processes, one traveling to the spinal cord, the other to a peripheral target organ. The neurons in the **spiral ganglion** are bipolar cells, which have one dendrite and one axon extending to the CNS. **Purkinje's cells** in the cerebellar cortex, **pyramidal cells** in the cerebral cortex, and **motor neurons** of the anterior horn in the spinal cord are all multipolar neurons. This drawing shows each of these different kinds of neurons as they appear following a variety of staining methods.

Figure 6-1. Types of Neurons
Various stains • High magnification

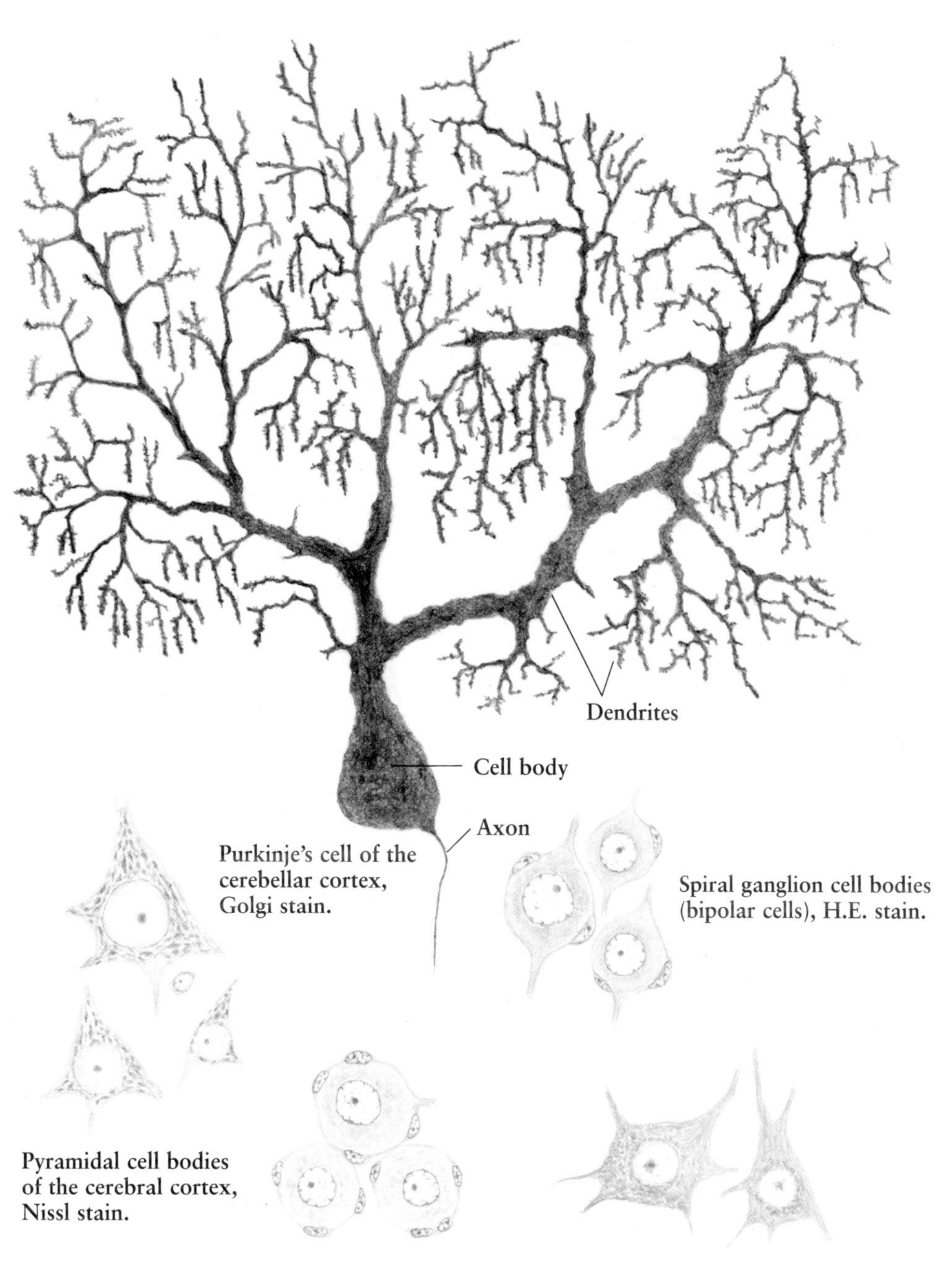

Fig. 6-2.

Neuron: Cell Body

The **cell body** of the neuron consists of its nucleus and surrounding cytoplasm, which contains a variety of organelles such as rough-surfaced endoplasmic reticulum (rER), the Golgi complex, mitochondria, neurofilaments, microtubules, and lysosomes.

The **nucleus** has a "fish-eye" appearance. It is large and spherical, with finely dispersed **chromatin** and one or more pronounced **nucleoli,** indicating active synthesis of the protein in the cell (**Fig. 6-2A–D**). The nucleus is surrounded by a nuclear envelope, which consists of a double layer of unit membrane and shows numerous pores under the electron microscope.

In preparations stained with a basic dye, the rER appears as clusters of an intensely basophilic component, known as **Nissl bodies.** These are most striking in the motor neurons of the spinal cord and pyramidal cells of the cerebral cortex. Nissl bodies can be found in the cytoplasm of the cell body and **dendrites,** but not in **axons** or **axon hillocks** (**Fig. 6-2A**). Injury to the neurons leads to the reduction or disappearance of Nissl bodies. If the injured neuron survives, Nissl bodies can reappear.

When prepared by silver impregnation, the **Golgi complex** of the neuron is demonstrated as *dictyosomes,* arranged around the nucleus or scattered throughout the cell body (**Fig. 6-2B**). Ultrastructurally, the Golgi complex consists of saccules, vesicles, and vacuoles, forming a three-dimensional distribution of the components, which has a curved appearance. The Golgi complex is involved in the accumulation and concentration of protein-rich secretory products.

In the cytoplasm of a neuron prepared with the Cajal stain, numerous **neurofibrillae** are visible under the light microscope (**Fig. 6-2C**). They appear as parallel bundles, extending into the **dendrites** and the **axon.** Neurofibrillae are considered to be artifacts of neurofilaments and microtubules clumped by silver impregnation. Neurofilaments (diameter of 10 nm) and microtubules (diameter of 24 nm) extend throughout the perikaryon and the processes. The functions of neurofilaments and microtubules are associated with transportation within the neuron.

Lipofuscin granules are often found in large neurons. These pigment granules are yellow-brown and thought to be secondary lysosomes involved in the degradation of unsaturated lipids and the detoxification of end products of metabolism. The number of lipofuscin granules increases with age. **Figure 6-2D** shows lipofuscin granules in the human spinal ganglial cells, which are surrounded by **satellite cells** that act as neuroglia, supporting the neurons.

The cell body of the neuron receives stimuli from other neurons, and acts as the trophic center, providing its processes with organelles and nutrients.

Figure 6-2. Neuron: Cell Body
Various stains • High magnification

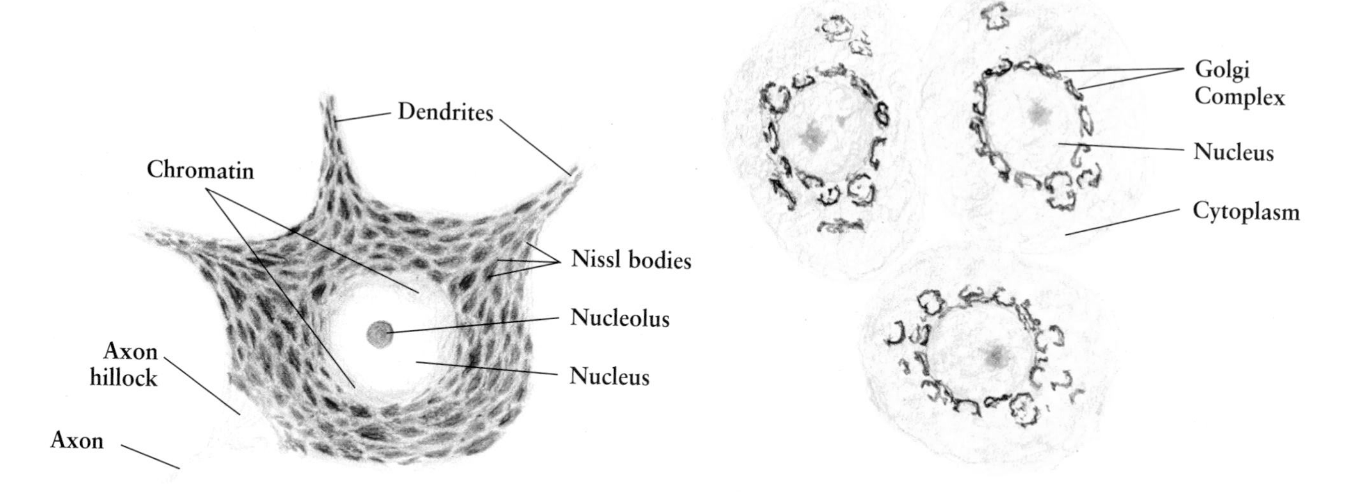

A. Nissl Bodies
Motor neuron of anterior horn of human spinal cord, Nissl stain.

B. Golgi Complex
Human spinal ganglion, Silver impregnation.

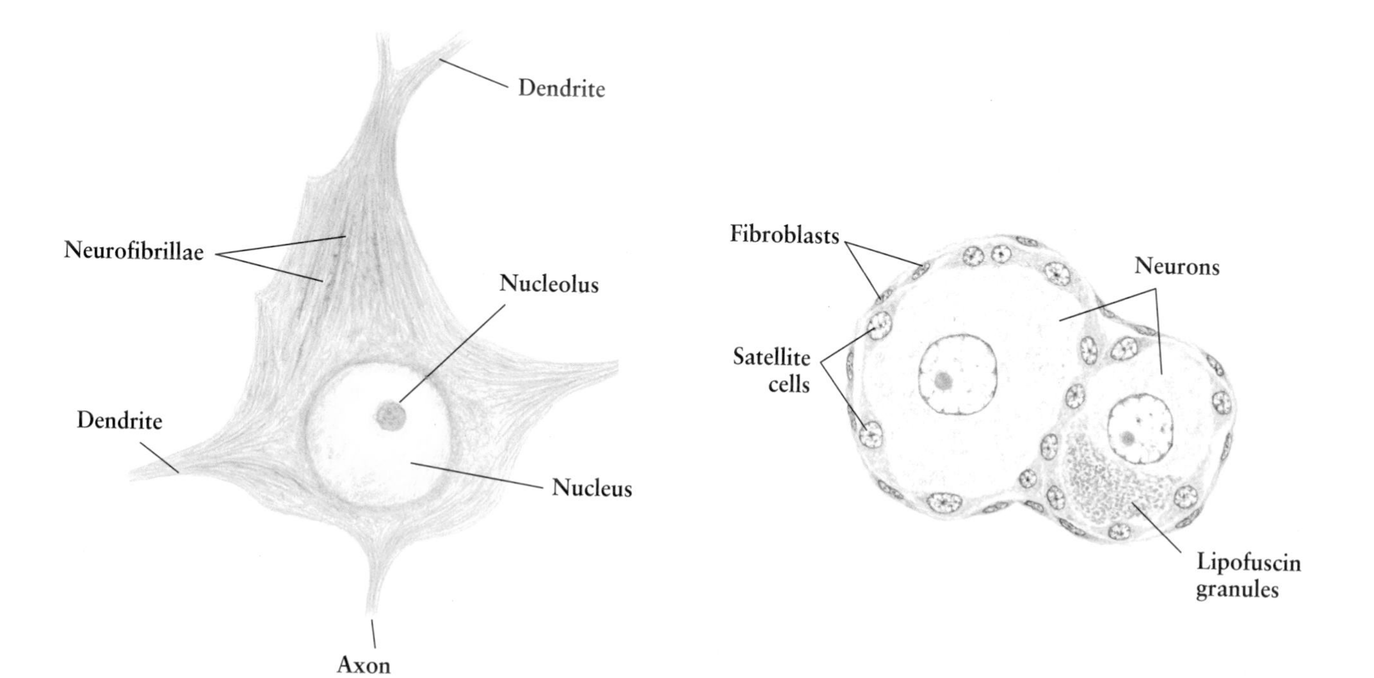

C. Neurofibrillae
Pyramidal cell of cat cerebral cortex, Cajal's stain

D. Lipofuscin Granules
Human spinal ganglion, H.E. stain.

Fig. 6-3. Myelinated Nerve Fibers

These four drawings illustrate some of the characteristic features of myelinated nerve fibers, demonstrated by very simple procedures that may be performed in the histology laboratory.

Fig. 6-3A. Take a small piece of rat spinal nerve, dissect it on a glass slide in physiological saline solution. The **axons, myelin sheaths, nuclei of Schwann cells,** and **node of Ranvier** may be seen under the light microscope.

Fig. 6-3B. Mount the nerve with tap water. The **contents of the myelinated nerve fibers** in the cut stump leak out and form a bubble-like mass.

Fig. 6-3C. To show the node of Ranvier, incubate the dissected spinal nerve in 0.5% silver nitrate for 24 hours, then add a reducing solution and mount with glycerin when the color turns brown. The **node of Ranvier** appears as a **silver cross.**

Fig. 6-3D. Incubate the spinal nerves in 1.0% osmium tetroxide solution for 24 hours, then place them on a glass slide and dissect further in glycerin. The **myelin sheath** is stained brown or black and is clearly visible under the light microscope.

Figure 6-3. Myelinated Nerve Fibers
Dissected fresh rat spinal nerve fibers • High magnification

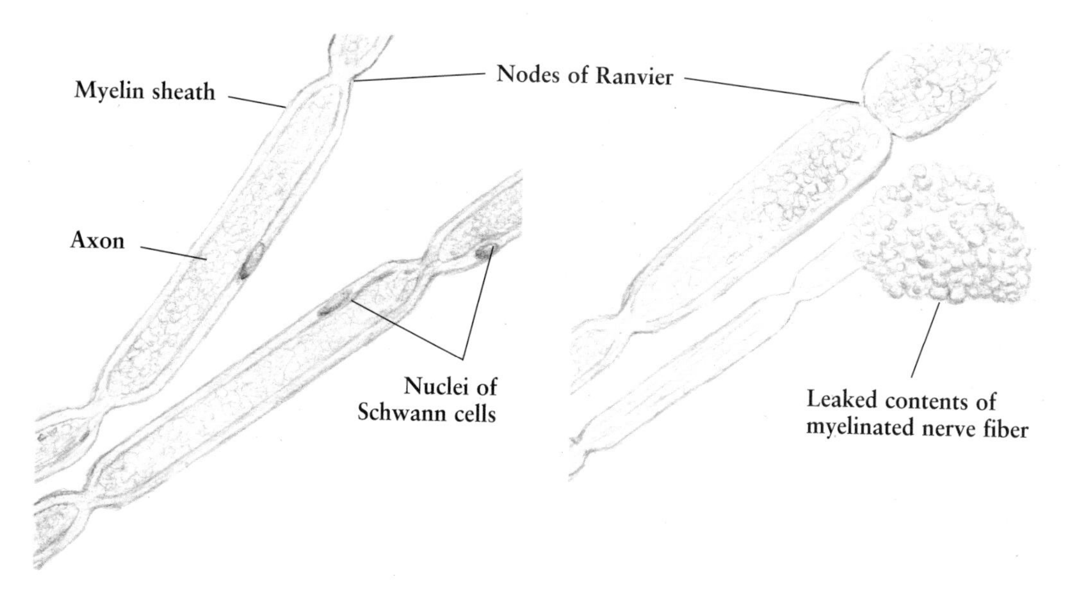

A. 0.85% saline solution added.

B. Tap water added.

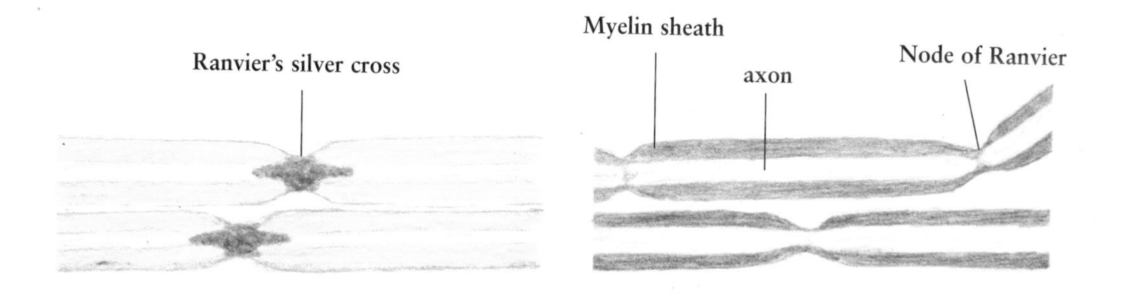

C. Incubated in 0.5% silver nitrate.

D. Incubated in osmium tetroxide.

Fig. 6-4.

Myelinated Nerve Fibers

Electron microscopic studies have revealed that the myelin sheath of myelinated nerve fibers in the peripheral nervous system consists of a system of concentric lamellar membranes derived from Schwann cell membranes. One **Schwann cell** winds around a single axon to form a **myelinated nerve fiber**. The gap between two adjacent Schwann cells along the axis of the axon is called the **node of Ranvier.** The distance between two nodes is called the **internode,** covered by a single Schwann cell.

In fixed preparations, the oblique incisions occurring in the myelin sheath are called **Schmidt-Lanterman clefts (Fig. 6-4B)**. These clefts are produced by inclusions of Schwann cell cytoplasm within the myelin. They probably serve as channels for exchanges of nutrients and gases between the axon and Schwann cell (myelin sheath). Around the myelinated nerve fiber is a very thin layer of loose connective tissue called **endoneurium.** In paraffin sections stained with hematoxylin and eosin, the myelin sheath only stains lightly, because the myelin has been dissolved by an organic solvent during the preparation.

In the central nervous system, axons become myelinated by oligodendrocytes in a manner similar to that of Schwann cells in the peripheral nervous system. In contrast to Schwann cells, a single oligodendrocyte myelinates several axons with its processes. **Figure 6-4** shows differences in myelinated nerve fibers following paraffin (**A**) or epon plastic (**B**) preparations.

Figure 6-4. Myelinated Nerve Fibers

Rabbit sciatic nerve • High magnification

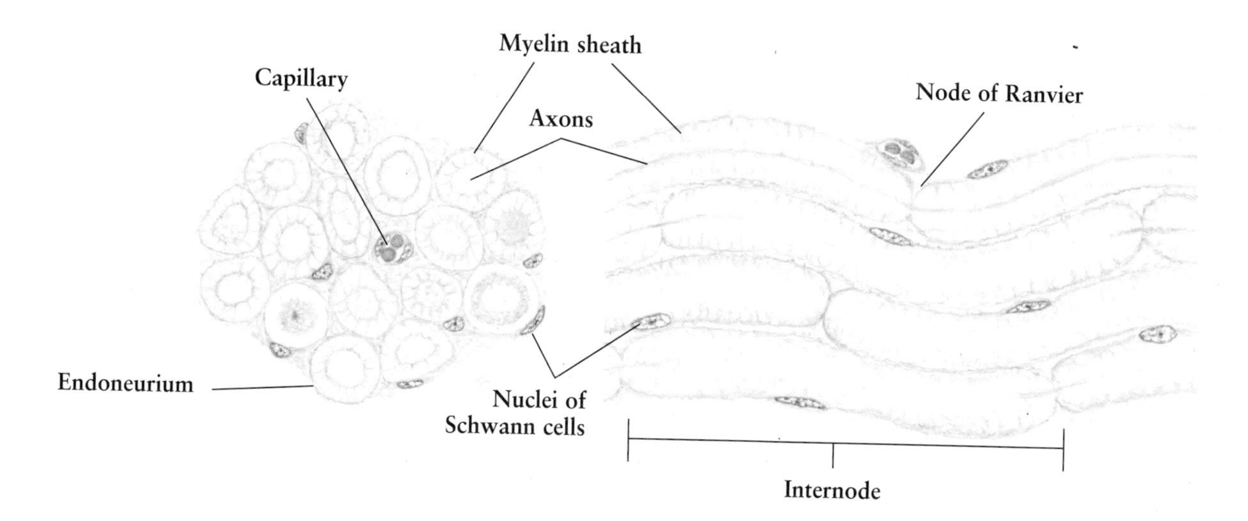

A. Paraffin sections. Left: cross section, Right: longitudinal section, H.E. stain.

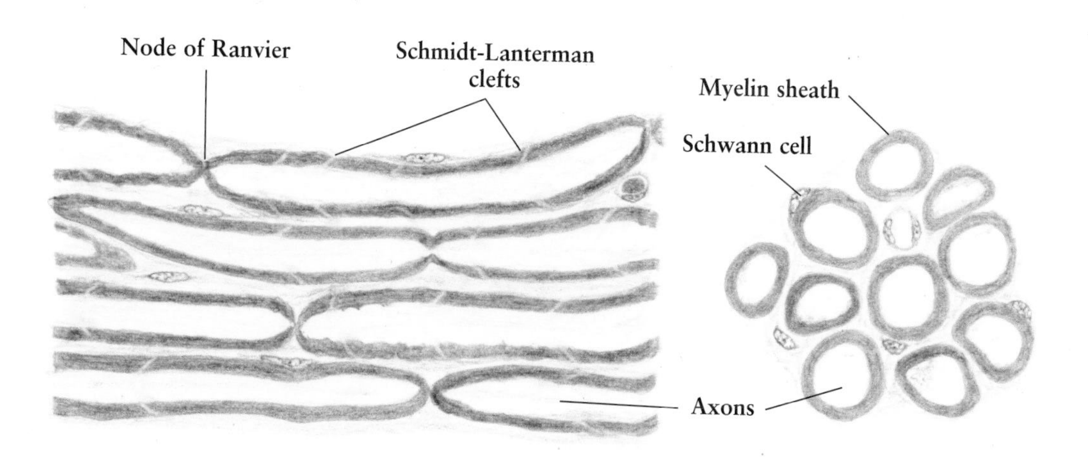

B. Epon sections. Left: longitudinal section, Right: cross section, Toluidine blue stain.

Fig. 6-5.

Unmyelinated Nerve Fibers

Unmyelinated nerve fibers in the central nervous system are completely naked axons, but in the peripheral nervous system they are surrounded by Schwann cell cytoplasm. A single Schwann cell may have several axons invaginated into its cytoplasm.

In **Fig. 6-5**, drawings **A, B,** and **C** show fresh bovine intrasplenic nerve. When 0.85% saline solution is added to dissected fresh nerve, the **nerve fibers** and **Schwann cell nuclei** can be recognized (**A**). When 1% acetic acid is added to the nerve fibers, the Schwann cell nuclei shrink and become more apparent (**B**). After the fresh unmyelinated nerve fibers have been incubated in hematoxylin solution for 24 hours, the **Schwann cell nuclei** stain light purple (**C**).

In permanent preparations, the unmyelinated nerve fibers are seen in parallel alignment with **Schwann cell nuclei** and **capillaries** interspersed among the fibers (**F**). In cross section, the **nerve fibers** without myelin sheaths are compacted tightly; Schwann cell nuclei and capillaries are scattered among them (**D, E**).

Figure 6-5. Unmyelinated Nerve Fibers

Bovine intrasplenic nerve • High magnification

A-C: dissected fresh specimen; D-F: permanent preparation

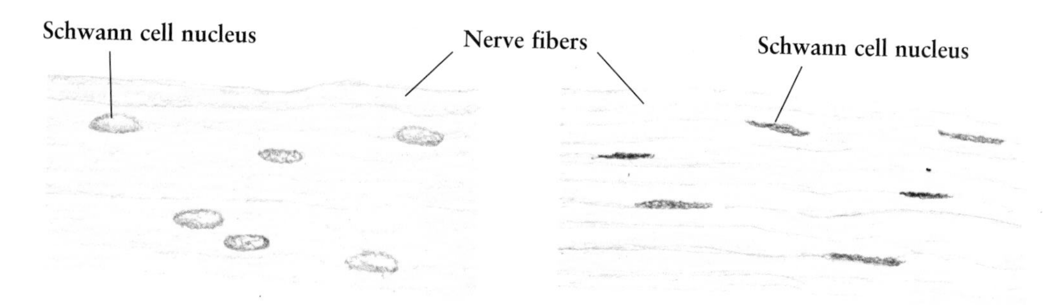

A. 0.85% saline solution added.

B. 1% acetic acid added.

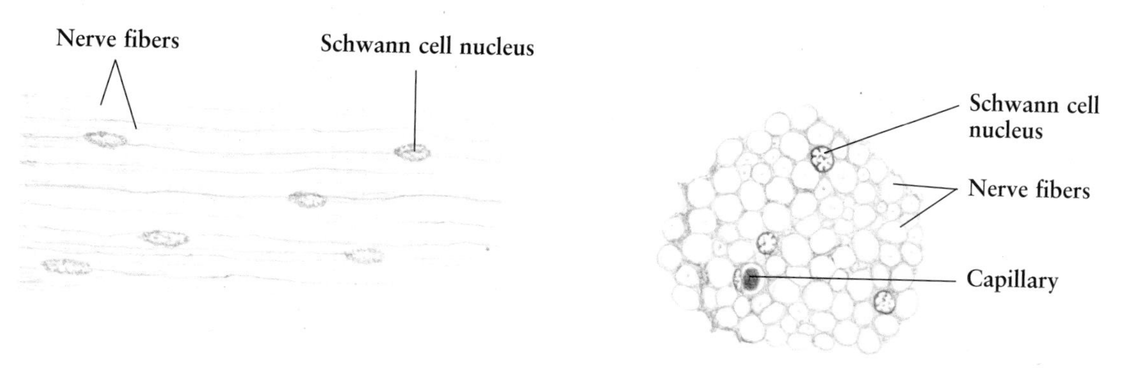

C. Stained with hematoxylin.

D. Epon preparation, cross section, toluidine blue stain.

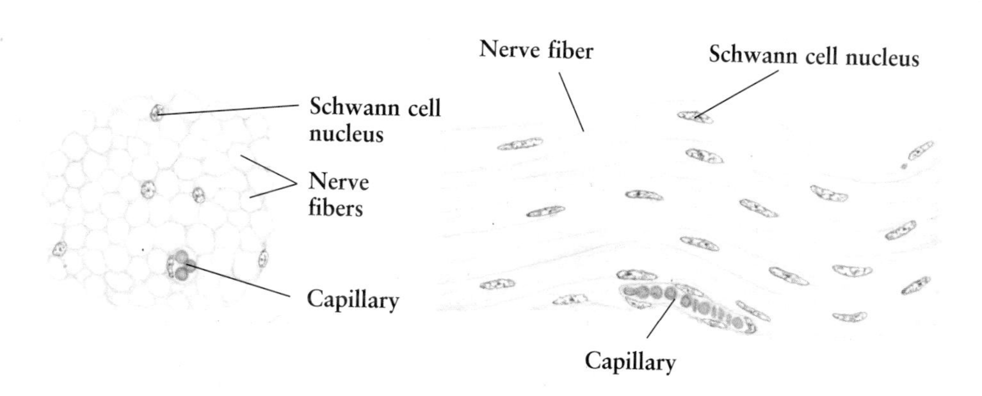

E. Cross section, H.E. stain.

F. Longitudinal section, H.E. stain.

Fig. 6-6.

Spinal Cord

The spinal cord is oval in cross section, divided into two halves by the **anterior median fissure** and the **posterior median sulcus.** It consists of gray matter and white matter.

The **gray matter,** composed of neurons, has a butterfly or H shape and is surrounded by peripheral white matter. The limbs of the H are called **anterior horns** (columns) and **posterior horns** (columns). The **motor neurons,** which control limb and trunk muscles, are located in the anterior horns. The interneurons, which serve important integrative functions, lie in the posterior horns. In segments from T1 to L3 and from S2 to S4, there is a **lateral horn** between the anterior and the posterior horns, where neurons of the autonomic nervous system are situated. In addition, there is a small **central canal,** lined by ependyma (see **Fig. 6-19**), occurring in the horizontal bar of the H.

The **white matter** of the spinal cord is composed of ascending tracts of sensory nerve fibers from dorsal root ganglia and descending motor tracts from the brain and the brain stem. Most of these fibers are myelinated. On each side of the spinal cord, the white matter can be divided into **dorsal (posterior) funiculus, lateral funiculus,** and **ventral** (anterior) **funiculus.**

These three drawings of spinal cord in cross section represent **cervical, thoracic,** and **lumbar segments,** respectively. Different stains were used for different emphasis. In **Fig. 6-6A,** note that gray matter contains a variety of different sizes of neurons. In **Fig. 6-6B,** note the nerve fibers in cross section in the white matter. In **Fig. 6-6C,** note the reticulated appearance of nerve fibers in the gray matter.

Figure 6-6. Spinal Cord
Cat • Very low magnification

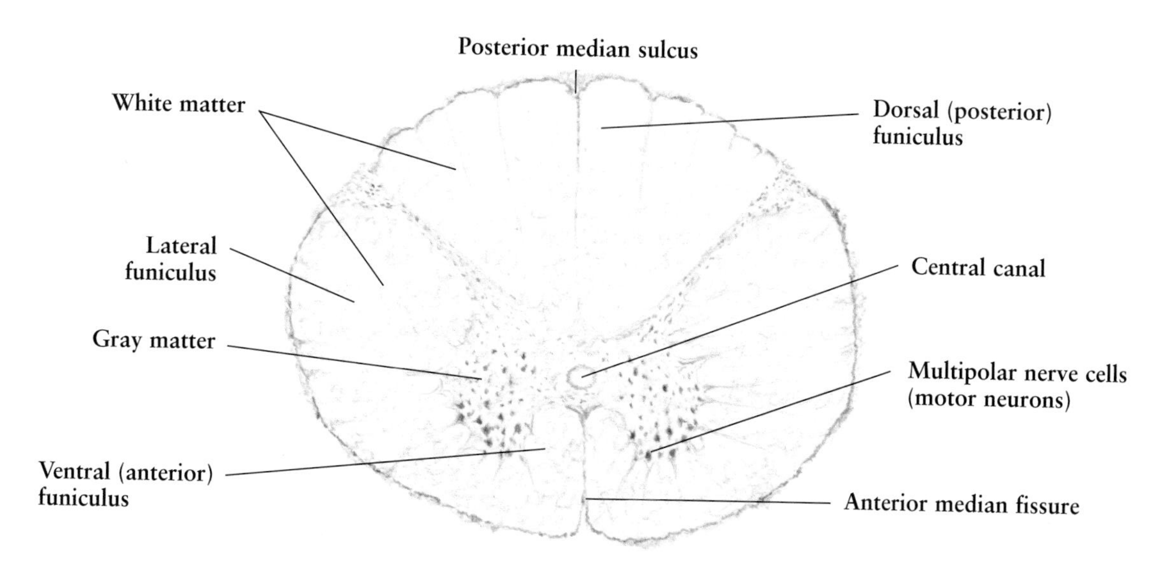

A. Cervical segment, Nissl stain.

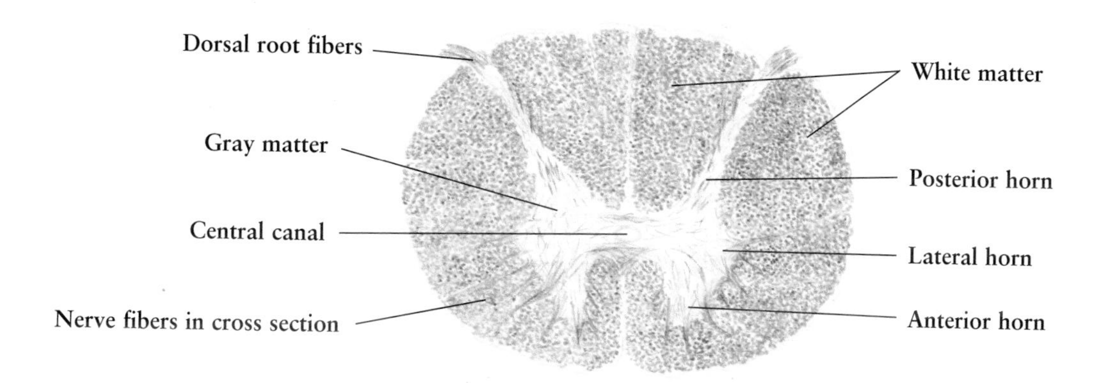

B. Thoracic segment, Weigert stain.

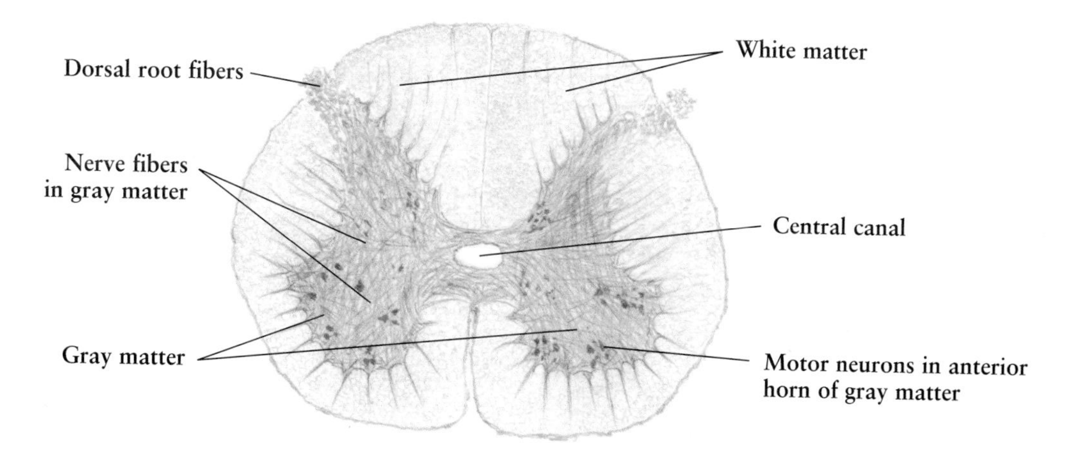

C. Lumbar segment, Bodian stain.

Fig. 6-7.

Spinal Cord

Figure 6-7 shows a **lumbar segment** of the human spinal cord stained with hematoxylin and eosin. The **anterior median fissure** and the **posterior median sulcus** divide the spinal cord into two halves. The **pia mater,** a very thin layer of loose connective tissue, attaches to the surface of the spinal cord. The **blood vessels** at the entry of the anterior median fissure are branches of the anterior spinal artery and vein, which supply the spinal cord. In the gray matter, the neurons are present in groups, and nerve fibers enter and leave, forming a dense network. The **dorsal root fibers** enter the **posterior horn** of the spinal cord through the **posterolateral sulcus,** and **ventral root fibers** leave the spinal cord through the **anterolateral sulcus.**

Fig. 6-8.

Spinal Cord

Figure 6-8 is an enlargement of the boxed area in **Fig. 6-7,** showing details of part of the anterior horn and the white matter. The **motor neurons** are multipolar cells with a large nucleus and prominent nucleolus. Nissl bodies are present in the cell body and dendrites, but not in the axons. Bundles of **dendrites** extend from the gray matter to the white matter, where the **myelinated nerve fibers** are seen in cross section. The small nuclei in both gray and white matter belong to the various **glial cells,** which cannot be classified in H.E.-stained preparations. In addition, **blood vessels** travel to the gray matter, forming the blood–brain barrier with the perivascular feet of astrocytes, which are not visible in this drawing.

Figure 6-7. Spinal Cord
Lumbar segment • Human • H.E. stain • Very low magnification

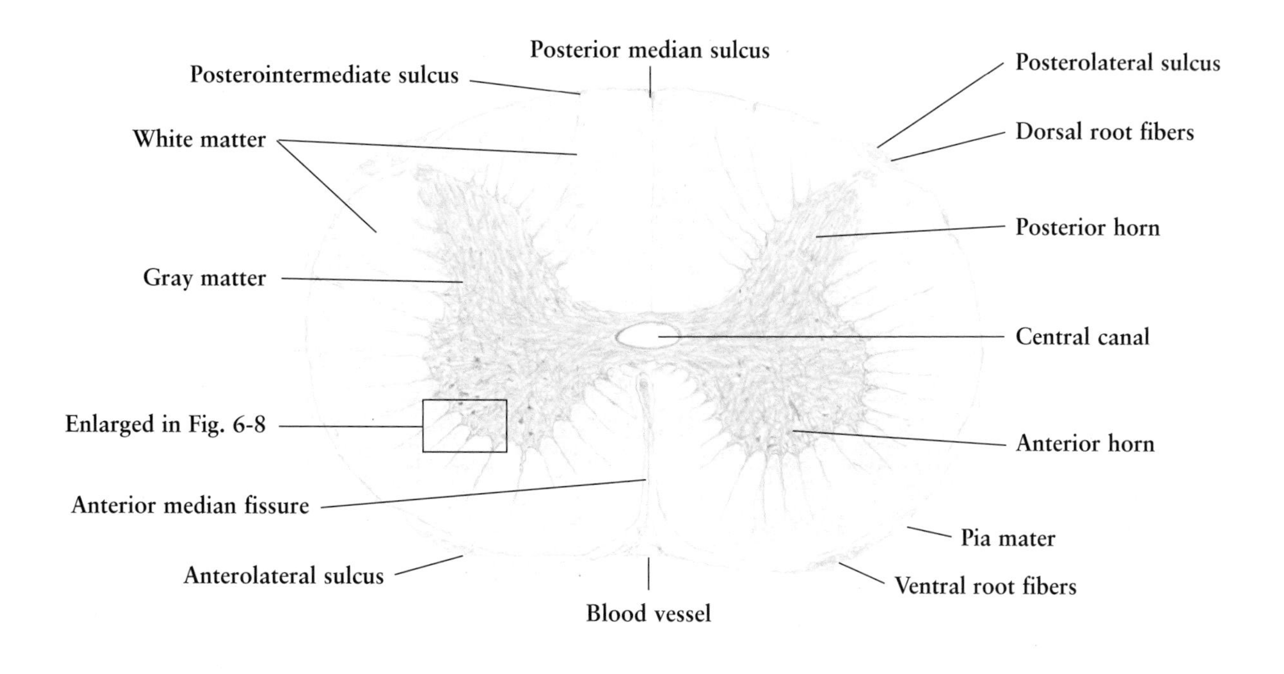

Figure 6-8. Spinal Cord
High magnification of the boxed area in Fig. 6-7

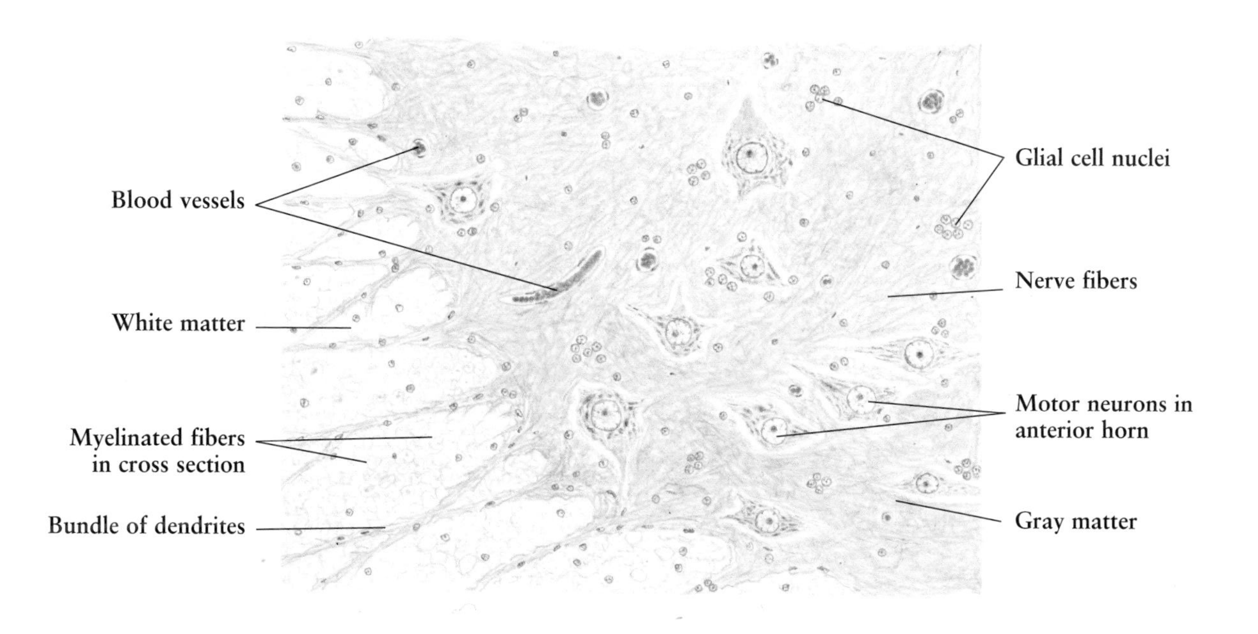

Figs. 6-9 and 6-10.

Cerebellar Cortex

The **cerebellar cortex** consists of three layers containing five different types of neurons. From the surface, these are the *molecular layer,* the *Purkinje cell layer,* and the *granular layer.*

The **molecular layer** contains outer stellate cells and basket cells. The outer **stellate cells,** with small bodies, are found in the outer part of the cortex. Their axons are confined to the molecular layer and make synaptic contacts with the dendrites of Purkinje cells. The **basket cells,** located in the deeper part of this layer (near the Purkinje cell body), send their dendrites toward the Purkinje cell and form baskets (**basket fibers**) that surround the Purkinje cell bodies.

The **Purkinje cell layer** is formed by a single layer of Purkinje cell bodies. Purkinje cells are large, flask-shaped multipolar neurons and have a large, round nucleus with a conspicuous nucleolus and basophilic cytoplasm. The dendrites arborize toward the surface of the cerebellar cortex and form a flattened, fanlike dendritic tree in the molecular layer, oriented at right angles to the long axis of the convoluted folds, or folia. Purkinje cell axons are myelinated fibers, passing through the granular layer and white matter, and terminating at the deep cerebellar nuclei.

The **granular layer** is composed of numerous granule cells and Golgi cells. The **granule cells** are small multipolar neurons with closely packed chromatic nuclei. They have four or five short dendrites, which join other nerve endings to form the so-called glomeruli. The axons are unmyelinated, travel up to the molecular layer, and bifurcate into branches parallel to the long axis of the convolution. **Golgi cells** are scattered in the upper part of the granular layer. Their dendrites enter the molecular layer, and axons contribute to the formation of the cerebellar glomeruli. The **cerebellar glomeruli** are irregularly dispersed spaces free of granule cells within the granular layer, consisting of one mossy fiber rosette, dendritic terminals of granule cells, axonal terminals, and dendrites of Golgi cells.

The unmyelinated **climbing fibers,** derived from the inferior olivary nuclei in the medulla, enter the cortex from the white matter and divide into numerous branches in the molecular layer while climbing the dendritic arborization of the Purkinje cells. The mossy fibers that form the fiber rosette of the cerebellar glomeruli come from the spinocerebellar system, pontocerebellar system, and vestibulocerebellar system.

Figure 6-9. Cerebellar Cortex
Dog • Bodian stain • High magnification

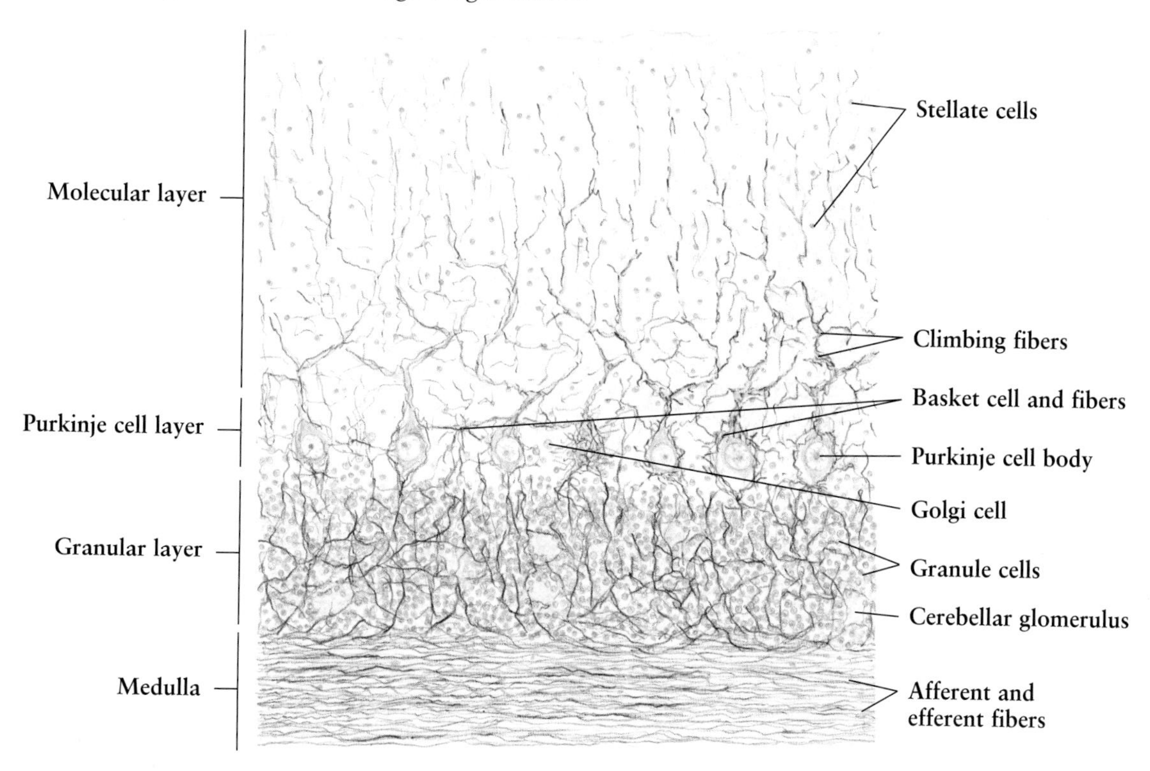

Figure 6-10. Cerebellar Cortex
Dog • Klüver-Bavera stain • High magnification

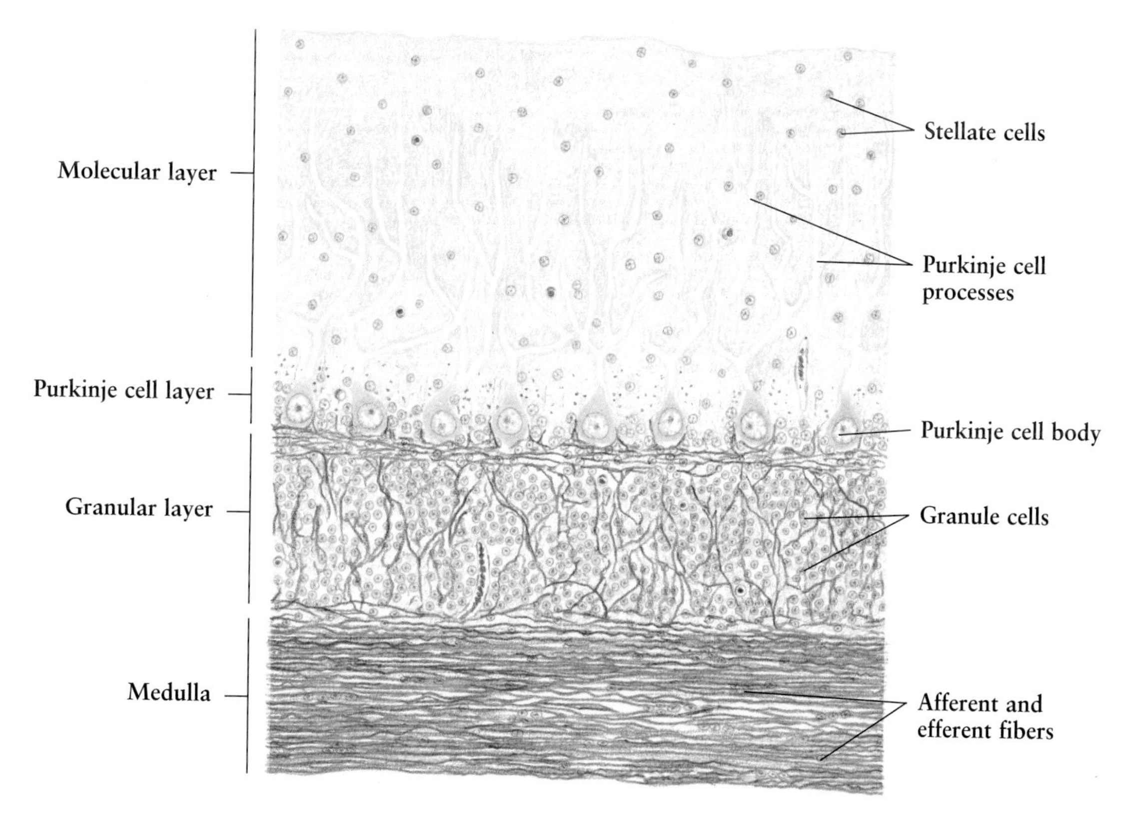

Fig. 6-11.

Cerebral Cortex: Areas 4 and 3

Histologically, the neurons in the cerebral cortex can be divided into five characteristic types with different morphological features, which constitute distinct layers of the cortex. These cell types are *horizontal cell, stellate cell, Martinotti cell, pyramidal cell,* and *fusiform cell,* the pyramidal cell and stellate cell being the most common cell types. In general, most of the cerebral cortex (neocortex) has structurally six fundamental layers distinguished by the cell type, size, density, and arrangement, with regional variation. The six layers from the pia surface to the deep part of the cortex are described here.

I. **Molecular layer.** This layer contains a few horizontal cells and a number of fine nerve fibers including axons of Martinotti cells and dendrites of pyramidal and fusiform cells, forming a synaptic field.

II. **External granular layer.** The secondary layer is formed by numerous stellate and small pyramidal neurons with dendrites ascending up to layer I and axons descending down to the deeper layers.

III. **External pyramidal layer.** This layer is composed of medium-sized pyramidal neurons in the upper part and a few larger neurons in the deeper part. Layer III is concerned with association and commissural efferents and afferents.

IV. **Internal granular layer.** Numerous densely packed small stellate neurons contribute to the formation of this layer. In addition to the dendrites and axons of stellate neurons, this layer also contains many horizontally running myelinated fibers from various thalamic nuclei, being receptive of specific thalamic efferents.

V. **Internal pyramidal layer.** Composed of medium and large pyramidal neurons, as well as some Martinotti and stellate cells, layer V is characterized by the presence of giant neurons, the Betz cells, especially in the primary motor area. Axons of pyramidal neurons leave the cortex as a major source of projection fiber efferents including the corticospinal, corticobulbar, extrapyramidal, corticopontine, and corticothalamic tracts.

VI. **Multiform layer.** Layer VI contains numerous Martinotti neurons, fusiform neurons, and pyramidal cells, and fibers entering and leaving the cortex.

Figure 6-11 is drawn from a human cerebral cortex preparation stained by the Nissl method. It shows the different arrangement of neurons in the six layers of the cerebral cortex. In fact, the neuroglial cells occur throughout the cortex and white matter. **Figure 6-11A** shows a primary **motor cortex** (Area 4 in precentral gyrus), which possesses a much thicker layer V and contains many giant pyramidal neurons in comparison with the primary **sensory cortex** (Area 3 in the postcentral gyrus), shown in **Fig. 6-11B.**

Figure 6-11. Cerebral Cortex: Areas 4 and 3
Human • Nissel stain • Low magnification

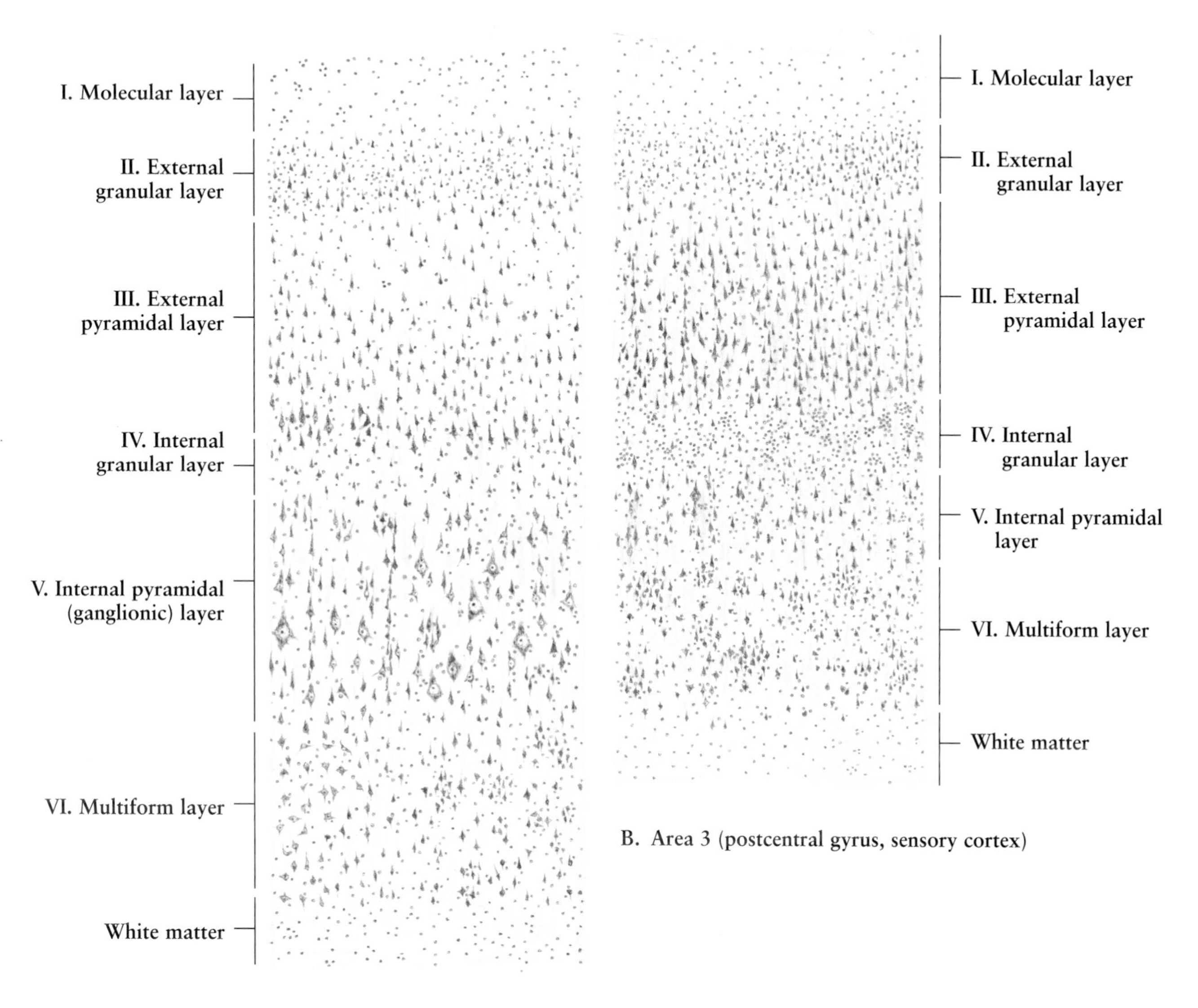

B. Area 3 (postcentral gyrus, sensory cortex)

A. Area 4 (precentral gyrus, motor cortex)

Fig. 6-12.

Cerebral Cortex: Areas 17 and 18

Area 17 of the cerebral cortex, located in the walls and floor of the calcarine fissure, is the primary visual area, while adjacent **area 18** is considered to be the secondary visual area. **Figure 6-12** shows the structures of both areas as well as the junction between them.

Histologically, both area 17 and area 18 have the six layers typical of the cerebral cortex. The **internal granular layer (layer IV)** of area 17 can be subdivided into three sublayers: **IVa, IVb,** and **IVc.** Sublayers IVa and IVb contain densely packed small granule cells. The **external pyramidal layer (layer III)** of area 18 is subdivided into three sublayers: **IIIa, IIIb,** and **IIIc.** Some larger pyramidal cells are present in sublayer IIIc. The border between areas 17 and 18 is characterized by the branching of the internal granular layer (layer IV) of area 18 into the three sublayers of layer IV in area 17, appearing in the shape of a tuning fork.

Figure 6-12. Cerebral Cortex: Areas 17 and 18
Human • Nissl stain • Low magnification

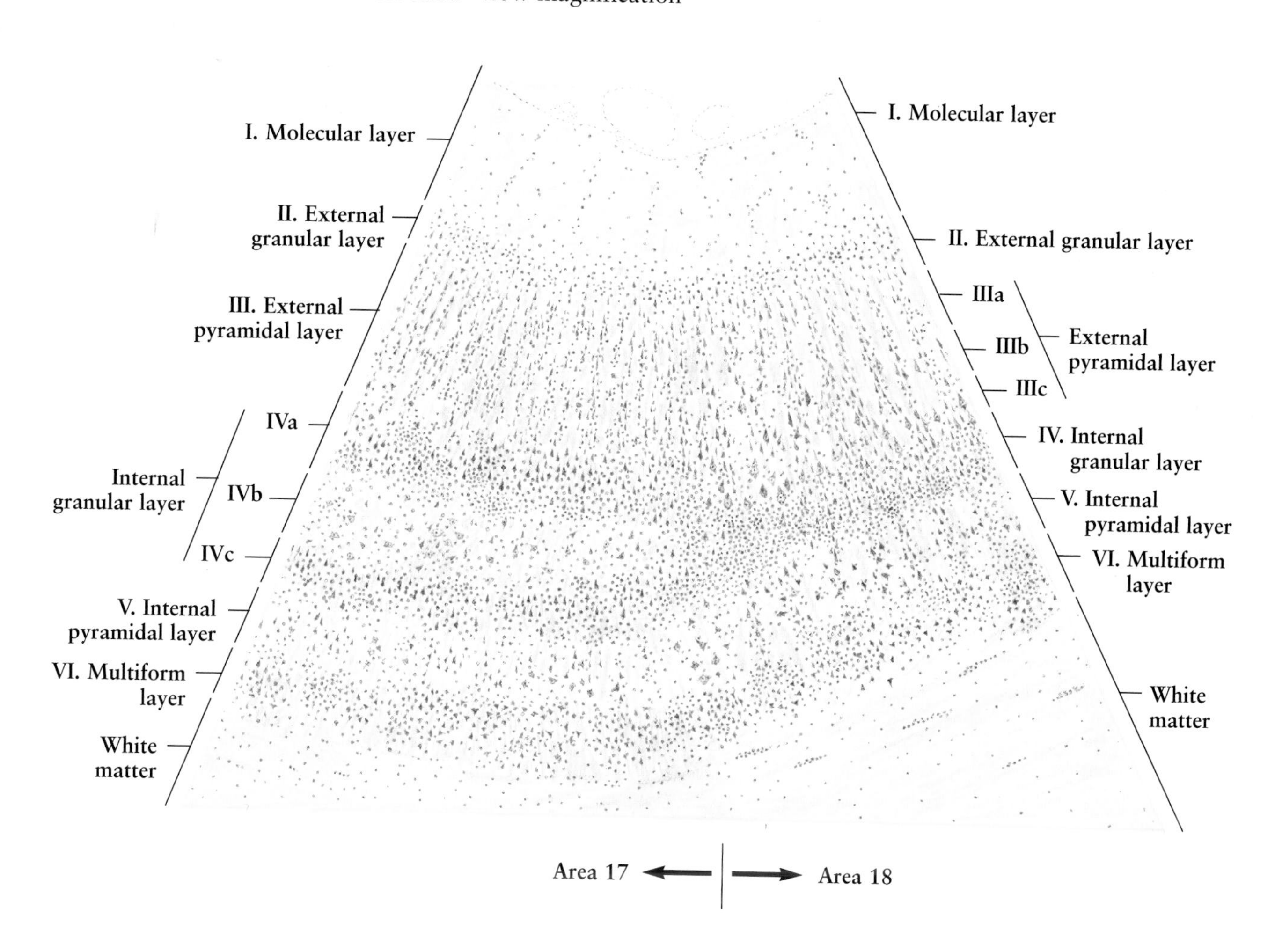

Figs. 6-13 and 6-14.

Cerebral Cortex: Pyramidal Cell

The **pyramidal cells** in the cerebral cortex are typical multipolar neurons. They are found in the external pyramidal layer (layer III), the internal pyramidal layer (layer V), and the multiform layer (layer VI). The size of these cells ranges from several micrometers to more than 100 μm in diameter. The neurons have a pyramidal perikaryon with a well-developed Golgi complex. In addition to their distinctive shape, pyramidal cells are characterized by the presence of abundant **Nissl bodies.** Their **nucleus** is large and round, with a conspicuous **nucleolus** and scarce chromatin, so that the entire nucleus appears to be a "fish-eye". Pyramidal cells have an **apical dendrite** and four or more **lateral dendrites** emerging from its lateral or basal side. Its **axon** leaves the cells from the central base. Like the axon itself, the initial portion of the axon, which is known as the **axon hillock,** lacks Nissl bodies.

Figure 6-13 is a Nissl-stained section showing the range of sizes among pyramidal cells and their prominent Nissl bodies. The small nuclei without visible cytoplasm stained by the Nissl method belong to **glial cells. Figure 6-14** is a Golgi silver-impregnated preparation showing whole neurons, including their dendrites and axons.

Figure 6-13. Cerebral Cortex: Pyramidal Cell
Human • Nissl stain • High magnification

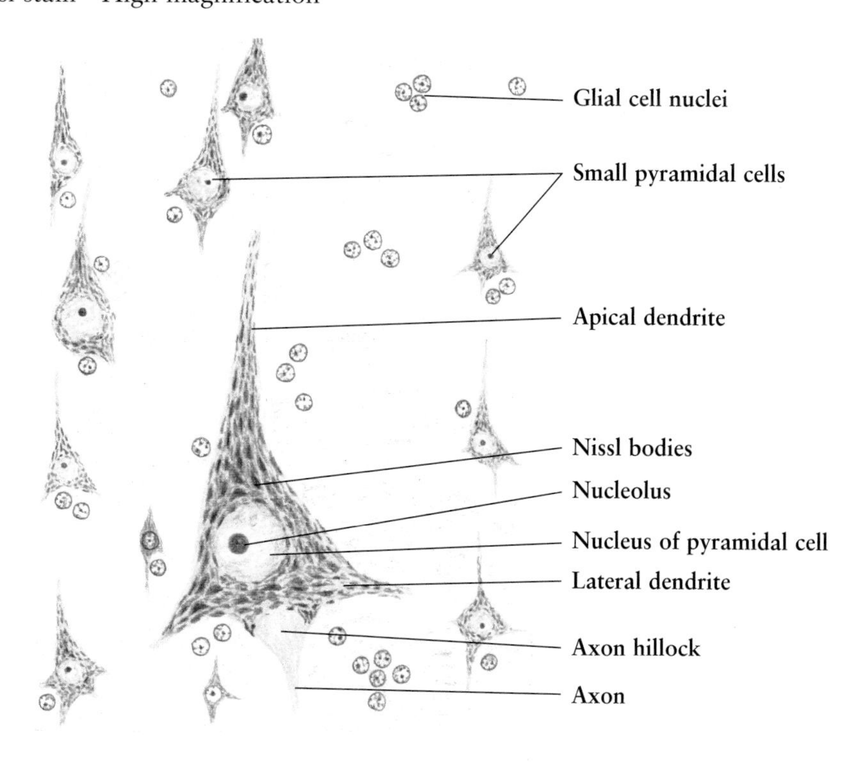

Figure 6-14. Cerebral Cortex: Pyramidal Cell
Human • Golgi stain • High magnification

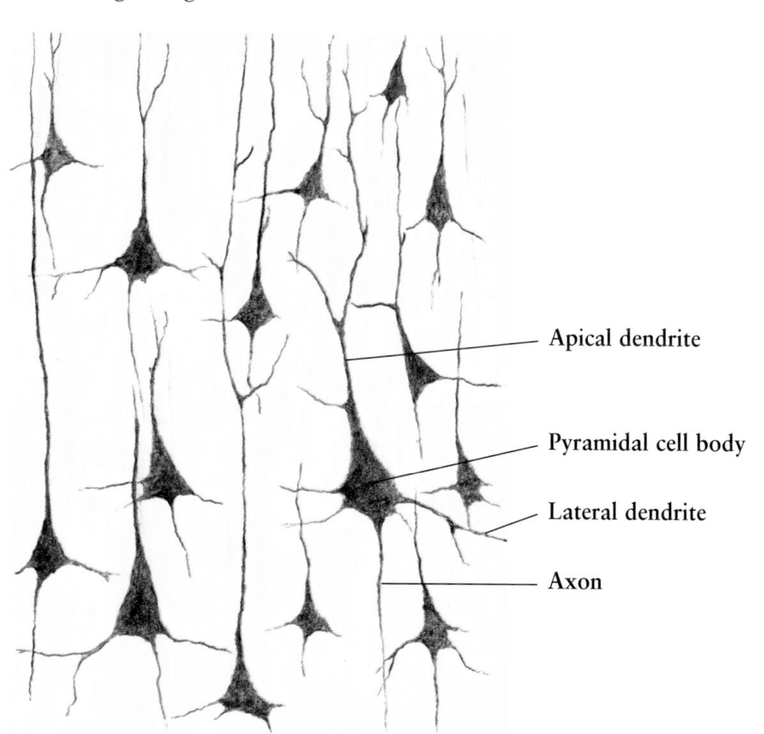

Figs. 6-15 and 6-16.

Neurosecretion: Supraoptic and Paraventricular Nuclei

The **neurons** in the **supraoptic nucleus** and the **paraventricular nucleus** produce the precursor molecules of hormones, in addition to conducting electrical impulses. They project through the supraoptic hypophyseal tract to the posterior lobe of the hypophysis. These neurosecretory products, including the precursors of oxytocin and vasopressin, are synthesized within the cell bodies as granules 120–200 nm in diameter. These granules are then transported down axons to the terminals in the hypophysis, where they are released into the blood.

Figure 6-15 shows a rat **supraoptic nucleus** stained by the aldehyde-thionine method, demonstrating blue-stained cell bodies and axons of the large neurosecretory cells. The **axons** filled with **endocrine products** appear to be a string of beads. At the base of this figure are the **tractus opticus** and **pia mater.**

Figure 6-16 shows a rat **paraventricular nucleus.** The left side of this picture shows the lumen of the **third ventricle,** with **ependymal cells** lining the wall of the ventricle. As in the supraoptic nucleus, the cell bodies and axons of the neurosecretory cells stain blue. **Blood vessels** can be recognized. The small nuclei belong to **neuroglial cells.**

Figure 6-15. Neurosecretion: Supraoptic Nucleus
Rat • Aldehyde-thionine stain • High magnification

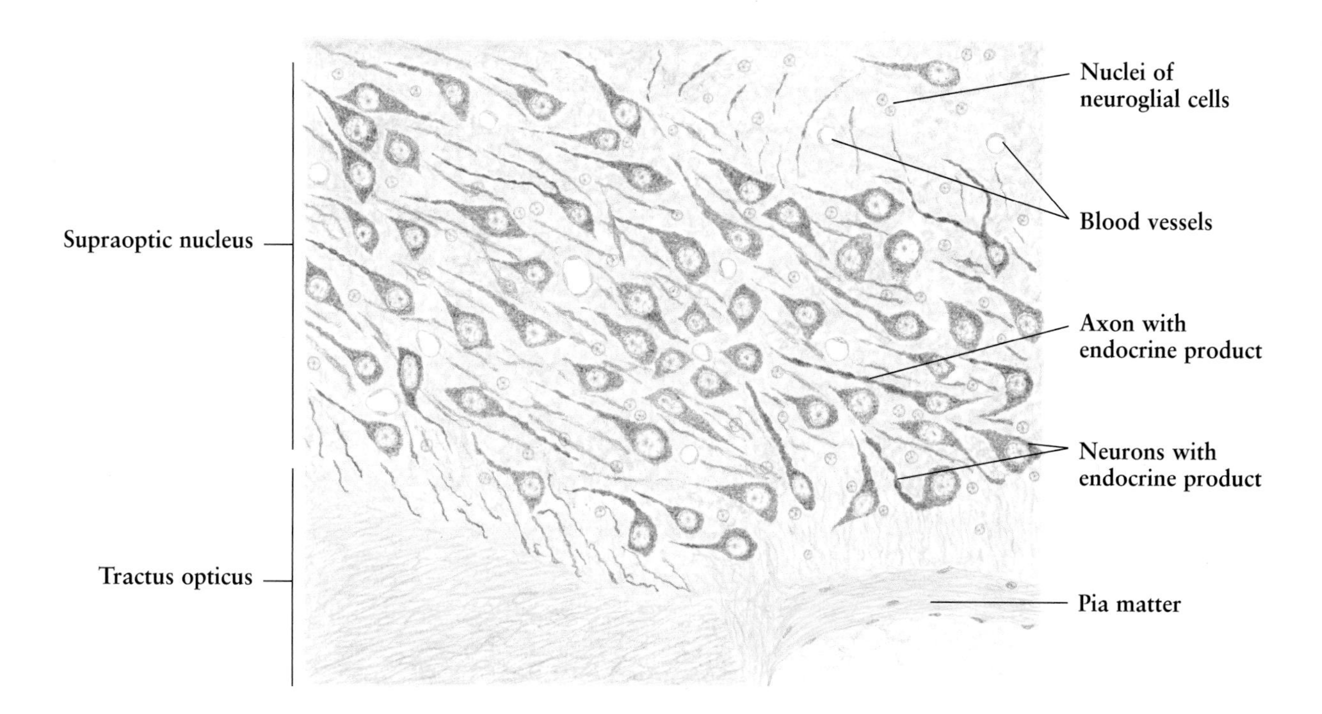

Figure 6-16. Neurosecretion: Paraventricular Nucleus
Rat • Aldehyde-thionine stain • High magnification

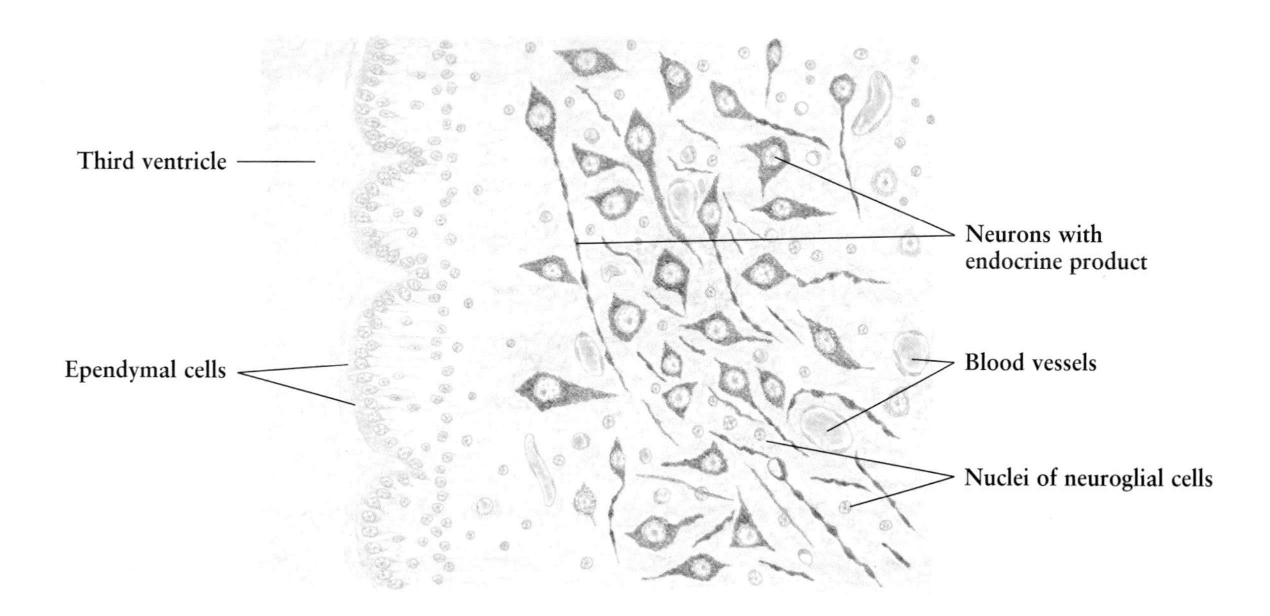

Fig. 6-17.

Hippocampal Formation

Anatomically, the **hippocampus proper,** the **dentate gyrus,** and the **subiculum** are referred to as the **hippocampal formation,** which is situated in the inferior horn of the **lateral ventricle.** In this drawing, the profile of the dentate gyrus, hippocampus proper, and part of the subiculum of the rabbit brain in frontal section can be appreciated. The hippocampus proper, which is continuous with the subiculum of the parahippocampal gyrus, extends to the inferior horn of the lateral ventricle, curves dorsally and medially, then curves again inward to form the dentate gyrus. The **hippocampal fissure** separates the dentate gyrus from the subiculum.

From the subiculum to the dentate gyrus, the cortical band of the hippocampus proper, can be divided into three fields: **CA1, CA2,** and **CA3.** Histologically, the cortex of the hippocampus proper consists of three fundamental layers: the **stratum radiatum,** the **pyramidal layer,** and the **stratum oriens.** The large and small pyramidal neurons in the pyramidal layer are the most conspicuous cells. They have complicated dendritic plexuses, which arise from both apical and basal poles of the cells and enter adjacent layers. The myelinated axons of the pyramidal neurons pass to the ventricular surface to form the **alveus,** a lamina of white matter, then collect into the **fimbria.**

The dentate gyrus also has three layers: the **molecular layer,** the **granular layer,** and the **polymorphic layer.** Its molecular layer is adjacent to that of the hippocampus proper at the end of the hippocampal fissure. The dentate gyrus is characterized by its granular layer, which is composed of densely packed small granule cells, arranged in a U-shaped configuration opening toward the fimbria. The axons of the granule cells pass through the polymorphic layer to terminate on the dendrites of pyramidal cells in the hippocampus proper (CA3). The dendrites of the granule cells mainly enter the molecular layer. The polymorphic layer contains several cell types, including pyramidal neurons.

The cellular organization of the subiculum shows a transitional change from a three-layered cortex to a six-layered cortex.

Figure 6-17. Hippocampal Formation
Rabbit brain • Frontal section • Klüver-Barvera stain • Low magnification

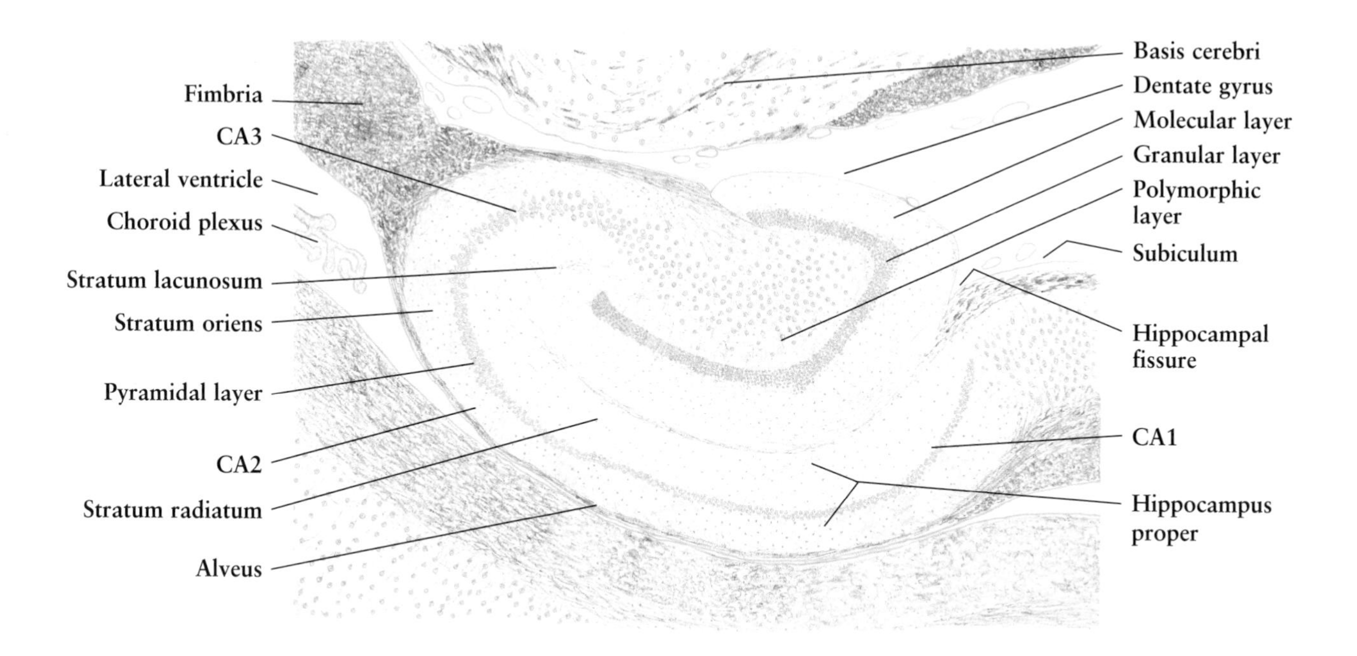

Fig. 6-18.

Neuroglia

Neuroglia comprise all the nonneural cells in the CNS. They are 10 times more numerous than the neurons, providing both mechanical and metabolic support to the neurons. Neuroglia are small cells, and only their nuclei can be recognized in routine preparations. The special procedures involving silver or other metal impregnation techniques are often used to demonstrate their cell bodies and processes. In addition, immunohistochemistry has been widely employed to localize them. Neuroglia include *astrocytes, oligodendrocytes, microglia,* and *ependymal cells.*

Astrocytes are large, star-shaped neuroglial cells, with numerous long branching cytoplasmic processes. They have a large ovoid or spherical nucleus with a nucleolus that is not obvious. Besides the ordinary organelles, the cytoplasm contains characteristic glial filaments which extend into the processes. Many of their cytoplasmic processes end in *pedicles* attaching to the walls of blood **capillaries,** known as **perivascular feet.** These perivascular feet as well as the endothelial cells of the capillary and the basement membrane constitute the structural basis of the *blood–brain barrier* in the CNS. The processes also extend to the surfaces of the brain and spinal cord, forming a layer beneath the pia mater. There are two types of astrocytes: protoplasmic and fibrous. The **protoplasmic astrocytes** are found mainly in the gray matter of the brain and spinal cord. They have thick processes with many branches and are often seen around neurons, synaptic areas, and blood vessels. The **fibrous astrocytes** are located chiefly in the white matter and have long, slender, smooth processes with few or no branches. When injury or damage occurs in the brain or spinal cord, the astrocytes play an important role in repair of CNS tissue.

Oligodendrocytes are smaller than astrocytes and have fewer, shorter processes. The cytoplasm is scanty, surrounding the nucleus, which contains condensed chromatin. Oligodendrocytes are present in both gray matter and white matter. In gray matter, they are seen lying around the neuronal perikarya, and in the white matter, they are responsible for the formation of the myelin sheath in the same manner as the Schwann cells in the PNS. Unlike a Schwann cell, however, each oligodendrocyte forms several myelin sheaths with its processes for adjacent axons.

Microglia are found in both gray and white matter. They are small, elongated cells with short processes. They have small irregular nuclei and relatively scanty cytoplasm. Microglia are thought to be the main source of phagocytic cells in the CNS.

Figure 6-18 is a summary drawing indicating the configuration and location of the neuroglia including protoplasmic astrocytes, fibrous astrocytes, oligodendrocytes, and microglia cells of the human brain stained with Cajal silver impregnation.

Figure 6-18. Neuroglia
Human cerebrum • Cajal stain • High magnification

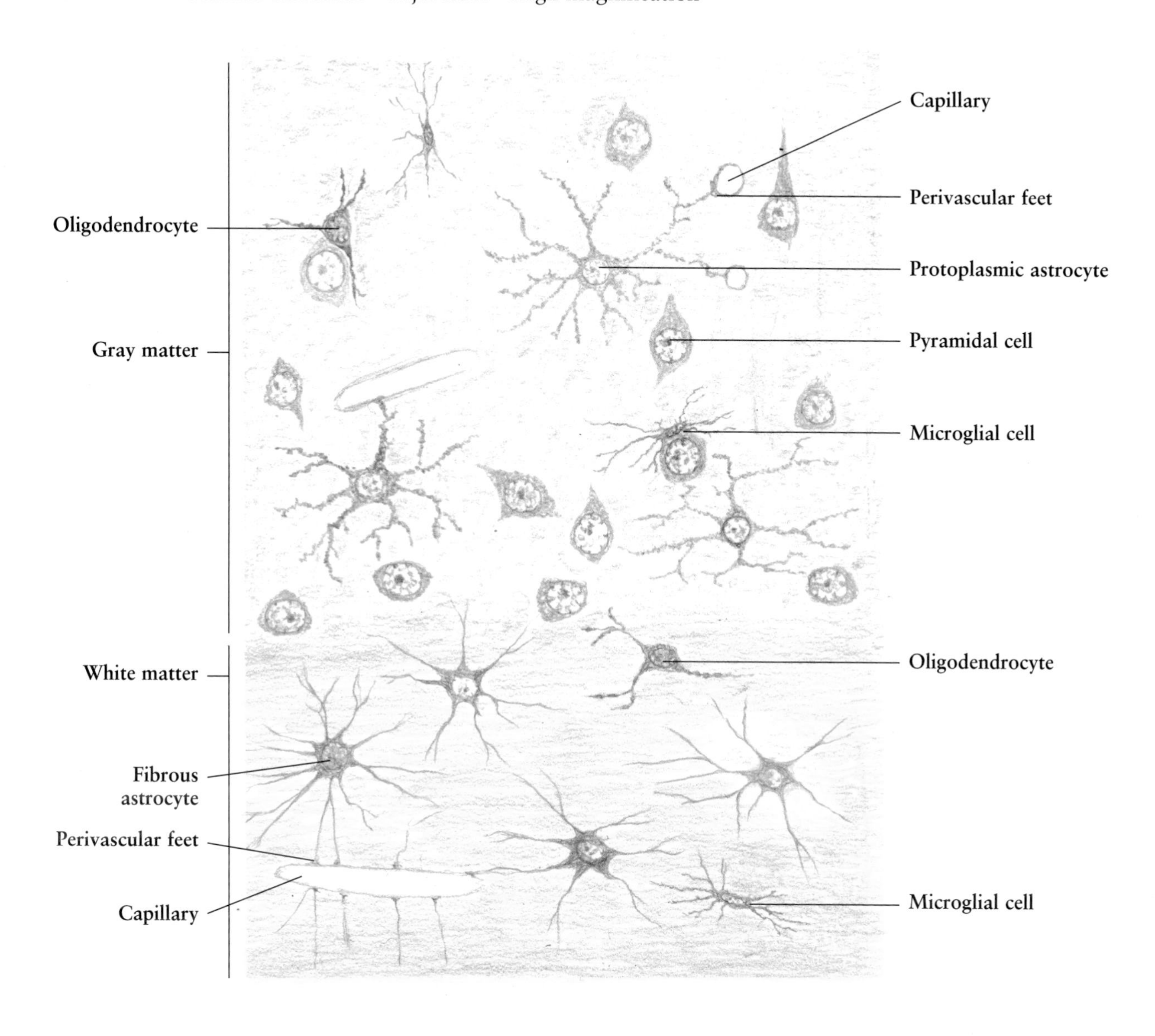

Fig. 6-19.

Ependyma

The **ependyma** is a simple epithelium lining the ventricles of the brain (see **Fig. 6-20**) and the central canal of the spinal cord. The epithelium is composed of simple cuboidal or columnar **ependymal cells** with cilia on their luminal surface. The nuclei are elliptical, and the cytoplasm contains a variety of organelles. Unlike other epithelium, there is no basement membrane under the ependymal epithelium. The adjacent ependymal cells are held together by junctional complexes. The ependymal cells belong to the neuroglia, which together with neurons are derived from the neural tube in the early embryo.

Figure 6-19 shows a section of dog spinal cord with Nissl stain. It highlights the **ependymal cells,** which line the **central canal** of the spinal cord. Each cell has a long process that extends from the basal part of the cell to the underlying subependymal cells. The other small nuclei belong to **neuroglial cells.** Two **neurons** at the bottom left of this picture can be recognized by their Nissl bodies.

Fig. 6-20.

Choroid Plexus

The **choroid plexus** is a highly ramified vascular structure projecting from the pia mater into the ventricles of the brain. Its main function is to produce cerebrospinal fluid, a modified tissue fluid which fills the ventricles of the brain, the subarachnoid space, and the central canal of the spinal cord.

The choroid plexus is covered by the choroid plexus epithelium, a simple **cuboidal epithelium** derived from the ependyma. The epithelial cells also participate in the production of cerebrospinal fluid. The underlying **lamina propria** is a somewhat loose connective tissue that contains numerous thin-walled **blood vessels,** some **fibroblasts,** and macrophages. Under the electron microscope the blood capillaries are seen to be the fenestrated type.

Figure 6-20 shows the choroid plexus of the third ventricle of the dog brain. Note the **ependymal cells** of the **third ventricle** at the lower left of this drawing.

Figure 6-19. Ependyma

Dog spinal cord • Nissl stain • High magnification

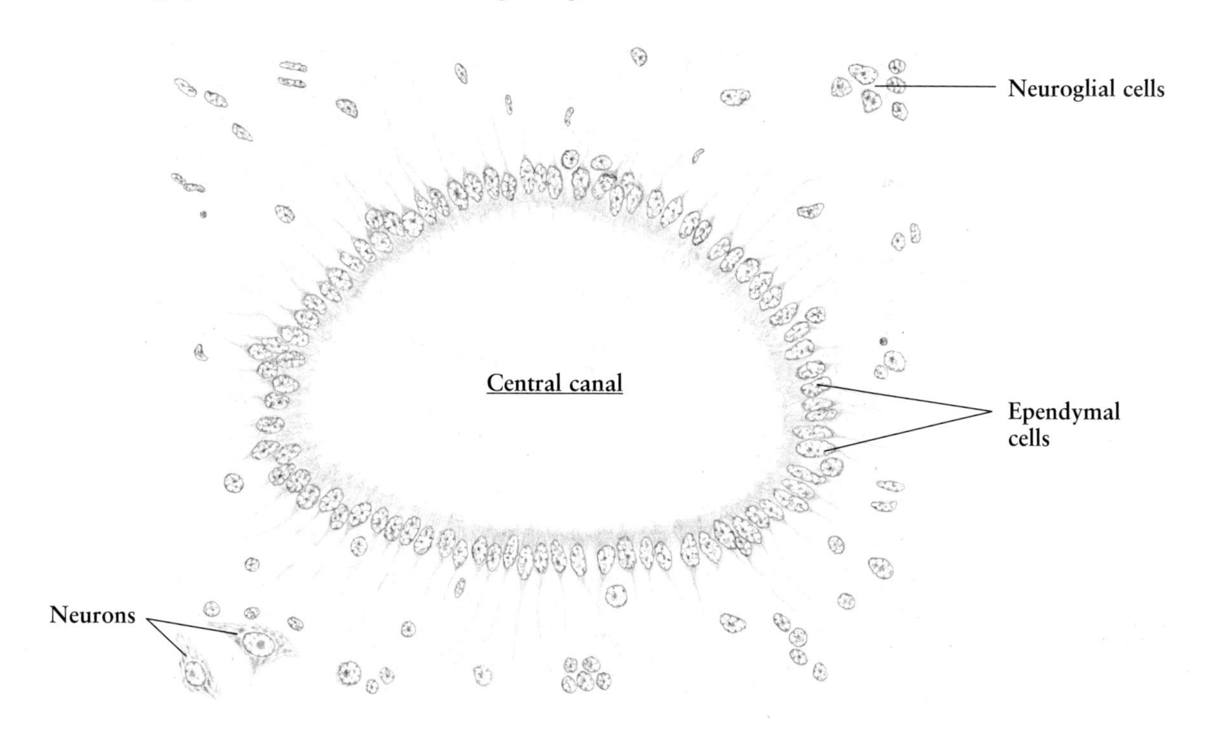

Figure 6-20. Choroid Plexus

Dog brain • H.E. stain • High magnification

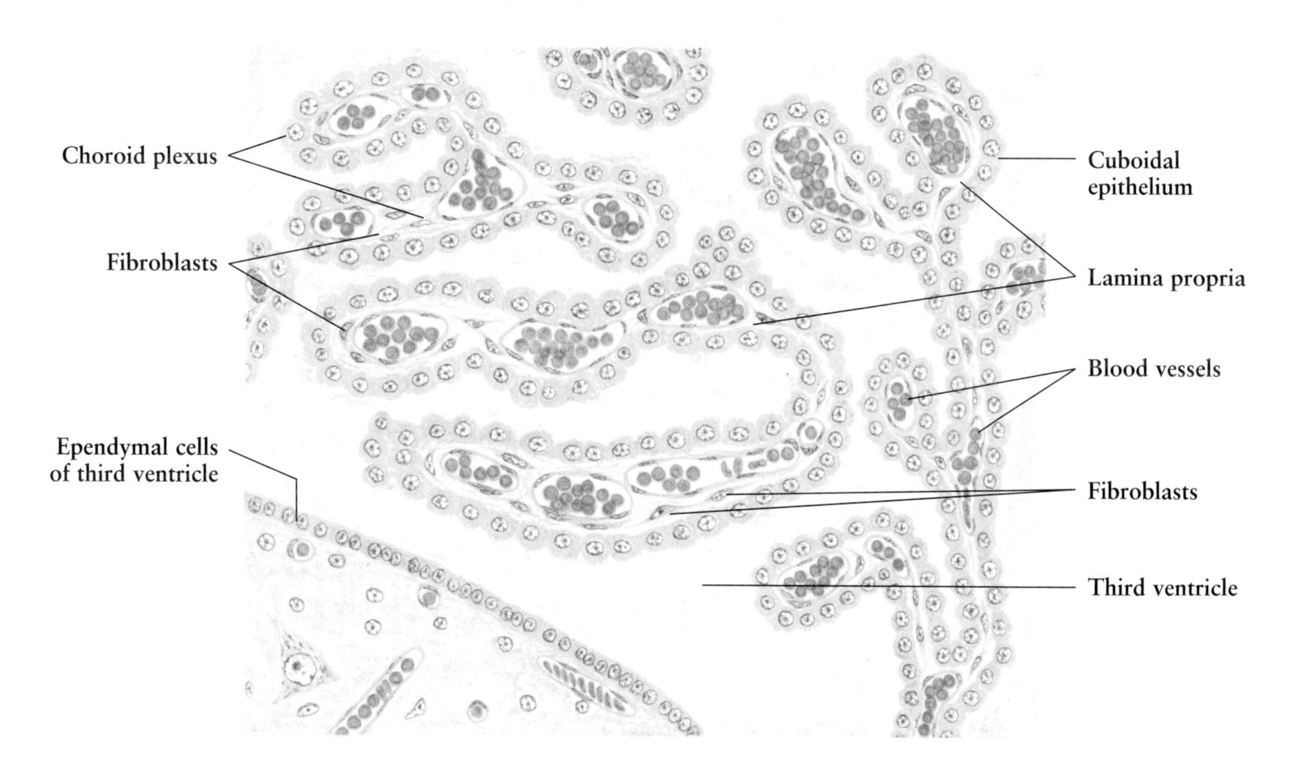

Fig. 6-21.

Spinal Ganglion

The **spinal ganglion** is located on the posterior nerve roots of the spinal cord. It contains the cell bodies of the pseudounipolar primary sensory neurons.

The ganglion is enclosed by a dense connective tissue **capsule,** which divides into **trabeculae** to provide a framework for the neuronal cells. The neurons of the spinal ganglion are large cells with a large nucleus. Their cell bodies appear round in section and display intense cytoplasmic basophilia. Each **ganglion cell body** is surrounded by a layer of flat **satellite cells,** which provide structural and metabolic support to the neurons. Within the ganglion, fascicle of **myelinated nerve fibers** in both cross and longitudinal sections can be observed. In addition, **blood vessels** occur throughout the ganglion.

Fig. 6-22.

Peripheral Nerve: Sciatic Nerve

Peripheral nerves, which include spinal nerves and cranial nerves, contain numerous afferent and efferent **nerve fibers** of the somatic and autonomic nervous systems.

In peripheral nerves, each individual axon is seen either enveloped by the myelin sheath (myelinated fibers) formed by Schwann cells, or surrounded by the cytoplasm of Schwann cells (unmyelinated fibers), which can be observed under the electron microscope. Between these nerve fibers is a delicate loose connective tissue, the endoneurium (see **Fig. 6-4A**), in close contact with the individual nerve fibers. The nerve fibers are grouped into bundles or fascicles, and covered by the **perineurium,** a layer of dense connective tissue composed of fibroblasts and collagen fibers. Each peripheral nerve is composed of one or more fascicles of nerve fibers and is surrounded by a layer of loose connective tissue, the **epineurium,** which extends from the outside and binds the fascicles together.

This figure is a rabbit's **sciatic nerve** in cross section, consisting of four fascicles of nerve fibers. Note that the **blood vessels** occur both outside and inside the fascicles as well as within the epineurium.

Figure 6-21. Spinal Ganglion
Human • H.E. stain • High magnification

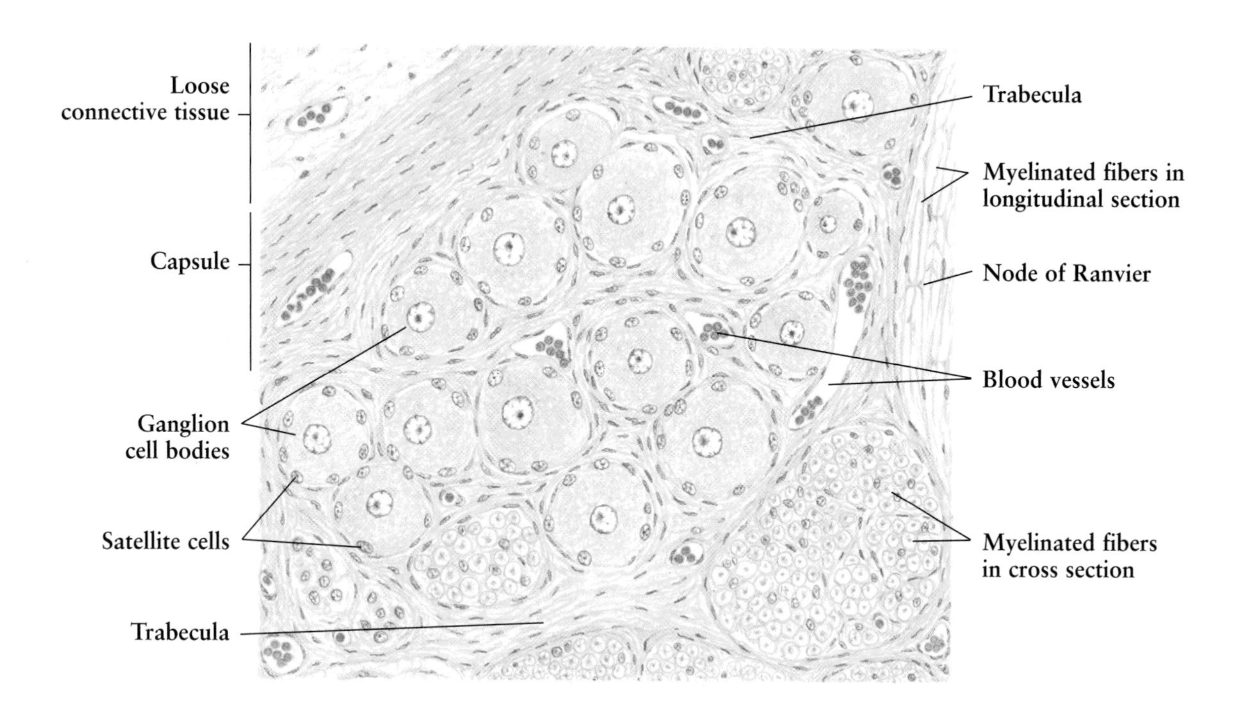

Figure 6-22. Peripheral Nerve: Sciatic Nerve
Rabbit • Cross section • H.E. stain • Low magnification

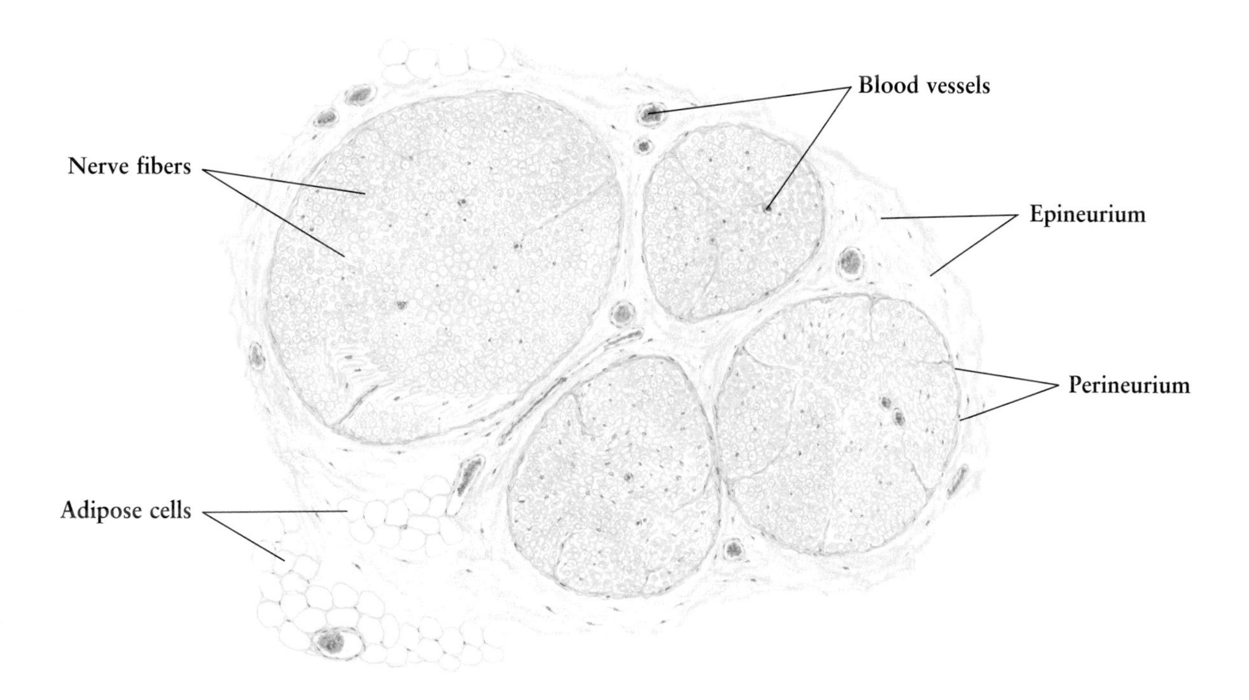

Fig. 6-23.

Motor End Plate

A **motor end plate** is a structure in which a motor axon terminal ends as small branches at the surface of a skeletal muscle fiber to form a *neuromuscular junction.* The motor end plate appears in whole mount as a flat, oval convex area, 40–60 µm in diameter. Approaching a **skeletal muscular fiber,** a **motor axon** loses its myelin sheath and ramifies into several **end branches,** which are covered by a thin cytoplasmic layer of Schwann cells. The neuromuscular junction is a kind of chemical synapse, commonly referred to as a *neuromuscular synapse.* The neurotransmitter is acetylcholine, and its hydrolytic enzyme is acetylcholinesterase.

In conventional preparations, the motor end plate cannot be recognized. This figure is a silver impregnation section of monkey tongue, displaying a motor axon and its ramified end branches. In this area, a number of **nuclei** of skeletal muscle fiber can be seen.

Fig. 6-24.

Muscle Spindle

The **muscle spindle** is a fusiform structure found between and in parallel with ordinary skeletal muscle fibers. It is typically up to 7 mm in length and less than 1 mm in diameter. The organ is enveloped by a **connective tissue capsule** consisting of fibroblasts and dense collagen fibers. Within it there are two distinctive types of muscle fibers, known as **intrafusal fibers.** One is the **nuclear bag fiber** characterized by having an expanded midregion containing numerous nuclei. The other is the **nuclear chain fiber,** with nuclei lined up in a row along its midregion.

This figure illustrates a muscle spindle in cross section. The connective tissue capsule separates the organ from the surrounding tissue. Intrafusal fibers are much thinner than **extrafusal fibers** or ordinary muscle fibers. The nuclear bag fibers show several nuclei in a cross-section profile. In contrast, the nuclear chain fibers show only one nucleus in a cross-section profile. A thin layer of fine connective tissue surrounds each individual intrafusal fiber. Inside the spindle are **capillaries** and unmyelinated **nerve fibers.**

The muscle spindle is a muscle stretch receptor organ, innervated with both sensory and motor nerves. The sensory nerve fibers terminate as annulospiral endings around the nuclear bag and the chain fibers and as a flower-spray ending near equators of chain fibers. The motor fibers (γ motor fibers) terminate as small, modified motor end plates on all intrafusal fibers.

Figure 6-23. Motor End Plate
Monkey tongue • Silver impregnation • High magnification

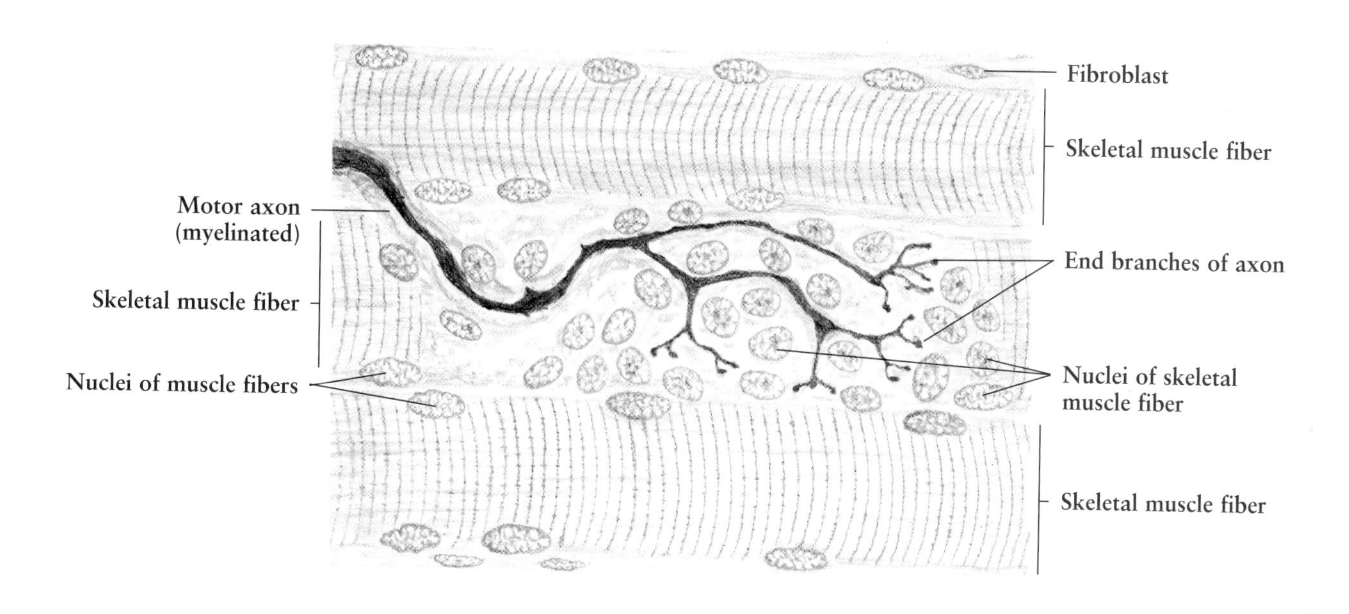

Figure 6-24. Muscle Spindle
Skeletal muscle of monkey • H.E. stain • High magnification

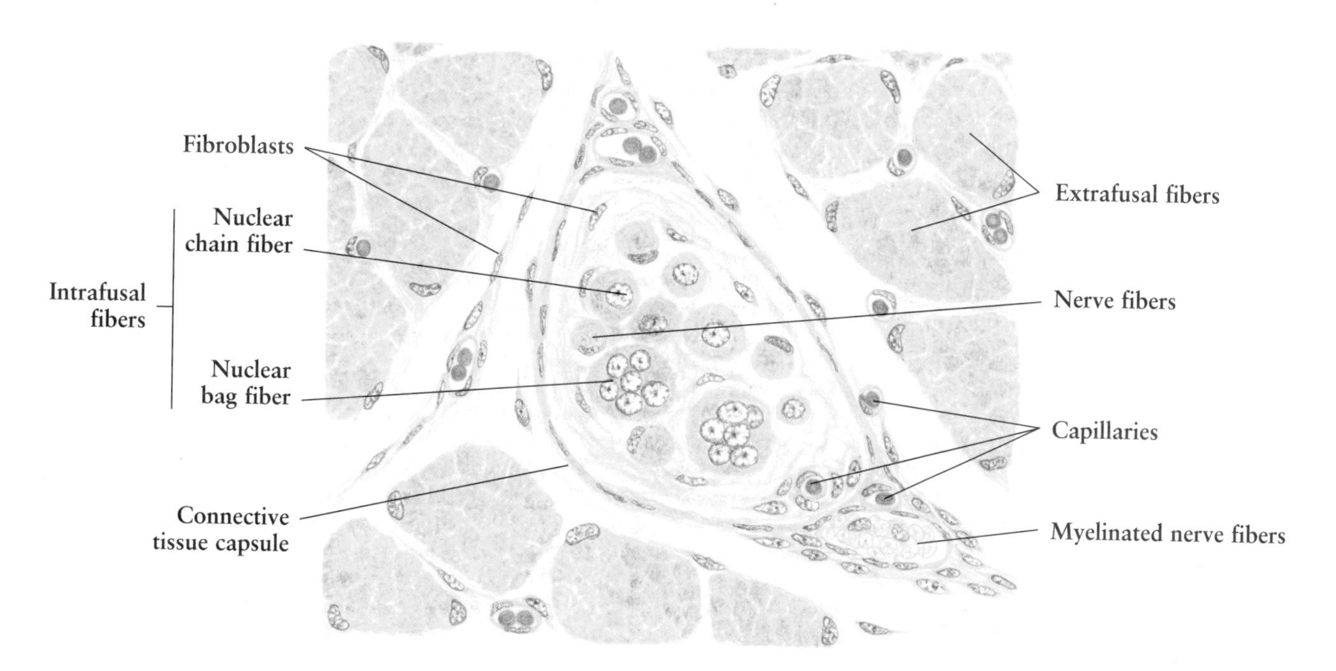

Fig. 6-25.

Free Nerve Endings

Free nerve endings cannot be recognized in the routine preparations with hematoxylin-eosin stain. This figure is a silver impregnation section of rat skin, showing the **free nerve endings** within the **epidermis** of the skin. Before entering the epidermis, the **sensory afferent nerve fiber** loses its investment. It passes and branches between the **epithelial cells** as a naked nerve fiber and terminates as free nerve endings. Free nerve endings are associated with the sensations of pain and light touch.

Fig. 6-26.

Meissner's Corpuscle

Meissner's corpuscles, also named *tactile corpuscles,* are sensitive mechanoreceptors responsible for touch sensitivity. They are distributed in the dermal papillae of skin, particularly of the fingertips (see **Figs. 15-1** and **15-2**), lips, nipples, and genitalia.

Meissner's corpuscles are elliptical in shape, 120 μm long and 70 μm wide. Their long axes are oriented perpendicular to the skin surface. The whole corpuscle is enclosed by a connective tissue **capsule.** Within the corpuscle, there is a stack of **pear-shaped cells,** regarded as modified Schwann cells, which dispose transversely, parallel to the skin surface. Losing their myelin sheaths, the **sensory nerve fibers** enter the corpuscle, ramify, and terminate between these pear-shaped cells.

Figure 6-26 consists of two drawings of Meissner's corpuscles from a human dermal papilla. **Figure 6-26A** is an H.E.-stained specimen demonstrating the connective tissue **capsule** and **pear-shaped cells** lying in a stack. **Figure 6-26B** shows a silver impregnation section, revealing a trace of a **sensory nerve fiber** and its **branches** and terminals.

Figure 6-25. Free Nerve Endings
Rat skin • Silver impregnation • High magnification

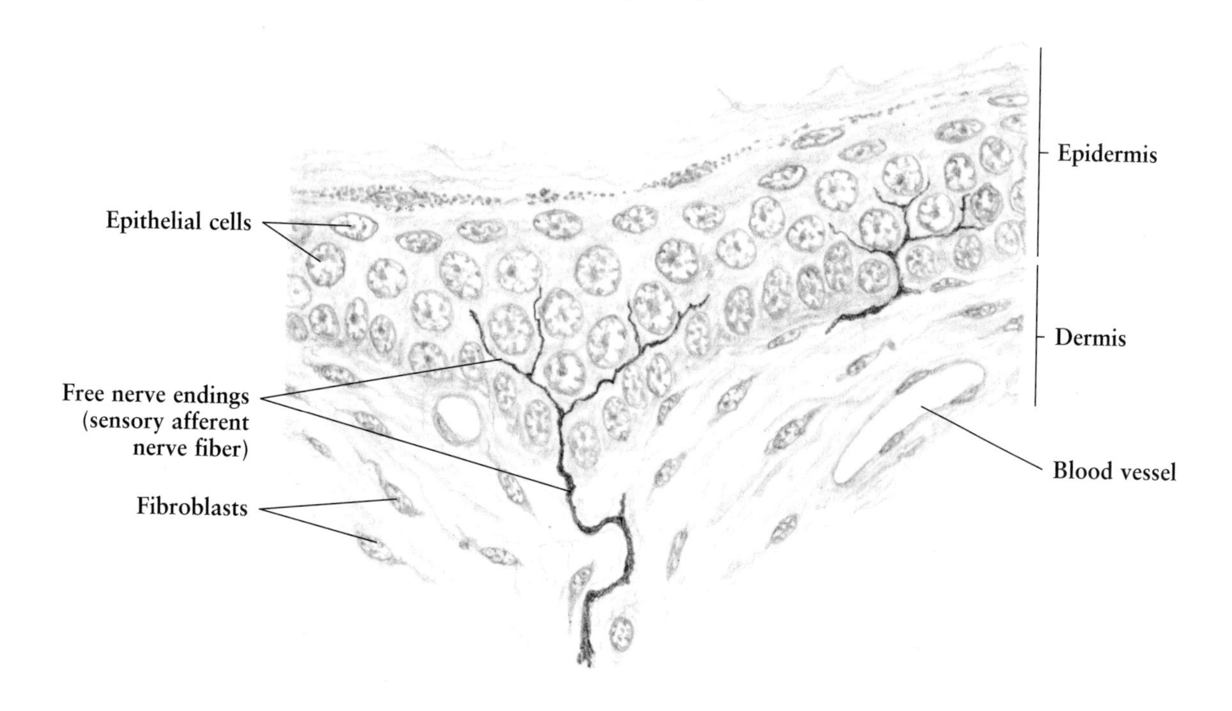

Figure 6-26. Meissner's Corpuscle
From human dermal papilla • High magnification

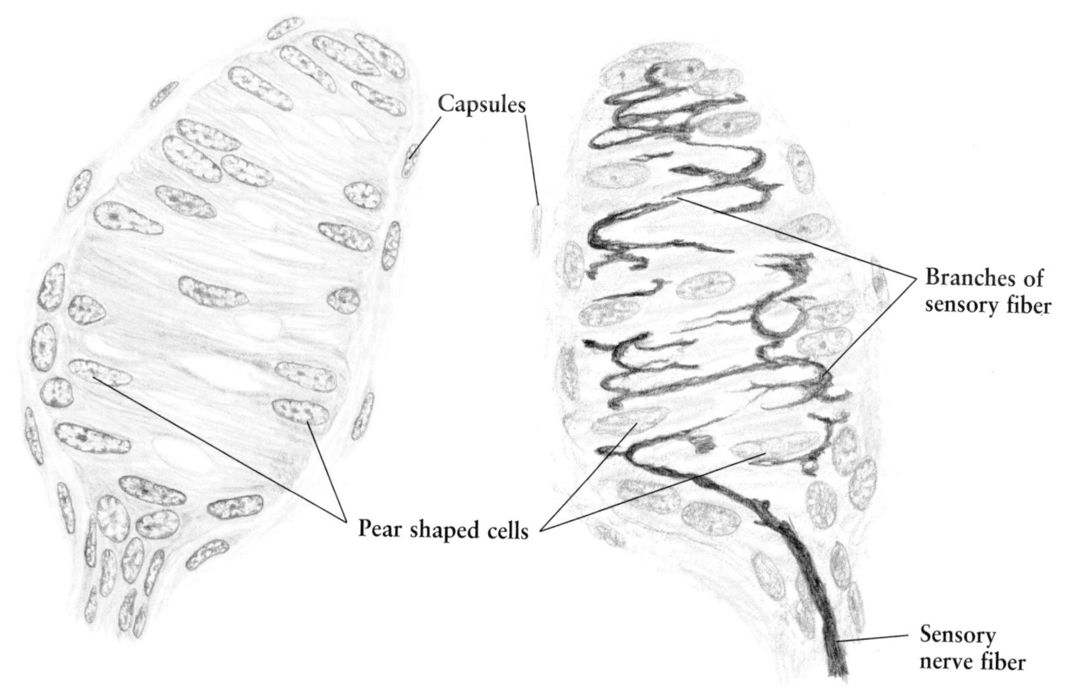

A. H.E. stain

B. Silver impregnation

Fig. 6-27.

Lamellar Corpuscle (Vater-Pacinian Corpuscle)

The **lamellar corpuscles** (**Vater-Pacinian corpuscles**) are oval bodies, 2 mm in length and 0.5–1 mm in diameter. They are found widely throughout the dermis and subcutaneous tissue (see **Fig. 15-1**), and are also present in tendons, ligaments, mesentery, and the pancreas.

The **capsule** of the corpuscle is a layer of delicate connective tissue. It is continuous with the perineurium of the nerve, which penetrates the corpuscle as an unmyelinated nerve ending. The corpuscle has an onion-like appearance, which consists of 10–60 concentric **lamellae** formed by flattened **fibroblasts**. Between them are a lymphlike fluid, collagen fibrillae, and some **capillaries.** In the center of the corpuscle is an **inner bulb,** containing a single unmyelinated sensory nerve ending surrounded by ramified and interdigitating Schwann cells.

The Vater-Pacinian corpuscle is a sensory receptor responsive to pressure and vibration because of fluid pressure between the lamellae of fibroblasts.

Figure 6-27. Lamellar Corpuscle (Vater-Pacinian Corpuscle)
Dermis of human skin • H.E. stain • High magnification

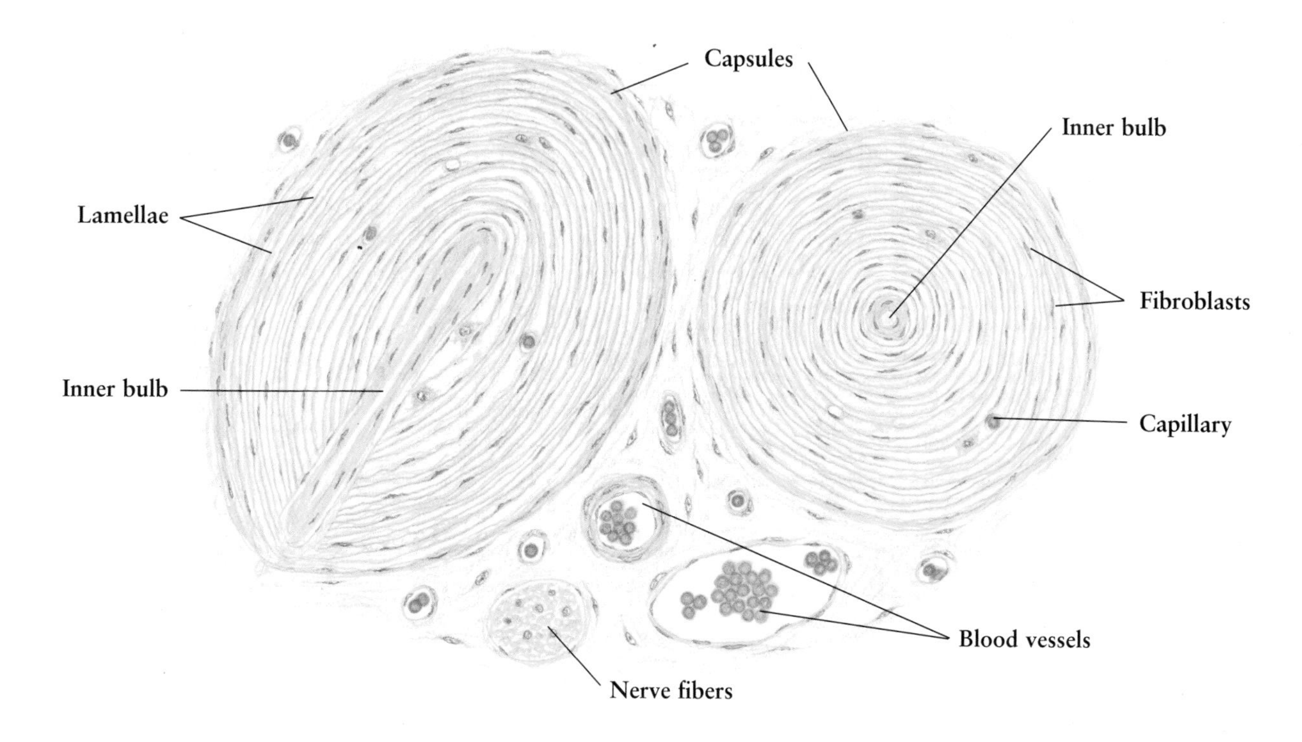

Figs. 6-28 and 6-29.

Palisade-shaped Nerve Endings

All hair follicles are innervated with a plexus of sensory nerve endings, known as peritrichial or **palisade-shaped nerve endings,** situated in the **dermis.** Approaching the **hair follicle,** the **myelinated nerve fibers** lose their myelin sheathes, divide into branches below the **duct of the sebaceous gland,** and then run up parallel to the long axis of the **hair shaft** and terminate at the **glassy membrane** of the hair follicle, giving rise to a palisade appearance. The nerve endings are sandwiched between two layers of Schwann cell cytoplasm. In transverse section, the palisade-shaped nerve endings are found just surrounding a half circle of a hair follicle. The hair shaft acts as a lever, and any slight movement of the shaft readily stimulates the nerve endings in the hair follicle.

Figure 6-28 shows the palisade-shaped nerve endings attaching to the outer surface of a hair follicle. Note the relationship between hair follicle and sebaceous gland. **Figure 6-29** is a high-magnification picture of a toluidine blue-stained epon section taken from a rat snout, illustrating how the sandwiched palisade-shaped nerve endings in cross section lie at the glassy membrane of the hair follicle and form a half circle of palisade.

Figure 6-28. Palisade-shaped Nerve Endings
Rat skin • Silver impregnation • High magnification

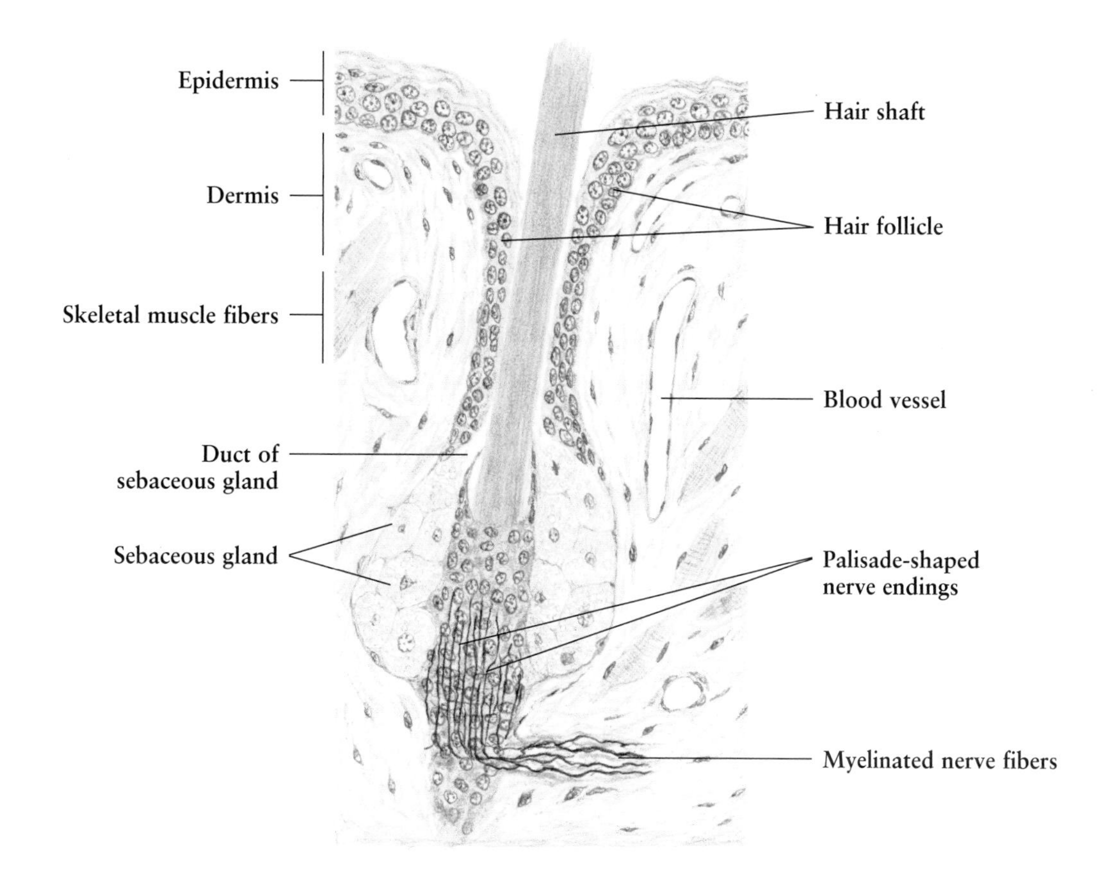

Figure 6-29. Palisade-shaped Nerve Endings
Rat snout • Epon section • Toluidine blue stain • High magnification

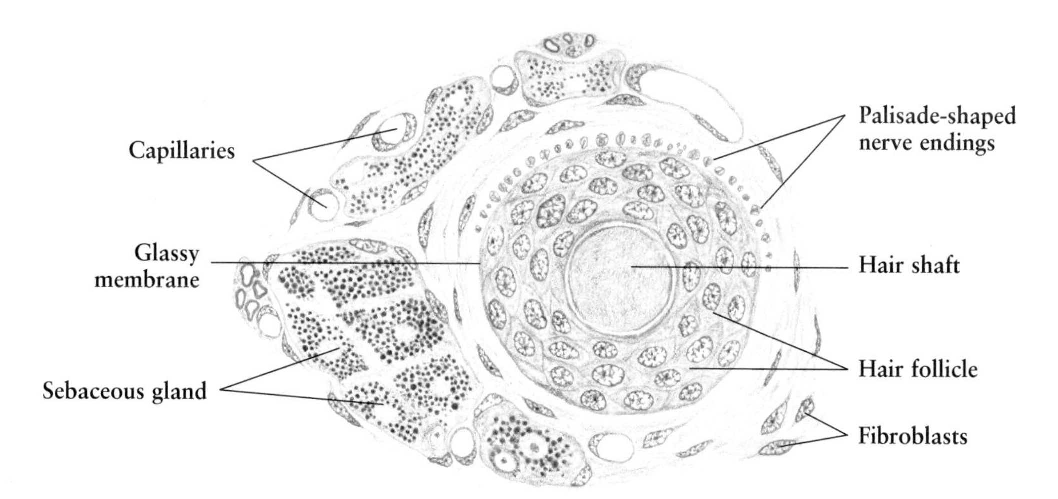

Fig. 6-30.

Taste Bud

Taste buds are small barrel-shaped chemoreceptors responsible for the sense of taste. They are mainly distributed in the circumvallate papillae and foliate papillae of the tongue, but also elsewhere in the oral cavity, palate, and epiglottis.

Taste buds are found embedded in the **stratified squamous epithelium**, extending from its **basement membrane** to the surface where they open into the oral cavity via a small pore, the **taste pore**. The taste bud is generally composed of three cell types: gustatory cells, supporting cells, and basal cells. Both **gustatory** and **supporting cells** are fusiform, with long **microvilli** on the free end of the cells at the taste pore where the cells are slightly depressed. The gustatory cells stain light, while supporting cells stain dark. The **basal cells** are small undifferentiated cells located at the base of the taste bud, and believed to be the stem cells for the other cell types. Ultrastructural studies have shown unmyelinated nerve fibers to be associated with both gustatory and supporting cells.

Four fundamental taste sensations can be detected: *sweet, bitter, sour,* and *salty.* Each sensation may be optimally perceived in a specific region of the tongue: sweet and salty at the tip, sour at the side, and bitter in the area of the circumvallate papillae. However no structural discrepancy in taste buds from different areas has been found. It has been demonstrated that a single taste bud can respond to all four basic taste qualities.

Fig. 6-31.

Olfactory Epithelium

Olfactory epithelium represents the receptor for the sense of smell. It is located in the olfactory mucosa in the roof of the nasal cavity in humans. This is a yellowish pseudostratified columnar epithelium, composed of three cell types: *olfactory cells, supporting cells,* and *basal cells.*

The **supporting cells** are tall and slender, with ovoid nuclei in their upper half. The bases are tapered, and rest on the **basement membrane,** while the apices are relatively broad with many long microvilli protruding into the overlying film of mucus. The **basal cells** are small conical cells, believed to be stem cells that are able to differentiate into supporting cells. The **olfactory cells** are fusiform bipolar neurons with nuclei in the lower half of the cells. A single dendrite terminates as a small swelling at the free surface of the epithelium, radiating to 6 or 10 **olfactory cilia.** At the base of the cell, each olfactory neuron gives rise to a single fine afferent axon, which join together to form olfactory nerve fibers in the lamina propria (see **Fig.** 9-3).

The **lamina propria** is a loose connective tissue beneath the olfactory epithelium. It contains numerous **blood vessels** and serous glands known as **Bowman's glands** (see **Fig.** 9-3). Secretions are expelled through the **excretory duct** to the surface of the epithelium to dissolve the odoriferous substance.

Figure 6-30. Taste Bud
Human tongue • H.E. stain • High magnification

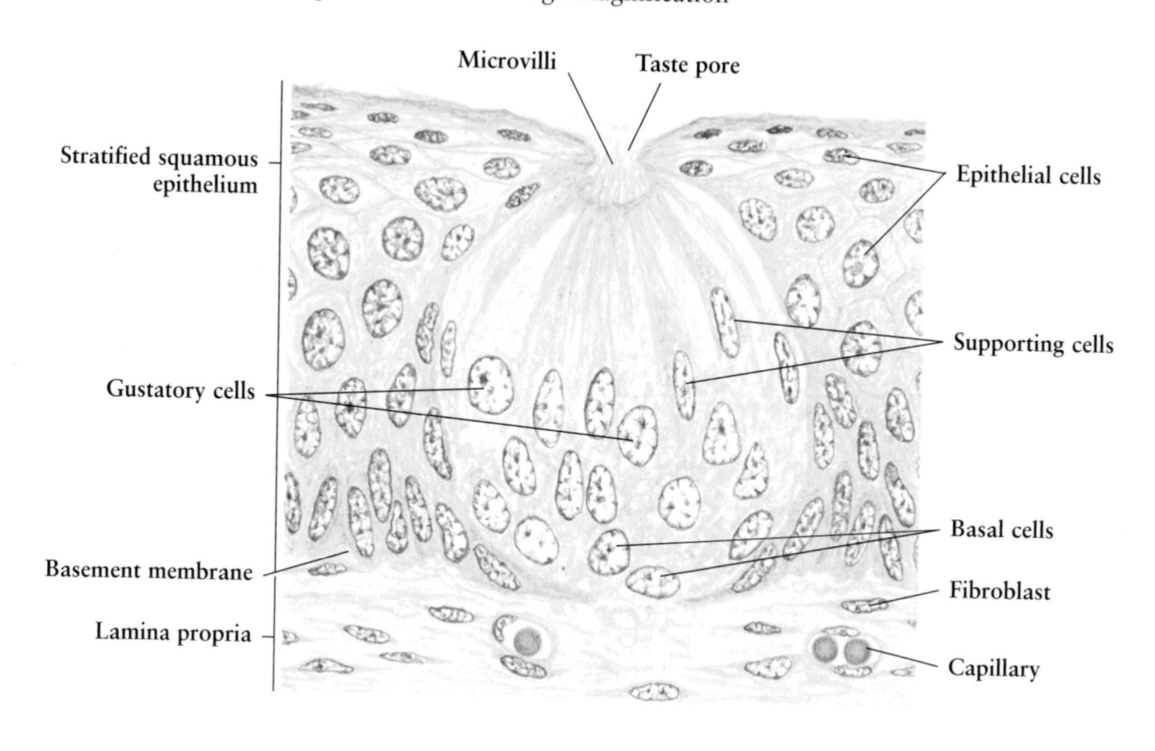

Figure 6-31. Olfactory Epithelium
Monkey olfactory mucosa • H.E. stain • High magnification

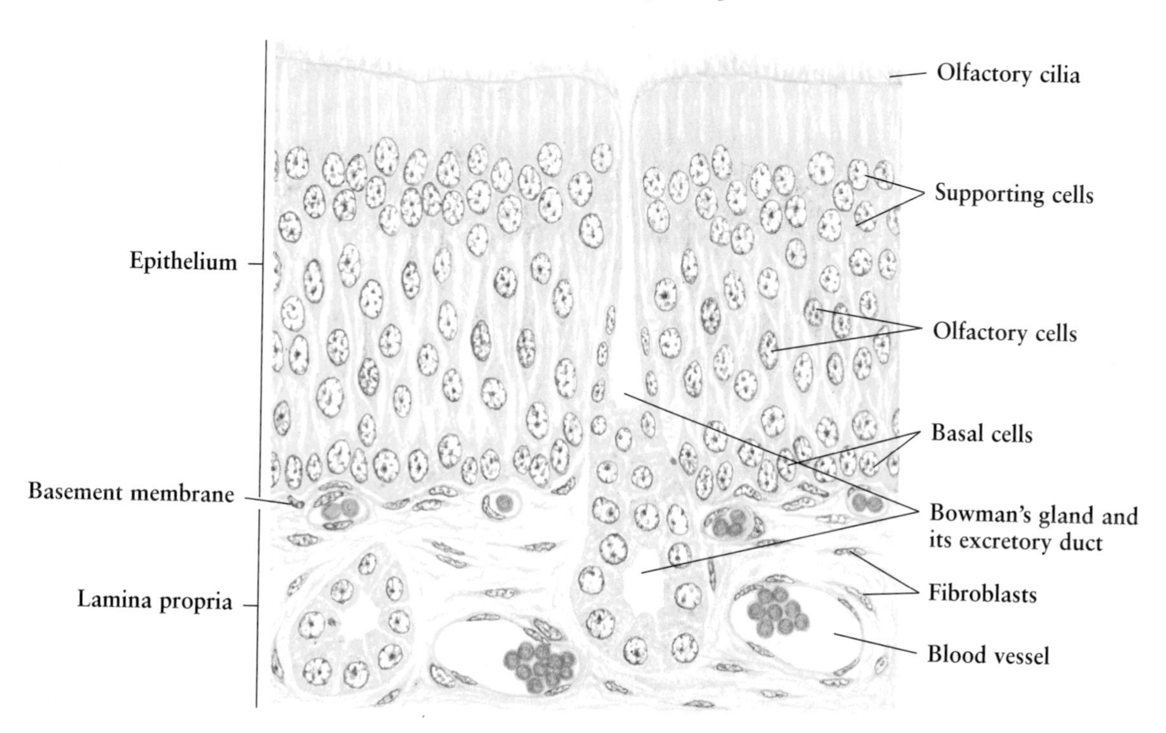

7. Circulatory System

The **circulatory system** consists of the blood vascular system and the lymphatic vascular system. The **blood vascular system** comprises (1) the **heart,** a muscular organ, which pumps blood into arteries; (2) **arteries,** thick-walled vessels that direct blood to capillary beds; (3) **capillaries,** a fine meshwork of anastomosing tubules, thin-walled vessels across which the exchange of substances between blood and tissues occurs; and (4) **veins,** which return blood from the capillaries to the heart.

The **lymphatic vascular system** consists of **lymphatic capillaries** and various-sized **lymphatic vessels.** The lymphatic capillaries are blind-ended tubules, which collect fluid (lymph) from tissue spaces. Lymph, after passing through lymphatic organs, is carried back to the blood vascular system via lymphatic vessels.

The components of the circulatory system have a number of structural features in common. Basically, their walls are composed of three layers: (1) **tunica intima,** an innermost layer, which minimally consists of a single layer of endothelial cells supported on a basement membrane, with or without an underlying delicate connective tissue layer; (2) **tunica media,** an intermediate layer with variable amounts of muscle, elastic, and reticular fibers; and (3) **tunica adventitia,** an outermost layer of connective tissue rich in collagen and elastic fibers.

In great vessels, because the wall is too thick to be nourished by diffusion from blood within the lumen, the **vasa vasorum** (i.e., vessels of the vessel) give rise to a blood capillary network in the adventitia and the outermost media. These vessels are derived either from the arteries they supply or from neighboring arteries.

Frequently, **nerve fibers,** which are unmyelinated sympathetic nerve fibers supplying the smooth muscle of the vessels, are found in company with arteries and veins.

Fig. 7-1. Heart: Wall of Ventricle

The **heart** is a four-chambered muscular organ composed of two atria and two ventricles. Although the wall of the ventricle is thicker than that of the atrium, they both consist of three layers: (1) an internal layer or **endocardium** (see **Fig.** 7-2); (2) an intermediate layer or **myocardium**; and (3) an external layer or **epicardium** (see **Fig.** 7-3).

The **myocardium** is composed of **cardiac muscle fibers,** arranged in layers that enclose the atria and ventricles in a complex spiral fashion. From the transverse section of the ventricle wall in this figure, it is difficult to distinguish the different layers of cardiac muscle fibers. Normally, most of these fibers are attached to the cardiac skeleton, a fibrous central supporting structure of the heart. The space between muscle fibers is occupied by connective tissue containing abundant blood vessels and lymphatic vessels. Projecting into the **ventricular cavity** are the **papillary muscles.** They are myocardium extensions covered by endocardium, which stabilize the cusps of the mitral and tricuspid valves via chordae tendineae (see **Fig.** 7-5).

Figure 7-1. Heart: Wall of Ventricle
Human • H.E. stain • Very low magnification

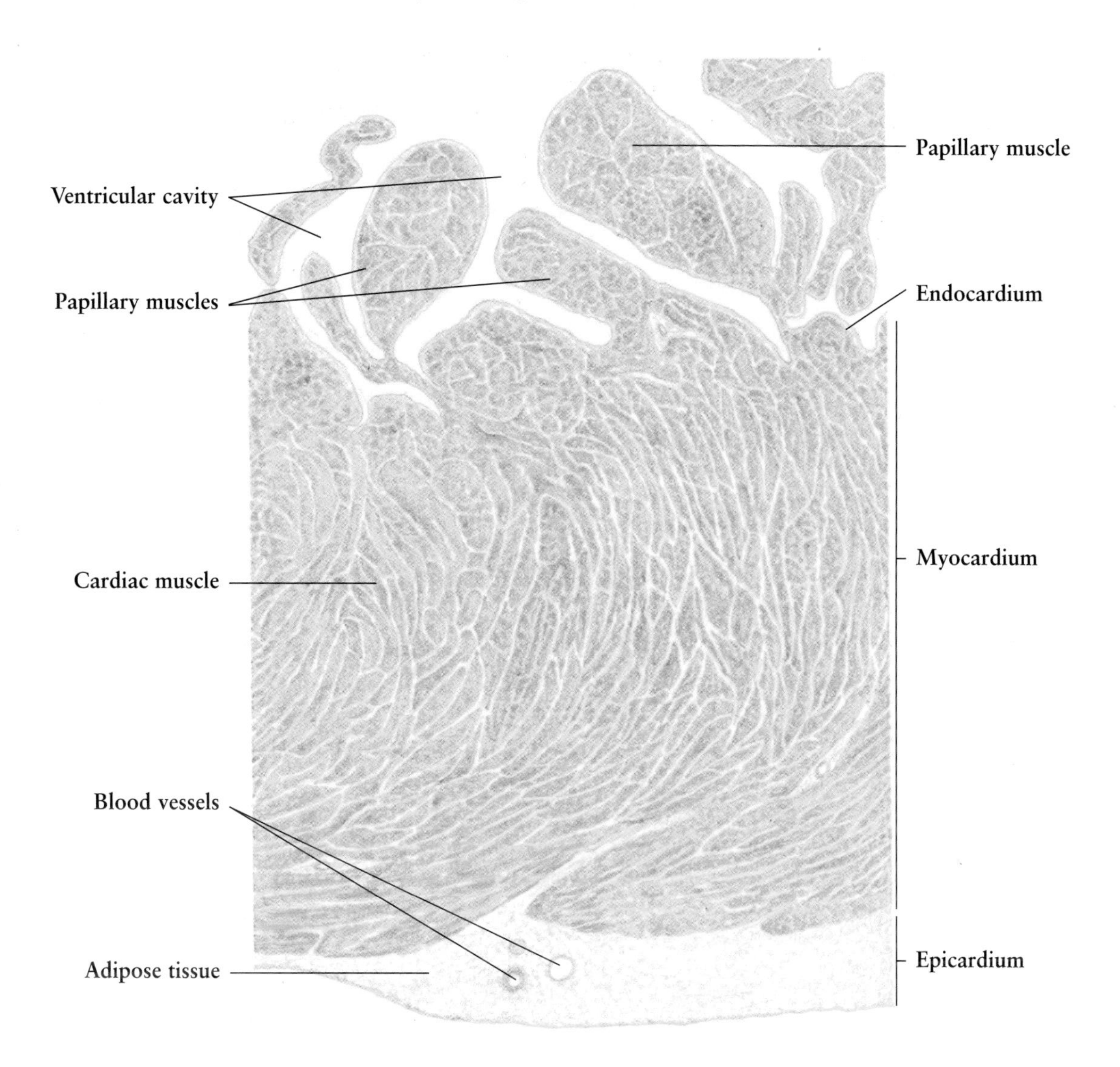

Fig. 7-2.

Heart: Endocardium

The **endocardium**, homologous to the tunica intima of blood vessels, consists of *endothelium,* a *subendothelial layer,* and a *subendocardial layer.*

The **endothelium** is a single layer of flattened epithelium, which covers all internal surfaces of the heart chambers, including the papillary muscle and chordae tendineae. It is continuous with that of the blood vessels entering and leaving the heart. Just beneath the endothelium is the **subendothelial layer** of connective tissue containing collagen and elastic fibers, **fibroblasts,** and some **smooth muscle fibers.** The subendothelial layer contains no blood vessels, and is nourished by direct imbibition from the circulating blood. Binding the endocardium to the underlying myocardium is a **subendocardial layer** of **loose connective tissue,** which contains numerous blood vessels, nerves, and branches of the conduction system of the heart, such as the **Purkinje fibers** (see **Fig.** 5-7).

Fig. 7-3.

Heart: Epicardium

The **epicardium** is a serous membrane, and forms the visceral pericardium. It is covered by a single layer of flattened epithelial cells, the **mesothelium,** which also lines the opposite parietal pericardial surface. The mesothelial cells secrete a small amount of serous fluid, which lubricates the sliding friction of the epicardium against the parietal pericardium during contraction and relaxation of the heart. Beneath the mesothelium is a layer of **connective tissue** containing numerous elastic fibers. This layer is bound to the **myocardium** by a broad layer of **adipose tissue.** Within the adipose tissue are embedded the **arteries** and **veins** of the coronary system and autonomic **nerve fibers,** which supply the myocardium.

Figure 7-2. Heart: Endocardium
Human • H.E. stain • High magnification

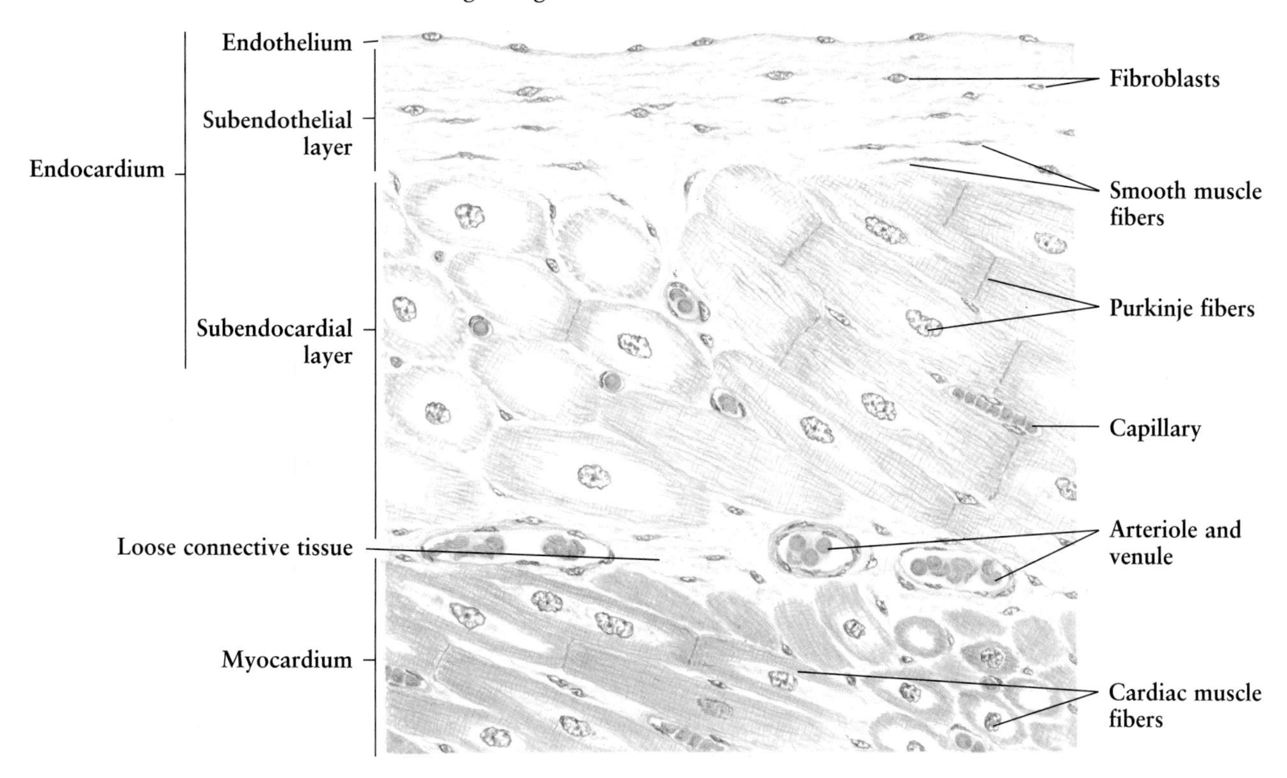

Figure 7-3. Heart: Epicardium
Human • H.E. stain • Low magnification

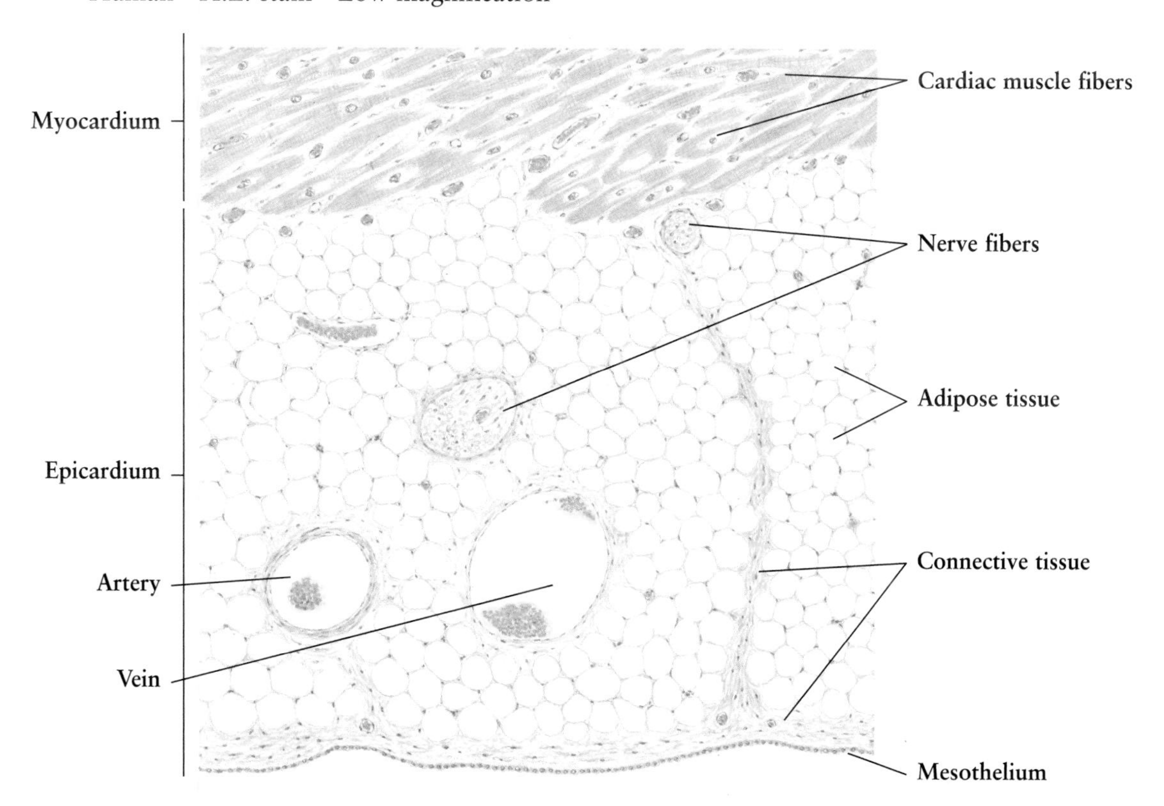

Fig. 7-4. Heart: Interventricular Septum

The **interventricular septum** is composed of a **membranous septum** and a **myocardial septum.** Located in the lower part of the interventricular septum, the myocardial septum is continuous with the myocardium of the heart, and is covered on both sides by the **endocardium.** The membranous septum consists of tough collagenous fibrous connective tissue, lying on the upper part of the interventricular septum. This constitutes part of the cardiac skeleton and serve as an attachment for some cardiac muscle fibers. In this illustration of the interventricular septum, the **cardiac muscle fibers** in the left of the myocardial septum are seen attaching to the membranous septum via a **tendon.** The **aortic valve, wall of the aorta, myocardium of the right atrium,** and the **tricuspid valve** are also shown.

Fig. 7-5. Heart: Atrioventricular Valve

Atrioventricular valves, the *tricuspid* and *mitral,* consist of a plate of tough fibrous connective tissue, covered by endocardium continuous with that of the heart chambers. Beneath the **endothelium** on the atrial side is a thick layer of **elastic connective tissue** composed of elastic fibers and **fibroblasts.** Between the elastic connective tissue and the endothelium of the ventricular side lies a thick layer of **dense connective tissue** that consists of bundles of **collagen fibers** and a small amount of fibroblasts. This dense connective tissue is penetrated by the fibrous **chordae tendineae,** which connect the valve flaps to the papillary muscles of the ventricles. The core of the chordae tendineae is composed of bundles of **collagen fibers** parallel to the long axis, with scattered **fibroblasts.** The surface of chordae tendineae is invested by a layer of **endothelium.**

Figure 7-4. Heart: Interventricular Septum
Human • H.E. stain • Very low magnification

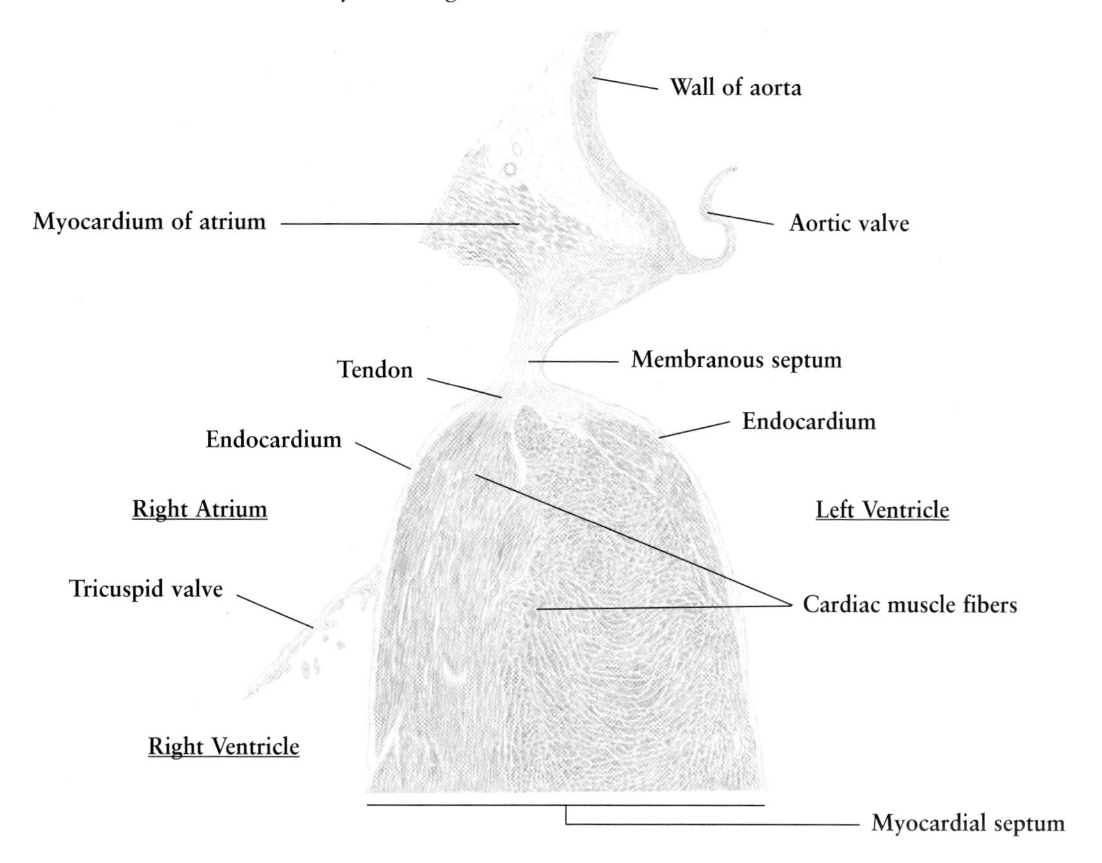

Figure 7-5. Heart: Atrioventricular Valve
Human • H.E. stain • Medium magnification

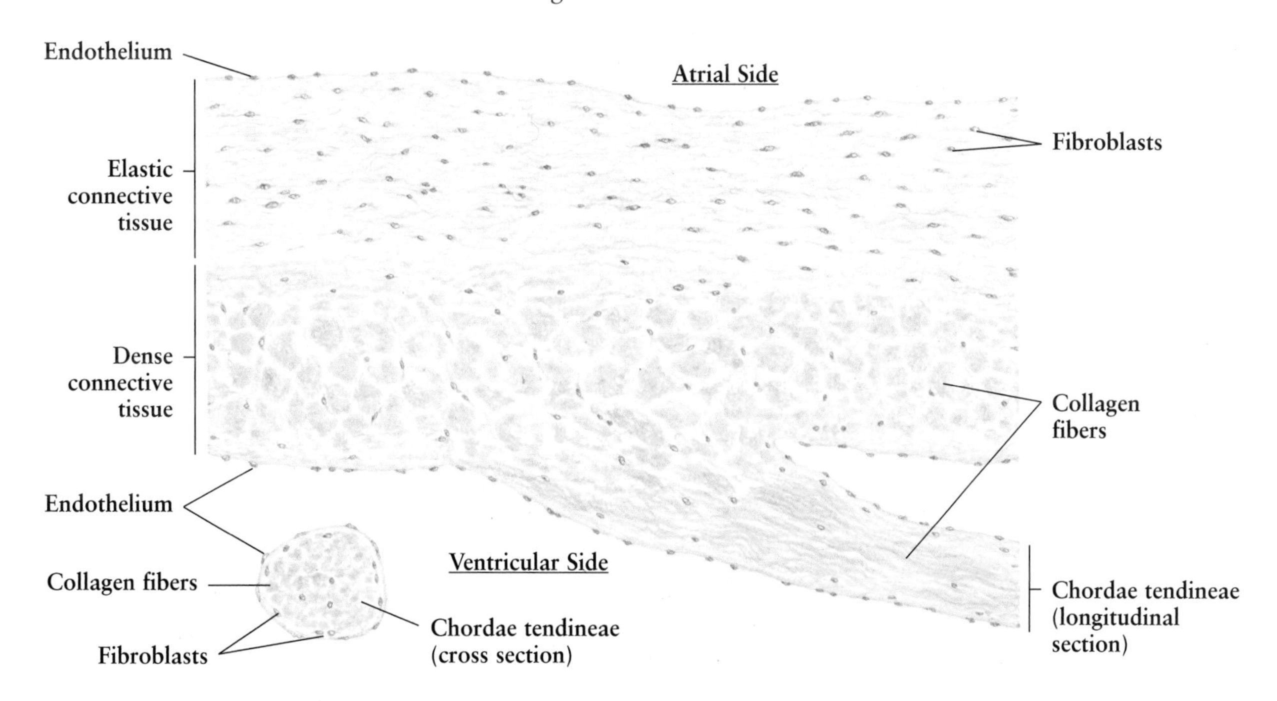

Figs. 7-6 and 7-7.

Elastic Artery: Aorta

The **aorta** is a typical elastic artery. Like most blood vessels, its wall consists of three layers: tunica intima, tunica media, and tunica adventitia.

The **tunica intima** is composed of **endothelium** and **subendothelial connective tissue.** The endothelium is formed of a single layer of endothelial cells, which are polygonal and flat, with the nucleus causing a local luminal protrusion of the cell. The endothelial cells are supported by a thin layer of basement membrane, which can be recognized under an electron microscope. The subendothelial connective tissue contains collagen and elastic fibers, scattered **fibroblasts,** and circularly arranged **smooth muscle fibers.** In the deeper portion of the intima, longitudinally arranged **smooth muscle fibers** are present. A distinct **internal elastic lamina** is difficult to identify, because it consists of two or three laminae that blend with similar membranes of the intima and media.

The **tunica media** is particularly thick, forming approximately four-fifths of the thickness of the wall. It is characterized by 40–60 layers of fenestrated **elastic laminae,** arranged concentrically. The space between adjacent elastic laminae is filled by amorphous ground substance, a fine elastic fiber network, **smooth muscle fibers,** and **fibroblasts.** At the outer limit of the media is an indistinct **external elastic lamina.**

The **tunica adventitia** consists of a thin layer of loose **connective tissue** containing **nerve fibers,** lymphatic vessels, some **adipose cells,** and **vasa vasorum,** which also penetrate the outer portion of the media.

Figure 7-6. Elastic Artery: Aorta
Human • Cross section • H.E. stain • Low magnification

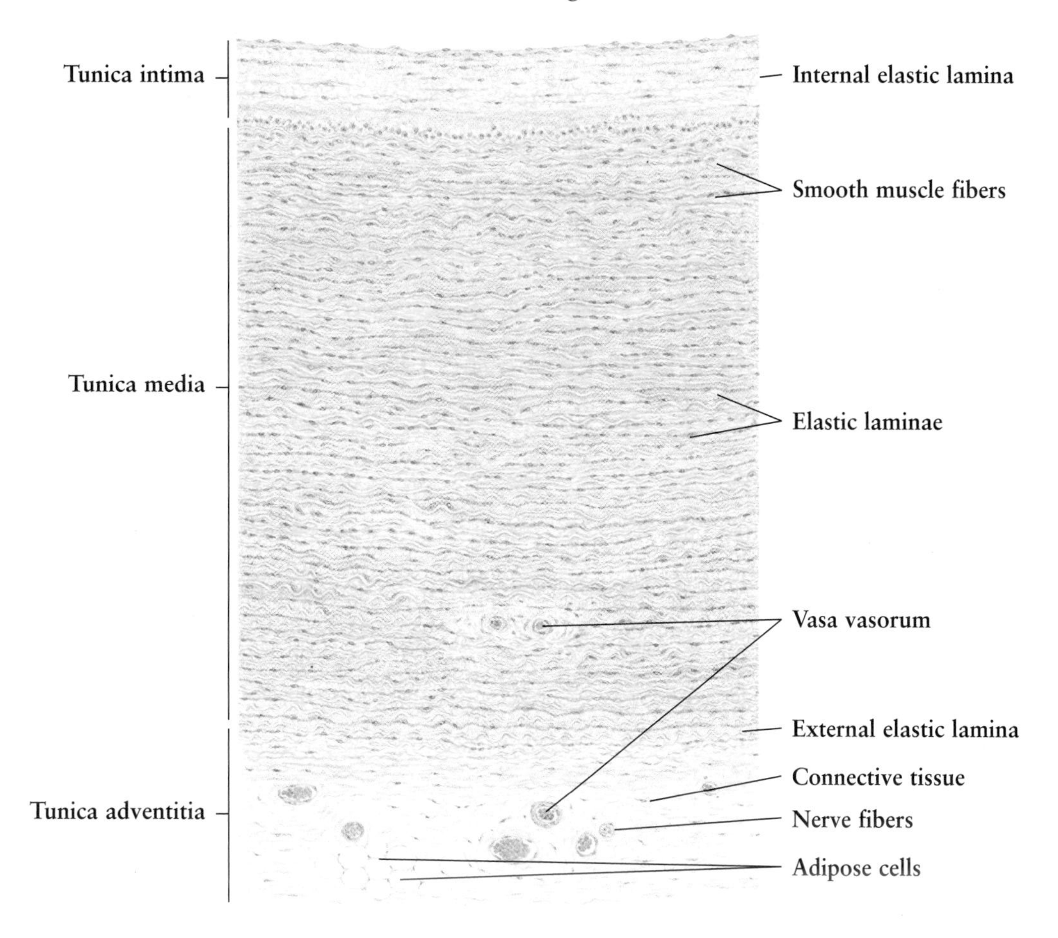

Figure 7-7. Elastic Artery: Aorta
Human • Cross section • H.E. stain • High magnification

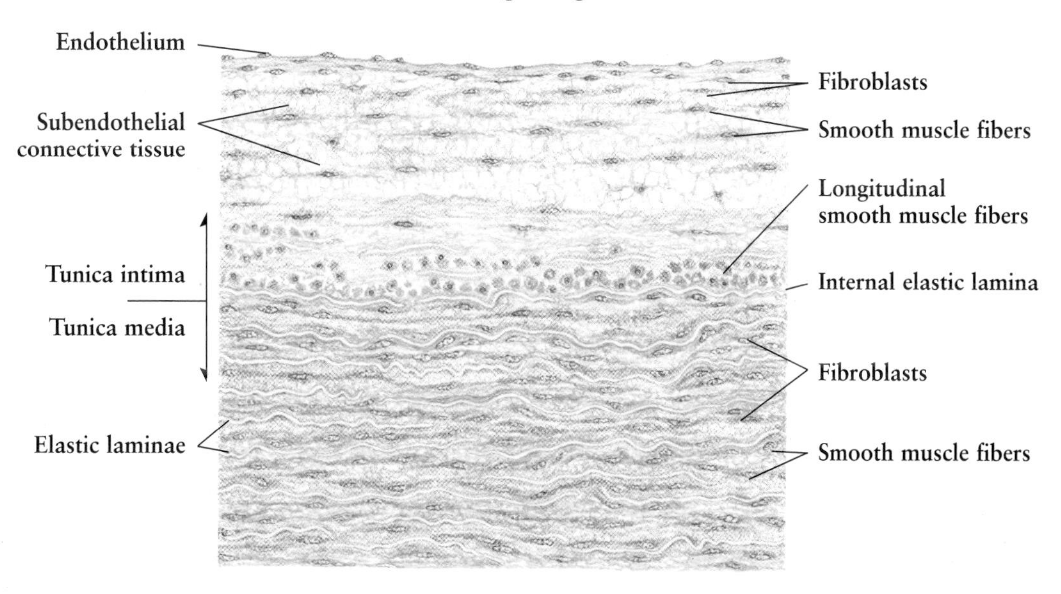

Figs. 7-8 and 7-9.

Elastic Artery: Aorta

The use of resorcin fuchsin stain demonstrates the distribution of elastin in the three layers of the aortic wall. Under high magnification, **elastic fibers** are found in the **subendothelial connective tissue** of the **tunica intima,** especially in its deeper portion. The **internal elastic lamina** is clearly recognized. Numerous **fenestrated elastic laminae** are seen in the **tunica media.** Between fenestrated elastic laminae are **elastic fibers.** The **external elastic lamina** is not easy to identify. The nuclei in the media belong to **smooth muscle fibers** as well as fibroblasts.

Figure 7-8. Elastic Artery: Aorta
Human • Cross section • Resorcin fuchsin stain • Low magnification

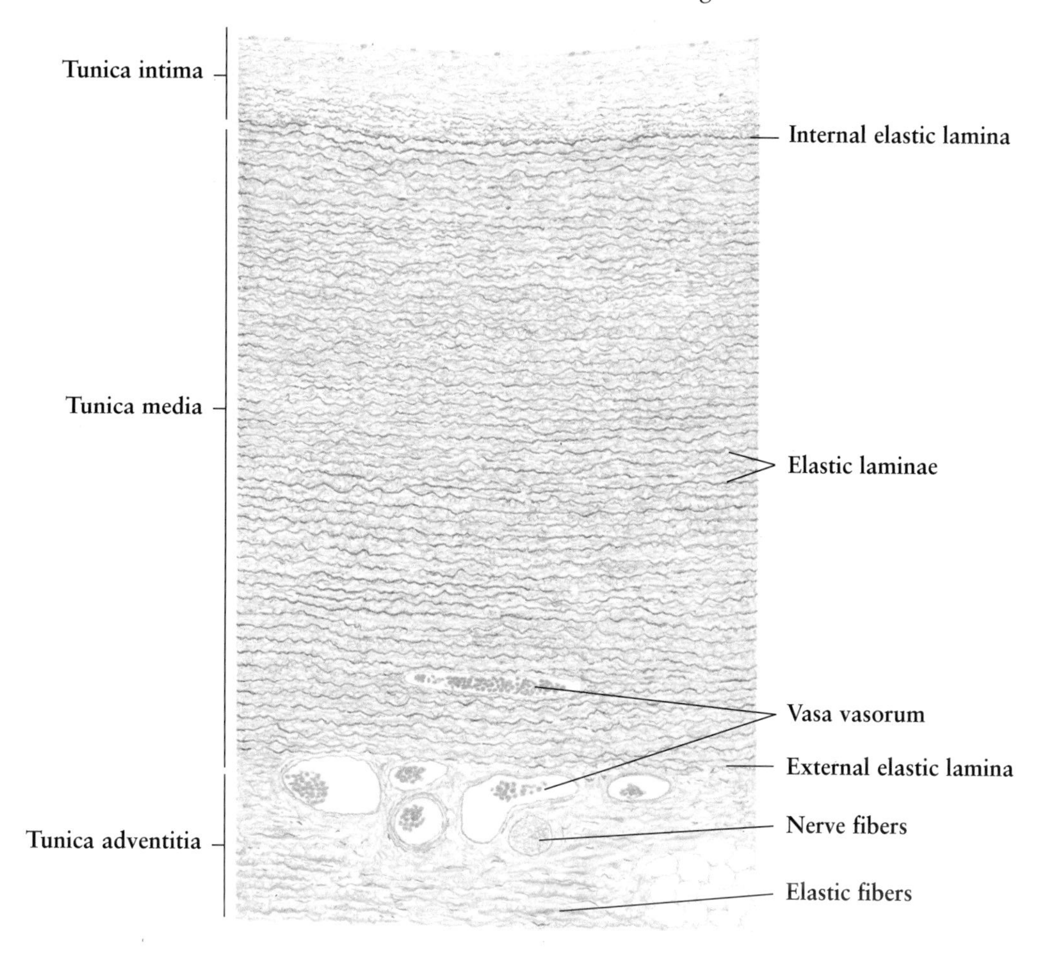

Figure 7-9. Elastic Artery: Aorta
Human • Cross section • Resorcin fuchsin stain • High magnification

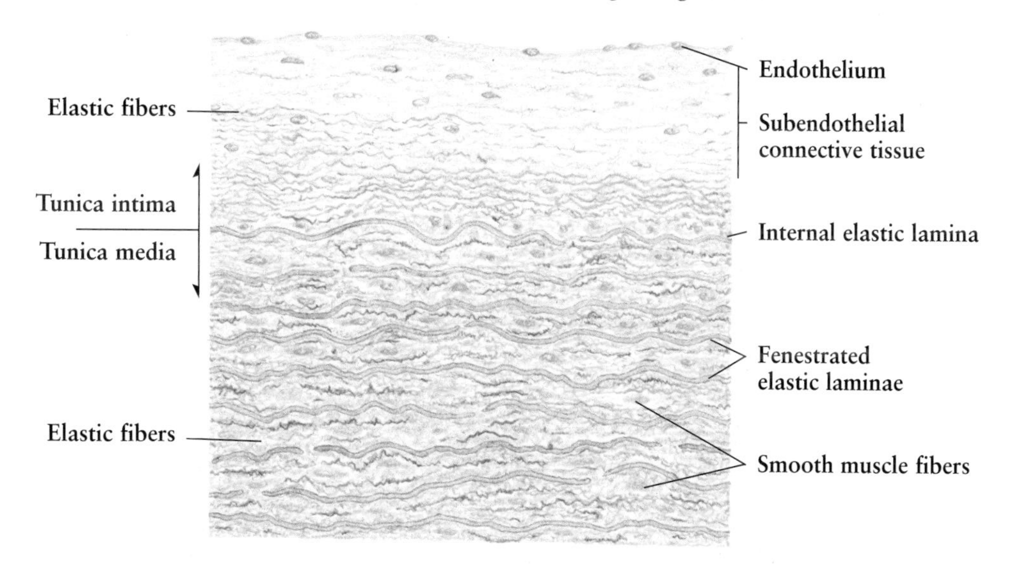

Figs. 7-10 and 7-11.

Muscular Artery

The **muscular arteries** are the largest class, and except for the elastic arteries, all named arteries of gross anatomy, including many of the large arteries and all medium-sized arteries and small arteries, belong to this group. The muscular artery possesses a basic structure similar to that of the elastic artery; however, its elastic tissue content is reduced while its smooth muscle content is increased. **Figures 7-10** and **7-11** show the characteristic features of the subclavian artery and radial artery, respectively.

The **tunica intima** comprises three layers: endothelium, subendothelial layer, and internal elastic lamina. The **endothelium** is similar to that of elastic artery. The **subendothelial layer** is a thin layer of connective tissue containing collagen and elastic fibers and a few **fibroblasts.** As the size of the vessel decreases this layer becomes thinner. The **internal elastic lamina** is well marked, and is always seen as a refractile wavy band in paraffin preparations.

The **tunica media** is characterized by a thick muscular layer with 20–40 layers of **smooth muscle fibers,** which are arranged circularly or spirally. Between the layers of muscle, there are small amounts of connective tissue. In the larger muscular artery, **elastic fibers (laminae)** are still prominent between the layers of smooth muscle fibers. The outer border of the media is marked by the **external elastic lamina,** which is a less conspicuous fenestrated layer of elastin. No vasa vasorum can be seen in this coat.

The **tunica adventitia** is as thick as the media. It consists of loose connective tissue containing **vasa vasorum, nerve fibers, adipose cells, fibroblasts,** and elastic fibers. Longitudinally arranged **collagen fibers** are found in the deeper portion of the adventitia.

Figure 7-10. Muscular Artery: Subclavian Artery
Human • Cross section • H.E. stain • Medium magnification

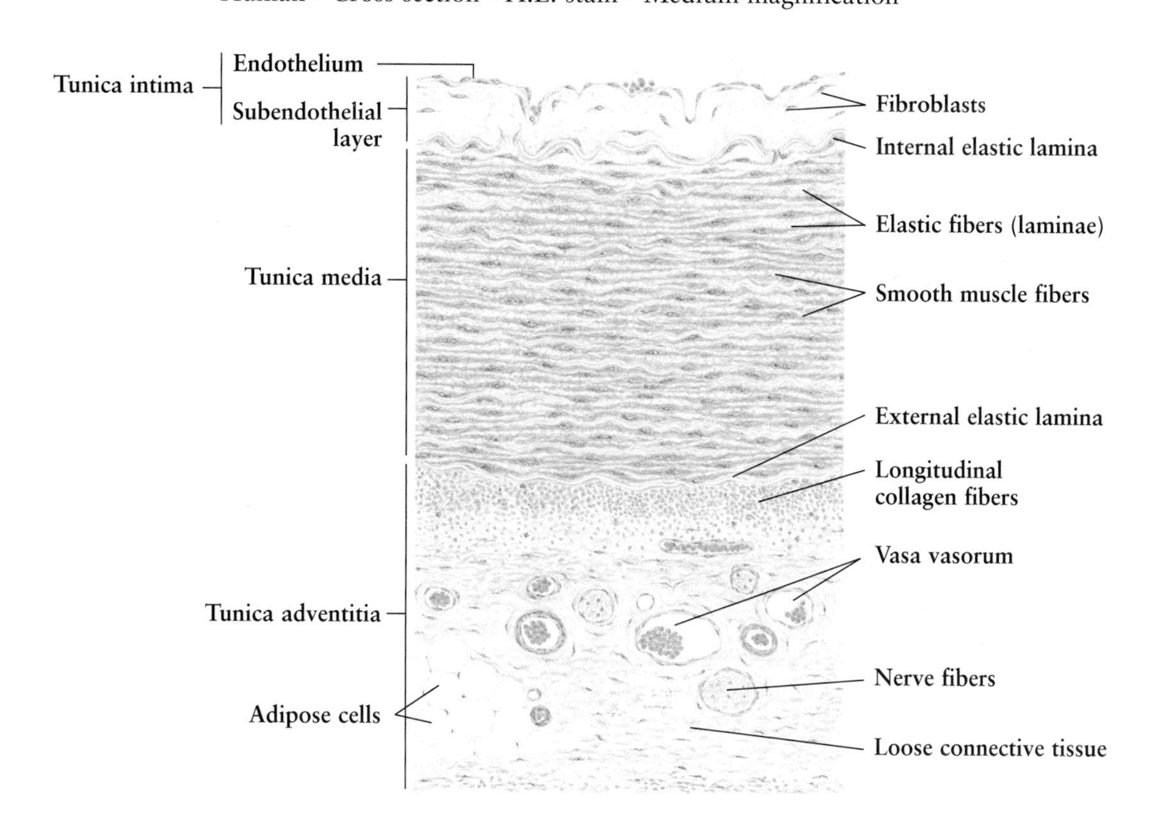

Figure 7-11. Muscular Artery: Medium-sized Artery
Human radial artery • Cross section • H.E. stain • Medium magnification

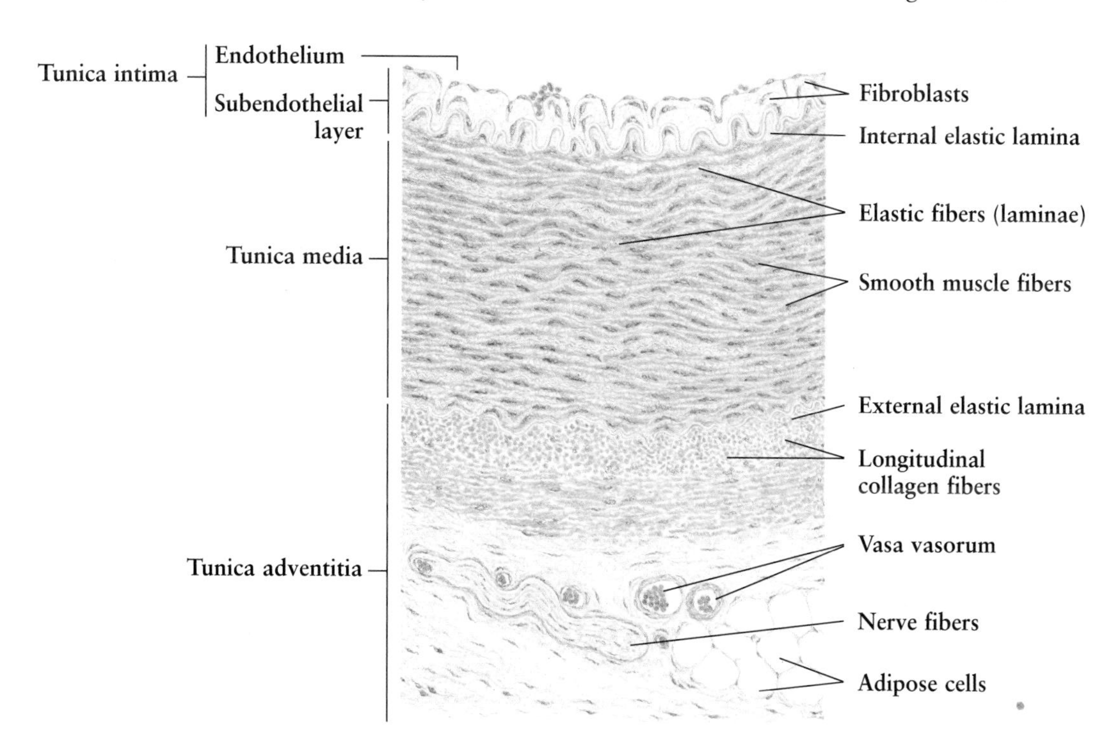

Fig. 7-12.

Muscular Artery: Small Artery

The **small artery,** 0.1–2 mm in diameter, belongs to the muscular group, and has its typical three-layer structure. The **tunica intima** consists of a single layer of **endothelium,** which rests directly on the **internal elastic lamina,** because the subendothelial connective tissue layer has disappeared. The **tunica media** consists of 5–10 layers of **smooth muscle fibers** arranged circularly or spirally, with interspersed elastic fibers. The external elastic lamina is absent or indistinct. The **tunica adventitia** is composed of a thin layer of connective tissue composed of **collagen fibers** arranged longitudinally and a few **fibroblasts.** There are no vasa vasorum in the tunica adventitia.

Fig. 7-13.

Small Artery and Small Vein

The structure of the **small artery** in this picture is almost the same as that in **Fig. 7-12,** except for its smaller diameter and fewer layers of smooth muscle fibers in the tunica media.

Small veins have a diameter ranging from 0.1 to 1 mm. The **tunica intima** is thin, consisting of a layer of **endothelium** and a very thin basement membrane. The **tunica media** is composed of two or three layers of circularly arranged **smooth muscle fibers** interspersed by collagen and elastic fibers. The **tunica adventitia** is a layer of loose connective tissue thicker than the tunica media. It contains small amounts of longitudinally aligned **collagen fibers** and **fibroblasts.**

In this figure, other vessel types are also shown. These are described later.

Figure 7-12. Muscular Artery: Small Artery
Human • Cross section • H.E. stain • High magnification

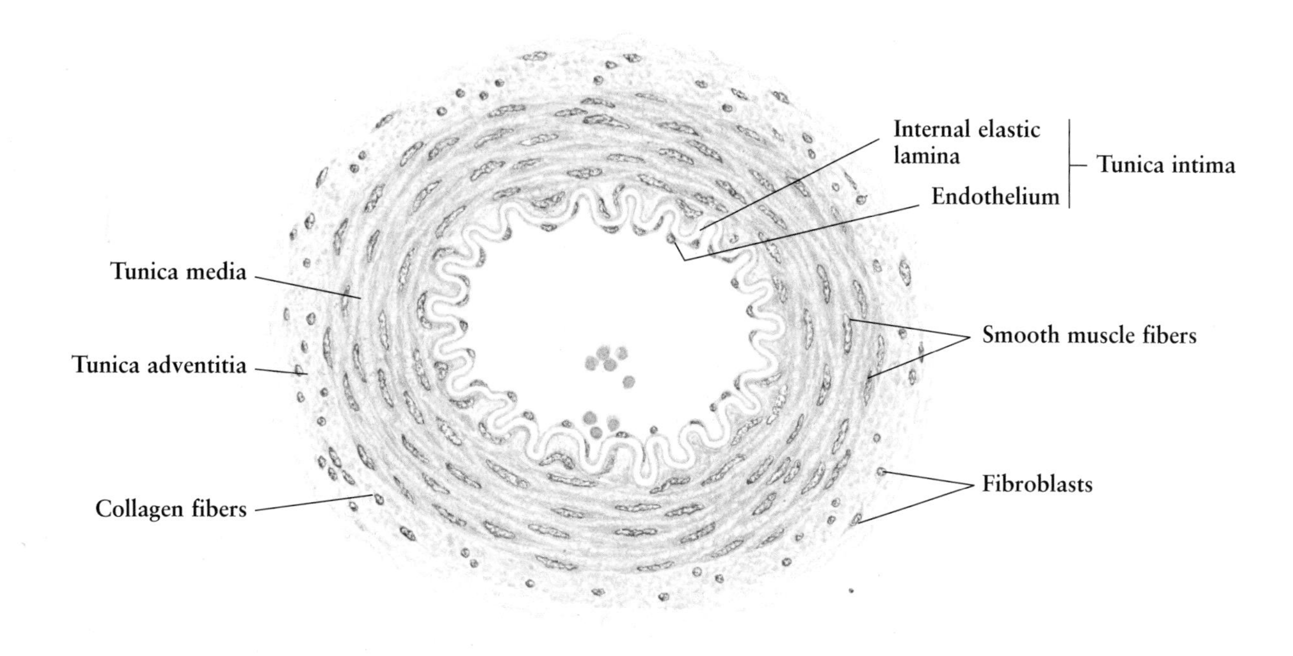

Figure 7-13. Small Artery and Small Vein
Human • Cross section • H.E. stain • High magnification

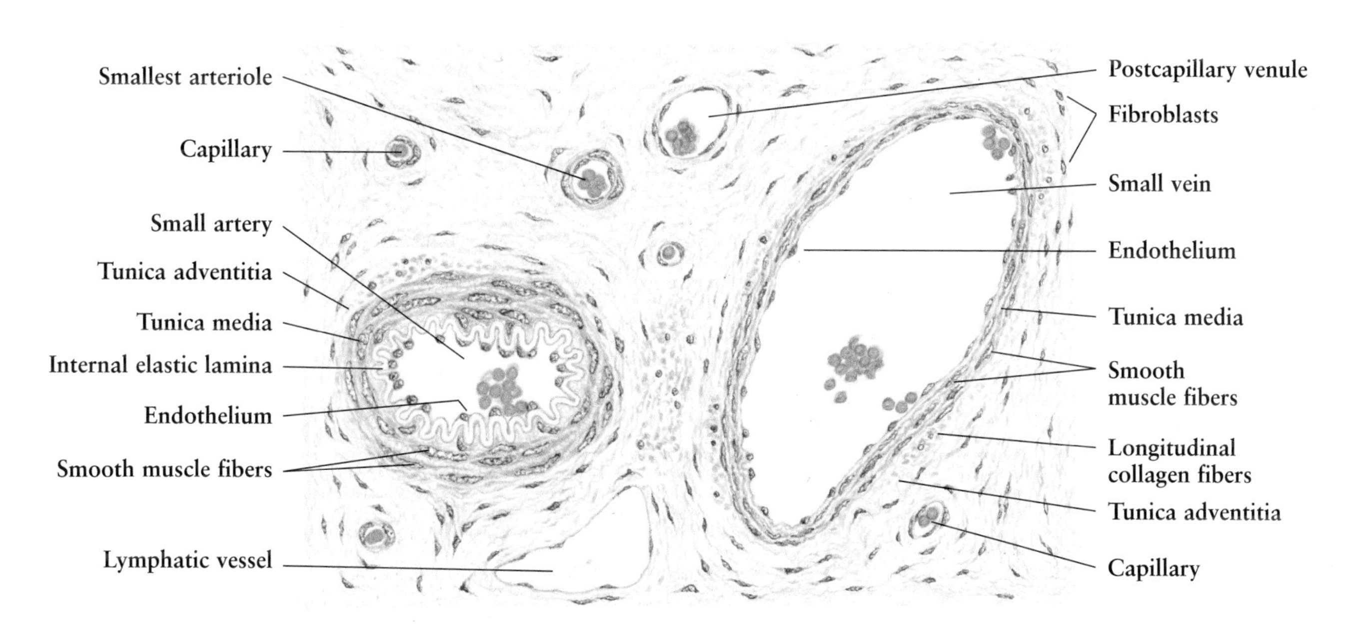

Fig. 7-14. Arteriole and Muscular Venule

The **arteriole** is a category of microvessels with a diameter of 30–300 μm. The **endothelial cells** of the tunica intima are elongated and arranged longitudinally, resting directly on the **internal elastic lamina**, which become thinner as the diameter of the vessel is reduced. The **tunica media** is composed of one to three layers of **smooth muscle fibers** arranged circularly or spirally. The **tunica adventitia** consists of a thin connective tissue sleeve containing longitudinally arranged collagen and elastic fibers and a few fibroblasts.

Muscular venules range in diameter from 50 μm to 1 mm, connecting collecting venules with small veins. The wall of the muscular venules contains a layer of **endothelium**, one or two layers of flattened **smooth muscle fibers** as its tunica media, and the tunica adventitia with **loose connective tissue** continuous with the surrounding tissue.

Fig. 7-15. Arteriole and Collecting Venule

In this figure there are several profiles of **arterioles** of different sizes. Their basic structure has been described in **Fig. 7-14.** The **smallest arterioles** lack an internal elastic lamina, with the endothelium resting directly on the basement membrane. The media of the smallest arterioles has only one layer of **smooth muscle fibers.** At the left part of this figure is an arteriole in longitudinal section showing ramification.

The **collecting venule** has a diameter ranging from 30 to 50 μm, and conducts blood from the postcapillary venule to the muscular venule. It is characterized by a complete layer of **pericytes** as its media. The endothelium of the collecting venule is thin and of the continuous type.

Figure 7-14. Arteriole and Muscular Venule
Human • Cross section • H.E. stain • High magnification

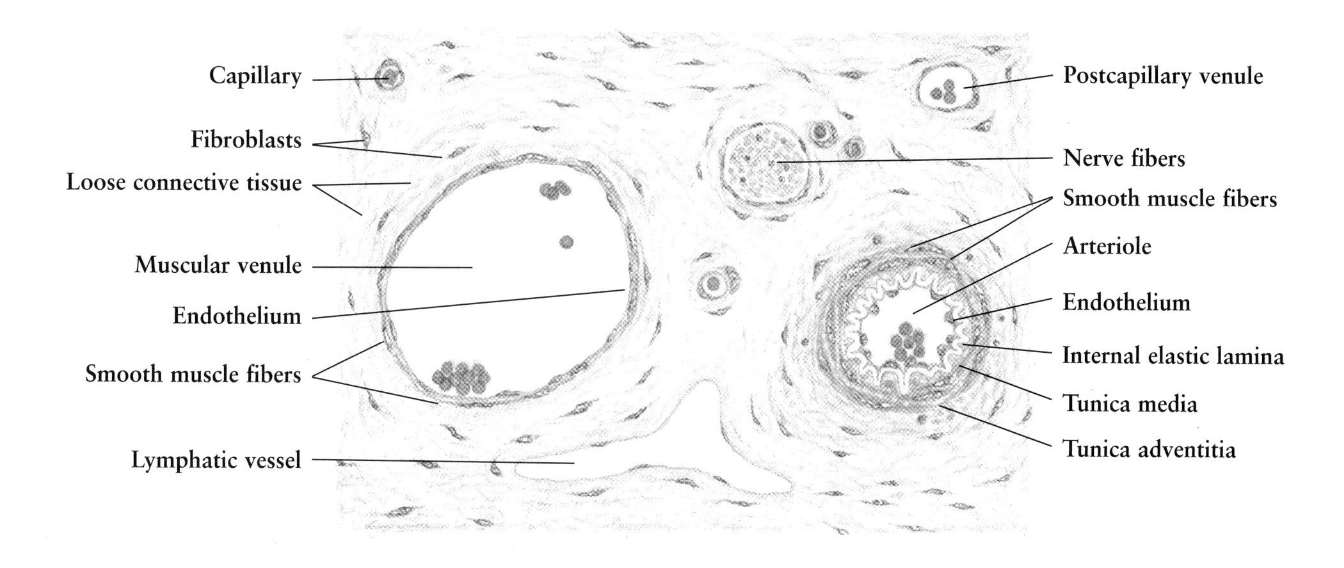

Figure 7-15. Arteriole and Collecting Venule
Human • H.E. stain • High magnification

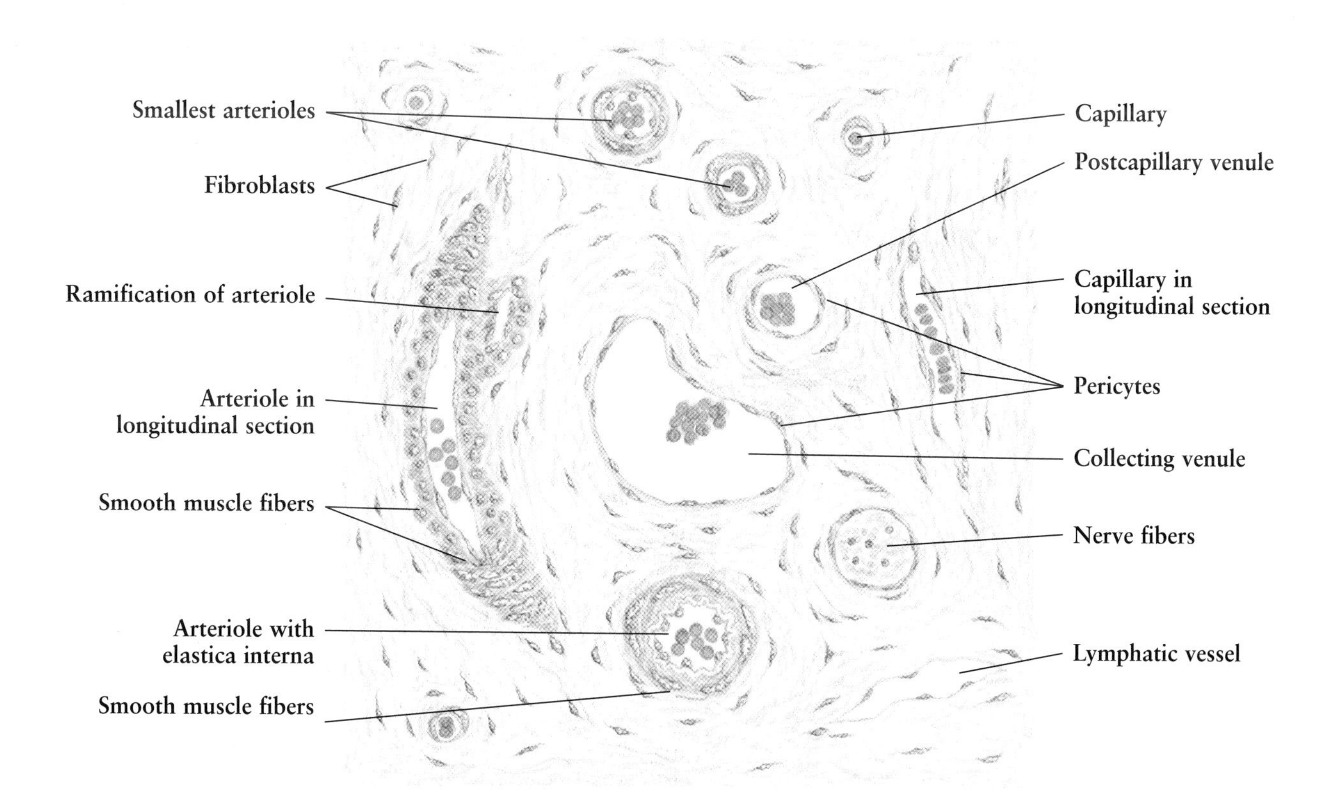

Fig. 7-16. Capillaries and Postcapillary Venules

Capillaries are small thin-walled blood vessels with an average luminal diameter of 5–10 μm, which form a continuous network of narrow canals between the smallest arterioles and the postcapillary venules, providing tissue fluid, nutrients, and oxygen for the cells of the various body tissues. The wall of a capillary is thin, composed of a single layer of **endothelial cells** that rest on the basement membrane, and occasional **pericytes** surrounding the endothelium. Such structural features of the capillary are well-suited for its role in material exchange. According to the ultrastructure of the endothelial cells by electron microscopic studies, capillaries have been classified into three major types: *continuous, fenestrated,* and *sinusoidal.* At the level of light microscopy, it is difficult to distinguish between the continuous and fenestrated types. Sinusoid capillaries are shown in **Figs. 10-46, 10-47,** and **10-48.**

Postcapillary venules are continuous with capillaries, with a diameter ranging from 8 to 30 μm. The wall of postcapillary venules is similar to that of capillaries, consisting of a layer of flat **endothelial cells,** a basement membrane, and an incomplete layer of **pericytes** surrounded by a thin loose connective tissue containing longitudinal collagen and elastic fibers. It is interesting that the endothelial cells of postcapillary venules in lymphatic tissue are cuboidal, not flat (see **Fig. 8-5**). In addition to their exchange function in a manner similar to that of capillaries, the postcapillary venules appear to be the main point where the white blood cells enter and leave the circulation.

In this figure, both longitudinal and transverse sections of capillaries and postcapillary venules are demonstrated in detail.

Fig. 7-17. Arteriovenous Anastomosis

Arteriovenous anastomosis is the direct communication between the arterial and venous circulation, occurring in the skin of the palm, fingertip, sole, lip, nose, and ear, as well as in tissues such as the thyroid gland, gastrointestinal tract, and erectile tissue. It also forms an essential part of the aortic, carotid, and coccygeal bodies.

The arteriovenous anastomosis forms a short, convoluted, and twisted vessel in outline, surrounded by a **connective tissue capsule**, and the whole structure is known as a **glomus** (see **Fig. 15-1**). Histologically an arteriovenous anastomosis is the direct connection of **arteriole** to **venule**, with a abrupt change of the wall without any transitional zone.

The **endothelial cells** in both **arterial** and **venous segments** are thin and flat. However, the arterial segment has lost its internal elastic lamina, and the **smooth muscle fibers** are modified in shape and become **epithelioid cells** in appearance. In the connective tissue surrounding the anastomotic portion there is a rich supply of **nerve fibers** belonging to the autonomic nervous system. The smooth muscle fibers (epithelioid cells) in the arteriovenous anastomosis, innervated by those nerve fibers, play the role of a muscular shunt in regulating blood circulation.

When the smooth muscle fibers of the arterial segment contract, all the blood in the arteriole passes through the capillaries and warms the periphery; when they dilate, the blood can flow directly into the venules, bypassing the capillary network.

Figure 7-16. Capillary and Postcapillary Venule
Human • H.E. stain • Higher magnification

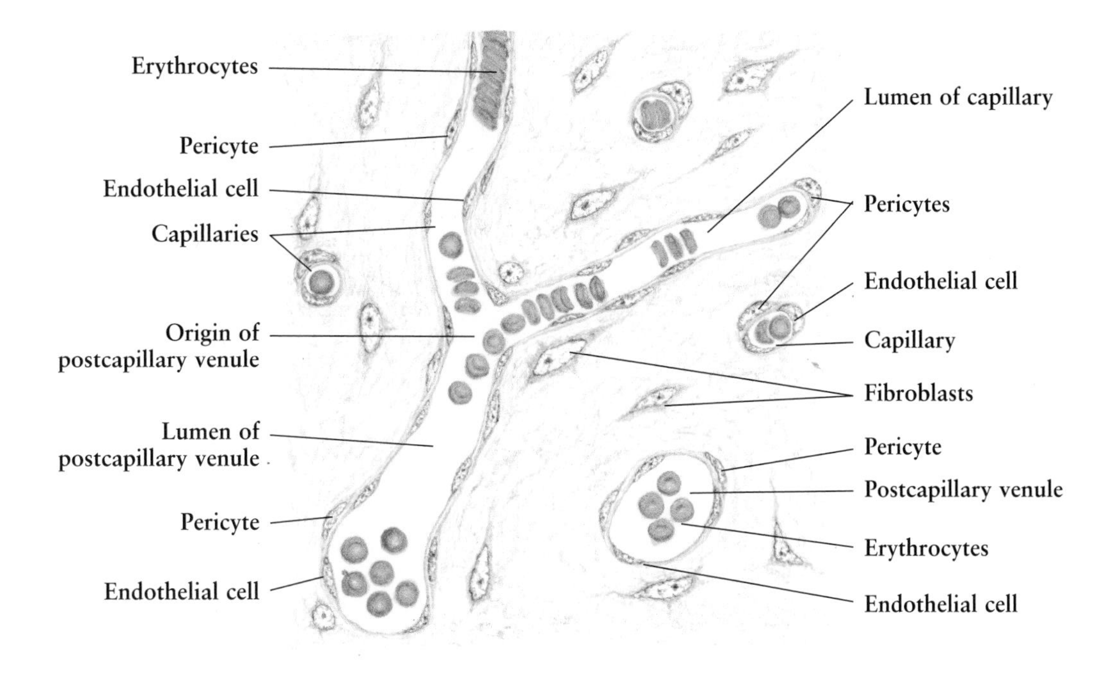

Figure 7-17. Arteriovenous Anastomosis
Glomus in human finger • H.E. stain • High magnification

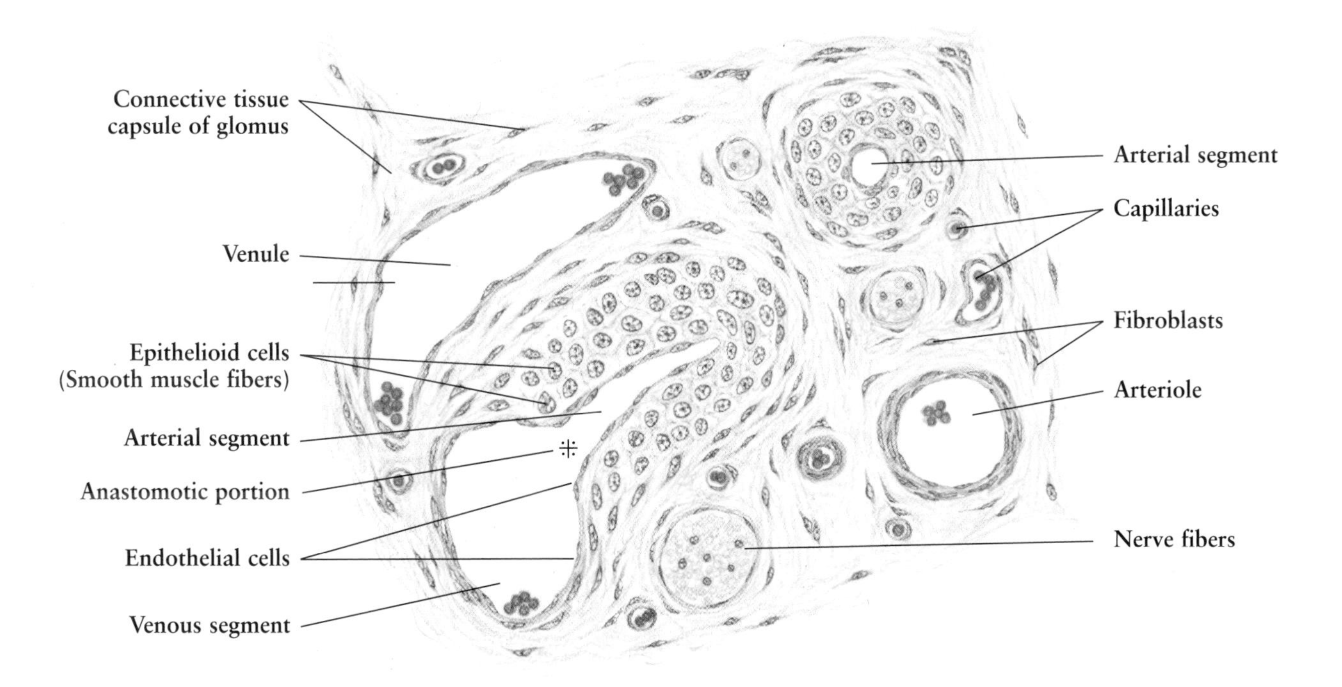

Fig. 7-18. Small Vein

This figure demonstrates a **small vein** in the tissue of the human lip. The **tunica intima** consists of a layer of **endothelium** and **subendothelial connective tissue.** Note that there is no internal elastic lamina in this coat. The **tunica media** typically consists of two or three layers of circular **smooth muscle fibers** with collagen and elastic fibers between them. The thickest of the layers is the **tunica adventitia.** It is composed of loose **connective tissue** containing a few **fibroblasts,** longitudinal **collagen fibers,** elastic fibers, and scattered bundles of **smooth muscle fibers** arranged longitudinally.

The small vein conducts blood from the muscular venule to the medium-sized vein. According to the function, situation, and position, some small veins possess venous valves (see **Fig.** 7-22).

Fig. 7-19. Medium-sized Vein: Saphenous Vein

Anatomically the **medium-sized vein,** with a diameter of 1–10 mm, is often found in company with a medium-sized artery, both of them having their own special names. Like that of the medium-sized artery, the wall of the medium-sized vein also comprises three layers: tunica intima, tunica media, and tunica adventitia.

The **tunica intima** is formed of flattened **endothelium** with a basement membrane and a thin layer of **subendothelial connective tissue** containing longitudinally oriented **smooth muscle fibers.**

The **tunica media** is composed of several layers of **circular smooth muscle fibers** intermingled with fibroblasts and collagen fibers. This layer is better developed in the veins of the lower limbs, such as the saphenous vein.

The **tunica adventitia** is well developed and constitutes the bulk of the wall. This layer consists of a loose **connective tissue** containing collagen fibers, **fibroblasts,** and **vasa vasorum.** In certain medium-sized veins such as the saphenous vein, the adventitia possesses thick bundles of **smooth muscle fibers** arranged longitudinally along the vessel.

Figure 7-18. Small Vein
Human • Cross section • H.E. stain • High magnification

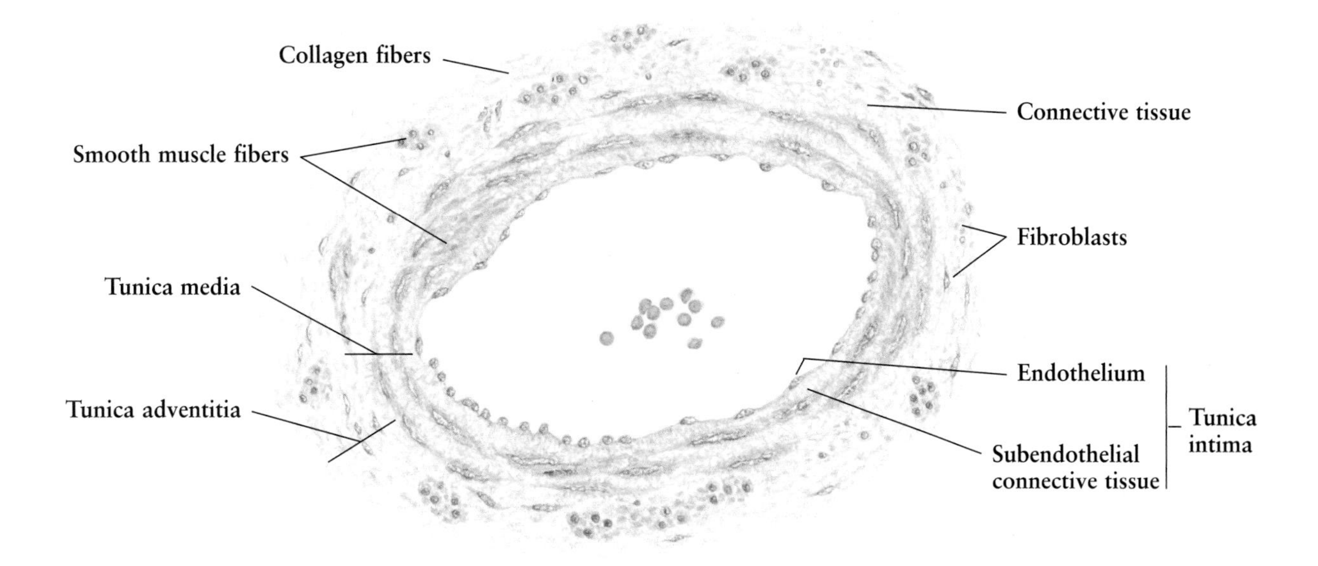

Figure 7-19. Medium-sized Vein: Saphenous Vein
Human • Cross section • H.E. stain • High magnification

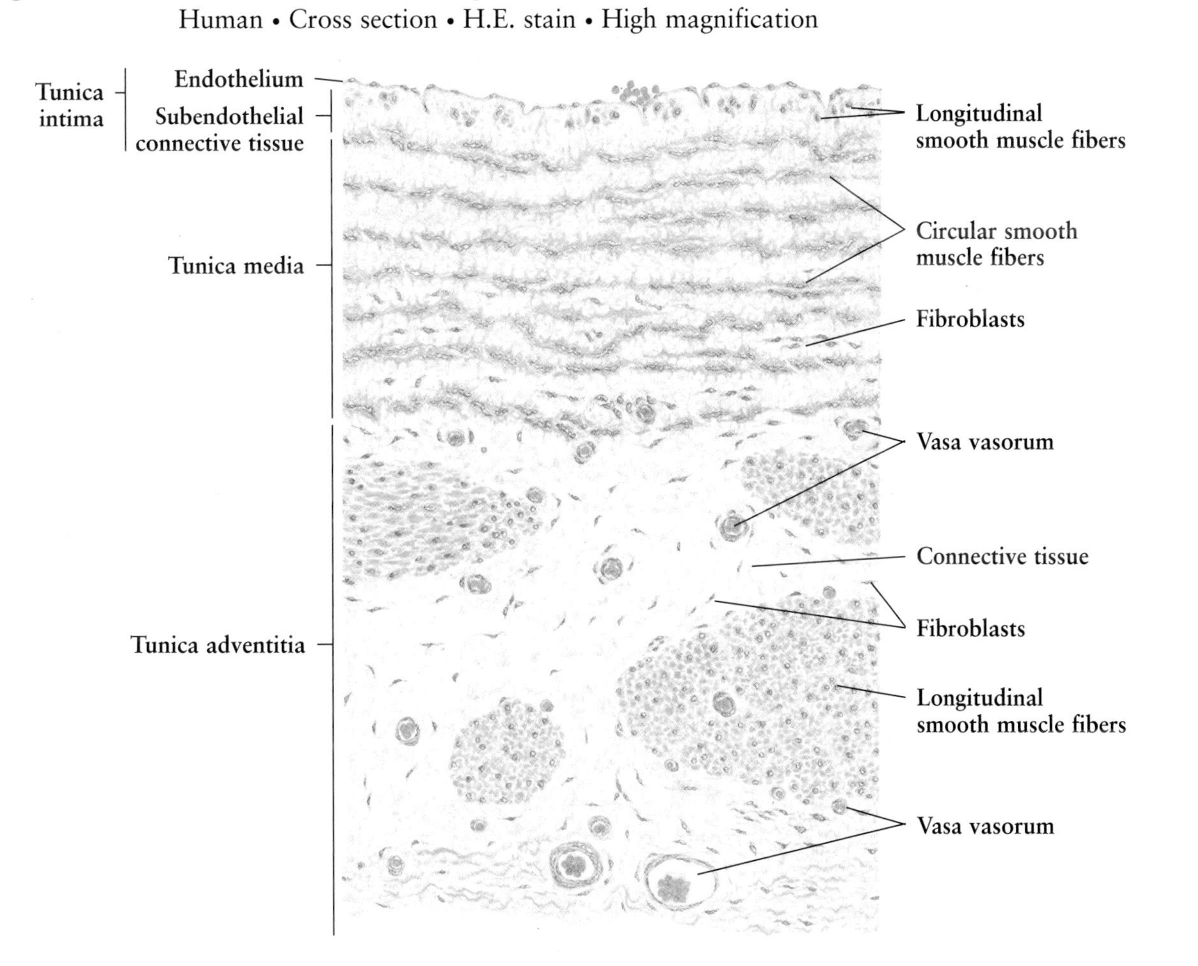

Figs. 7-20 and 7-21.

Large Veins: Venae Cavae

The **large veins** have a diameter greater than 10 mm. To this group belong the *venae cavae, pulmonary, portal, brachiocephalic, azygos, renal, splenic,* and *superior mesenteric veins.* The details of the human **superior** and **inferior venae cavae** are demonstrated in these figures.

The **tunica intima**, similar to that found in small or medium-sized veins, consists of a single layer of **endothelium** and a thin layer of **subendothelial connective tissue** containing scattered longitudinal smooth muscle fibers. Both the basement membrane and the internal elastic lamina are indistinct.

The **tunica media** is poorly developed and quite thin in most large veins. In the inferior vena cava it contains only 2 or 3 layers of **circular smooth muscle fibers.** In the superior vena cava, however, it may have more than 10 layers of **circular smooth muscle fibers** intermingled with **fibroblasts** and collagen fibers.

The **tunica adventitia** is extremely thick and well-developed, with three zones. The *innermost zone,* adjacent to the media, is a dense fibroelastic **connective tissue** with coarse collagen and elastic fibers as well as fibroblasts. The *middle zone* mainly contains longitudinally oriented bundles of **smooth muscle fibers**. The most conspicuous structural difference between the superior and inferior venae cavae occurs in this zone. The inferior vena cava contains numerous thick longitudinal bundles of smooth muscle fibers, which characterizes this large vein, while the superior vena cava has fewer smooth muscle fibers in this coat. The *outermost zone* is composed of a coarse network of collagen and elastic fibers. The **vasa vasorum** are found in the entire layer of the adventitia, and even penetrate the deepest portion of adventitia, very close to the lumen of the vena cava (see **Fig. 7-21**).

Figure 7-20. Large Vein: Superior Vena Cava
Human • Cross section • H.E. stain • Low magnification

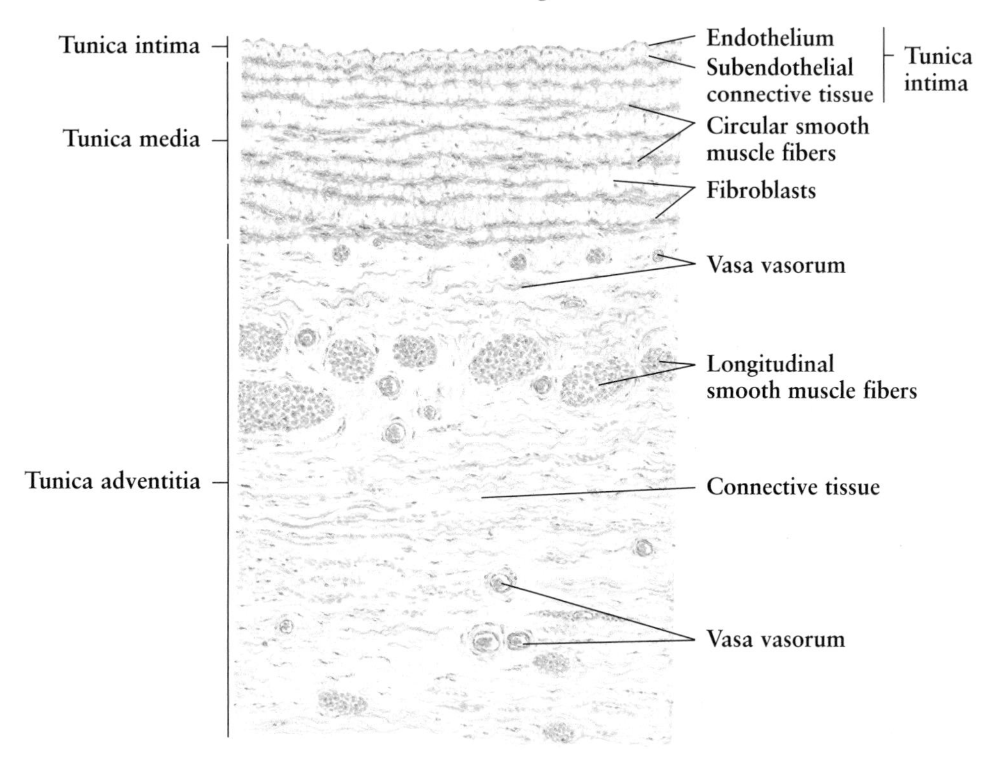

Figure 7-21. Large Vein: Inferior Vena Cava
Human • Cross section • H.E. stain • Low magnification

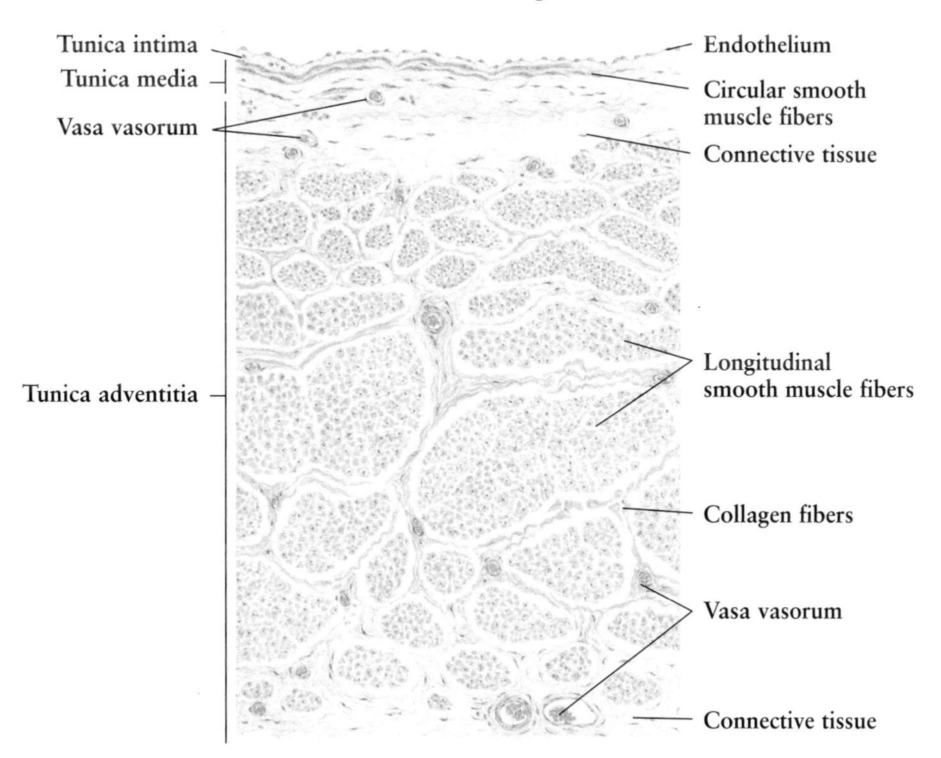

Figs. 7-22 and 7-23.

Venous Valves

Venous valves occur in many veins, both small and large, that conduct blood flow against gravity. They prevent the backflow of blood from the heart. The valves are slender, semilunar, pocket-like flaps formed by local folding of the intima. Each **valve** is usually composed of two **leaflets,** positioned opposite each other, with their free edges directed toward the heart. When blood passes through the lumen between the leaflets, they flatten out against the wall of the vein. When blood begins to regurgitate the pockets fill up, causing contact of the two leaflets and resulting in closing up of the lumen of the vein.

In **small veins** the leaflets of the valves consist of thin fibroelastic connective tissue covered on both sides by **endothelium. Figure 7-22** shows the details of valves in small veins, with one in transverse section and the other in longitudinal section. The *arrow* indicates the direction of blood flow.

In large veins such as the **femoral vein** shown in **Fig. 7-23,** valve **leaflets** are much thicker. The connective tissue core, completely covered by **endothelium** on both sides, contains longitudinal **collagen fibers** and scattered **fibroblasts,** with a well-developed **network of elastic fibers** just beneath the endothelium of the luminal surface directed against the blood flow. The **tunica intima,** together with **media,** protrude into the lumen of the vein, forming the **base of the valve,** which connects with the leaflets of valves.

Venous valves are not found in cerebral veins, the superior vena cava, pulmonary veins, umbilical veins, or veins of viscera and bone marrow.

Figure 7-22. Venous Valves: Small Vein
Human • H.E. stain • High magnification

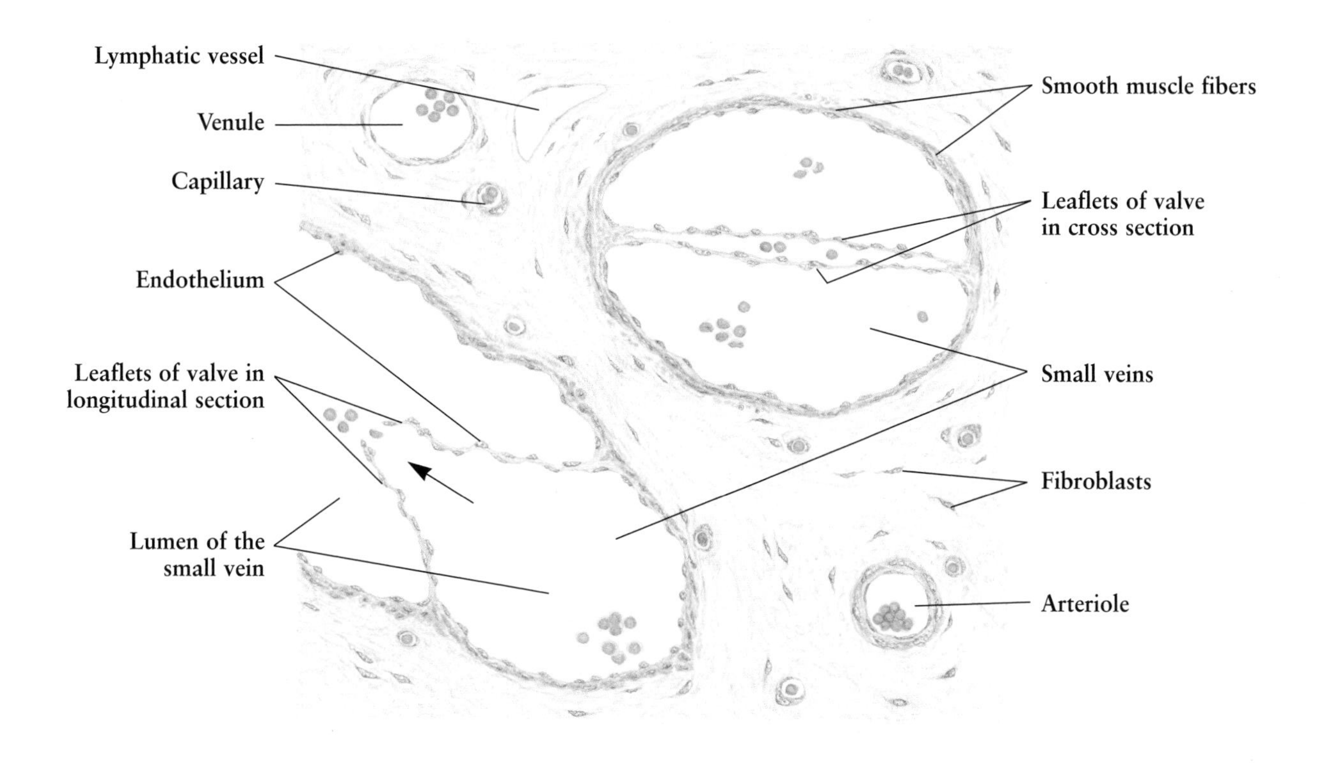

Figure 7-23. Venous Valves: Femoral Vein
Human • Cross section • H.E. stain • Low-and High magnification

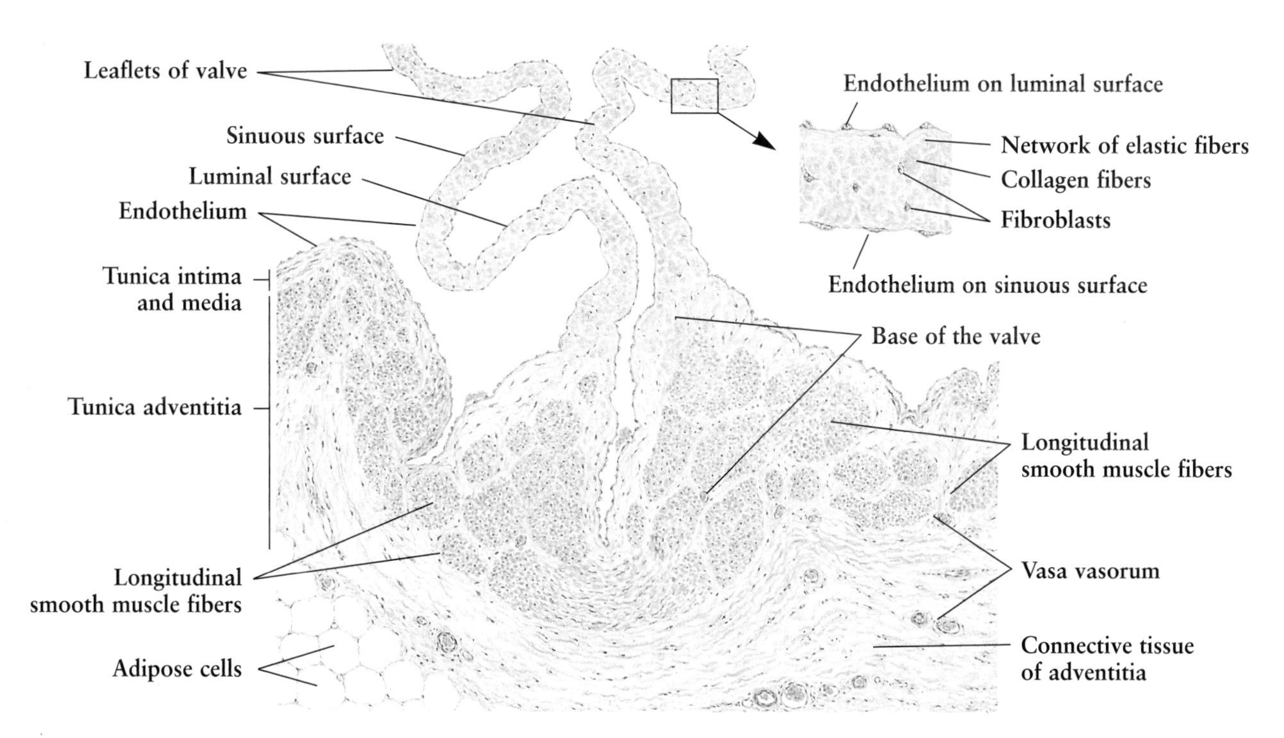

Fig. 7-24. Lymphatic Capillary and Lymphatic Vessel

Lymph drains from the network of *lymphatic capillaries* to the *lymphatic vessels* and then either to the *thoracic duct* or *right lymphatic duct,* which empty into the large veins.

The **lymphatic capillary** has a broader and irregular lumen with a diameter ranging from 10 to 50 μm. The wall is very thin and consists only of a continuous **endothelium**, lacking a basement membrane and pericytes.

The wall of the **collecting lymphatic vessel** is composed of a continuous layer of endothelium, an incomplete layer of smooth muscle fibers enclosing them, and a thin layer of connective tissue as its adventitia. There are many **valves** in these vessels. Like those in the venous system, each valve in a lymphatic vessel has two leaflets. Each is a fold of intima formed of two layers of endothelium separated by a narrow plate of only fine collagen and elastic fibers, without fibroblasts. The free edges are oriented toward the heart.

The lymphatic capillaries and vessels are found in most tissues and organs; the exceptions are the central nervous system, cartilage, bone, bone marrow, teeth, spleen pulp, thymus, lens and sclera of the eye, internal ear, and the fetal part of the placenta.

Fig. 7-25. Thoracic Duct

The **thoracic duct** structurally resembles a large vein. The *tunica intima* consists of an **endothelial lining,** a subendothelial connective tissue containing some **longitudinal smooth muscle fibers**, and a thin, inconstant internal elastic lamina. The *tunica media* is composed of **longitudinal** and **circular smooth muscle bundles** separated by **connective tissue**. The *tunica adventitia* is an irregular layer of connective tissue with scattered bundles of **longitudinal** and **circular smooth muscle fibers** and **vasa vasorum**. However, in contrast to a vein, the layers of the thoracic duct are less distinct.

Figure 7-24. Lymphatic Capillary and Lymphatic Vessel

Human • H.E. stain • Low magnification

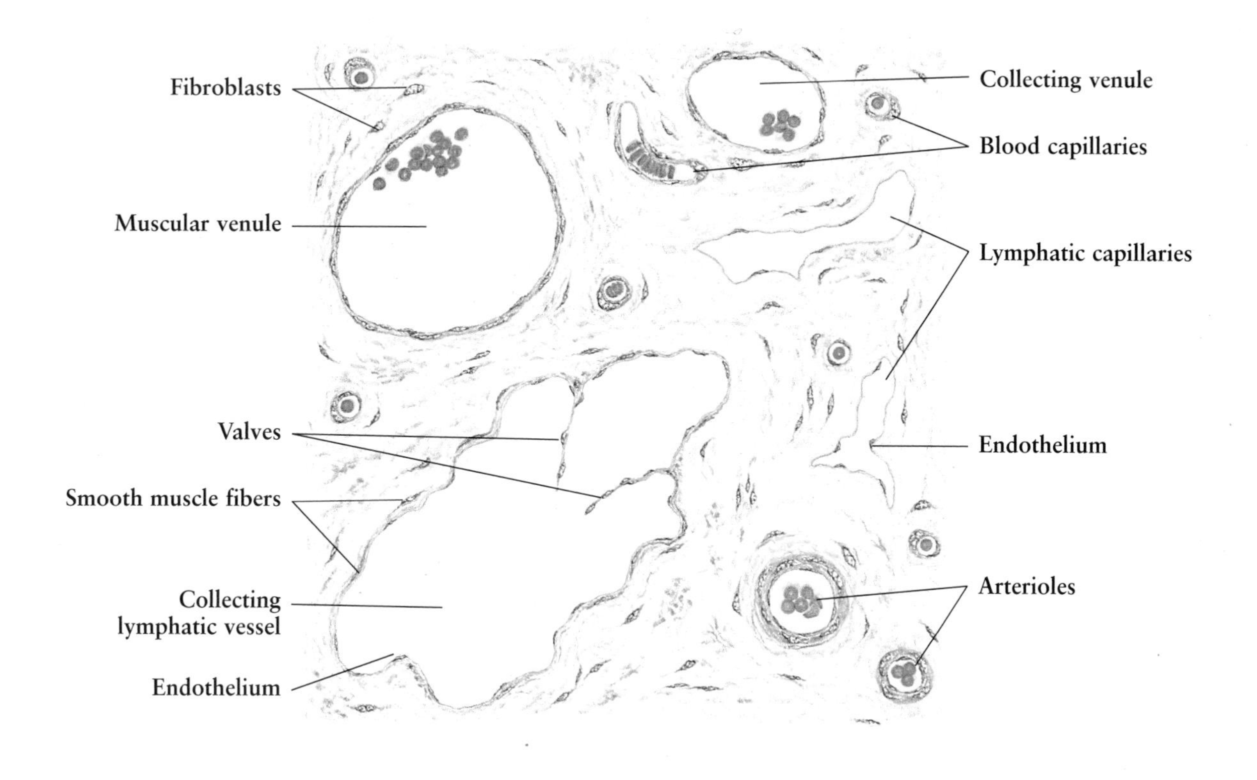

Figure 7-25. Thoracic Duct

Human • H.E. stain • Low magnification

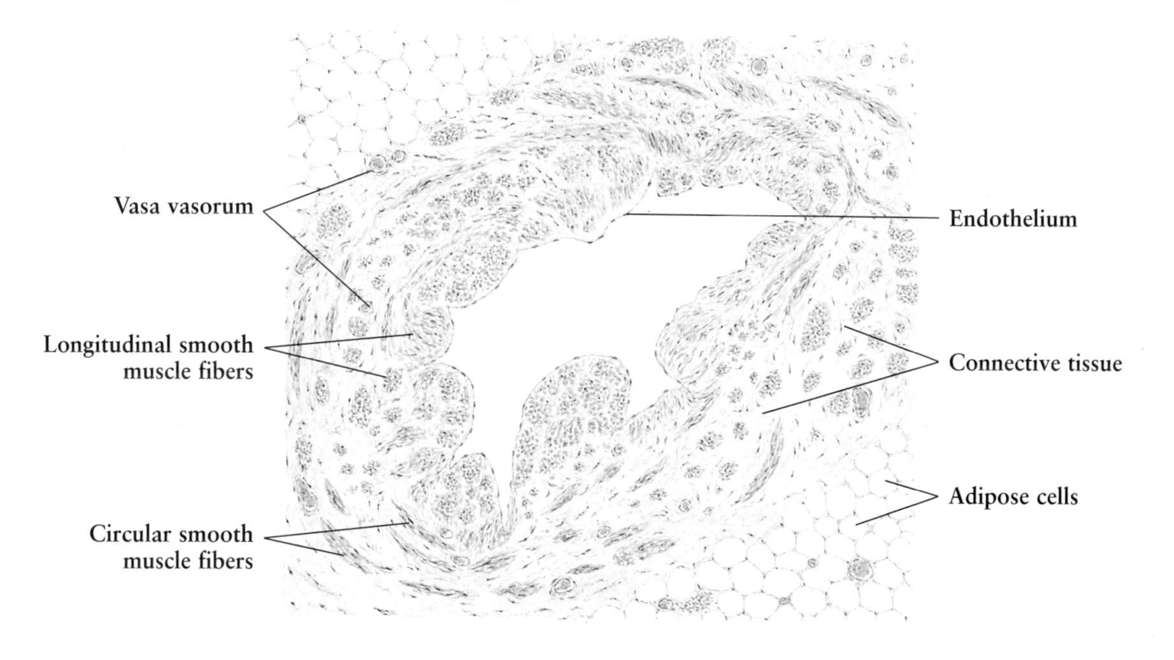

8. Lymphatic Organs

The **lymphatic organs** include the **thymus, lymph nodes, tonsils,** and **spleen**, and are involved in lymphocyte production, immune responses, or both. Lymphatic organs are composed of **lymphatic tissue**, aggregates of **lymphocytes** and **macrophages** supported by a framework of **reticular tissue**, which is formed of **reticular fibers** and **reticular cells**.

The lymphatic tissue, which forms the basis of the immune system of the body, is classified as **diffuse** and **solitary lymphatic tissue**, the **lymphatic nodule**. Diffuse lymphatic tissue occurs throughout the body, especially in the lamina propria of the digestive and the respiratory tracts. Lymphatic nodules can be found singly or aggregated as in lymph nodes, tonsils, and spleen, and also widely distributed along the digestive tract, where they are called the **Peyer's patches**.

The lymphatic organs are classified as **central** and **peripheral organs**. The central lymphatic organs include the **thymus** and **bursa of Fabricius** in birds; in mammals, they are the **thymus** and **bone marrow**. The latter is considered as the equivalent of the bursa of Fabricius. The peripheral lymphatic organs are the lymph nodes, tonsils, spleen, and lymphatic tissues in the gastrointestinal tract. Of these lymphatic organs, only the lymph nodes are directly connected with lymphatic vessels, and thus filter lymph.

Functionally lymphocytes are divided into **B lymphocytes**, mediators of the humoral immune response, and **T lymphocytes**, mediators of the cell-mediated immune response. In mammals all lymphocytes originate in the bone marrow. T lymphocytes, however, leave the bone marrow and mature in the thymus; B lymphocytes remain and mature within the bone marrow. Both mature T and B lymphocytes migrate to the peripheral lymphatic organs where they proliferate and function.

Fig. 8-1.

Lymphatic Tissue: Diffuse and Solitary

Lymphatic tissue consists of two principal components: *lymphocytes* and *reticular tissue*. Lymphocytes lie within the interstices of the reticular tissue, which is composed of a framework of reticular cells and fibers.

Diffuse lymphatic tissue is an infiltration of lymphocytes, which is not demarcated from the surrounding connective tissue. It is often a component of the lamina propria of the mucosa, especially in the digestive and respiratory tracts. **Figure 8-1** demonstrates diffuse lymphatic tissue located in the **lamina propria** of the esophageal mucosa.

A **lymphatic nodule** is a dense spherical mass of lymphatic tissue, with or without a **germinal center**. It may occur singly as a solitary lymphatic nodule, or in groups as aggregated lymphatic nodules present in Peyer's patches, or in specific lymphatic organs such as lymph node, tonsil, and spleen. This illustration also shows a **solitary lymphatic nodule** in the lamina propria of the esophageal mucosa.

Fig. 8-2.

Aggregated Lymphatic Nodules

Aggregated lymphatic nodules occur in special lymphatic organs such as lymph nodes, tonsils, and spleen. They may also be aggregated into less highly organized structures than lymphatic organs, forming the unencapsulated Peyer's patches in the lamina propria of the ileal mucosa.

The **Peyer's patches** shown in this picture are a huge accumulation of lymphatic nodules, each with a germinal center. They occupy the **lamina propria** of the **ileal mucosa,** without a connective tissue capsule. Because of the infiltration of enormous numbers of lymphocytes, the Peyer's patches interrupt the **muscularis mucosa** to enter the **submucosa,** and bulge into the intestinal lumen, resulting in the disappearance of the intestinal **villi.**

With the main features of a lymph node, such as a diffuse cortex, lymphatic nodules with germinal centers, and specialized postcapillary venules as the entrance of lymphocytes into the lymphatic tissue, the Peyer's patches are considered as modified lymph nodes.

Figure 8-1. Lymphatic Tissue: Diffuse and Solitary
Human esophagus • Cross section • H.E. stain • Low magnification

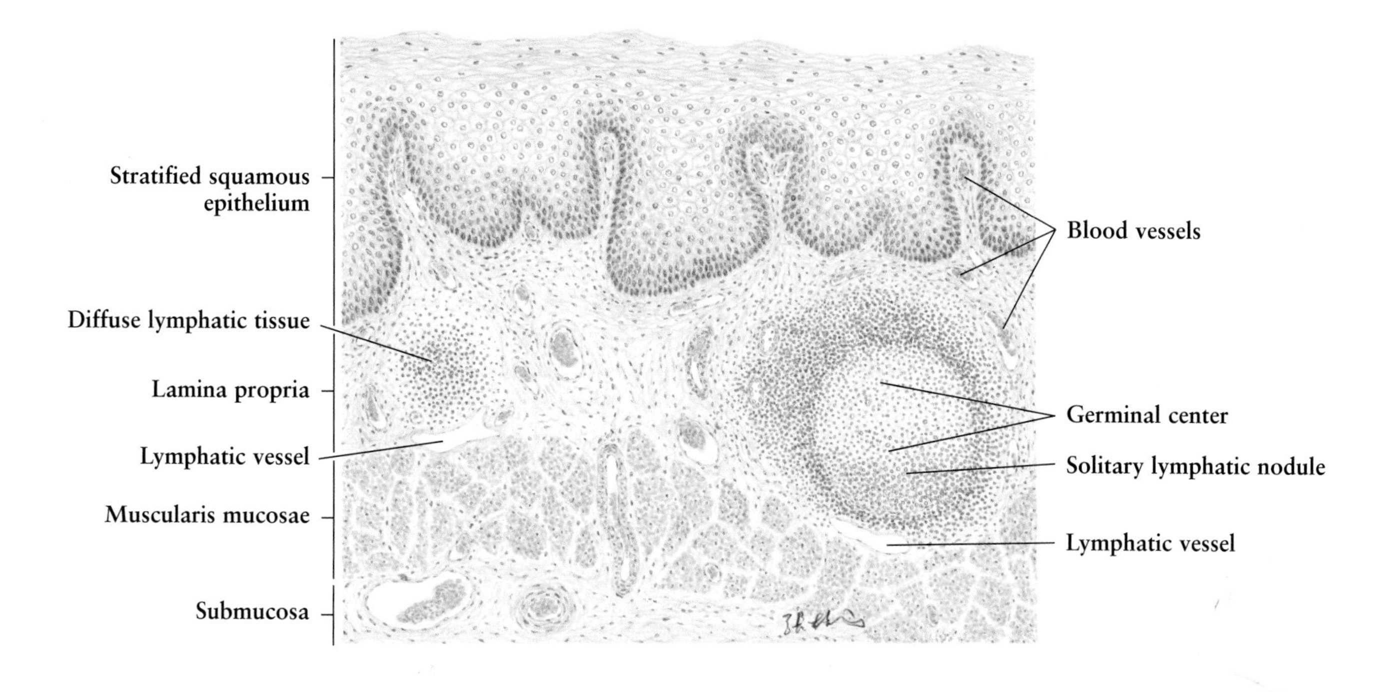

Figure 8-2. Aggregated Lymphatic Nodules
Human ileum • Longitudinal section • H.E. stain • Low magnification

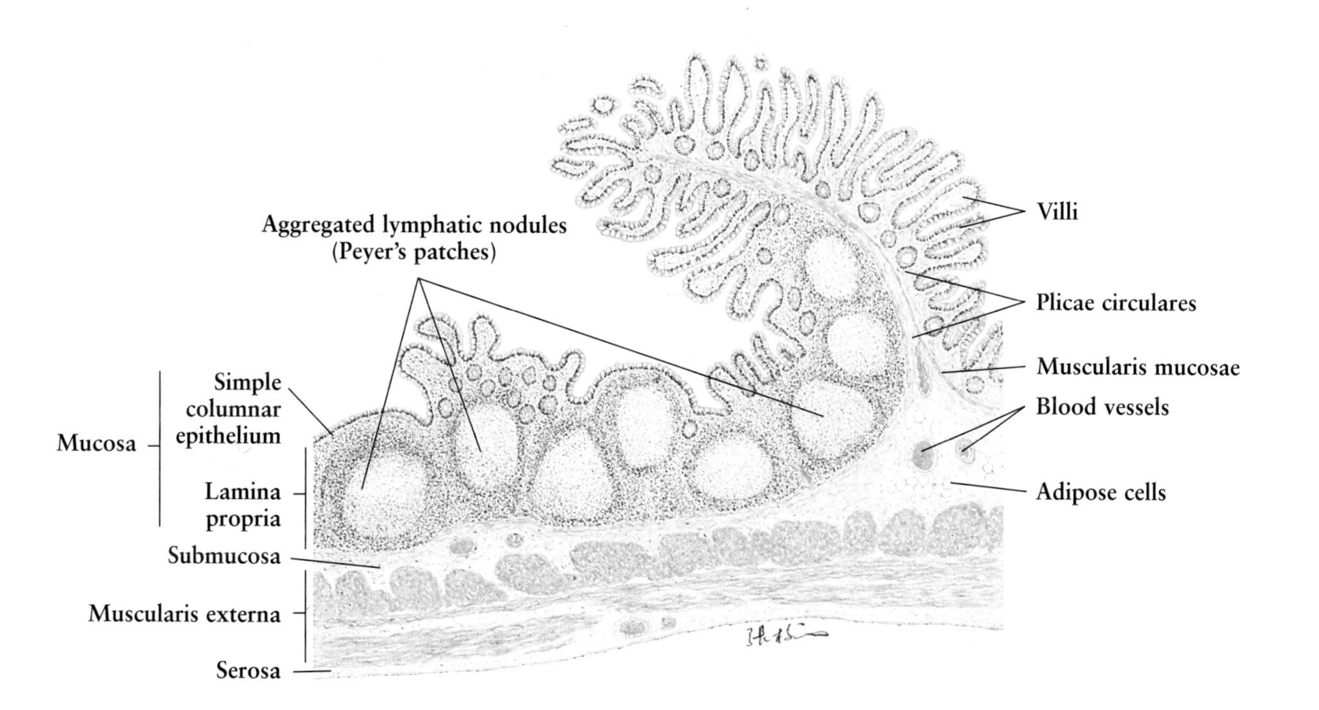

Fig. 8-3.

Lymphatic Nodule

Histologically, a typical **lymphatic nodule** is composed of a germinal center and a cap at one pole, invested by an outer boundary. The whole structure is embedded by surrounding **diffuse lymphatic tissue.**

The crescent-shaped **cap,** overlying the germinal center, consists of densely packed small lymphocytes. In general, the cap is always directed toward the marginal sinus in a lymph node (see **Figs. 8-6** and 8-7), and toward the overlying epithelium in tonsils and the digestive and respiratory tracts.

The **germinal center** displays two distinguishable zones: a light region and a dark region. The **light region** is consistently adjacent to the cap. Its cell component contains mainly loosely arranged medium-sized lymphocytes, a few large lymphoblasts, and plasma cells. The **dark region** is located on the opposite pole of the cap. It is composed of numerous closely packed small, medium, and large lymphocytes, and lymphoblasts. Macrophages containing residues of phagocytized degenerating lymphocytes are often seen in this region. The germinal center is believed to be involved in the functional differentiation of B lymphocytes, which are associated with humoral immune response.

The **outer boundary** consists of a few layers of circularly arranged flattened reticular cells. External to the outer boundary, **blood capillaries** and **high endothelial postcapillary venules** can be observed in the surrounding diffuse lymphatic tissue.

Fig. 8-4.

Germinal Center: Dark Region

This illustration is an enlargement of the boxed area in **Fig. 8-3,** displaying the cell components of the **dark region** in the **germinal center.** Clearly recognized are **small lymphocytes, medium-sized lymphocytes,** and **large lymphocytes,** as well as **lymphocytes in mitosis.** Additionally, numerous **macrophages** containing phagocytic vacuoles in their cytoplasm are found in this region. Also present between lymphocytes are **reticular cells,** which form (together with reticular fibers, not shown in this illustration) a framework in the lymphatic tissue.

The outer **boundary** separating the germinal center from the **peripheral corona** is composed of a few layers of **reticular cells** in a regular arrangement surrounding the germinal center. The peripheral corona contains numerous small lymphocytes as well as some medium-sized lymphocytes and reticular cells.

Fig. 8-5.

High Endothelial Postcapillary Venule

High endothelial postcapillary venules (HEV; also see **Fig.** 8-3), 30–50 µm in diameter, are located in the inner cortex of lymph nodes and lymphatic tissue of tonsils. They are different in the endothelium from those found in other kinds of tissues and organs. The endothelium of a high endothelial postcapillary venule, shown in this illustration, is composed of **cuboidal endothelial cells** surrounded by a layer of **pericytes.** The **migrating lymphocytes,** which pass from the blood into perivascular thymus-dependent lymphatic tissue, are often found between the endothelial cells. The high endothelial postcapillary venule is a major pathway through which B lymphocytes and T lymphocytes enter the lymphatic tissue from the blood.

Figure 8-3. Lymphatic Nodule
Human lymph node • H.E. stain • Low magnification

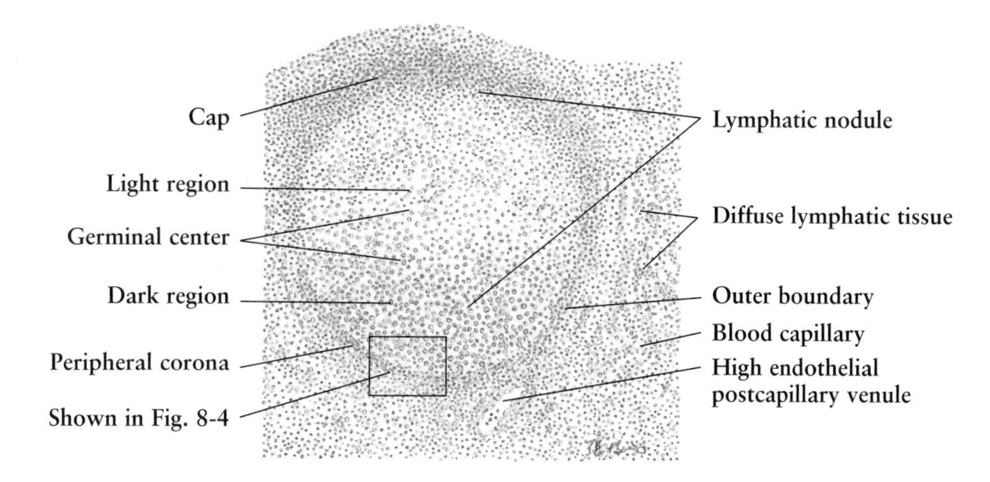

Figure 8-4. Germinal Center: Dark Region
Human lymph node • H.E. stain • High magnification

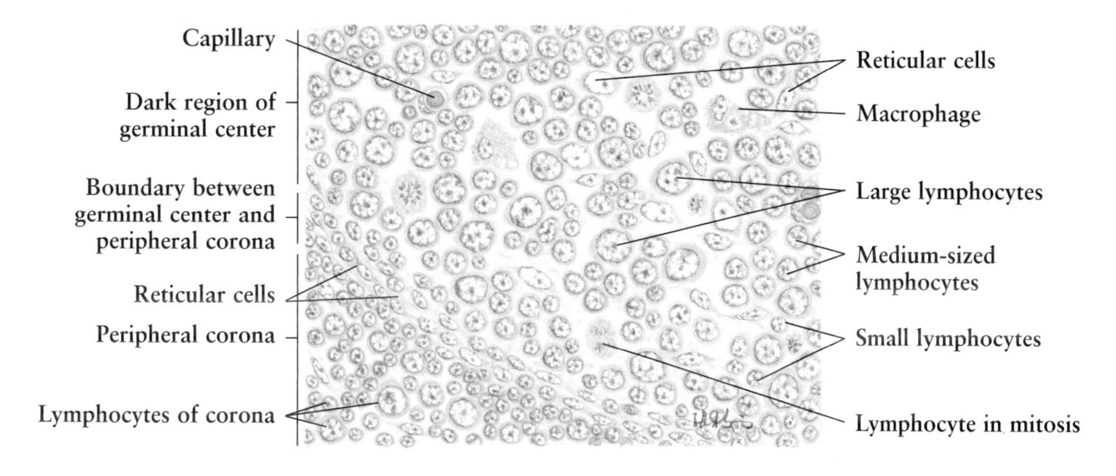

Figure 8-5. High Endothelial Postcapillary Venule
Human lymph node • H.E. stain • High magnification

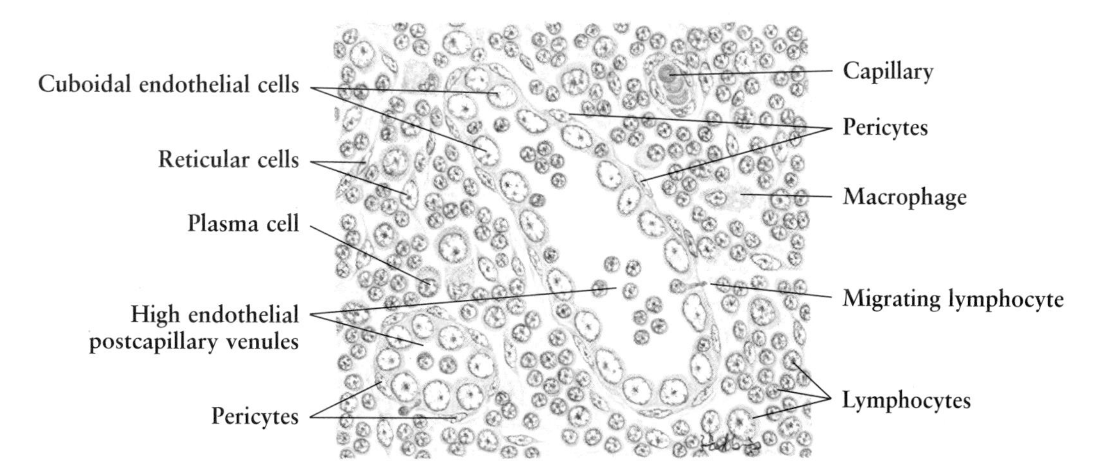

Fig. 8-6. Lymph Node

Lymph nodes are small, bean-shaped organs, ranging from 3 to 25 mm in diameter, that occur in chains or groups along the course of lymphatic vessels. Each lymph node is covered by dense connective tissue, the **capsule,** which extends into the substance of the organ, forming a number of **trabeculae.** The lymph node has a convex contour, and on one side a concave region, the **hilum,** where the blood vessels (**artery** and **vein**)enter and leave the organ. **Afferent lymphatic vessels** enter the node at multiple points on the convex surface while **efferent lymphatic vessels** leave only at the hilum.

The parenchyma of the lymph node can be classified into two regions: an outer portion or the cortex, and an inner portion or the medulla. The **cortex** is further divided into an outer cortex and an inner or deep cortex. The **outer cortex** contains numerous lymphatic nodules, most of which show **germinal centers.** This is the region of B lymphocytes. The **deep cortex,** also called the paracortex, is particularly rich in T lymphocytes. In the **medulla** the lymphatic tissue is arranged as strands, or **medullary cords,** which contain B lymphocytes and numerous plasma cells.

The percolation of lymph through a lymph node involves afferent lymphatic vessels, a system of lymph sinuses within the node, and efferent lymphatic vessels. The *system of lymph sinuses* consists of three parts: **subcapsular (marginal) sinus,** the space between outer cortex and capsule; **intermediate (cortical) sinus,** the space between cortical nodule and trabecula that separates the cortex into incomplete compartments; and **medullary sinus,** the space between medullary cords.

Small **arteries** enter the node at the hilum, ramify, and run in the medulla and in the trabeculae, giving rise to extensive capillary networks in the cortex and medullary cords. From the capillaries, blood drains into the high endothelial postcapillary venule through which numerous lymphocytes enter the lymphatic tissue.

The *functions* of a lymph node include the percolation of lymph, the production of lymphocytes, and the formation of antibodies.

Figure 8-6. Lymph Node
Human • H.E. stain • Very low magnification

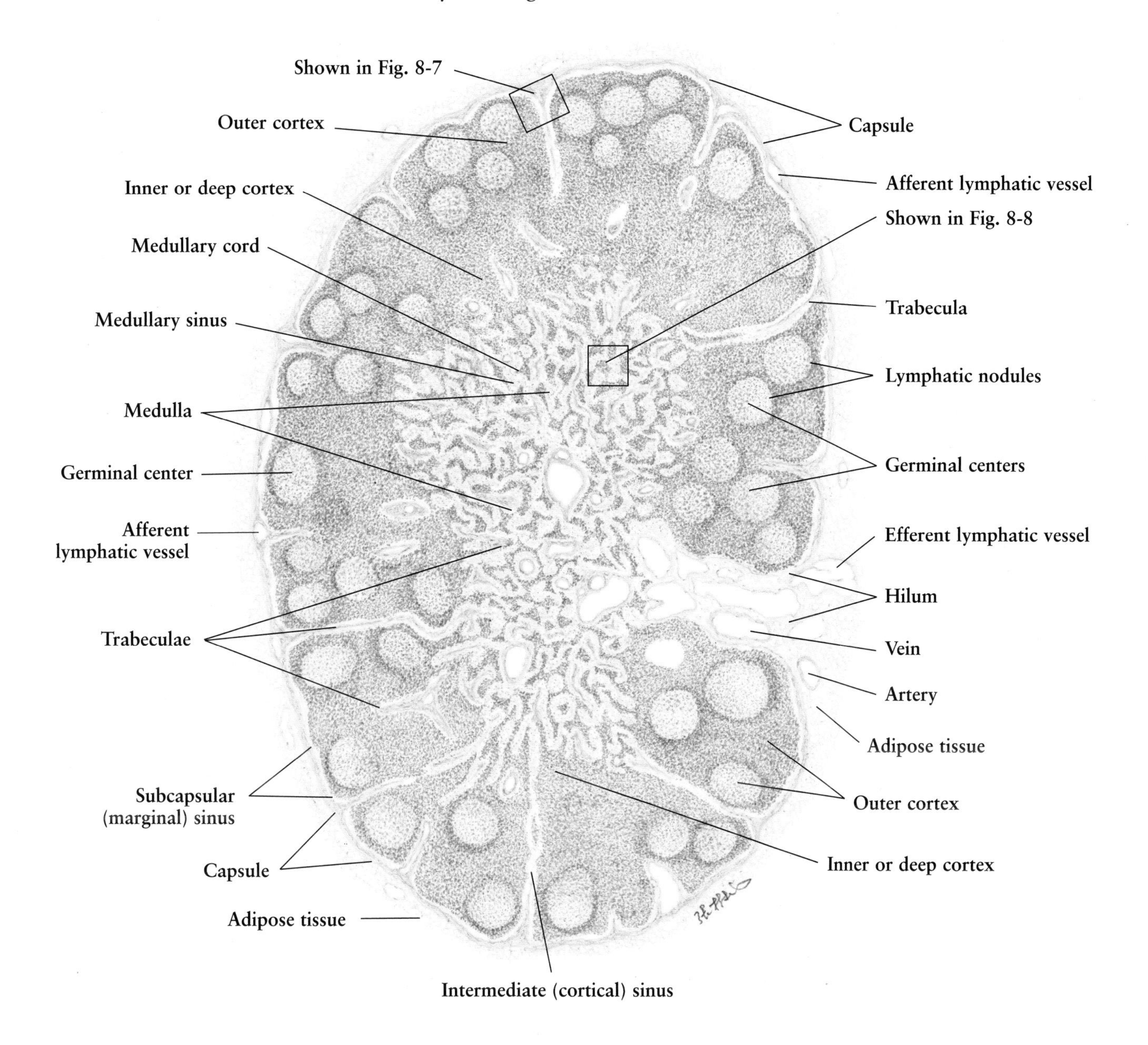

Fig. 8-7. Lymph Node: Subcapsular Sinus

This figure is an enlargement of the boxed area in **Fig. 8-6,** giving details of the capsule of the lymph node and the lymphatic sinuses.

The fibrous **capsule** consists of dense connective tissue with enormous numbers of collagen fibers and fibroblasts. The connective tissue extends into the substance of the node, forming **trabeculae,** which separate the cortex into incomplete compartments. Within the capsule **afferent lymphatic vessels** with **valves, arteries, veins,** and **capillaries** are present. **Adipose cells** also can be seen external to the capsule.

Beneath the capsule is the **subcapsular sinus,** which becomes continuous with the **intermediate sinus.** The sinuses are lined by a layer of squamous **littoral cells,** which are endothelial in nature. The littoral lining on the nodule side is readily traversed by lymphocytes to enter the sinus from the lymphatic tissue. The sinuses are bridged by numerous **reticular cells,** which are interconnected by their thin cytoplasmic processes to form a three-dimensional network. Within the lumen of the sinuses, there is a great number of **macrophages** in addition to lymphocytes. Macrophages engulf bacteria, foreign materials, dead cells, and cell debris from the afferent lymph. It is believed that macrophages are involved in the processing of many antigens and the transmission of information to lymphocytes for the production of antibodies.

Fig. 8-8. Lymph Node: Medulla

The **medulla,** located in the inner part of the lymph node (see **Fig. 8-6**), consists of medullary cords, medullary sinuses, and trabeculae.

The **medullary cords** are branching strands of lymphatic tissue, separated by irregular medullary sinuses. The cords are filled with numerous **lymphocytes** (B lymphocytes, lymphoblasts), **plasma cells,** and a few **macrophages. Blood vessels** are also present within the cords. Some medullary cords are continuous with the deepest portion of the paracortex and appear as extensions of it into the underlying medulla.

The **medullary sinuses** are irregular spaces between the medullary cords. Like subcapsular and intermediate sinuses, the medullary sinuses are lined by a layer of **littoral cells,** and have **reticular cells** in a three-dimensional network. The lumen of the medullary sinus is full of flowing lymph with **small** and **medium-sized lymphocytes** and **macrophages.**

The **trabeculae,** extending from the fibrous connective tissue of the capsule, may be found throughout the medulla. The structure of the trabeculae demonstrates a large number of collagen fibers and **fibroblasts,** as well as **arteries** and **veins.**

Figure 8-7. Lymph Node: Subcapsular Sinus
Human • H.E. stain • High magnification

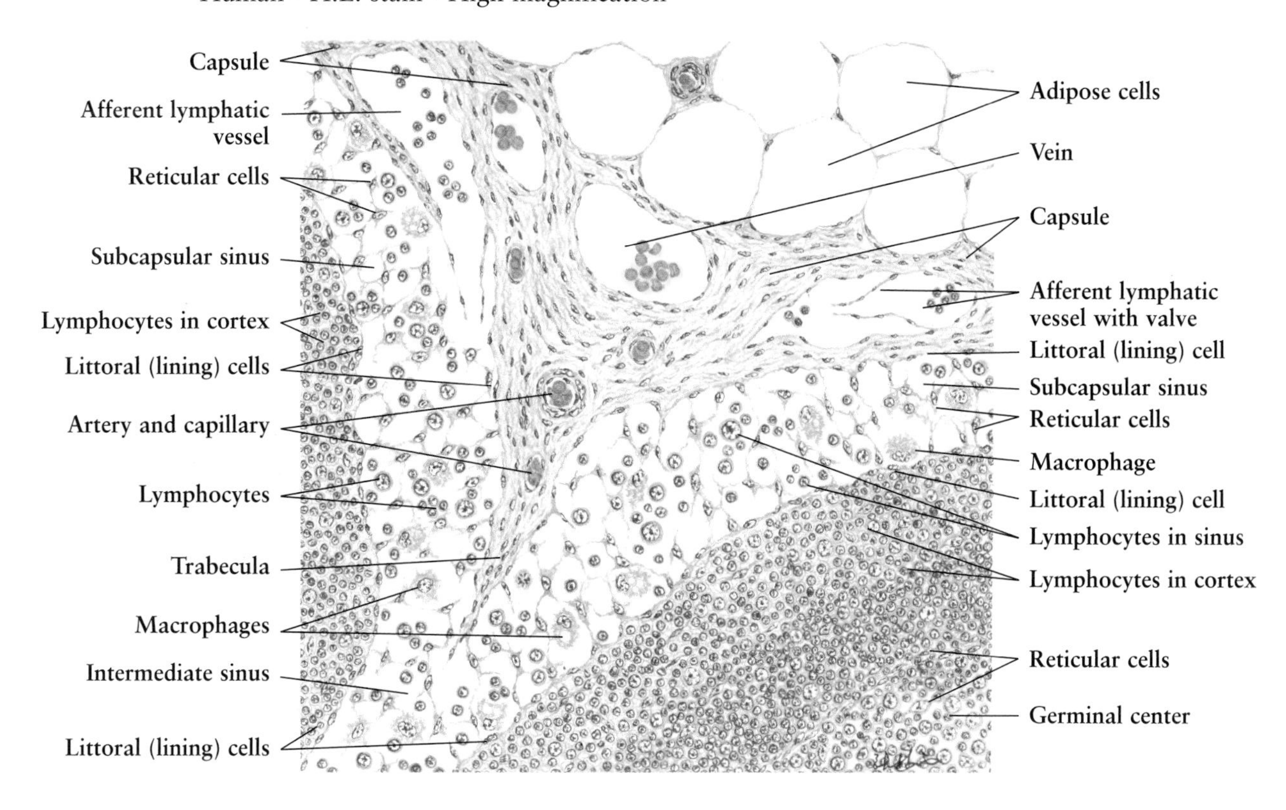

Figure 8-8. Lymph Node: Medulla
Human • H.E. stain • High magnification

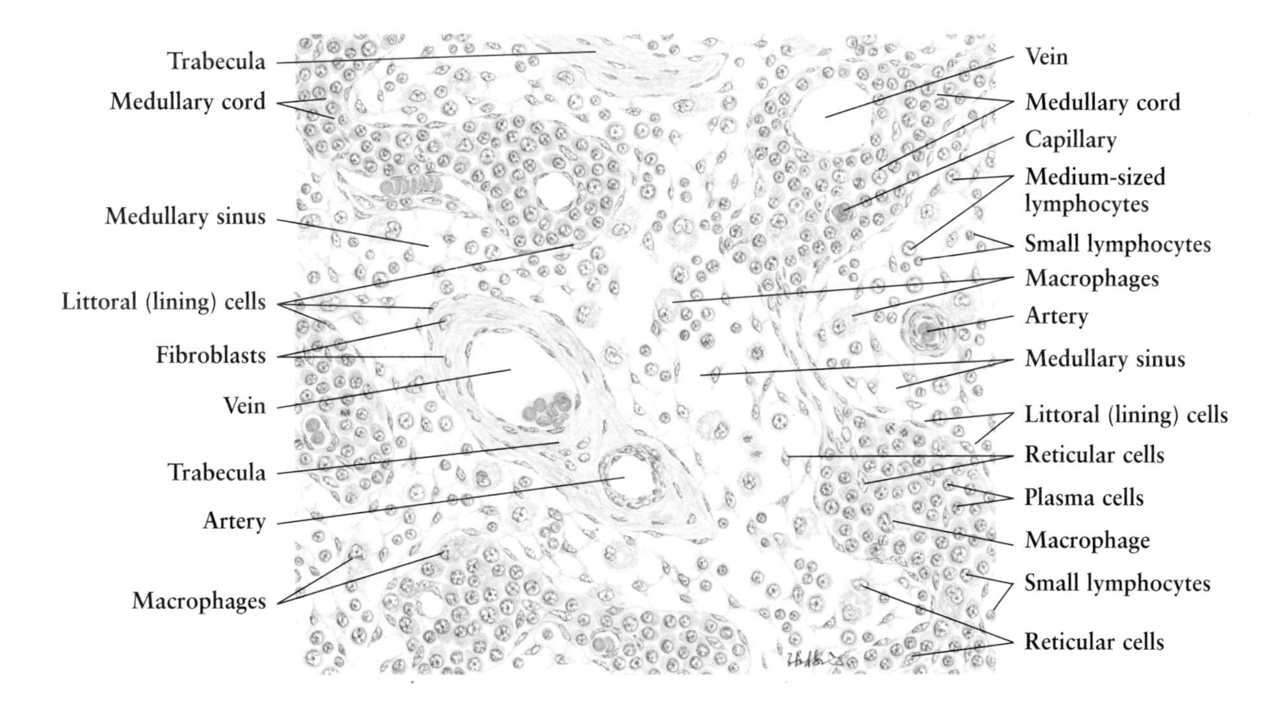

Figs. 8-9 and 8-10.

Lymph Node: Reticular Fibers

Reticular fibers are invisible in routine preparations; the silver impregnation method, however, has been used to demonstrate the *reticular fibers* in a human *lymph node.* In these two figures, the reticular fibers are stained black while the nuclei of lymphocytes and other cell types appear brown.

Distributed throughout the lymph node, **reticular fibers** form a delicate three-dimensional network, supporting the huge mass of lymphocytes within the cortex and medulla. In the **cortex,** the reticular network is generally dense, but it is relatively sparse in the **germinal center.** In the **medulla,** the reticular fibers are densely concentrated in the **trabeculae** and around the **blood vessels** in the **medullary cords.** In the **subcapsular, intermediate,** and **medullary sinuses,** reticular fibers run along with the fine processes of reticular cells, bridging the lumen of the sinuses.

Reticular fibers are also present in the connective tissue of the **capsule.** They surround the **blood vessels** and **lymphatic vessels,** and even penetrate the **valves** of the **afferent lymphatic vessels.**

Figure 8-9. Lymph Node: Reticular Fibers of Cortex

Human • Silver impregnation • Low magnification

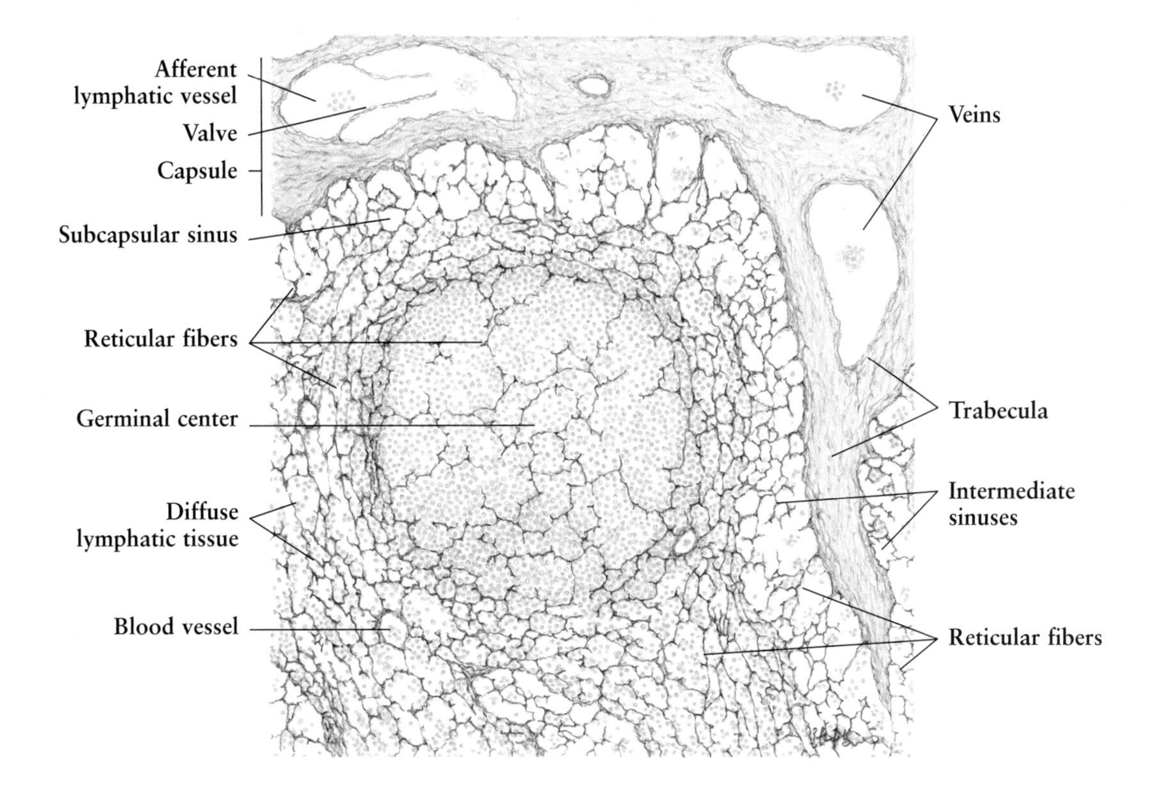

Figure 8-10. Lymph Node: Reticular Fibers of Medulla

Human • Silver impregnation • Low magnification

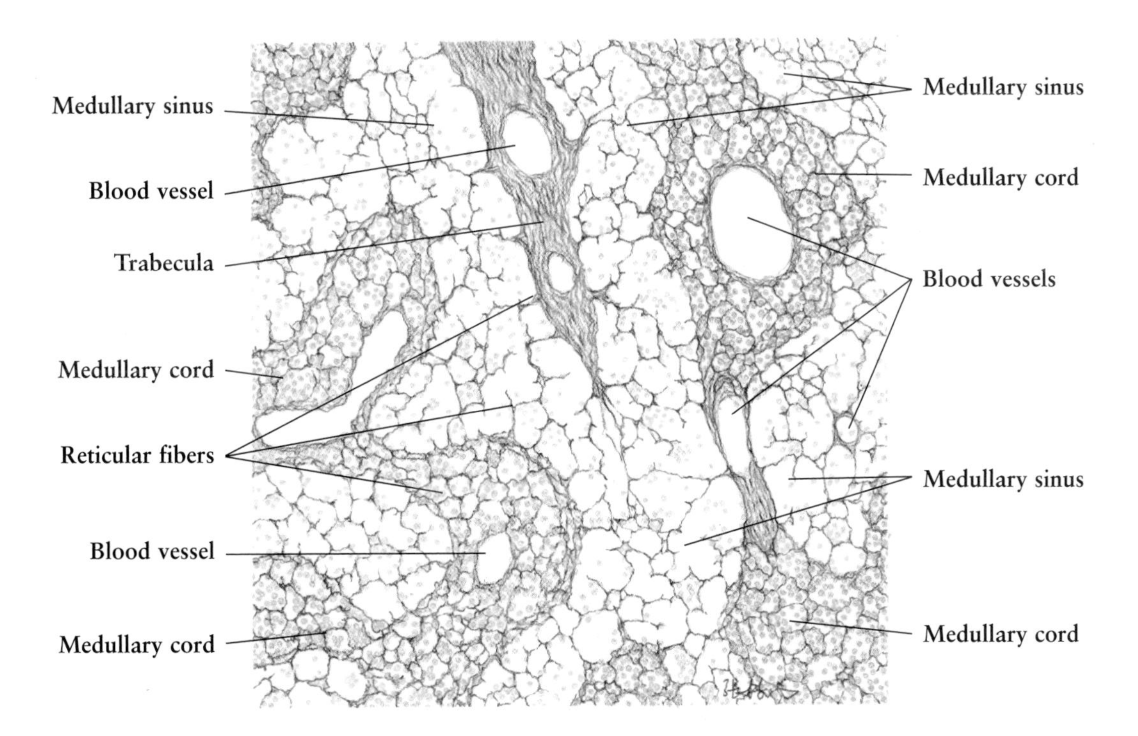

Fig. 8-11. Palatine Tonsil

The **palatine tonsils** are paired, large ovoid masses of aggregated lymphatic nodules, located bilaterally between the glossopalatine and pharyngopalatine arches. The base and sides are separated from the adjacent **skeletal muscles** by a dense **connective tissue capsule,** which extends into the lymphatic tissue to form the **internodular septa.**

The luminal surface of the tonsil is covered by nonkeratinized **stratified squamous epithelium,** which is similar to and continuous with that of the oral cavity. The surface epithelium is deeply invaginated into the lymphatic tissue to form 15 or more tonsillar **crypts.** The lumen of the crypts is narrow and contains mucus with living and degenerating lymphocytes, desquamated epithelial cells, neutrophils, detritus, and microorganisms. Some of the crypts may become extended by the accumulation of a **cheesy plug.**

The palatine tonsils consist of numerous **lymphatic nodules** and **diffuse lymphatic tissues** immediately beneath the epithelium along the sides of the crypts. These lymphatic nodules possess prominent germinal centers, and are arranged in rows, separated by a thin layer of connective tissue, the internodular septa.

The palatine tonsils and other two tonsillar groups, the lingual tonsils (see **Fig. 10-18**) and the pharyngeal tonsils, jointly form a ring of lymphatic tissue surrounding the pharynx, where nasal and oral passages cross each other. The tonsils are involved in defending the body against invading bacteria, viruses, and other foreign proteins by making early chemical or physical contact with the invaders, followed by the production of antibodies, especially IgA, which affords the body considerable protection against infection by the microorganisms.

Fig. 8-12. Palatine Tonsil: Epithelium of Crypt

This figure, an enlargement of the boxed area in **Fig. 8-11,** demonstrates the **epithelium** of the palatine tonsillar crypt. In the deeper portions of the crypts, the **stratified squamous epithelium** is extensively infiltrated with lymphocytes. In **heavily infiltrated epithelium,** the **boundary** between the epithelium and lymphatic tissue may become obscured; numerous epithelial cells disappear, leaving the spaces occupied by migrating **lymphocytes** and **plasma cells. Blood vessels** may also invade this area from the underlying connective tissue and lymphatic tissue.

In the **lumen of the crypt, lymphocytes** and **neutrophils** are shown mixing with the **mucus.**

Figure 8-11. Palatine Tonsil

Human • H.E. stain • Very low magnification

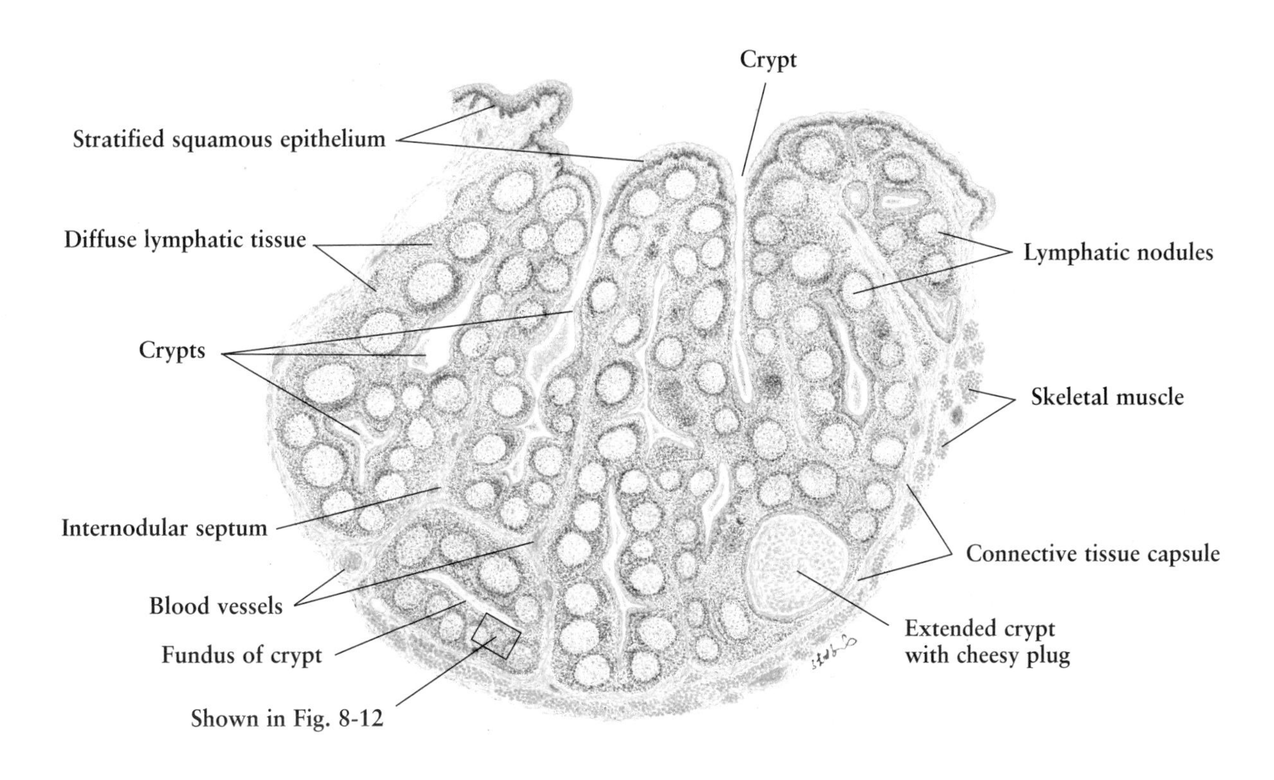

Figure 8-12. Palatine Tonsil: Epithelium of Crypt

Human • H.E. stain • High magnification

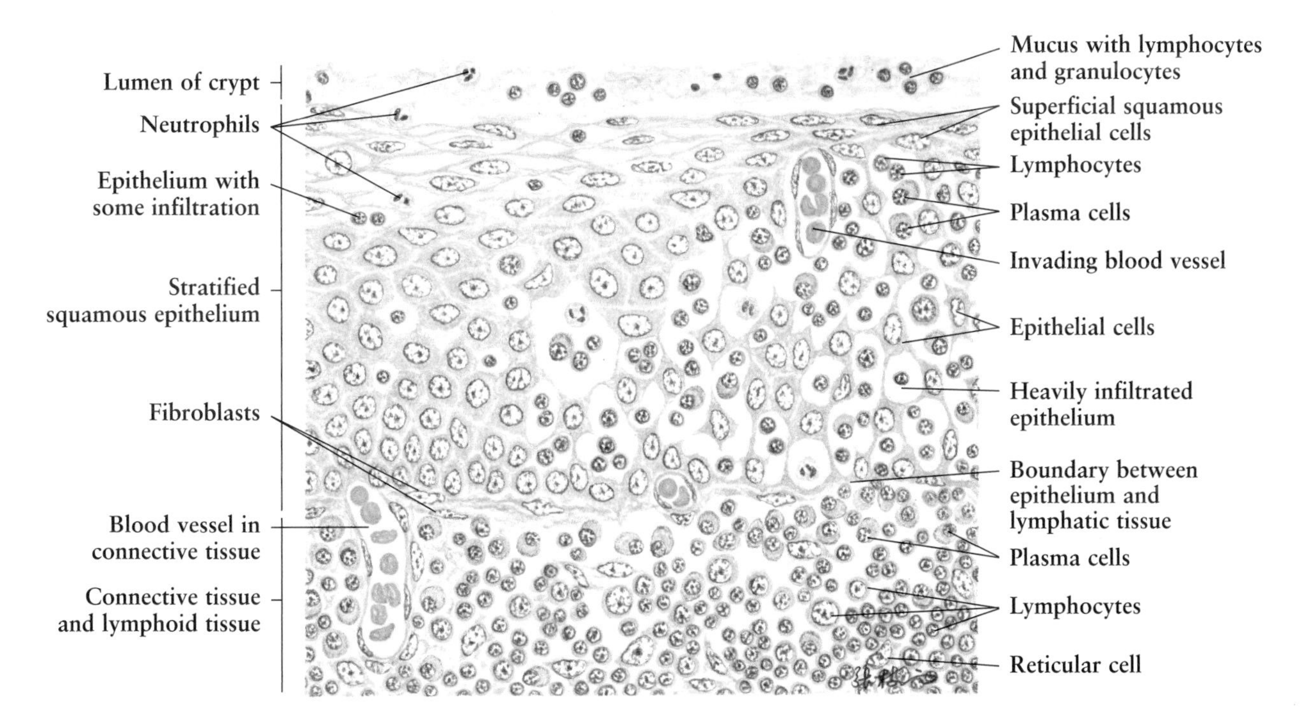

Fig. 8-13. Spleen

The **spleen,** the largest lymphatic organ, is located in the left upper part of the abdomen. It is invested by a connective tissue capsule, which penetrates the substance of the organ, forming trabeculae. The capsule becomes especially thick at the deep indentation, the hilum, where the splenic artery and vein enter and leave.

The **capsule** is very thick. It consists of dense collagen connective tissue with some elastic fibers and smooth muscle fibers. The luminal surface of the capsule is covered by a thin layer of flattened **mesothelium,** a component of the peritoneum. The capsule also contains blood and lymphatic vessels. The **trabeculae** have a composition similar to that of the capsule. They branch and anastomose repeatedly to form a complex three-dimensional network that supports the substance of the spleen and conveys the arteries and veins.

The parenchyma of the spleen is composed of white pulp and red pulp. The **white pulp** is typical lymphatic tissue that encloses and follows the arteries, forming **periarterial lymphatic sheath,** or thickening into ovoid lymphatic nodules known as **splenic nodules,** which contain prominent **germinal centers.** The **red pulp** is more abundant, and made up of splenic cords, strands of lymphatic tissue, and venous sinuses. The junction between red pulp and white pulp is called the **marginal zone,** a diffuse lymphatic tissue.

The splenic artery enters the spleen through the hilum. It branches several times into **trabecular arteries,** which run within the trabeculae, accompanied by the **trabecular veins.** The arteries leave the trabeculae, travel, and ramify in the red pulp as **central arteries** and then **penicillar arteries,** and terminate as arterial capillaries in the splenic cords. The blood is collected from the venous sinuses by pulp veins, then empties into trabecular veins.

The *main functions* of the spleen include destruction of erythrocytes, production of blood cells, chiefly lymphocytes, storage of blood (platelets), and production of antibodies.

Figure 8-13. Spleen
Human • H.E. stain • Low magnification

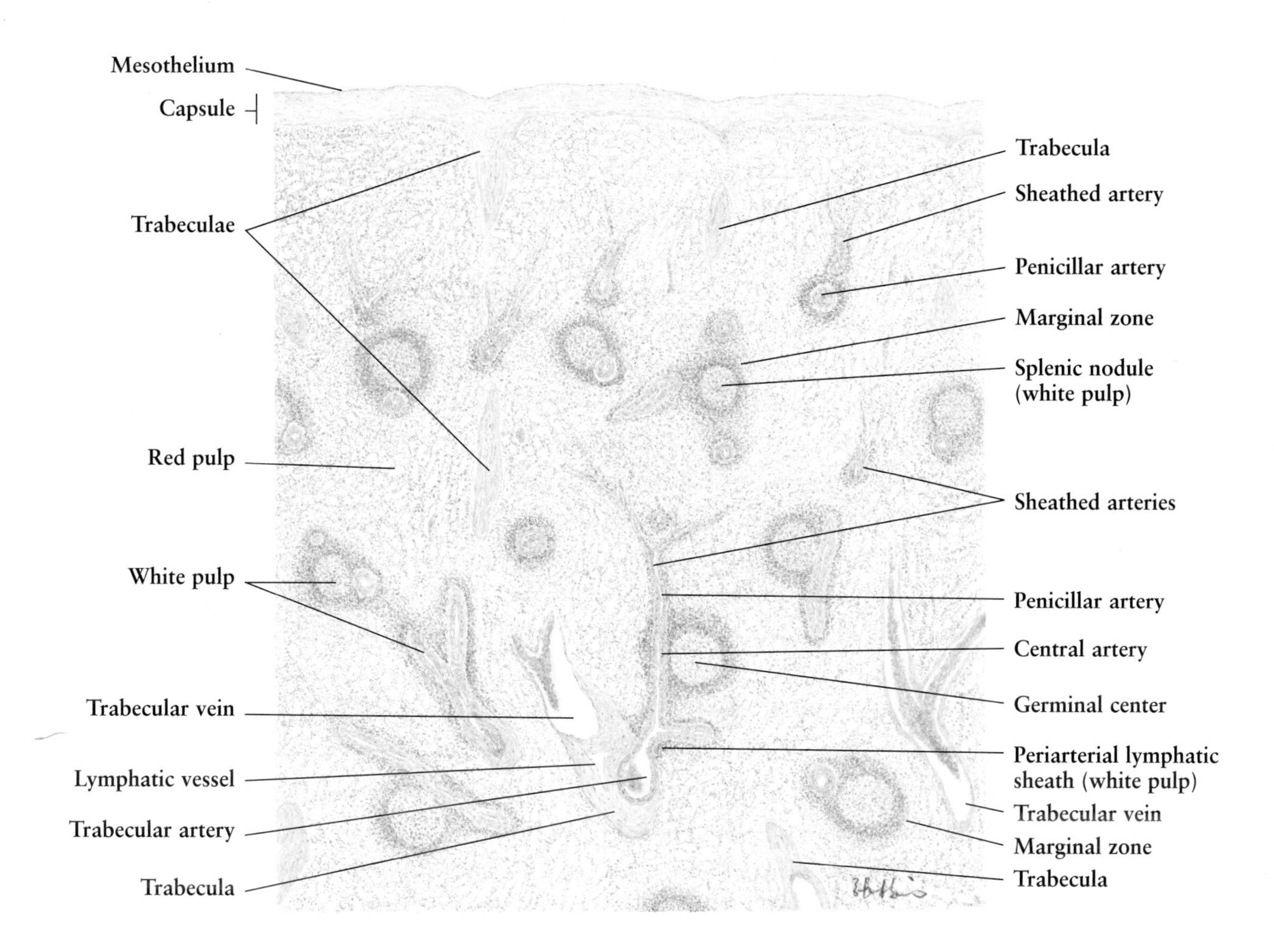

Fig. 8-14. Spleen: Blood Vessels

This figure demonstrates the **blood vessels** that conduct blood from the trabecular arteries, through the splenic pulp, and back to the trabecular veins in the substance of the spleen. The **trabecula** is a dense fibrous connective tissue, composed of numerous collagen fibers, fibroblasts, some elastic fibers, and smooth muscle fibers. It contains trabecular arteries, trabecular veins, and **lymphatic vessels,** as well as **nerve fibers.**

The **trabecular artery,** histologically a small muscular artery, travels within the trabecula always accompanied by the trabecular vein. Once it leaves a trabecula and enters the splenic pulp, the artery is enclosed by a lymphatic tissue sheath composed of lymphocytes, reticular cells, plasma cells, and macrophages. When the lymphatic tissue sheath swells to form a lymphatic nodule known as the **splenic nodule** (with a germinal center), the artery is called a **central artery.** A few capillaries are given off to the white pulp and the **marginal zone.** After passing through the nodules, the arteries enter the **red pulp** as the penicilli (**penicillar arteries**), which are divided into a number of vessels called the **sheathed arteries.** In turn, the sheathed arteries branch again to form two or three **arterial capillaries,** which may open onto the **splenic cord** or directly connect with the **venous sinuses.** Blood is collected by the venous sinuses, which form an anastomosing network penetrating the whole of the red pulp, and then drain away through the **pulp vein** into the trabecular vein.

Fig. 8-15. White Pulp: Splenic Nodule

The **white pulp** is typical lymphatic tissue and consists of two similar structures: periarterial lymphatic sheath and splenic nodule.

In the tissue section, the **splenic nodules** are scattered among the **red pulp.** They are denser accumulations of lymphocytes along the strands of white pulp and form the typical lymphatic nodules, which possess prominent **germinal centers.** Each splenic nodule has one or more **central arteries** passing through, which are arterioles and eccentric in position because they avoid the germinal centers. The central artery gives off some **capillaries** supplying the nodules and the marginal zone. The *B lymphocytes* are produced in the germinal center.

Between the white pulp and surrounding red pulp is the so-called **marginal zone** of diffuse lymphatic tissue. It is characterized by a wider network of reticular cells and contains few lymphocytes and numerous macrophages. Lymphocytes formed in the white pulp are shed into the marginal zone, which is a *B lymphocyte area.*

The structure of the *periarterial lymphatic sheath* is similar to that of the splenic nodule, but the former possesses no typical germinal center (see **Fig. 8-13**). *T lymphocytes* are concentrated in the periarterial lymphatic sheath.

Figure 8-14. Spleen: Blood Vessels
Human • H.E. stain • Low magnification

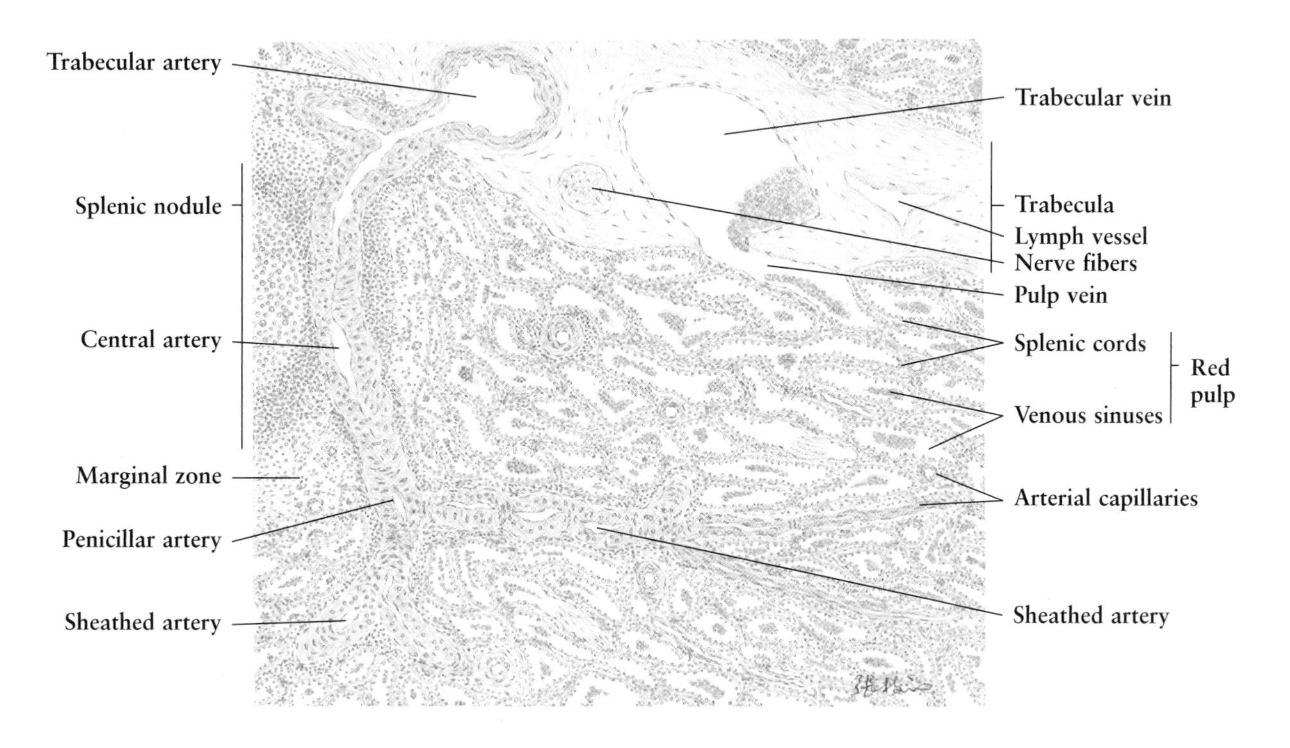

Figure 8-15. White Pulp: Splenic Nodule
Human • H.E. stain • Medium magnification

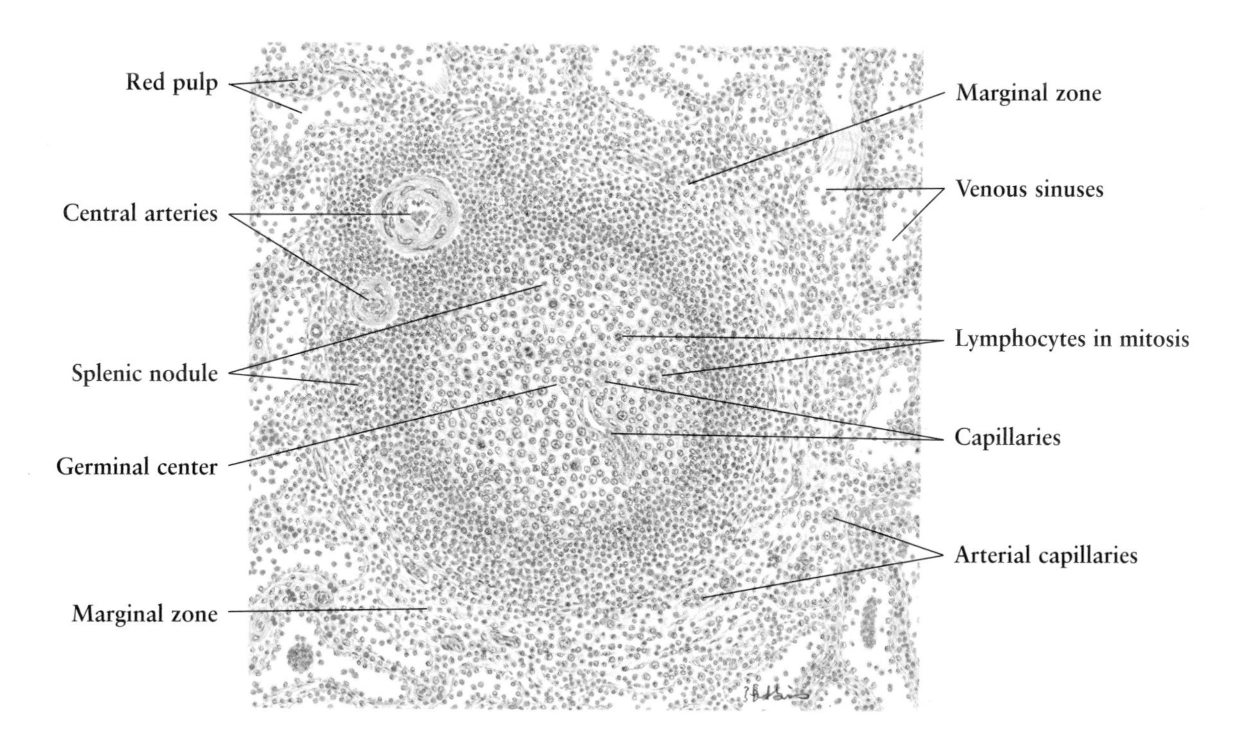

Fig. 8-16. Red Pulp: Venous Sinus and Splenic Cord

The **red pulp** consists of venous sinuses and splenic cords. The splenic cords are located between venous sinuses, which occupy more spaces of the red pulp.

The **venous sinuses** vary in shape and size, and anastomose with one another. The **endothelial cells** are highly elongated fusiform, up to 100 μm in length. They taper toward each end, but the cytoplasmic portion containing the nucleus bulges into the lumen of the sinuses. The endothelial cells rest on an incomplete layer of **basement membrane,** and are arranged longitudinally, contacting each other by transverse processes. The clefts between the parallel endothelial cells are wide enough to permit the blood cells to leave and reenter the sinuses. Externally the sinus is enclosed and supported by a network composed of reticular fibers (see **Fig. 8-17**).

The **splenic cords** are formed by the three-dimensional framework of the **reticular cells** with numerous reticular fibers. The meshes of this framework are filled with **large, medium-sized,** and **small lymphocytes, plasma cells, macrophages,** and the formed elements of blood, among which the **erythrocytes** are the most prominent. The macrophages are responsible for the phagocytosis of foreign materials, the destruction of aged erythrocytes, and the salvage of hemoglobin iron. One macrophage is shown containing a phagocytosed erythrocyte *(arrow)*. Also, the **sheathed arterioles** and **arterial capillaries** can be seen in the splenic cords.

Fig. 8-17. Red Pulp: Reticular Fibers

The network of reticular fibers in the red pulp of the human spleen is demonstrated by the silver impregnation method of Bielschowsky. The **reticular fibers** are stained black, while the nuclei of all cell types appear brown. The reticular fibers in the **splenic cord** are shown as a complicated network. Around the **endothelial cells** of the **venous sinus,** however, the reticular fibers are arranged transversely in a manner reminiscent of the steel bands that hold the staves of a wooden barrel together. Note that the reticular fibers are most dense around the **sheathed arterioles** and **arterial capillaries.**

Figure 8-16. Red Pulp: Venous Sinus and Splenic Cord
Human spleen • H.E. stain • High magnification

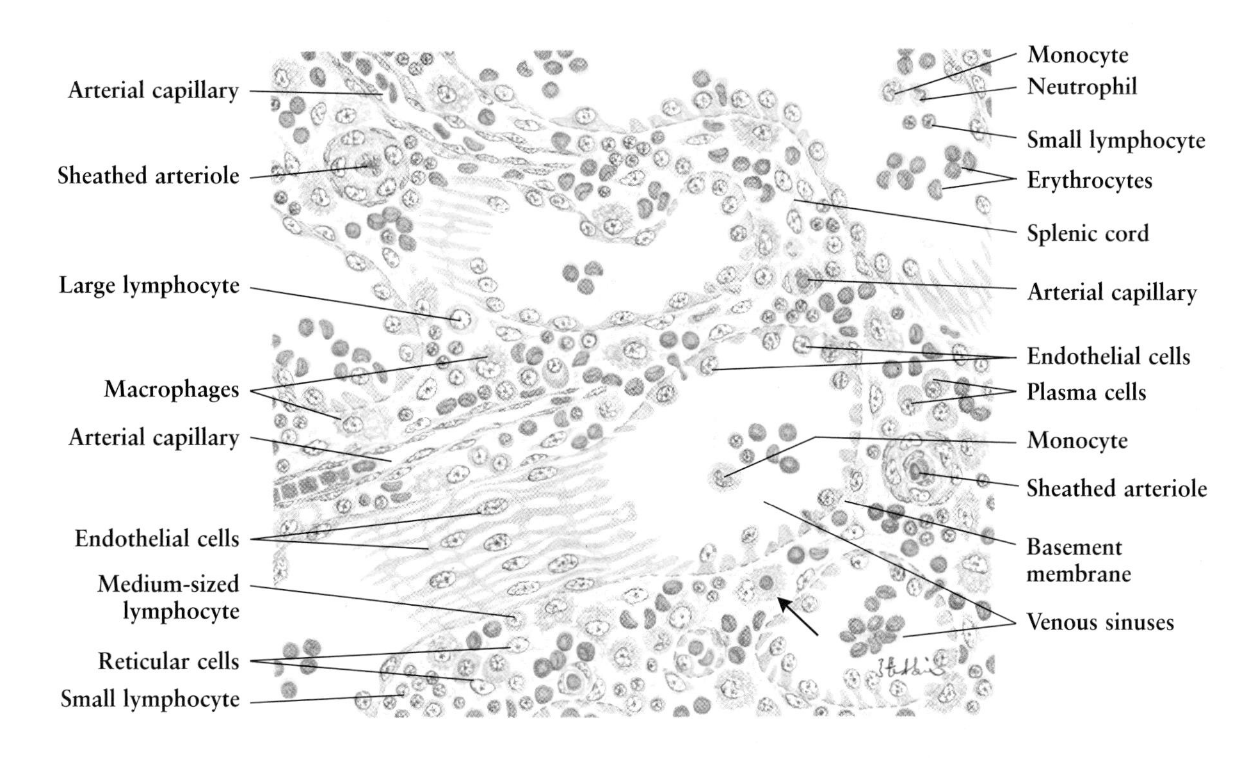

Figure 8-17. Red Pulp: Reticular Fibers
Human spleen • Silver impregnation • Gomori • High magnification

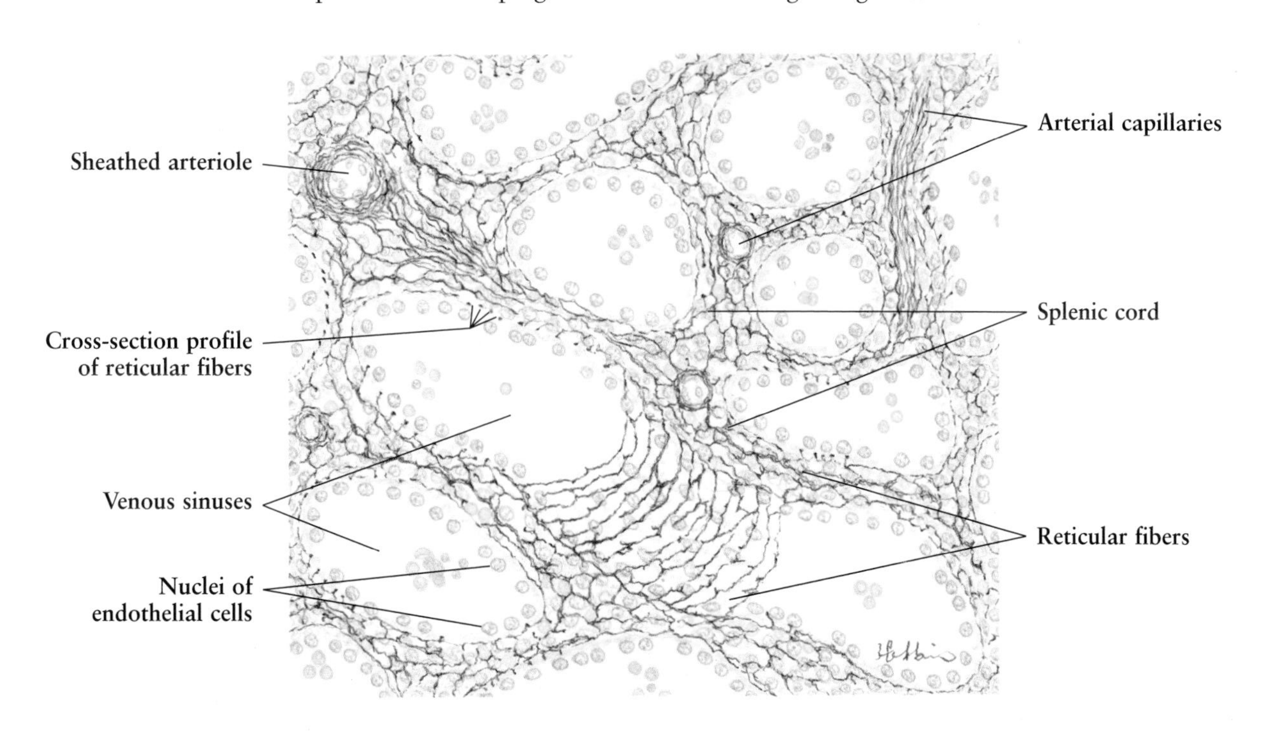

Fig. 8-18. Thymus: Child

The **thymus** is a large lymphatic organ, located in the anterior mediastinum. It consists of two main lobes and is enclosed by a thin connective tissue **capsule,** from which the connective tissue penetrates the substance of the organ as the **interlobular septa,** subdividing it into numerous incomplete **lobules.** The interlobular septa also convey the **blood vessels** from the capsule to the parenchyma of the thymus.

In routine preparations, the lobules are divided into two parts, a darkly stained peripheral **cortex,** which is composed of densely packed lymphocytes, and a lightly stained, central **medulla,** which is composed of loosely packed lymphocytes and other cell types. The medulla is characterized by the eosinophilic **Hassall's corpuscles.** The medulla of some lobules is continuous with that of other lobules because of the incomplete separation of the lobules by the interlobular connective tissue septa.

The thymus is developed from the epithelial outgrowth of the third branchial pouches in embryo and varies in size with the age of the individual. This organ develops until the age of puberty (as shown in this figure), after which involution begins (see **Fig. 8-19**). The *principal function* of the thymus is associated with the production of T lymphocytes, which are able to recognize the foreign proteins, and the production of thymic hormone, the *thymosine.*

Fig. 8-19. Thymus: Adult

In humans, the thymus begins to involute gradually after puberty. These changes include a decrease in size and weight, diminution in the number of lymphocytes, shrinking of the parenchyma, and replacement of the cortex by adipose tissue. This last change is the most prominent histological one. Normally the thymus remains functional in the adult, despite continuous involution throughout life.

Figure 8-19 is an adult **thymus,** illustrating that the mesenchyma, especially of the **cortex,** shrinks prominently. It is replaced by **adipose tissue,** which occupies the spaces between mesenchyma and **interlobular septa** or **capsule.** Note that the **medulla** contains numerous **Hassall's corpuscles.**

Figure 8-18. Thymus: Child
Human • H.E. stain • Low magnification

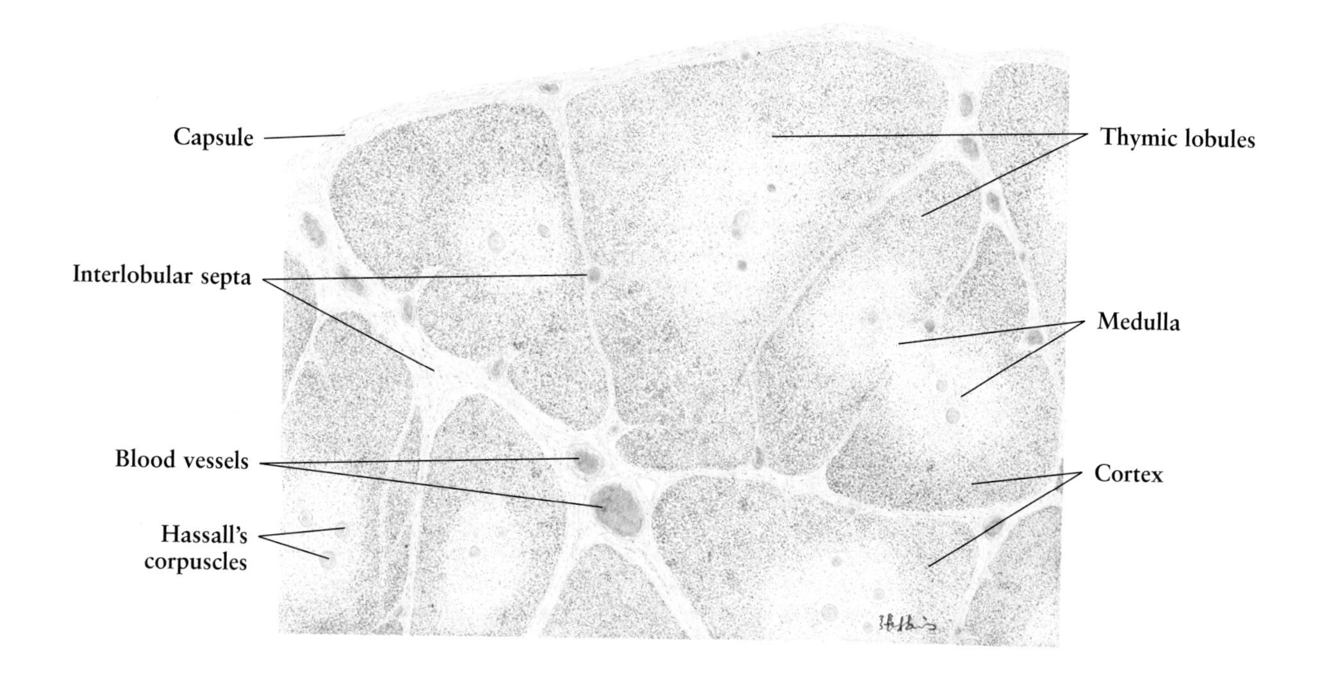

Figure 8-19. Thymus: Adult
Human • H.E. stain • Low magnification

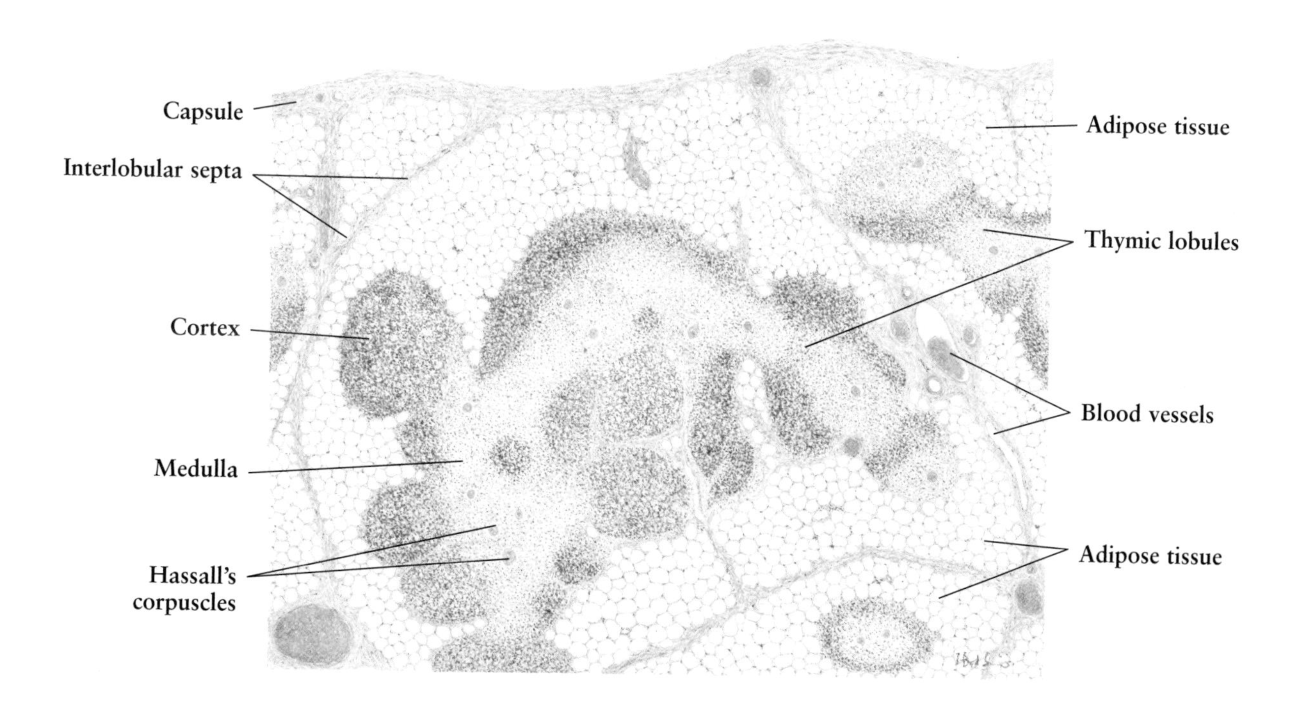

Fig. 8-20.

Thymic Lobule: Cortex

Unlike that in other lymphatic organs, lymphatic tissue in the thymus does not form distinct lymphatic nodules with germinal centers. The **thymic cortex** is characterized by numerous **small lymphocytes,** also known as *thymocytes,* which are densely and irregularly packed. These small lymphocytes are derived from stem cells that originally migrated from the bone marrow to the thymus, where they developed into mature T lymphocytes. While the majority of the thymocytes die by apoptosis (programmed cell death) within the thymus, the remainder mature and leave the thymus via the postcapillary venules in the medulla (see **Fig. 8-21**), and eventually take up residence in the T lymphocyte (or thymus-dependent) zone of other lymphatic organs, such as the lymph nodes and spleen. In addition to the small lymphocytes, **medium-sized** and **large lymphocytes** and occasional **plasma cells** are also present in the cortex. The mitotic figures of dividing lymphocytes (**in mitosis**) are prominent in the thymic cortex. Some **lymphocyte with dense marginated nuclear chromatin** may indicate a cell death in apoptosis.

The lymphocytes are found in the meshwork formed by **epithelial reticular cells,** which anastomose with each other by means of their long, thin cytoplasmic processes. The flattened epithelial reticular cells also line on the inner surface of the **capsule** and invest the **capillaries** and interlobular connective tissue septa.

There are numerous **macrophages** in the cortex. They have a dense nucleus and a large cell body with phagocytic vacuoles within the cytoplasm. The macrophage actively engulf and destroy degenerating or dead lymphocytes within the thymus.

Also shown in this drawing is the connective tissue **capsule,** which consists of numerous collagen fibers and **fibroblasts,** containing **blood vessels.** Note that a **capillary** penetrates the thymic cortex from the capsule, invested by an incomplete layer of reticular cells.

Fig. 8-21.

Thymic Lobule: Medulla

One of the characteristic features of the **thymic medulla** is that the **small lymphocytes** are much less abundant. In addition to the small lymphocytes, **medium-sized lymphocytes** and large lymphocytes are present in the medulla. A number of **macrophages** and a few **eosinophils** are also found. Because lymphocytes are much less numerous in the medulla, the **epithelial reticular cells** are easily recognized.

Another characteristic feature of the medulla is the presence of **Hassall's (thymic) corpuscles.** These spherical bodies, 20–100 μm in diameter, are composed of concentrically arranged epithelial reticular cells, which look like keratinized epithelial cells in the skin. The central cells are huge and often stain red with eosin, showing evidence of degeneration and hyalinization. The function of these corpuscles is unknown.

The medulla is rich in blood vessels. Found at the junction of cortex and medulla are the **postcapillary venules.** These are lined by a layer of flattened endothelial cells, rather than the cuboidal endothelial cells of the high endothelial postcapillary venules of lymph nodes. Via the postcapillary venules, T lymphocytes leave the thymus and enter the circulation.

Figure 8-20. Thymic Lobule: Cortex
Human • H.E. stain • High magnification

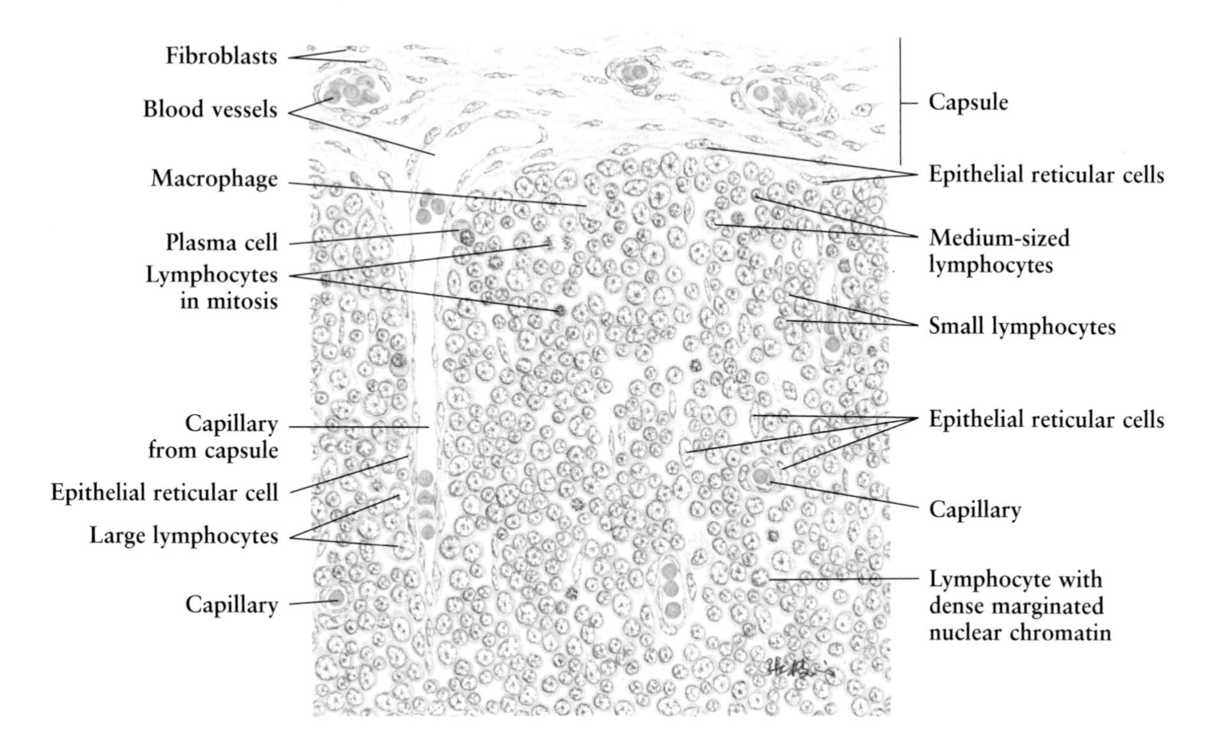

Figure 8-21. Thymic Lobule: Medulla
Human • H.E. stain • High magnification

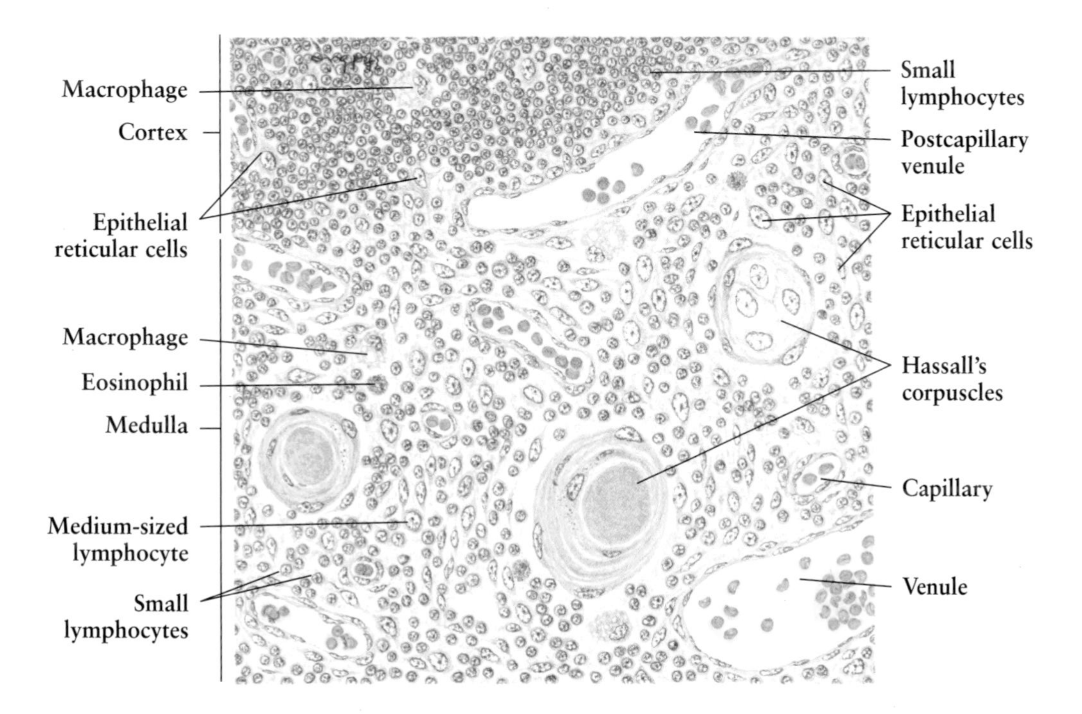

9. Respiratory System

Anatomically, the **respiratory system** consists of a series of passages: the **nose, nasopharynx, pharynx, larynx, trachea** and **bronchi,** and two **lungs.** Its main *function* is to provide for an intake of oxygen by the blood and to eliminate carbon dioxide from the blood.

Functionally, the respiratory system can be divided into a conducting portion and a respiratory portion. The **conducting portion,** which includes the nose, nasopharynx, larynx, trachea, bronchi, bronchioles, and terminal bronchioles, is responsible for the transport of inspired and expired air. As the inspired air passes through the conducting portion, it is filtered, moistened, and warmed. The **respiratory portion,** where gaseous exchange takes place, consists of **respiratory bronchioles, alveolar ducts, alveolar sacs,** and **alveoli.** The alveoli, which make up the bulk of the lung tissue, are thin-walled saclike structures enveloped by a rich network of capillaries. This arrangement provides a structural basis for gaseous exchange.

The nasal cavity, nasopharynx, and larynx are also called the **upper respiratory tract.** These structures are lined by a layer of ciliated pseudostratified columnar epithelium, **respiratory epithelium,** which is involved in filtering, moistening, and warming the inspired air. The **olfactory mucosa,** present in the nasal cavity, is sensitive to odors in the air. The paired **vocal folds** responsible for phonation lie in the larynx.

The trachea and all its branches are known as the **lower respiratory tract.** They are rigid, flexible, hollow air tubes that continuously branch, becoming smaller in diameter. In general, the wall of the trachea is composed of **mucosa, submucosa,** and **adventitia,** which contains a series of C-shaped **hyaline cartilages.** With branching, they show the features of changing structure. The epithelium of the mucosa transforms from the ciliated pseudostratified columnar to the simple cuboidal or squamous. The glands in the submucosa decrease in number and disappear later. The cartilage, which keeps the airways open, gradually becomes irregular plates, and finally disappears.

Fig. 9-1. Ala of Nose

The **nose** is divided by a median nasal septum into *right* and *left nasal cavities* that are supported by bone, cartilage, and muscle. Each nasal cavity consists of three parts: *vestibule,* covered with stratified squamous epithelium; *respiratory part,* covered with ciliated pseudostratified columnar epithelium (see **Fig. 9-2**); and *olfactory part,* covered with olfactory mucosa (see **Fig. 9-3**).

This illustration demonstrates an **ala** of a human nose; its internal surface, belonging to the vestibule, is covered with **keratinized stratified squamous epithelium.** The **skin** covering the outer surface of the ala is characterized by many unusually large **sebaceous glands** and **hair follicles** with very fine hair. The keratinized stratified squamous epithelium of the skin extends through the nares into the vestibule, where it contains some **sweat glands,** many **sebaceous glands,** and **hair follicles** with a few stiff, thick hairs projecting into the airway, helping to filter large dust particles from the inspired air. In the core of the ala are found **hyaline cartilage, adipose tissue, skeletal muscle fibers, larger blood vessels,** and **nerve fibers.** The size of the ala can be changed by the contraction of those skeletal muscle fibers.

In the posterior region of the vestibule the keratinized epithelium becomes a nonkeratinized and nonhairy type, indicating the beginning of the respiratory part of the nasal cavity.

Figure 9-1. Ala of Nose
Human • H.E. stain • Low magnification

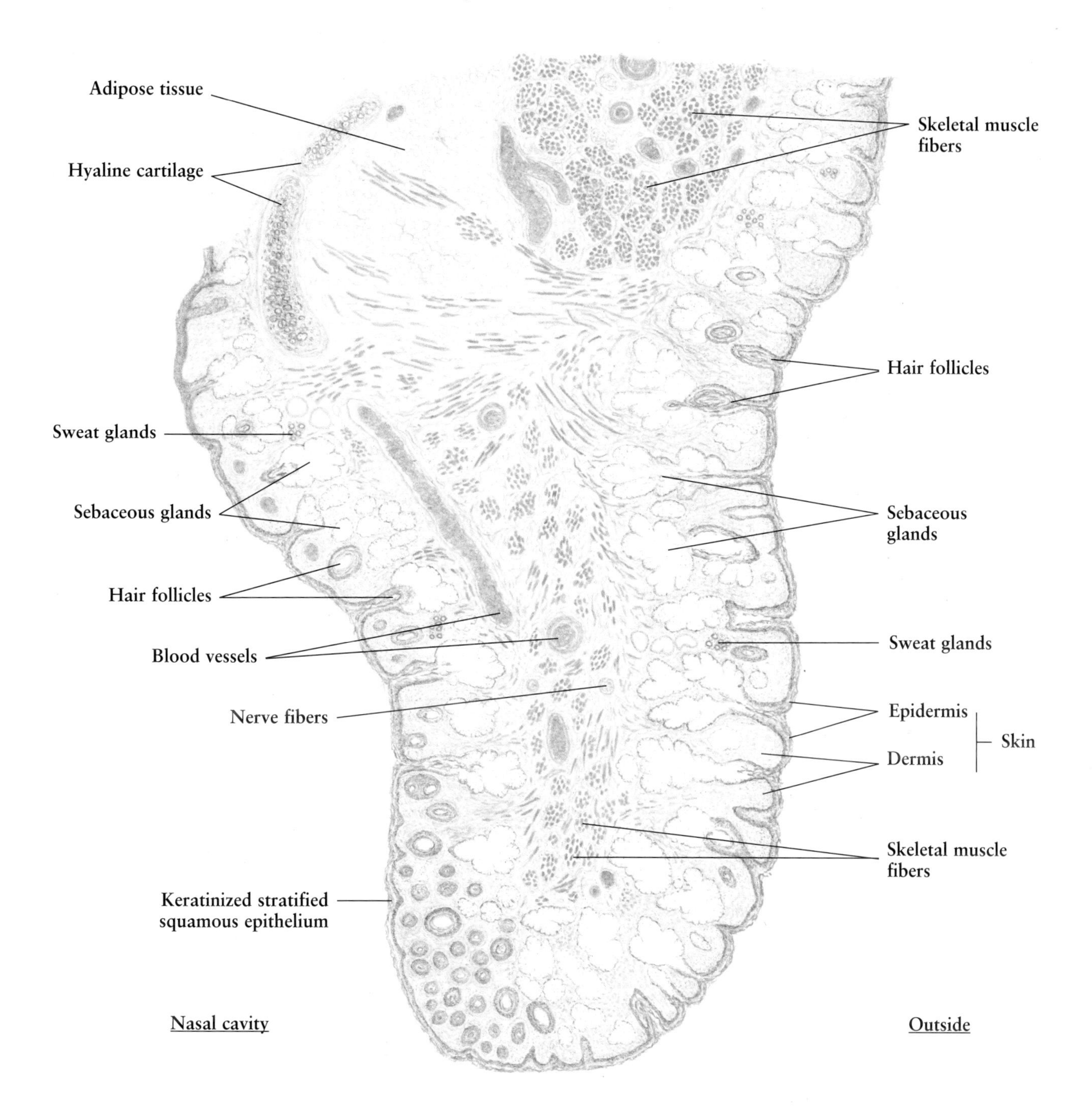

Fig. 9-2. Nasal Cavity: Respiratory Area

The major part of the **nasal cavity,** with the exception of the vestibule and the upper one-third, is lined with **respiratory mucosa,** which is made up of *ciliated pseudostratified columnar epithelium* and *lamina propria.*

The **ciliated pseudostratified columnar epithelium** rests on a **basement membrane** and consists of ciliated cells, goblet cells, intermediate cells, and basal cells (see **Fig. 1-8**). **Intraepithelial glands** may be found among these epithelial cells. A thin layer of surface mucus, secreted by the goblet cells and intraepithelial glands, traps dust particles in inspired air. The mucus with trapped particles is propelled by the wavelike movement of the cilia toward the pharynx, where it is either swallowed or expectorated.

The **lamina propria** is composed of unusually thick but **loose** fibroelastic **connective tissue** characterized by numerous glands and a large venous plexus. The **glands** are of **mixed** type and are mainly located beneath the epithelium, with **excretory ducts** opening into the nasal cavity. Their secretions, passing via excretory ducts that open into the nasal cavity, are involved in cleaning and humidifying the inspired air. The **venous plexus** is formed of thin-walled veins, and is located in the middle of the lamina propria. The **smooth muscle fibers** attached to the wall of the vein may form **sphincters.** The primary function of the venous plexus is to warm the air as it passes by. However, when an allergic reaction or inflammation occurs in the nasal cavities, the venous plexus may become engorged, resulting in swelling of the nasal mucosa and restriction of the air passage. Small **arteries, capillaries, lymphatic vessels,** and **nerve fibers** are also present in the lamina propria. Additionally, a number of **plasma cells** are found near the epithelium and around the mixed glands. The deepest layer of the lamina propria is firmly attached to the **periosteum** or perichondrium.

A similar but much thinner mucosa also lines the paranasal sinuses, nasopharynx, and auditory tubes.

Figure 9-2. Nasal Cavity: Respiratory Area
Human • H.E. stain • Low magnification

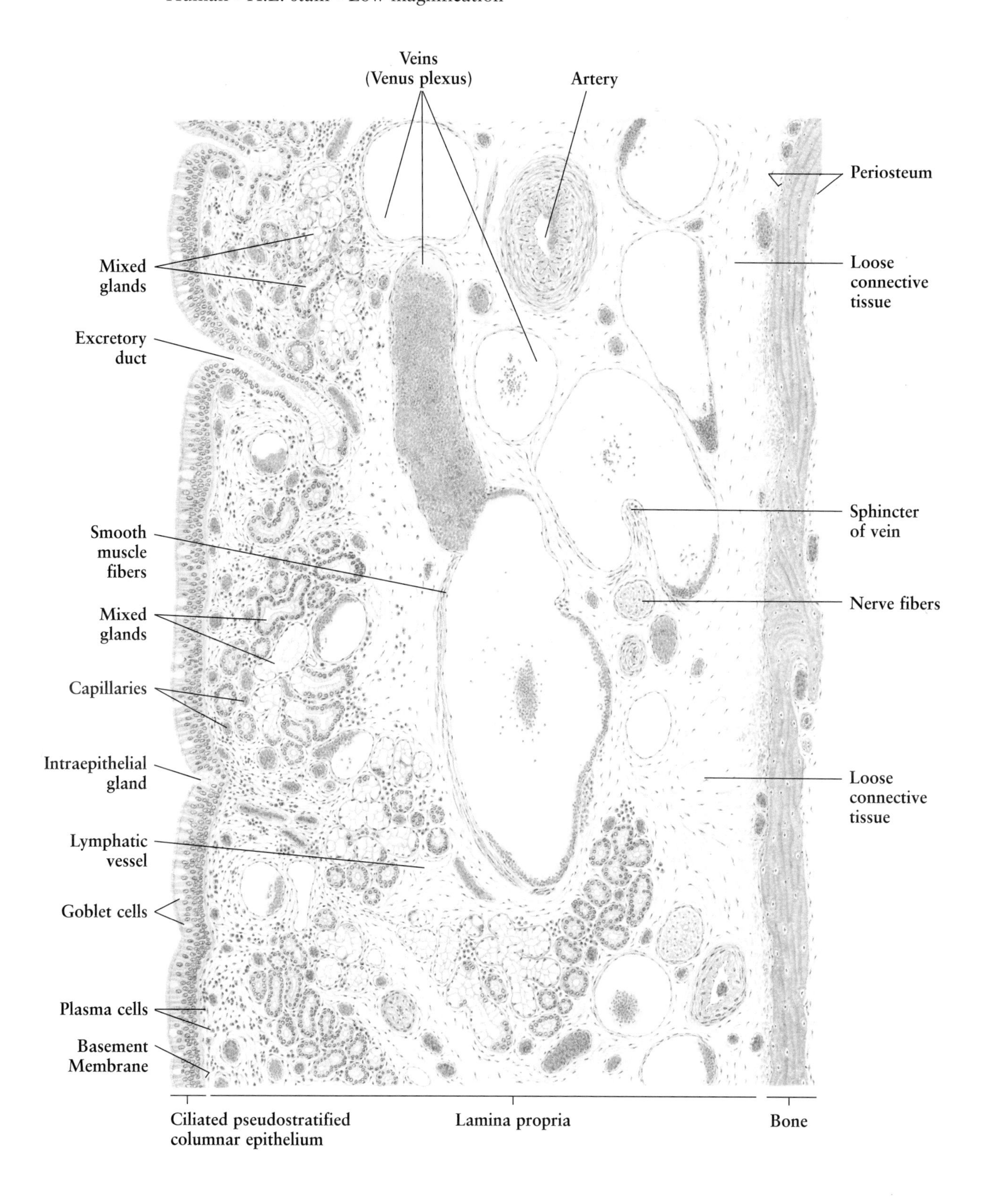

Fig. 9-3. Nasal Cavity: Olfactory Mucosa

The **olfactory mucosa** is yellowish brown in the fresh state, and is present in the olfactory part of each nasal cavity, including the roof of the nasal cavity, the upper third of the nasal septum, and the superior nasal concha. It is composed of *olfactory epithelium* and *lamina propria.*

The **olfactory epithelium** (for details, see **Fig. 6-31**) is a type of pseudostratified columnar epithelium that lacks goblet cells. It is formed of three cell types: **olfactory cells,** slender bipolar chemoreceptor neurons; **supporting cells,** tall columnar cells with numerous microvilli at their free surface; and **basal cells,** precursors of columnar supporting cells.

The **lamina propria** consists of a loose network of fibroelastic connective tissue; its deepest layer is continuous with the **periosteum.** There are many branched tubuloacinar serous glands in the lamina propria, called **Bowman's glands.** They have narrow **excretory ducts** that penetrate the olfactory epithelium and open into the nasal cavity. The secretions of Bowman's glands serve as a solvent for odoriferous substances, eliminate remnants of such substances, and humidify the olfactory mucosa. The axons of the olfactory cells are collected in the lamina propria into bundles of **olfactory fibers.** These fibers pass superiorly through the fine canals of the cribriform plate of the ethmoid bone to enter the olfactory bulb of the brain. The lamina propria is also rich in **blood vessels** and **lymphatic vessels,** but the large venous plexuses are less prominent.

Figure 9-3. Nasal Cavity: Olfactory Mucosa
Monkey • H.E. stain • High magnification

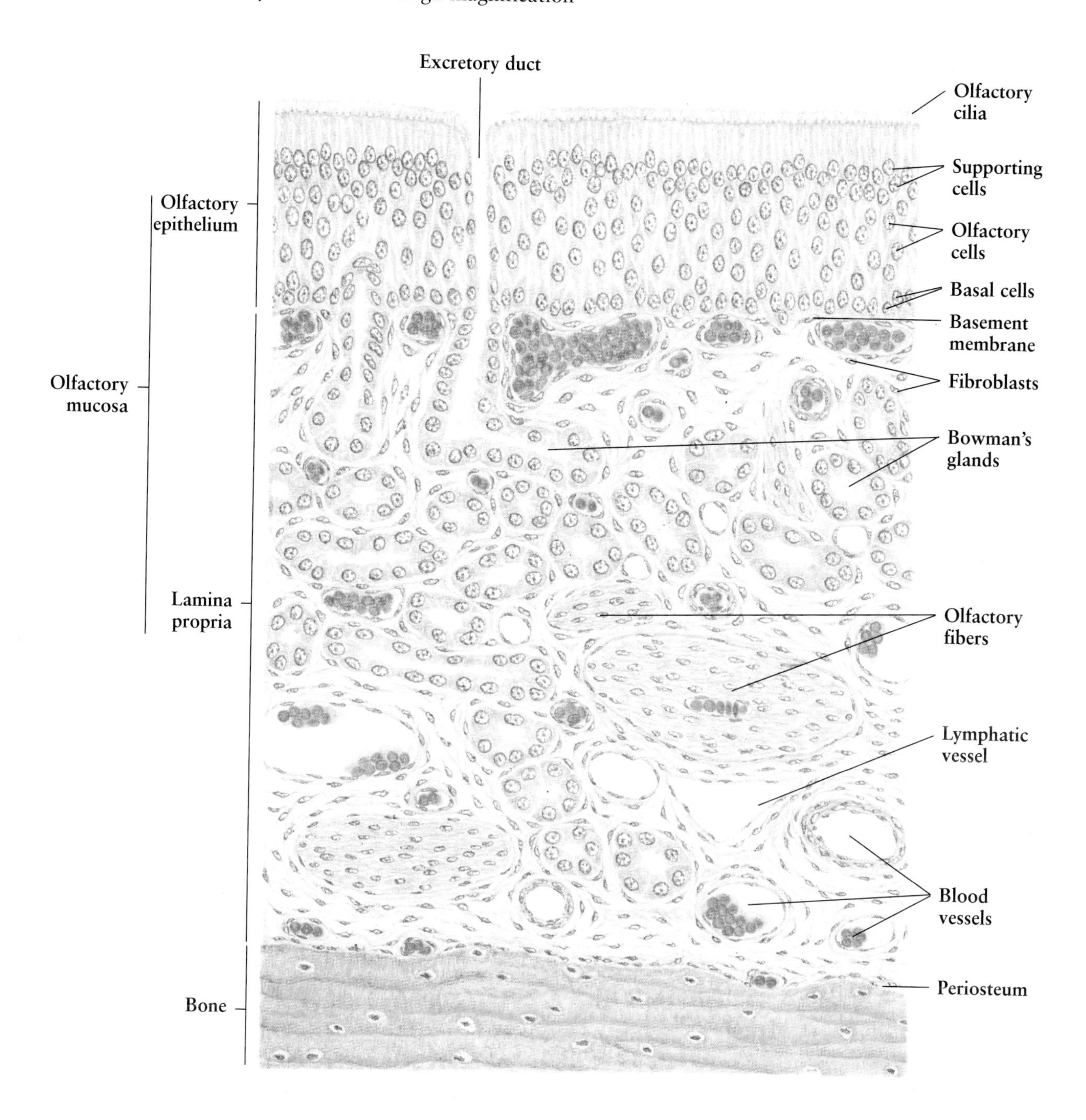

Fig. 9-4. Epiglottis

The **epiglottis,** the uppermost portion of the larynx, is a flaplike upward extension of the anterior wall of the larynx. It is supported by a plate of elastic cartilage and covered by a mucosa containing stratified squamous epithelium.

Both oral and laryngeal sides are covered by the **mucosa,** which consists of **nonkeratinized stratified squamous epithelium** and the underlying **lamina propria.** On the **oral side,** the epithelium is thick with connective tissue papillae beneath it. On the **laryngeal side,** however, the epithelium is much thinner and has no papillae. The laryngeal side is also characterized by **taste buds** (for details, see **Fig. 6-30**) scattered among the epithelial cells. At the base of the epiglottis on the laryngeal side, the stratified squamous epithelium undergoes a transition into ciliated pseudostratified columnar epithelium (not shown in this illustration).

In the lamina propria, diffuse **lymphatic tissue** (infiltration) may be found just beneath the epithelium. The lamina propria is a loose connective tissue with blood and lymphatic vessels, **adipose cells,** and nerve fibers. The **epiglottis glands** are of mixed type, and predominantly located on the laryngeal side. The core of the epiglottis is occupied by a large area of **elastic cartilage** with its **perichondrium** bound to the deepest layer of the lamina propria.

During swallowing, the epiglottis is pressed by the base of the tongue toward the posterior wall of the pharynx, thus closing the larynx. As a result, the bolus of food is pushed to slide through the oral surface of the epiglottis and to enter the esophagus, not the trachea.

Figure 9-4. Epiglottis

Human • Longitudinal section • H.E. stain • Very low magnification

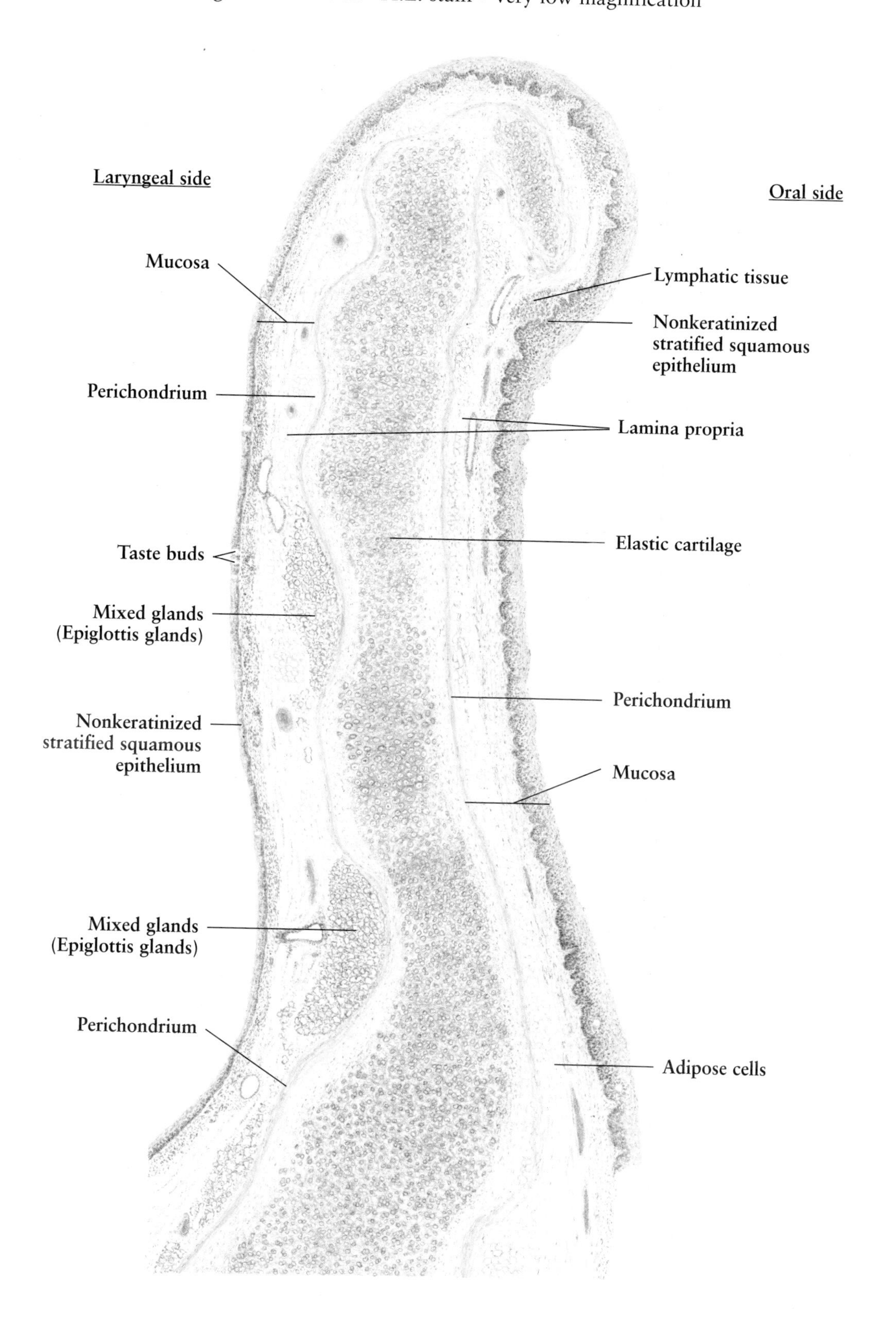

Fig. 9-5. Larynx

The **larynx** is a complex hollow organ that connects the pharynx above and the trachea below. Its lumen is covered by a *mucosa* lined with *ciliated pseudostratified columnar epithelium* and *stratified squamous epithelium.* The wall of the larynx is supported by a group of *cartilages* and a group of *skeletal muscles.*

The **mucosa** of the larynx forms two pairs of folds: the upper are the **vestibular folds** (false vocal cords) and the lower are the **vocal cords.** Between these two pairs of folds is the **laryngeal ventricle** with its narrow pouchlike prolongation, the ventricular recess. The vocal cord and vestibular fold are covered by nonkeratinized **stratified squamous epithelium,** while the rest is lined by **ciliated pseudostratified columnar epithelium** with goblet cells. The beating direction of the cilia is toward the pharynx, moving foreign particles, bacteria, and mucus toward the exterior. The **lamina propria** is a loose connective tissue containing blood vessels and diffuse **lymphatic tissue.** Within the lamina propria are numerous **mixed glands** except where the vocal cords are present. The core of the vocal cord is composed of vocal ligaments with bundles of elastic fibers (see **Fig. 9-6**).

Beside the vocal ligament is the **vocalis muscle,** which controls the tension on the vocal cords and is therefore associated with *phonation.* The other skeletal muscles surrounding the larynx include the **aryepiglottic muscle, thyroarytenoid muscles,** and **lateral cricoarytenoid muscle.** These muscles are involved in *breathing* and *swallowing.* Between the bundles of muscles is a thin layer of loose connective tissue that contains blood vessels and **adipose cells.** The major cartilages supporting the wall of the larynx are the *thyroid,* the *cricoid,* and the *arytenoids.* These are all of the hyaline type. Additionally there are several small elastic cartilages: the *corniculates, cuneiforms,* and the *tips of the arytenoid.* No cartilages are shown in this illustration; for details, see a relevant anatomy textbook or atlas.

The larynx is designed to *produce sound,* to *close the trachea* during swallowing to prevent food and saliva from passing down the airways to the lungs, and *to function as a part of the respiratory conducting system.*

Figure 9-5. Larynx
Human • Frontal section • H.E. stain • Very low magnification

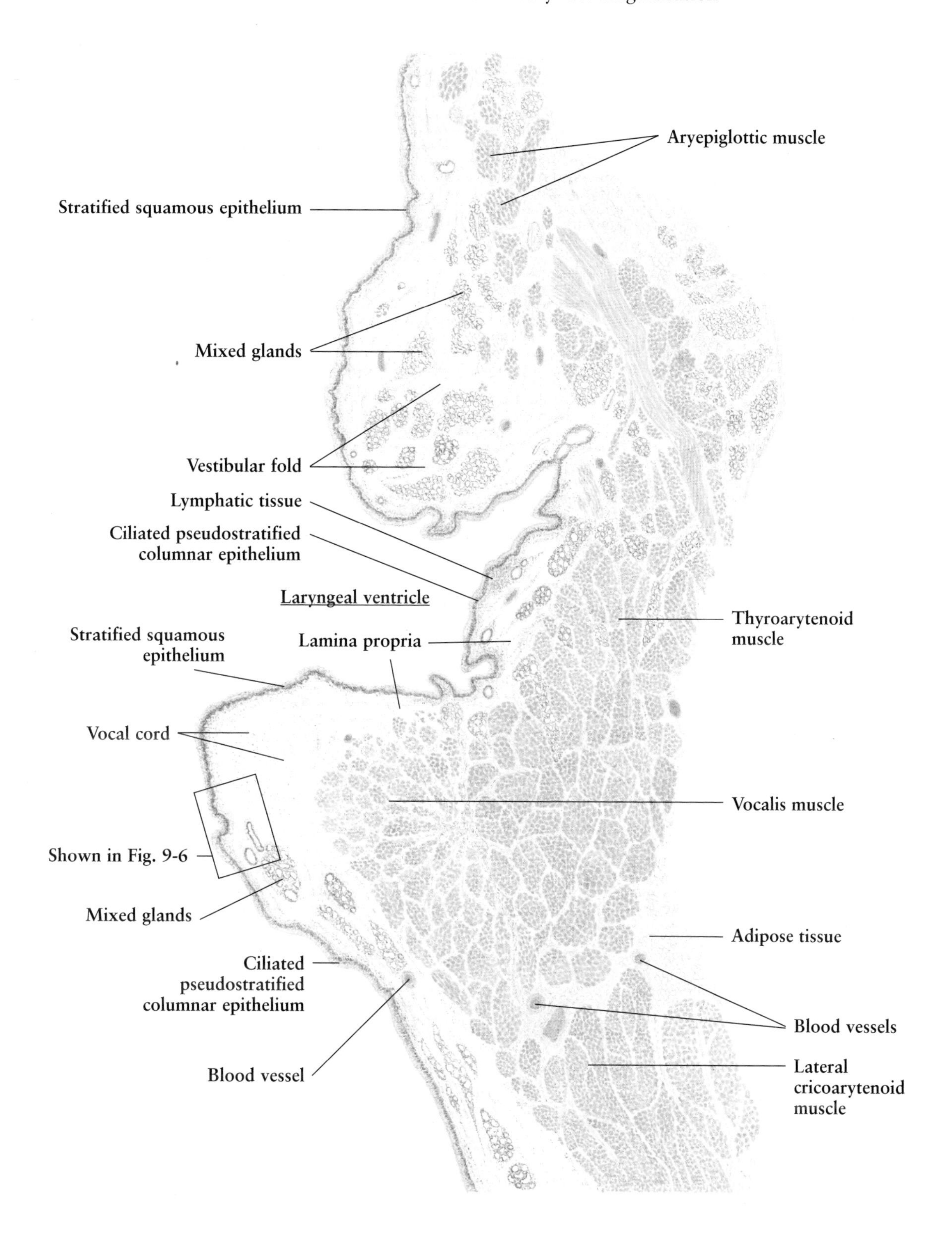

Fig. 9-6. Larynx: Mucosa

This illustration is an enlargement of the boxed area in **Fig. 9-5**, showing the *mucosa* around the *vocal fold.*

The **epithelium** on the left side of this figure is **ciliated pseudostratified columnar epithelium** with **goblet cells;** on the right side it is **stratified squamous epithelium** with **connective tissue papillae** beneath it. Between these two different types of epithelia is a **transitional region,** with the arrangement of the epithelial cells showing the combined characteristic features of these two epithelia, but without any connective tissue papillae under the epithelia.

The **lamina propria** is a **connective tissue** rich in **blood vessels.** There are numerous **fibroblasts** arranged in a streamlike appearance. The **glands** are of **mixed** type with **excretory ducts** opening into the lumen of the larynx. Their secretions moisten the mucosa of the larynx.

Also seen in this figure is the vocal ligament. As a part of the vocal fold, the **vocal ligament** is composed of a dense accumulation of elastic fibers stretched between the midline point of the thyroid and arytenoid cartilages. Sound is produced by the vibration of the free edges of the vocal cord (vocal ligament), as air passes between them.

Fig. 9-7. Trachea

The **trachea** is a rigid, but flexible, air tube about 10–12 cm long and 2–2.5 cm in diameter. It is continuous above with the larynx and below with two primary bronchi.

The **lumen** of the trachea is kept patent by a series of as many as 20 horseshoe-shaped **hyaline cartilage** rings with a posterior gap, oriented one above the other. Adjacent cartilage rings are interconnected by dense fibroelastic connective tissue, which is continuous with the perichondrium, giving the trachea added elasticity. The gap between the ends of each horseshoe-shaped cartilage is traversed by interlacing bundles of **smooth muscle fibers** and supported by **fibroelastic connective tissue.** The smooth muscle fibers are mainly arranged transversely; a few run longitudinally. The diameter of the trachea may be diminished as a result of contraction of these smooth muscle fibers.

The **mucosa** lining the trachea consists of **ciliated pseudostratified columnar epithelium** and a **lamina propria.** The ciliated pseudostratified columnar epithelium or respiratory epithelium, also lines the bronchi. The lamina propria is a layer of loose connective tissue and abundant elastin, contributing substantially to its elasticity. The **submucosa** consists of loose connective tissue and contains numerous small **mixed glands.** The mixed glands are also found even external to the smooth muscle in the gap between the ends of each incomplete cartilage ring.

External to the cartilage ring is **loose fibroelastic connective tissue,** containing **blood vessels** and autonomic nerve fibers that supply the trachea.

Figure 9-6. Larynx: Mucosa
Human • H.E. stain • High magnification of the boxed area in Fig. 9-5

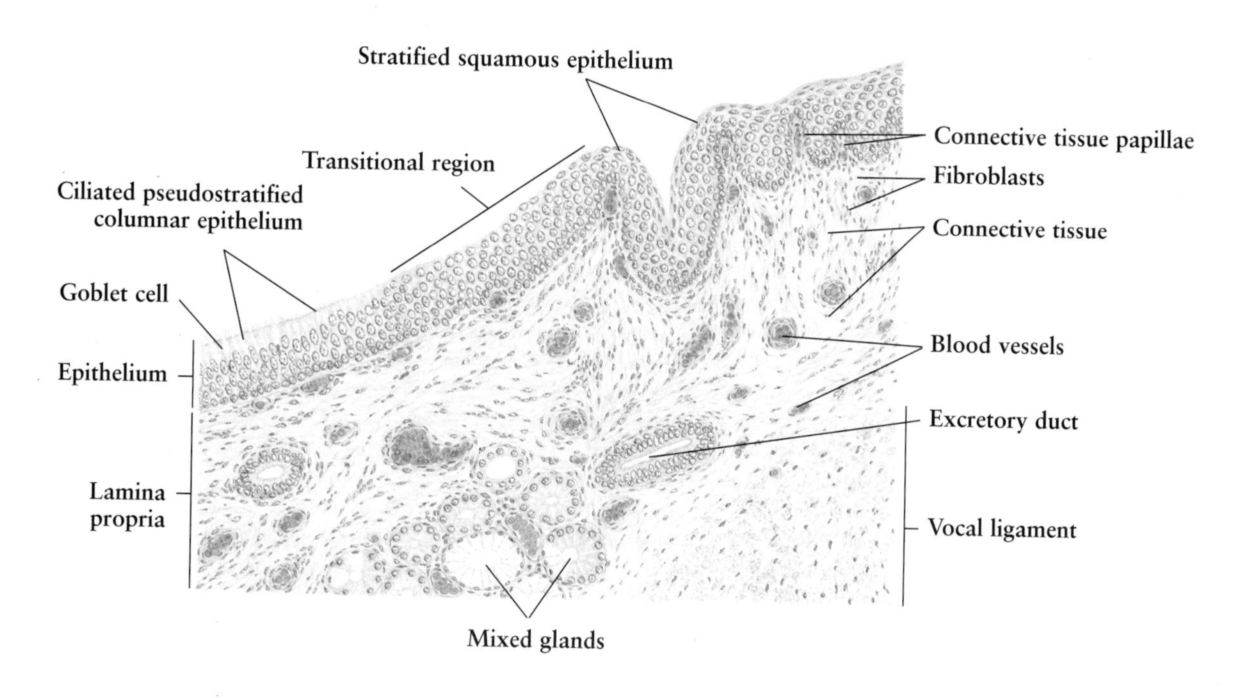

Figure 9-7. Trachea
Human • Cross section • H.E. stain • Very low magnification

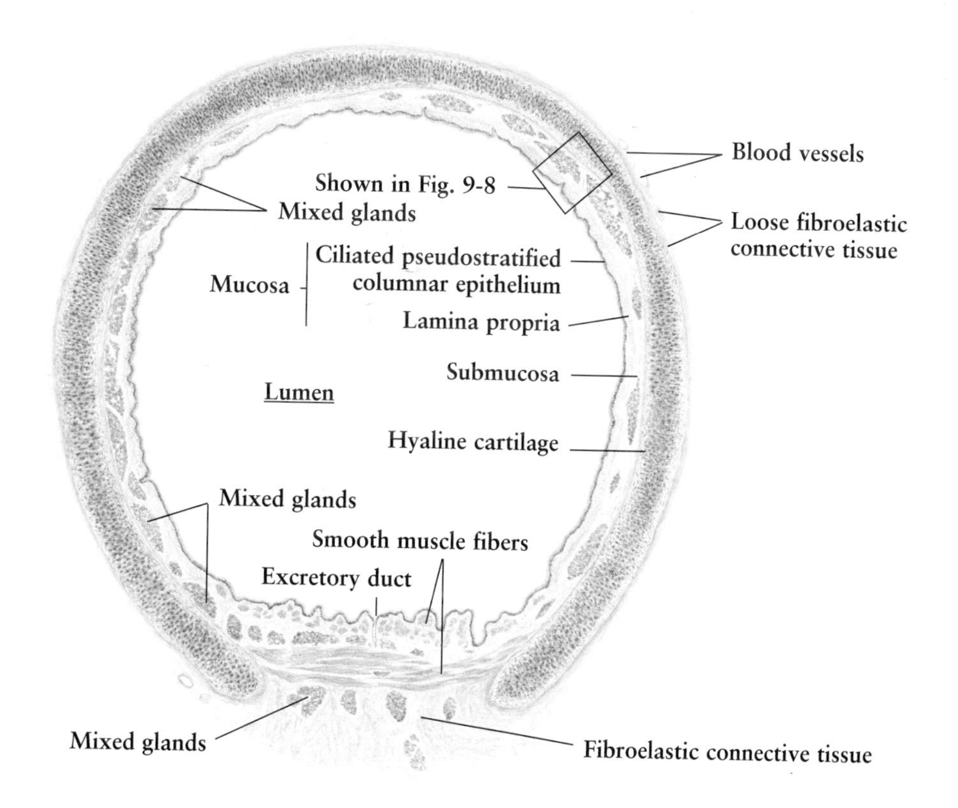

Fig. 9-8. — Tracheal Wall

Figure 9-8 is an enlargement of the boxed area in **Fig. 9-7,** showing the fine structure of the **tracheal wall.** The **mucosa** of the trachea consists of *respiratory epithelium* and *lamina propria,* adjacent to *submucosa.*

The respiratory epithelium, described as **ciliated pseudostratified columnar epithelium,** is composed of ciliated columnar cells, goblet cells, intermediate cells, and basal cells (see **Fig. 1-8**), with some APUD (amine precursor uptake and decarboxylation) cells. The **basement membrane** is particularly thick, formed of a thin basal lamina and a thick, dense network of reticular and collagen fibrils.

The **lamina propria** consists of highly vascular, loose connective tissue containing numerous **blood vessels, nerve fibers,** and **fibroblasts.** It becomes more condensed at its deeper portion to form a band of fibroblastic tissue.

Adjacent to the lamina propria is the **submucosa,** which is loose connective tissue filled with numerous **mixed glands** and some **adipose cells.** The **excretory ducts** of the mixed glands travel through the lamina propria and the epithelium, and open onto the surface of the respiratory epithelium. Their secretion is associated with the moisture of the mucosa and cleaning of the inspired air. The deepest portion of the submucosa contains abundant **collagen fibers** arranged circularly and longitudinally, and merges with the **perichondrium** of the underlying **hyaline cartilage.**

Figure 9-8. Tracheal Wall
Human • H.E. stain • High magnification

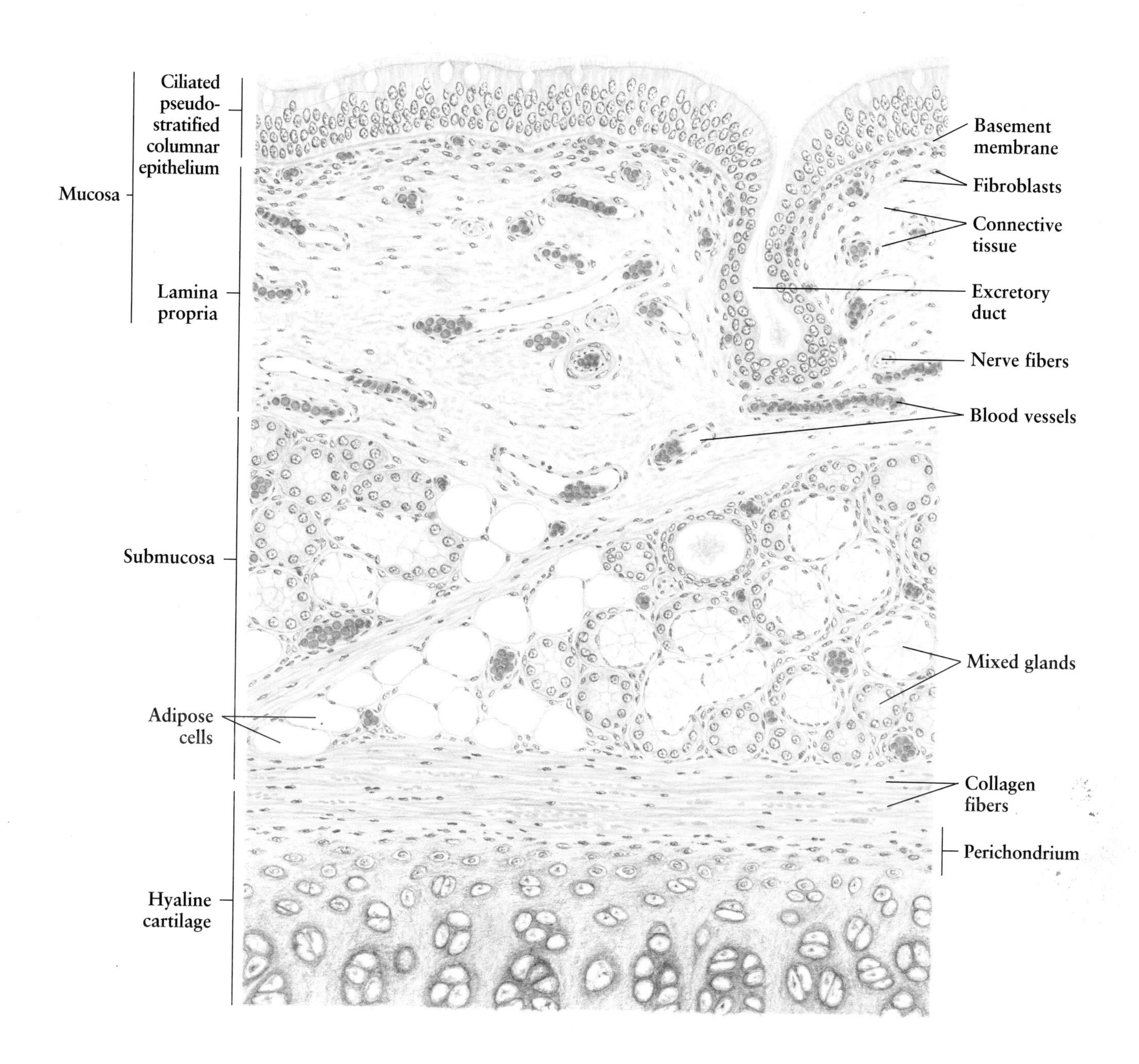

Fig. 9-9. Lung

The **lungs** are paired lobed organs situated within the thoracic cavity on each side of the centrally placed mediastinum. Each lung is surrounded by a serous membrane, the visceral pleura, from which connective tissue passes into the organ, dividing it into lobes and lobules. Each lung has a hilum at the mediastinum, where the bronchus, blood vessels, lymphatic vessels, and nerves enter and leave.

Lung *parenchyma* consists of *alveoli* to which air is conducted by a series of *tubes*. In the mediastinum, the *trachea* branches to form right and left *primary bronchi*. Each primary bronchus divides into *secondary bronchi,* which supply the lobes before branching into *tertiary bronchi* that enter the segments of each lobes. Then the tertiary bronchi give rise to the **bronchioles,** each supplying a *lobule,* the basic unit of the lung. Within each lobule, the bronchiole divides further into **terminal bronchioles** and **respiratory bronchioles.** The next subdivisions are the **alveolar ducts,** followed by the **alveolar sac,** which is the common opening of the **alveoli,** the most peripheral part of the lung.

The lungs have a dual blood supply. The *pulmonary vessels* carry deoxygenated blood (**pulmonary artery**) from the right ventricle to the lungs for respiratory exchanges, and take back oxygenated blood (pulmonary vein) to the left atrium. They run along with the bronchi in the connective tissue outside their walls, and finally form capillary anastomoses surrounding the alveoli. The *bronchial vessels* carry oxygenated blood for the nutrition of the bronchial tubes and lung tissue. They also follow the bronchi, running within their fibrous coats to be distributed to their glands and walls and to the connective tissue of the pleura.

Figure 9-9. Lung
Human • H.E. stain • Low magnification

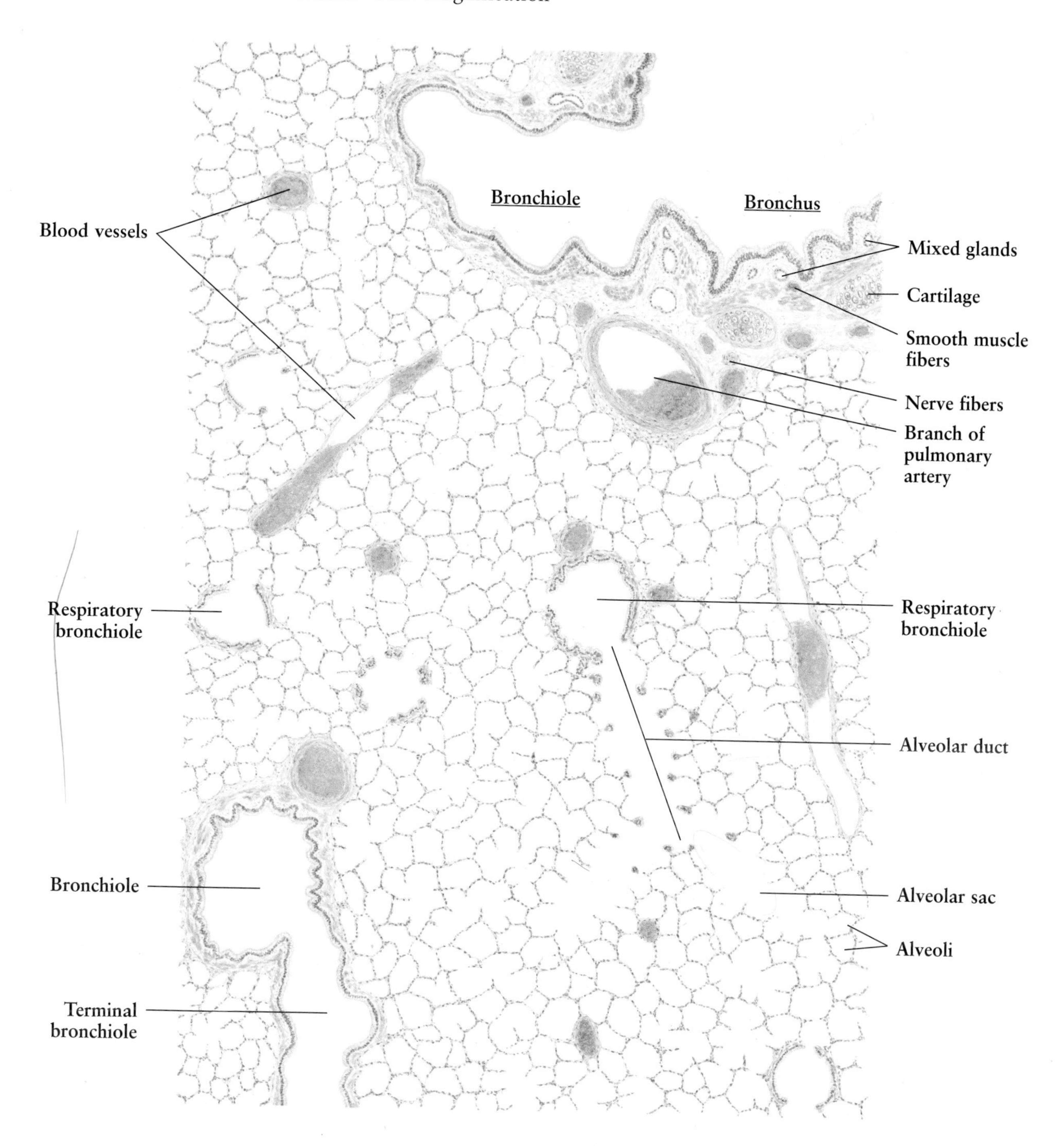

Fig. 9-10.

Lung: Small (Intrapulmonary) Bronchus

The extrapulmonary bronchi are histologically identical with the trachea. These airways differ only in their smaller diameter. The **small** or **intrapulmonary bronchi** consist of *mucosa, submucosa, cartilage,* and *adventitia.*

The epithelium of the **mucosa** is made up of **ciliated pseudostratified columnar epithelium** with **goblet cells,** resting on a **basement membrane.** The underlying **lamina propria** is a elastic connective tissue containing **blood vessels** and some lymphatic tissue. The **smooth muscle fibers** comprise a complete layer of **muscularis mucosae,** encircling the entire lumen. The **submucosa** is a loose connective tissue, which contains seromucous **glands,** blood vessels, lymphatic vessels, and **nerve fibers.** The **cartilages** become irregularly shaped plates, and function as a skeletal support maintaining the patency of the bronchus. Between the cartilage plates is connective tissue containing a few **adipose cells** or glands. The **adventitia** is a layer of dense fibroelastic membrane with blood vessels and nerve fibers.

Fig. 9-11.

Lung: Bronchiole and Terminal Bronchiole

A **bronchiole,** with a diameter of 0.5–1 mm, supplies a single lobule. Its lumen is lined by a lower **ciliated pseudostratified columnar epithelium** with some goblet cells. The **basement membrane** becomes thinner. The cartilage is absent, and only a few **glands** can be found. The prominent bundles of **smooth muscle** and elastic fibers comprise the lamina propria, supporting the wall of the bronchiole. Also seen in the wall of the bronchiole are **blood vessels** and **nerve fibers.**

The **terminal bronchioles** are the thin-walled branches of the bronchiole. The epithelium becomes **ciliated simple columnar** type, without goblet cells. The *Clara cells* can be observed among these epithelial cells under higher magnification. The cartilage and glands completely disappear. The connective tissue is much reduced, and the **smooth muscle fibers** form an incomplete layer. The terminal bronchiole gives rise to respiratory bronchioles.

Figure 9-10. Lung: Small (Intrapulmonary) Bronchus
Human • H.E. stain • Medium magnification

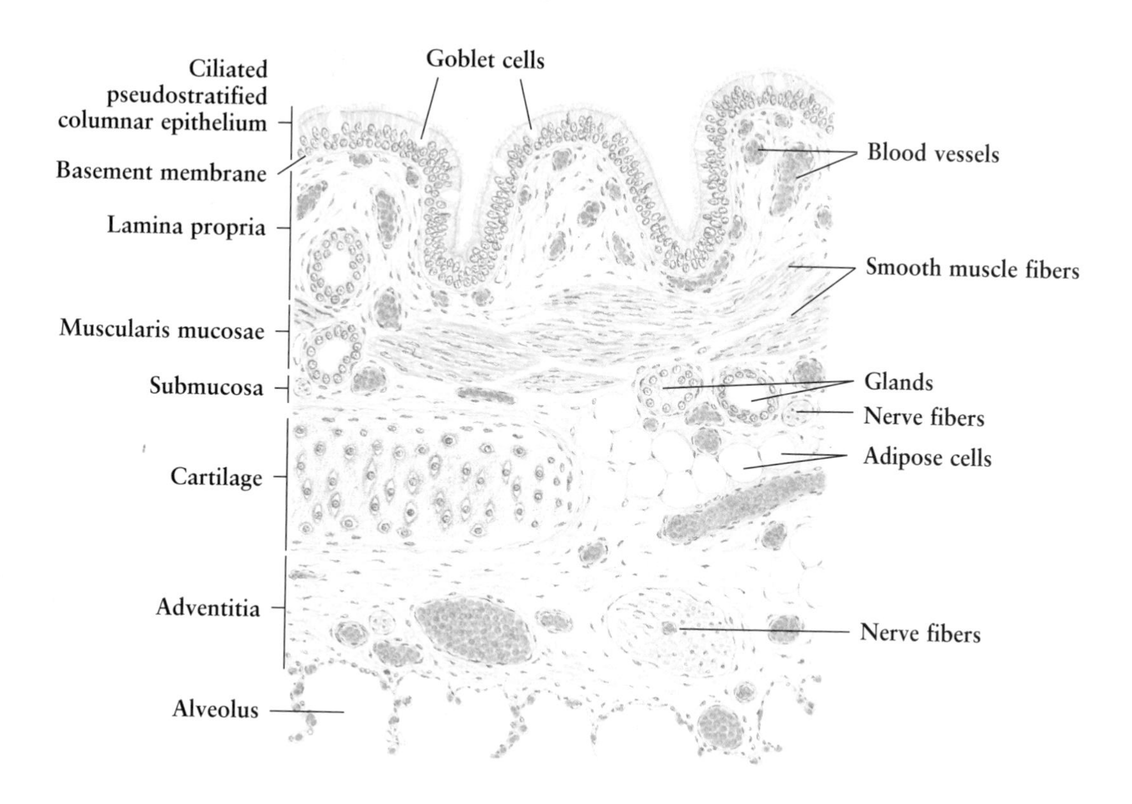

Figure 9-11. Lung: Bronchiole and Terminal Bronchiole
Human • H.E. stain • Low magnification

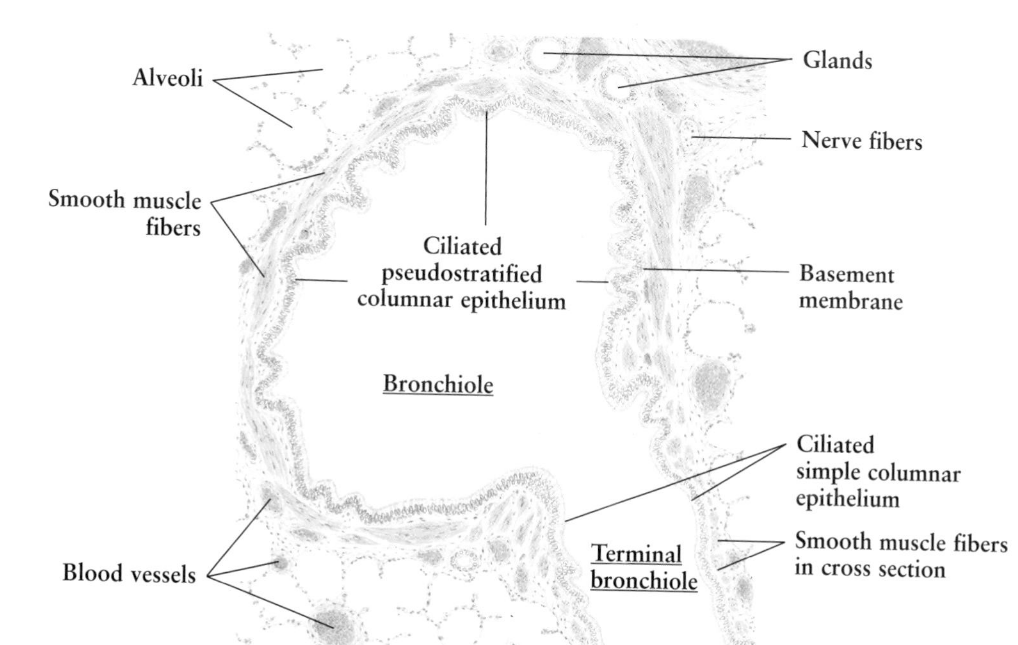

Fig. 9-12.

Lung: Respiratory Subdivisions

This illustration demonstrates the **respiratory subdivisions** of the lung, including the *respiratory bronchioles, alveolar ducts, alveolar sacs,* and *alveoli.*

The **respiratory bronchioles** are short branches of a terminal bronchiole. The lumen is lined by a **simple cuboidal epithelium** composed of **ciliated** cells and Clara cells. The wall contains an incomplete layer of **smooth muscle fibers** embedded in fibroelastic connective tissue. The most outstanding characteristic of the respiratory bronchioles is that the wall is interrupted by saccular outpocketings, the **alveoli,** where gaseous exchange first occurs. The respiratory bronchioles lead into alveolar ducts.

The **alveolar ducts** present enormous numbers of alveoli opening into the lumen, arranged side by side. They do not essentially possess a real wall. The so-called wall, in a longitudinal section of the alveolar duct, is composed of the lines of many small knoblike structures that are covered by a layer of **simple cuboidal epithelium,** with **smooth muscle fibers** and fibroelastic tissue in it. In three dimensions, the "wall" of the alveolar duct is a spiral structure in which the gyra are separated by alveoli. The alveolar duct continues into alveolar sacs.

An **alveolar sac** is a distended common air space at the end of an alveolar duct. Opening into it is a group of alveoli, which are clustered around it. There is no smooth muscle fiber in the wall of the alveolar sac.

The **alveoli** are very thin-walled spheres with a small opening on one aspect into the alveolar sac, alveolar duct, or respiratory bronchiole. Neighboring alveoli are separated by a common wall, the **interalveolar septum,** which is extremely thin and consists of a stroma containing numerous anastomosing capillaries, supported in a network of fine reticular and elastic fibers. There are about 150 million alveoli in each lung, with 100 square meters in total area.

Also seen in this illustration are the **pulmonary vessels,** running in the lung parenchyma, and the **visceral pleura,** which invests the lung. The **pleura,** lined by a layer of **mesothelium,** is relatively thick in humans, and consists of a layer of fibroelastic **connective tissue** containing collagen and elastic fibers, and occasionally smooth muscle fibers.

Figure 9-12. Lung: Respiratory Subdivisions

Human • H.E. stain • Medium magnification

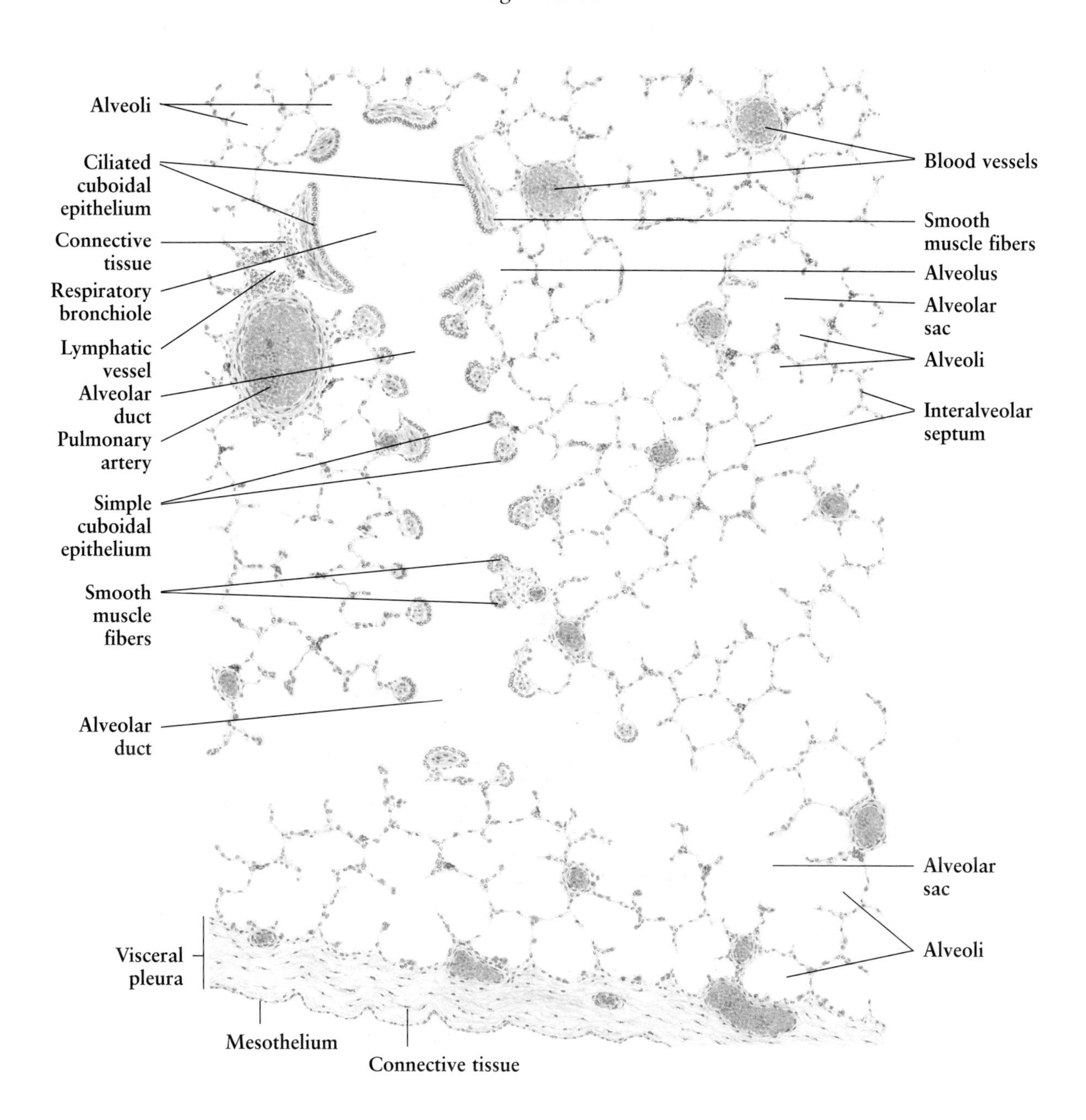

Fig. 9-13.

Lung: Alveoli

The **alveoli** are small polyhedral pockets opening on one side into the alveolar sac, alveolar duct, or respiratory bronchiole. Adjacent alveoli share a common interalveolar septum, which contains numerous blood capillaries and elastic and reticular fibers, as well as some fibroblasts and macrophages. The lining of the alveolus consists of two cell types: type I and type II cells.

Type I cells, the *squamous alveolar cells,* resting on a basement membrane, have an extremely large, attenuated cytoplasm about 50–100 μm wide, and a flattened nucleus located in a niche. Because the cell is so thin and long, the cytoplasm is difficult to recognize at the light microscopic level, and in thin section nuclei are seen only occasionally. Type I cells are associated with gaseous exchange.

Type II cells, the *great alveolar cells,* are scattered singly or are present in a small group among the squamous alveolar cells. They are cuboidal with a large rounded nucleus, and rest on a basement membrane. In routine preparations, the type II cell has a foamy appearance. At the electron microscopic level, the cell is characterized by multilamellar bodies within its cytoplasm, which contain phospholipid, protein, and glycosaminoglycans. The great alveolar cells are responsible for the secretion of surfactant, which prevents the alveoli from collapsing by reducing the surface tension of the squamous alveolar cells.

The **interalveolar septum** separating two adjacent alveoli is a very thin structure, consisting of epithelium and connective tissue. Both sides of the adjacent alveolar surfaces are lined by alveolar epithelium composed of type I and II cells. Between these two layers of epithelia is an interstitium containing alveolar **capillaries,** supported in a network of elastic and reticular fibers. Some **fibroblasts** are found in the septum. The alveolar **macrophage** or **dust cells** can be observed in the interstitium of the interalveolar septum or free in the alveolar spaces. At the light microscopic level, the dust cells are characterized by enormous numbers of black phagocytosed carbon particles deposited in their cytoplasm. The lungs of heavy smokers appear darker in color because of numerous dust cells with carbon particles.

The structures through which gaseous exchange occurs between air in alveoli and blood in alveolar capillaries are known collectively as the *blood–air barrier.* These structures, measuring about 0.1–0.2 μm in thickness, include (1) attenuated cytoplasm of pulmonary epithelial cells; (2) basement membrane of the epithelium; (3) basement membrane of the capillary; and (4) very thin cytoplasm of capillary endothelial cells. To reduce the thickness of the blood–air barrier, the two basement membranes are generally fused into one layer.

Alveolar pores, 10–15 μm in diameter, may be found in the interalveolar septum, connecting two adjacent alveoli. They may equalize pressure between alveoli, or help to prevent collapse or overdistension of the alveoli if a small bronchiole is occluded.

The illustration also shows (upper-right portion) a knoblike structure of the alveolar duct, which is lined by **simple cuboidal epithelium** and has some **smooth muscle fibers** beneath the epithelial cells.

Fig. 9-14.

Alveoli: Elastic Fibers

A special staining method using resorcin fuchsin and hematoxylin demonstrates the large amount of elastin, which forms a network of **elastic fibers** in the interalveolar septum. The elastic fibers, stained purple, are shown in the interstitia, surrounding the **capillaries** and **pulmonary artery,** and in the knoblike structures of the alveolar duct.

Figure 9-13. Lung: Alveoli
Human • H.E. stain • High magnification

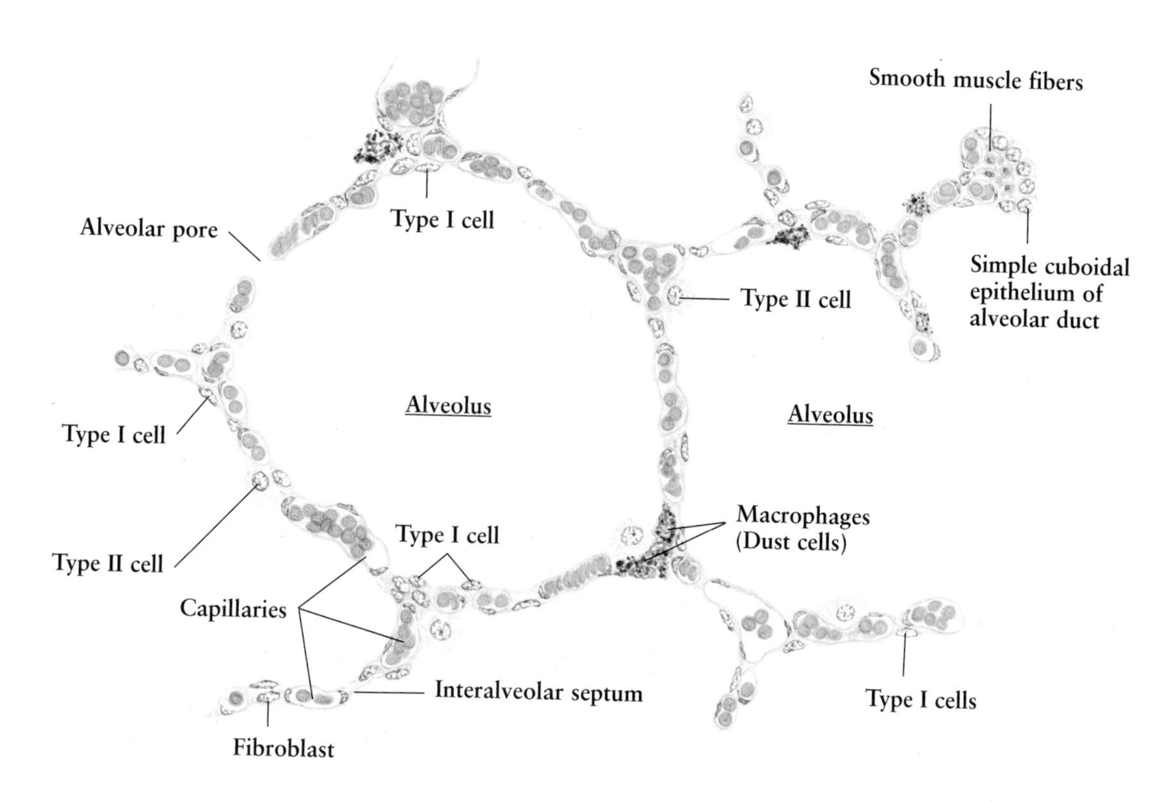

Figure 9-14. Alveoli: Elastic Fibers
Human lung • Resorcin fuchsin-hematoxylin stain • High magnification

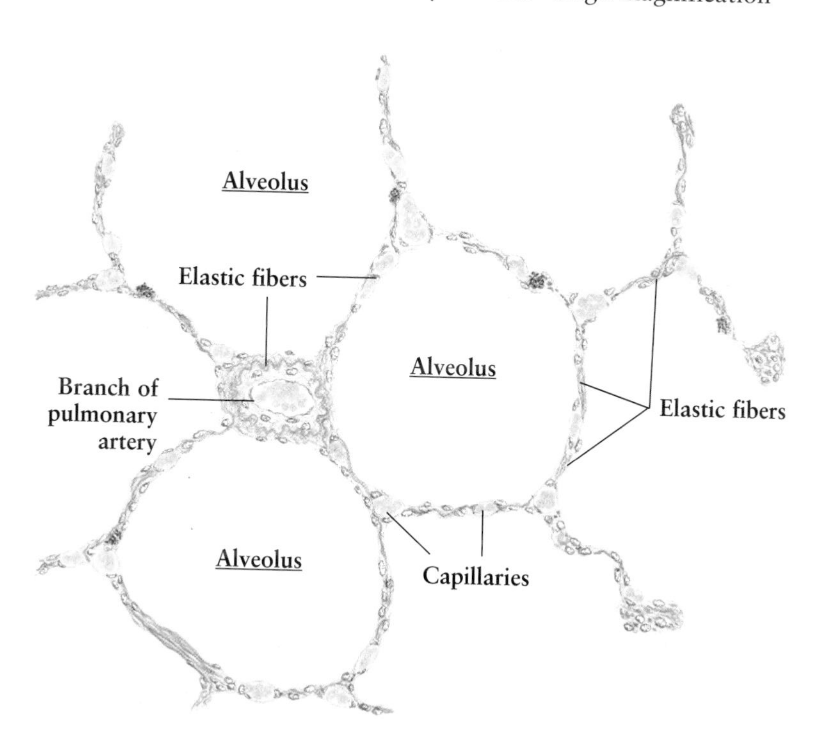

10. Digestive System

The **digestive system** consists of the **digestive** or **alimentary tract** and its **associated glands,** such as salivary glands, the liver, and the pancreas. It *functions* in the processing, digestion, and absorption of food, followed by elimination of the residue.

The digestive tract is divided into **oral cavity, esophagus, stomach, small intestine, large intestine, rectum,** and **anal canal.** The contents of the oral cavity include the **lips, teeth, tongue** and its **lingual papillae,** and the salivary glands, such as **parotid glands, submandibular glands,** and **sublingual glands.**

From esophagus to anal canal, the digestive tract shows certain common structural characteristics. The wall of the tract is composed of four principal layers: the *mucosa, submucosa, muscularis externa,* and *adventitia* or *serosa.*

The **mucosa** consists of an **epithelial lining,** a **lamina propria** of loose connective tissue, and the **muscularis mucosae,** which is a layer or two layers of smooth muscle fibers, separating the mucosa from the submucosa.

The **submucosa** is a layer of irregular loose connective tissue, containing large blood and lymphatic vessels, glands, lymphatic tissue, and a **submucous (Meissner's) nerve plexus.**

The **muscularis externa** is formed of thick bundles of smooth muscle fibers, which are arranged into two sublayers: circular in the internal sublayer and longitudinal in the external sublayer. Between the two muscle sublayers are **myenteric (Auerbach's) nerve plexus.**

The **adventitia** is a thin layer of loose connective tissue. It is continuous with adjacent tissue in the esophagus. In the gastrointestinal tract, the adventitia is replaced by the serosa, which is composed of a layer of **mesothelium** and the underlying loose connective tissue, rich in blood and lymphatic vessels and adipose cells.

The **liver** is the largest gland associated with digestion, formed of numerous **hepatic lobules,** which are the structural units of the liver. It is the principal metabolic organ and central detoxification center. The *bile salts* secreted by the liver emulsify fats, aiding their digestion and absorption.

The **pancreas** is the principal enzyme-producing gland, consisting of an exocrine portion and an endocrine portion. The **exocrine portion** is composed of numerous **acini,** and secretes a variety of digestive enzymes, involved in the digestion of protein, carbohydrates, fats, and nucleic acids. The **endocrine portion** consists of numerous **islets of Langerhans,** which are scattered among the acini of the exocrine portion. The islets of Langerhans produce *glucagon, insulin,* and *somatostatin.*

In addition, a great number of small glands are present in the lamina propria and in the submucosa along the digestive tract. These are responsible for the production of the mucus or digestive enzymes.

Fig. 10-1. — Lip

The **lip** is an organ that has skeletal muscle as its core, covered on the outside by skin and on the inside by an oral mucosa. From the outside to the oral cavity, the structures of *skin, subcutaneous tissue, muscularis, submucosa,* and *mucosa* can be recognized in order.

The **skin** covering the outside of the lip is the same as that of the greater part of the body. It is composed of *epidermis* and *dermis.* The **epidermis** is a keratinized stratified squamous epithelium with a stratum corneum at the surface. The **dermis** consists of a layer of fibroelastic connective tissue. Like other parts of the skin, the skin of the lip is also supplied with **hair follicles, sebaceous glands,** and **sweat glands.** The underlying **subcutaneous tissue (hypodermis)** is a loose connective tissue containing numerous **adipose cells.**

The **muscularis** is a layer of skeletal muscle consisting mainly of **orbicularis oris muscle,** which is seen in cross section in this figure, with a few strands of the compressor labii muscle cut longitudinally.

The **submucosa** is composed of fibroelastic connective tissue containing **labial glands,** the mixed glands whose secretion is released onto the surface of the oral cavity via ducts. In addition, the larger **blood vessels, nerve bundles,** and **adipose cells** can be found in this layer.

The **mucosa** consists of nonkeratinized stratified squamous epithelium (see **Fig. 1-10**) and its **lamina propria,** a fibroelastic connective tissue.

The **red portion** (vermilion border) at the lip margin is a transitional region from skin to oral mucosa. The epidermis appears thicker than that of skin, and is composed of a nonkeratinized, translucent, stratified squamous epithelium. The underlying dermis shows very high papillae with a rich vascular plexus, giving rise to the red color in this region. There are no hairs, sebaceous, or sweat glands in this region, but it is very rich in sensory nerve endings; because of these, the lip is also a very sensitive organ.

Figure 10-1. Lip
Human • H.E. stain • Very low magnification

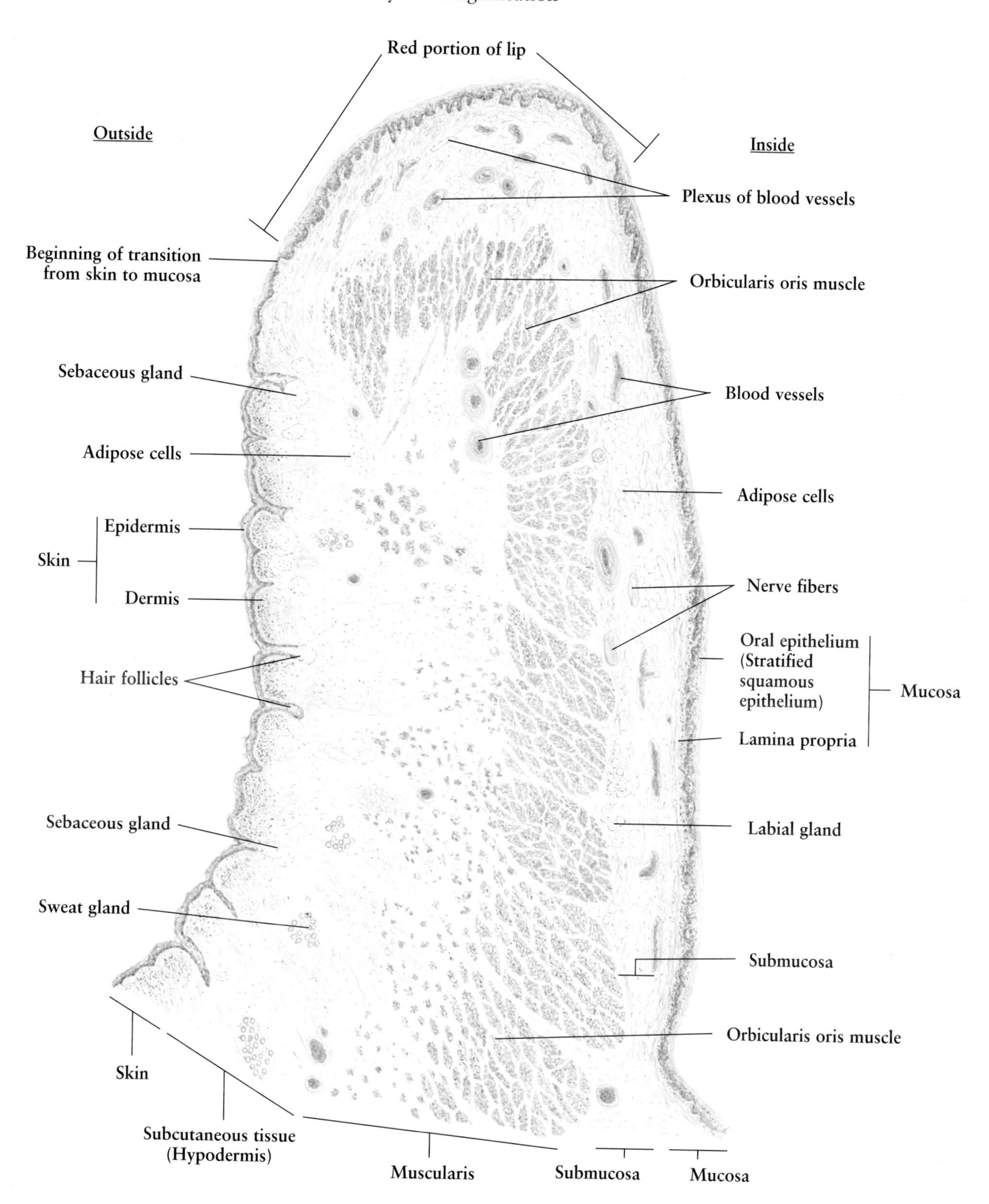

Fig. 10-2. Soft Palate and Uvula

The **soft palate** is continuous with the hard palate, separating the oral cavity from the nasal cavity. Its center is composed of **skeletal muscles** that belong to the musculi levator veli palatini, palatoglossus, and palatopharyngeus. In this illustration, the muscles are cut either longitudinally or transversely. The soft palate is covered by oral mucosa on the oral side, and by nasal mucosa on the nasal side. The oral **mucosa** consists of **stratified squamous epithelium** with a **lamina propria** of fibroelastic connective tissue. Between the lamina propria and muscularis is a loose connective tissue with abundant **mucous glands** and some interposed **adipose cells.** The nasal mucosa is composed of **ciliated pseudostratified columnar epithelium** and a lamina propria of fibroelastic connective tissue with lymphocytic infiltration. The underlying submucosa contains a large number of **mixed glands.**

The **uvula** is a small conical process extending downward from the center of the free edge of the soft palate. It has a core of **skeletal muscle** embedded in **loose connective tissue** and invested peripherally by typical oral mucosa. The loose connective tissue contains **mixed** or **mucous glands** and adipose cells. The lamina propria of the mucosa is often infiltrated by **diffuse lymphatic tissue.**

Figure 10-2. Soft Palate and Uvula
Human • Sagittal section • H.E. stain • Very low magnification

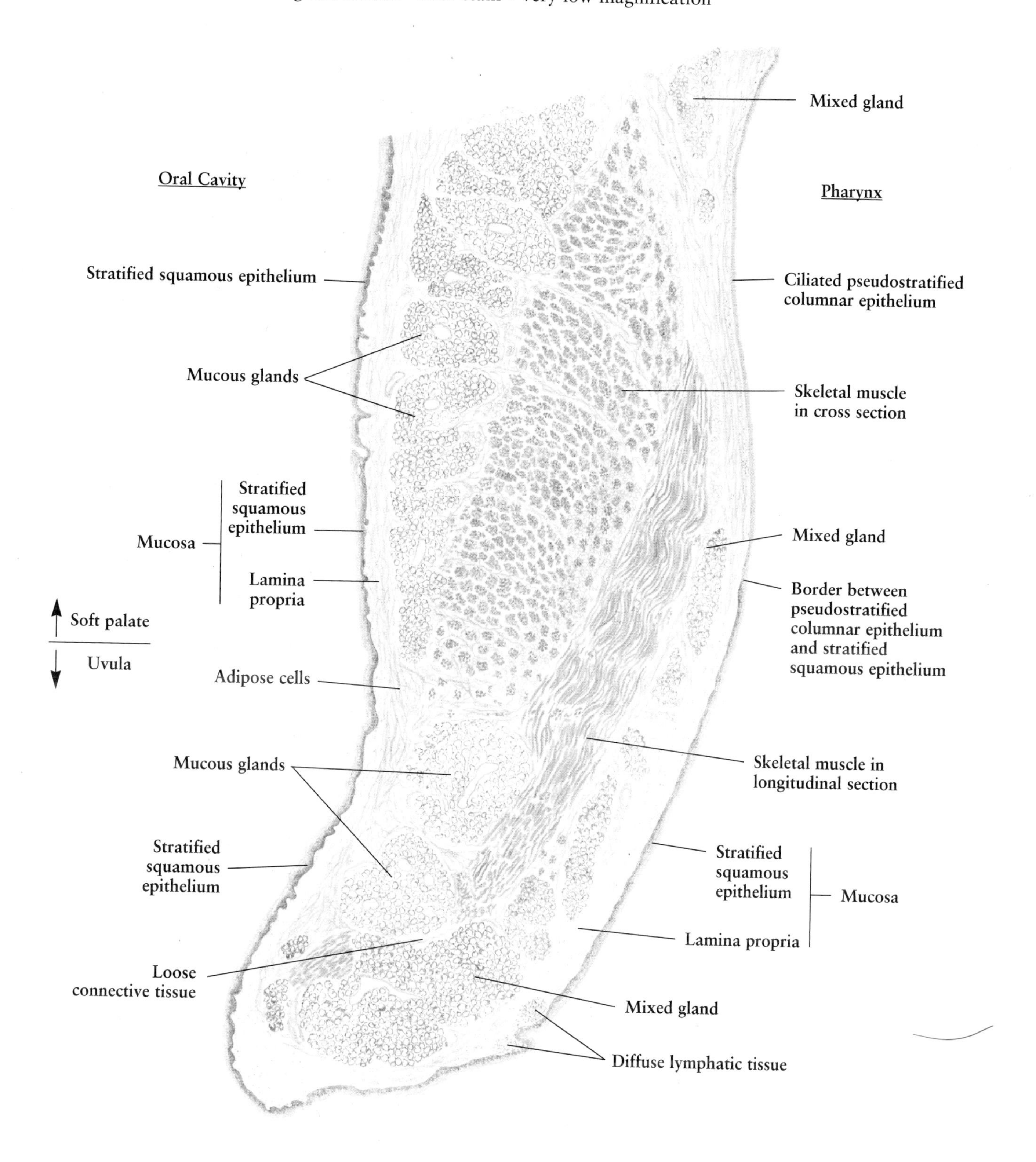

Fig. 10-3. Tooth: Ground Section of Incisor

The **teeth** are hard conical organs with a soft central connective tissue, the *dental pulp*. Each individual tooth can be grossly divided into two parts, the *crown* and the *root;* the *neck* intervenes between the two. The dentin forms the main bulk of an entire tooth, including both crown and root.

The **crown** is completely covered by **enamel,** which is a highly mineralized layer and is the hardest tissue in the body. In this figure of a ground section of a human incisor, the **stripes of Retzius** and **Schreger's lines** may be identified clearly in the enamel.

The **root** is covered peripherally by a thin layer of **cementum,** which is a calcified tissue structurally resembling bone. The next layer is composed of **dentin,** a material also similar to bone, but harder, and void of lacunae or bone cells. In the core of the tooth is a long narrow cavity, the **pulp cavity,** filled with dental pulp, which communicates with the periodontal space through the **root canal** and the **apical foramen,** at the apex of the tooth.

Figure 10-3. Tooth: Ground Section of Incisor
Human • Undecalcified and unstained • Very low magnification

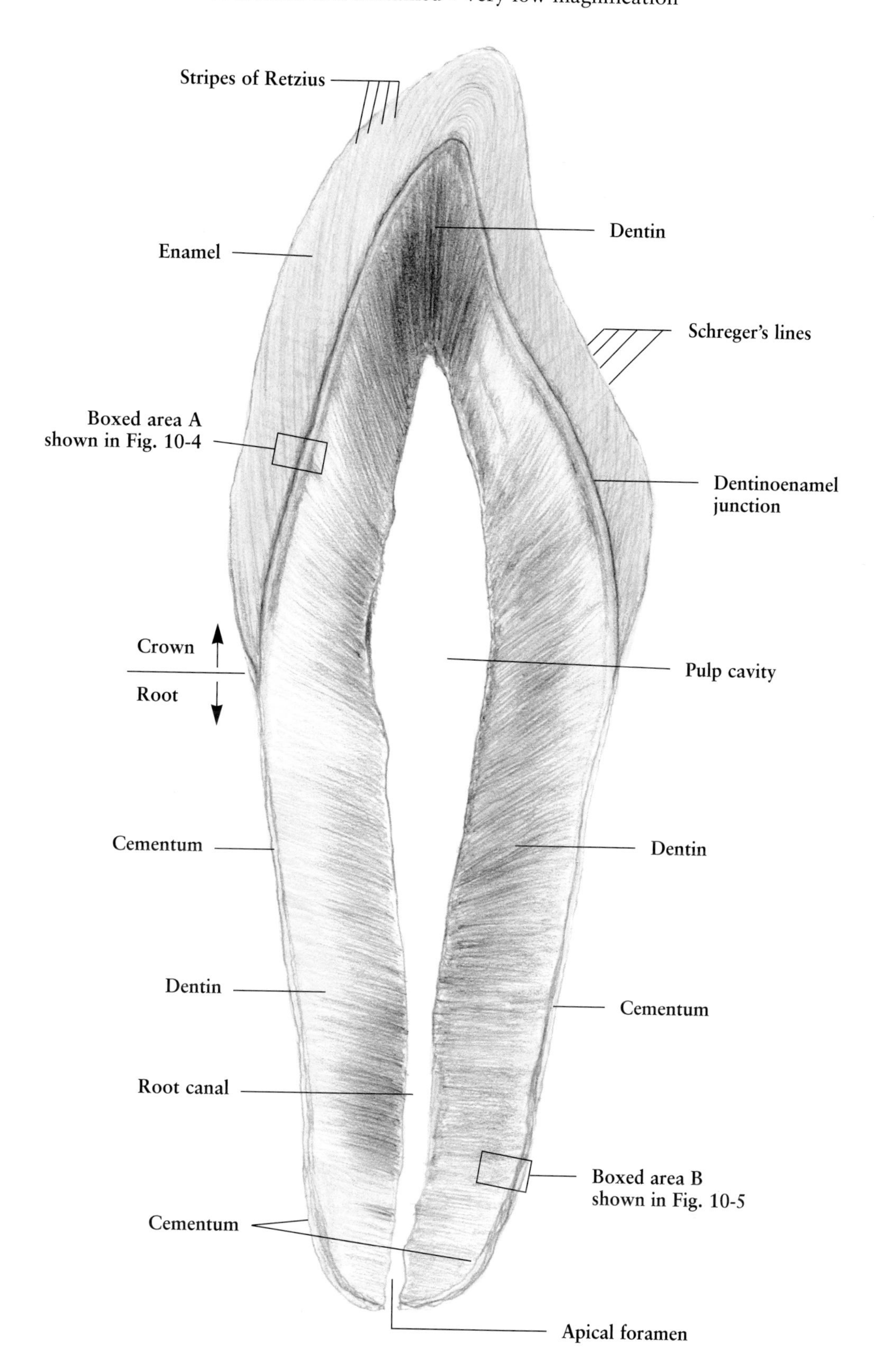

Fig. 10-4. Tooth: Crown of Incisor

This illustration is an enlargement of **boxed area A** in **Fig. 10-3,** showing the details of an **incisor crown,** which consists of dentin and enamel. The border between the enamel and the dentin is the **dentinoenamel junction.**

Enamel is composed of highly calcified rods or prisms with a diameter of 5 μm, separated by a small amount of interprismatic substance. The enamel rods radiate from the dentinoenamel junction toward the free surface of the crown. The enamel rods and the interprismatic substance are represented by the **Schreger's lines** at the level of light microscopy. The **stripes of Retzius** run obliquely from the dentinoenamel junction toward the surface of the enamel and represent successive stages in mineralization of enamel rods. The **enamel tufts** are groups of poorly calcified, twisted enamel rods embedded in an abundant matrix, appearing close to the dentinoenamel junction.

Dentin consists of a calcified component and an organic component that is chiefly collagen fibers. Dentin is traversed by numerous **dentinal tubules** radiating from the pulp to the periphery of the dentin. In the living state, each tubule houses a process of an odontoblast as well as tissue fluid. Near the dentinoenamel junction is a row of irregular **interglobular spaces,** which are incompletely mineralized dentinal areas, containing only organic matrix.

Fig. 10-5. Tooth: Root of Incisor

Boxed area B in **Fig. 10-3** is enlarged in this figure, showing the microscopic structure of dentin and cementum in the root of an incisor. These two layers are separated by the **junction** of dentin and cementum.

Dentin has been described in **Fig. 10-4.** Like the interglobular spaces in the crown, **Tome's granular layer** of dentin beneath the junction of dentin and cementum is formed as a result of incomplete calcification of ground substance.

Cementum is a special, avascular bonelike tissue that covers the layer of dentin of the root and the neck of the tooth. Histologically, it is similar to bone with coarse bundles of collagen fibers in a calcified matrix, but definite Haversian systems are usually absent. The upper third of the cementum is acellular; the lower two-third, however, has cementocytes located in **lacunae,** interconnected by **canaliculi.** The collagen fibers are continuous with Sharpey's fibers penetrating from the periodontal membrane.

Figure 10-4. Tooth: Crown of Incisor
Human • Undecalcified ground section • Unstained • High magnification

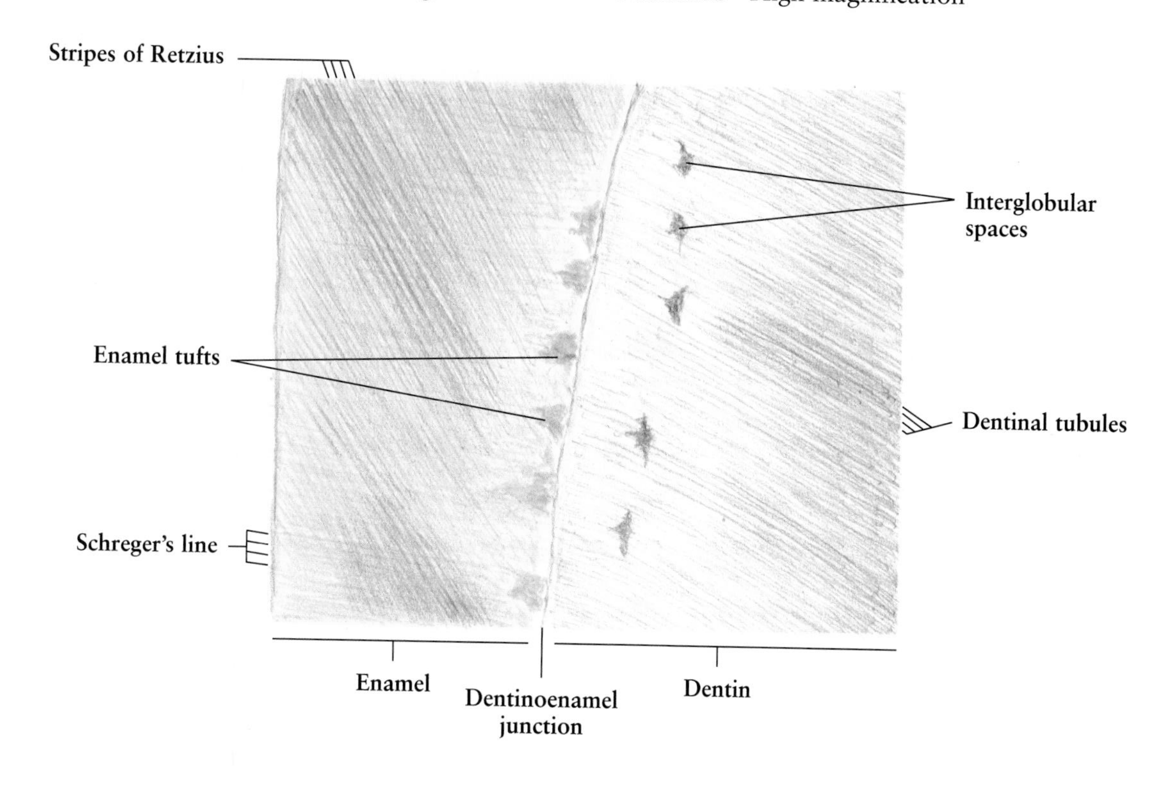

Figure 10-5. Tooth: Root of Incisor
Human • Undecalcified ground section • Unstained • High magnification

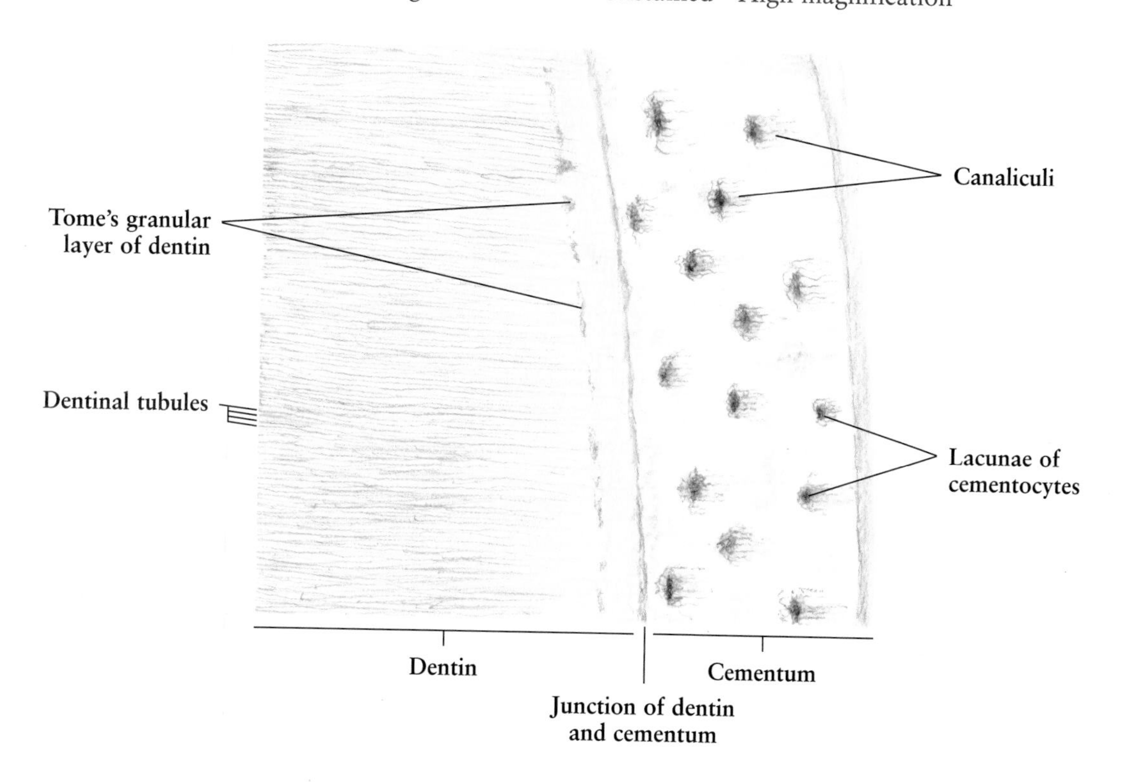

Fig. 10-6. Tooth: Incisor and Surrounding Structures

This drawing is of a decalcified section of a monkey incisor, showing the incisor and its surrounding structures.

Because the *enamel* contains very little organic substance and no collagen fibers, it is dissolved completely by acetic acid during decalcification. **Dentin** and **cementum,** however, remain intact. The **dentinal tubules** and **contour lines** are prominent. The imbricative contour lines are rather wide bands following the contour of the tooth. They are formed as a result of nonhomogeneous calcification during the appositional growth of dentin. Resembling primitive mesenchyme, the **pulp** that fills the pulp cavity is a loose type of connective tissue containing **odontoblasts,** fibroblasts, collagen and elastic fibers, and myelinated and unmyelinated **nerve fibers,** as well as **blood vessels.** The odontoblasts are arranged in a pseudostratified layer that separates the pulp from the dentin layer.

The tooth is suspended in the crypt of the **alveolar bone** or socket. The **periodontal membrane,** a type of special dense connective tissue, binds the cementum of the tooth to the bony wall of the socket (see **Fig. 10-8**), allowing limited movement of the tooth.

The **alveolar bone,** which surrounds the root just peripheral to the periodontal membrane, consists of a thin compact lamina of bone. The periodontal membrane is firmly cemented into this bony tissue as its periosteum. Vessels and nerves run through this alveolar bone and the **apical foramen** of the root to enter the pulp.

The **gingiva** is the modified oral mucous membrane covering the surface of the alveolar bone. It is composed of **stratified squamous epithelium** and its **lamina propria,** a dense connective tissue. The gingiva is connected to the periodontal membrane by the **gingival periodontal fibers,** and is attached to the tooth at the level of the base of the **clinical crown.** Between the enamel and the gingiva is a groove called the **gingival sulcus.**

Figure 10-6. Tooth: Incisor and Surrounding Structures
Monkey • Decalcified • Longitudinal section • H.E. stain • Low magnification

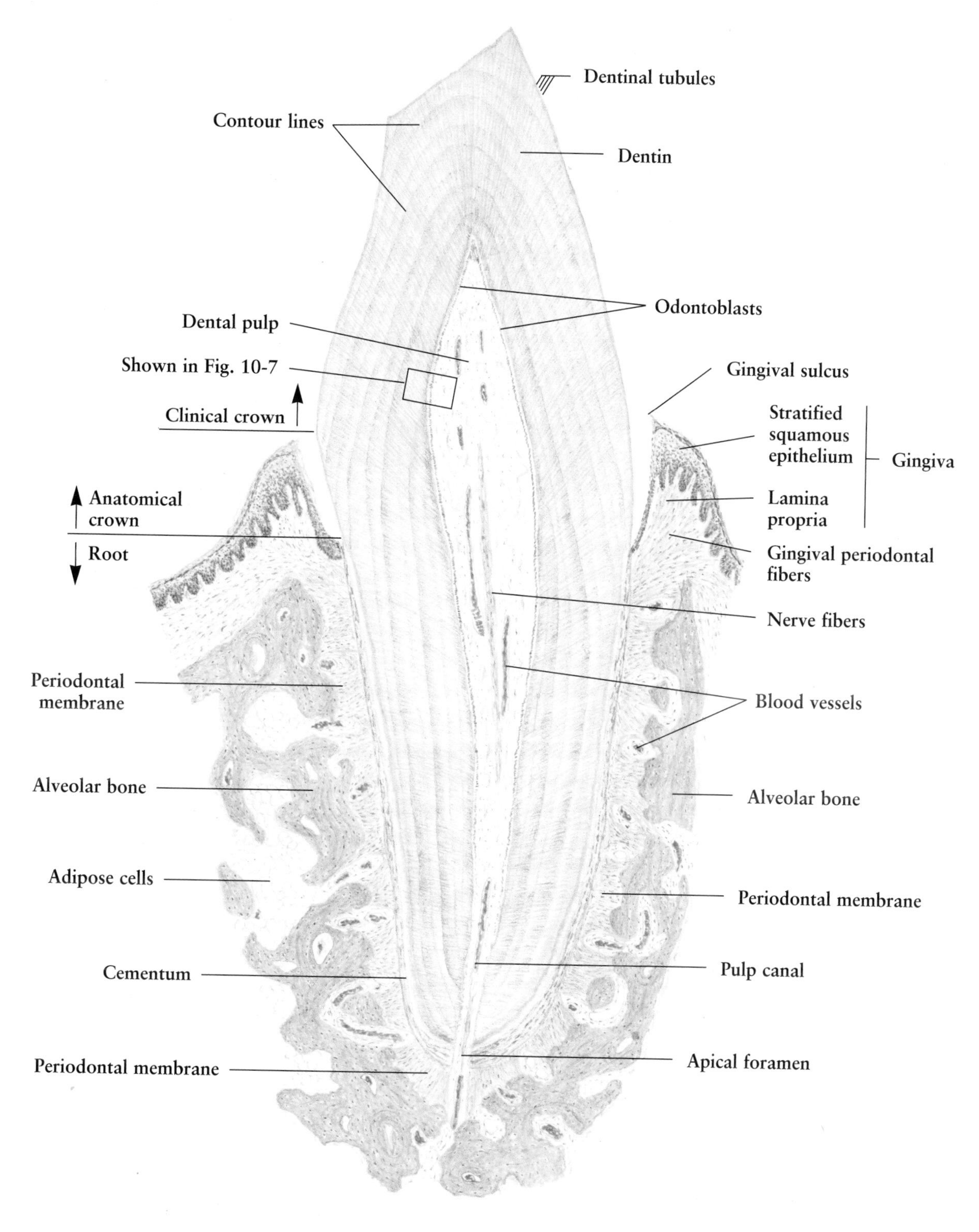

Note: The enamel has been dissolved by decalcification.

Fig. 10-7. Tooth Pulp and Odontoblasts

This illustration is an enlargement of the boxed area in **Fig. 10-6,** giving details of the tooth pulp and the dentin.

The **odontoblasts,** which secrete the components of dentin, are arranged in a pseudostratified layer on the dentinal surface of the pulp. These cells are fusiform in shape and have an ovoid nucleus with a prominent nucleolus, invested by a thin layer of cytoplasm. Each odontoblast has a long cytoplasmic **process** penetrating into a **dentinal tubule** in the dentin. The parallel dentinal tubules, each housing a process of odontoblasts, extend into the **mineralized dentin** through a narrow pale-staining zone of uncalcified dentin matrix known as **predentin.**

Tooth pulp also contains numerous stellate **fibroblasts** and **capillaries,** as well as nerve fibers, collagen, and reticular fibers, which are not visible in this figure.

Fig. 10-8. Periodontal Membrane

The **periodontal membrane** (ligament) is a special type of dense fibrous connective tissue composed of strong, thick bundles of **collagen fibers** embedded in an amorphous intercellular substance. One end of these collagen fibers penetrates the **cementum** of the tooth as Sharpey's fibers. The other end is anchored in the **alveolar bone.** The periodontal membrane acts as a sling suspending the tooth in the socket. It also functions as a periosteum of the alveolar bone and supports the gingiva. Between the thick bundles of collagen fibers there are small areas of **loose connective tissue** containing **blood vessels** and nerve fibers. In addition, the **osteoblasts** and **cementoblasts** are found in a row on the surfaces of the alveolar bone and the cementum, respectively.

The collagen fibers that suspend the tooth run a slightly wavy course between the alveolar bone and the cementum; this allows slight movements of the tooth in various directions, and thus the impact of occlusal forces is dampened. The turnover rate of collagen in the periodontal membrane is high; thus, this tissue is continuously remodeled. This is the mechanism that allows tooth movement during orthodontic treatment.

Figure 10-7. Tooth Pulp and Odontoblasts
Human • H.E. stain • High magnification

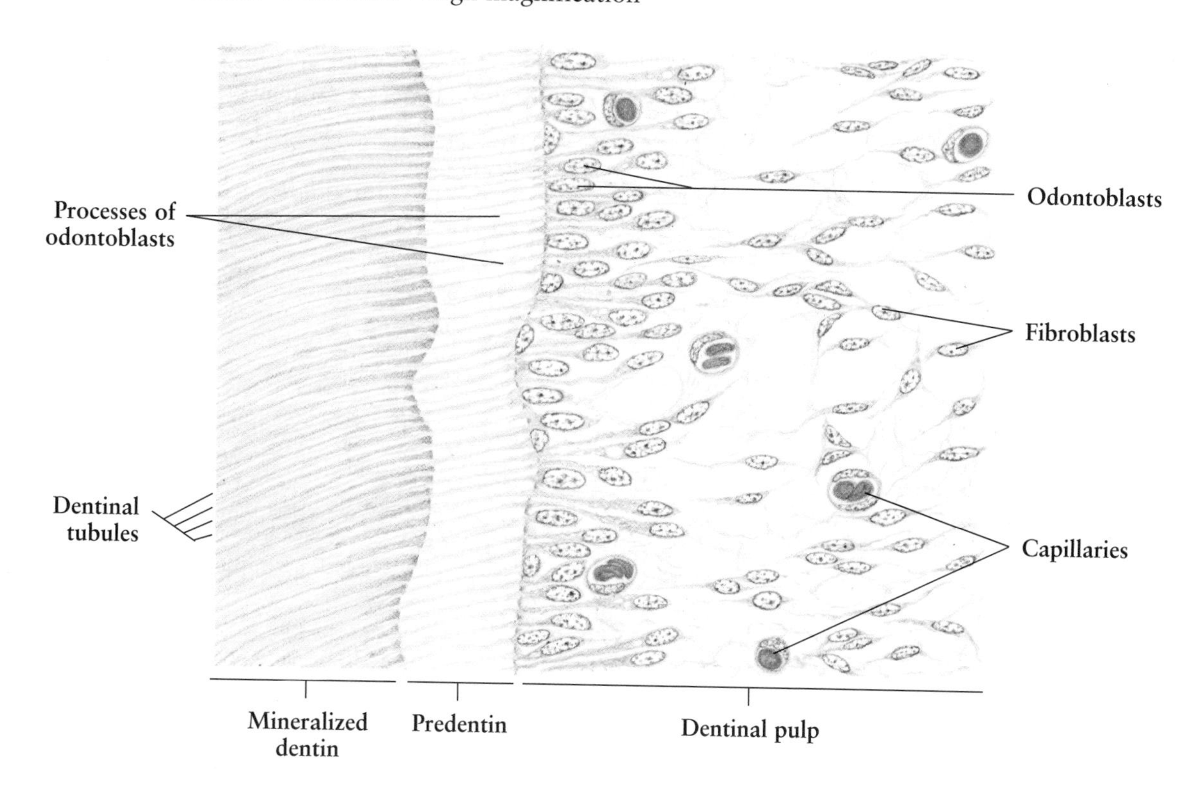

Figure 10-8. Periodontal Membrane
Monkey • H.E. stain • Decalcified • Cross section • Medium magnification

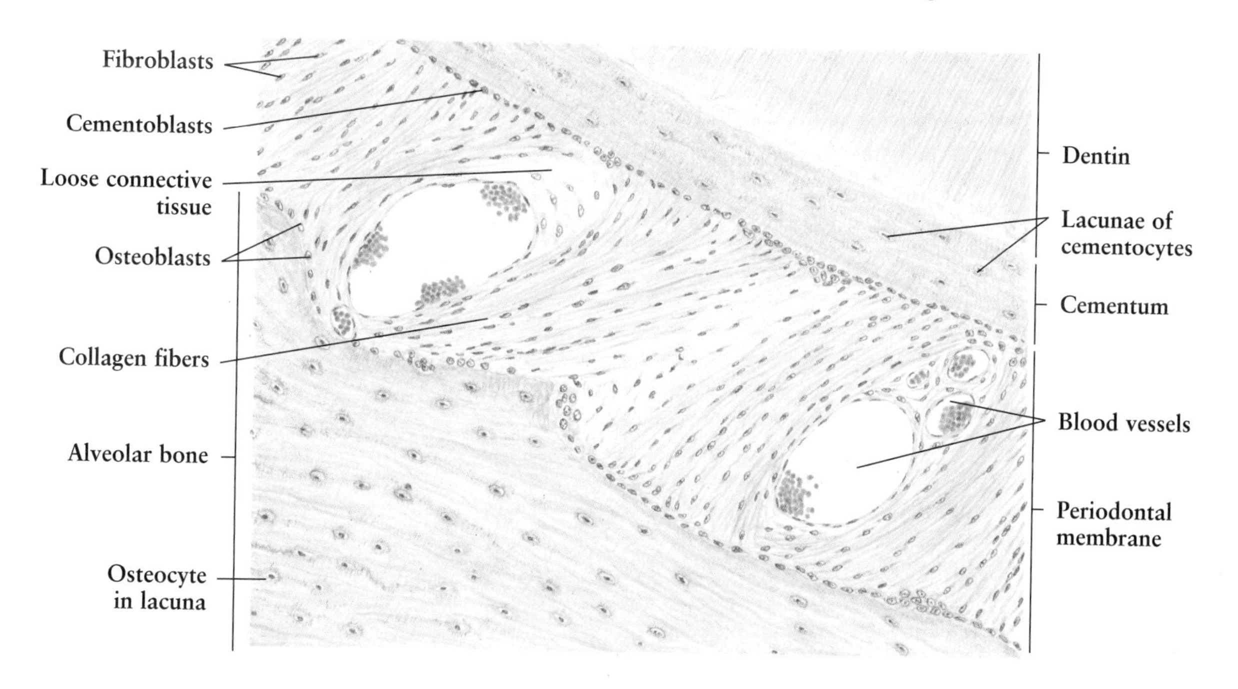

Fig. 10-9.

Tooth Development: Bud Stage

Each tooth has an ectodermal and a mesodermal component, the former giving rise to the enamel and the latter to the dentin, cementum, pulp, and the periodontal membrane. Tooth development, begining with the sixth fetal week, is divided into three main stages: *bud stage, cap stage,* and *bell stage.*

During the **bud stage,** the **oral epithelium** covering the edge of the jaw thickens and dips into the underlying mesoderm to form the **dental lamina,** the primordium of the tooth. As it grows into the **mesenchyme,** the primordium develops into a C-shaped epithelial cord, the **epithelial tooth bud.** Immediately beneath the epithelial downgrowth, the adjacent mesenchyme becomes **condensed** as the primordium of the dental papilla.

Fig. 10-10.

Tooth Development: Cap Stage

During the **cap stage,** as a result of fast proliferation and differentiation of the epithelial cells of the dental lamina, the end portion of the epithelial tooth bud dilates and transforms into a cap-shaped enamel organ composed of three layers of epithelial cells: the **inner enamel epithelium,** the **outer enamel epithelium,** and the intervening **stellate reticulum.**

The condensed mesenchyme beneath the enamel organ transforms into the **dental papilla** and is soon enclosed by the inner enamel epithelium. The mesenchymal cells surrounding the enamel organ and dental papilla become elongated, forming the **dental sac.** This structure envelops the enamel organ and dental papilla.

At the same time, another epithelial cord forms from the dental lamina between the oral epithelium and the enamel organ. This is the **bud of the permanent tooth.**

Figure 10-9. Tooth Development: Bud Stage
Human • H.E. stain • Low magnification

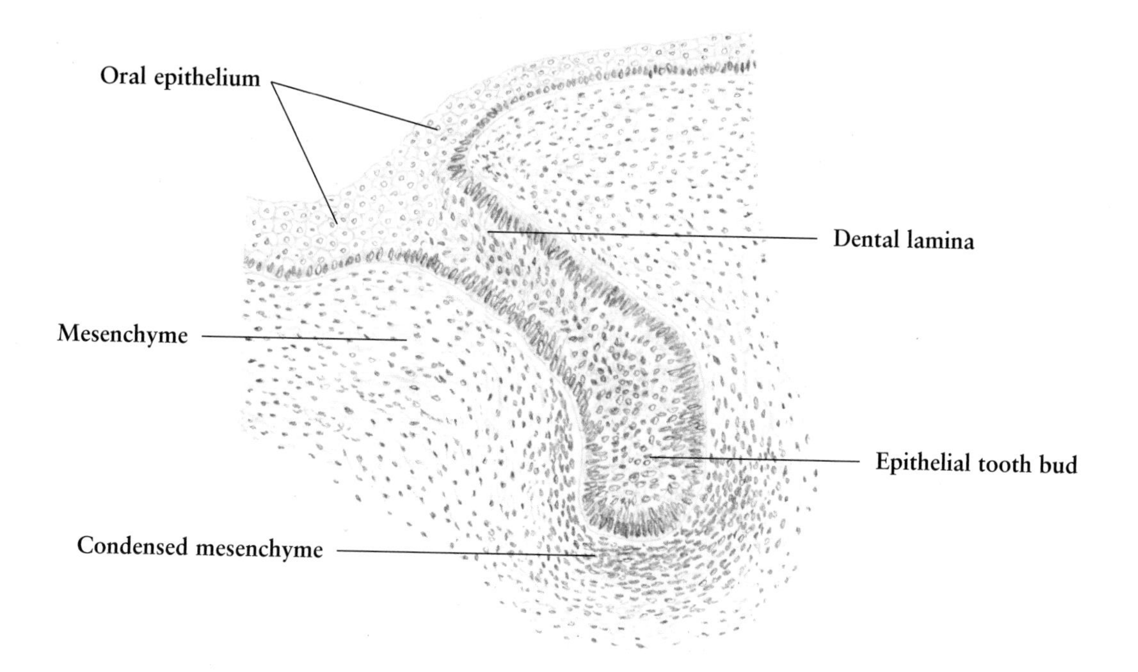

Figure 10-10. Tooth Development: Cap Stage
Human • H.E. stain • Low magnification

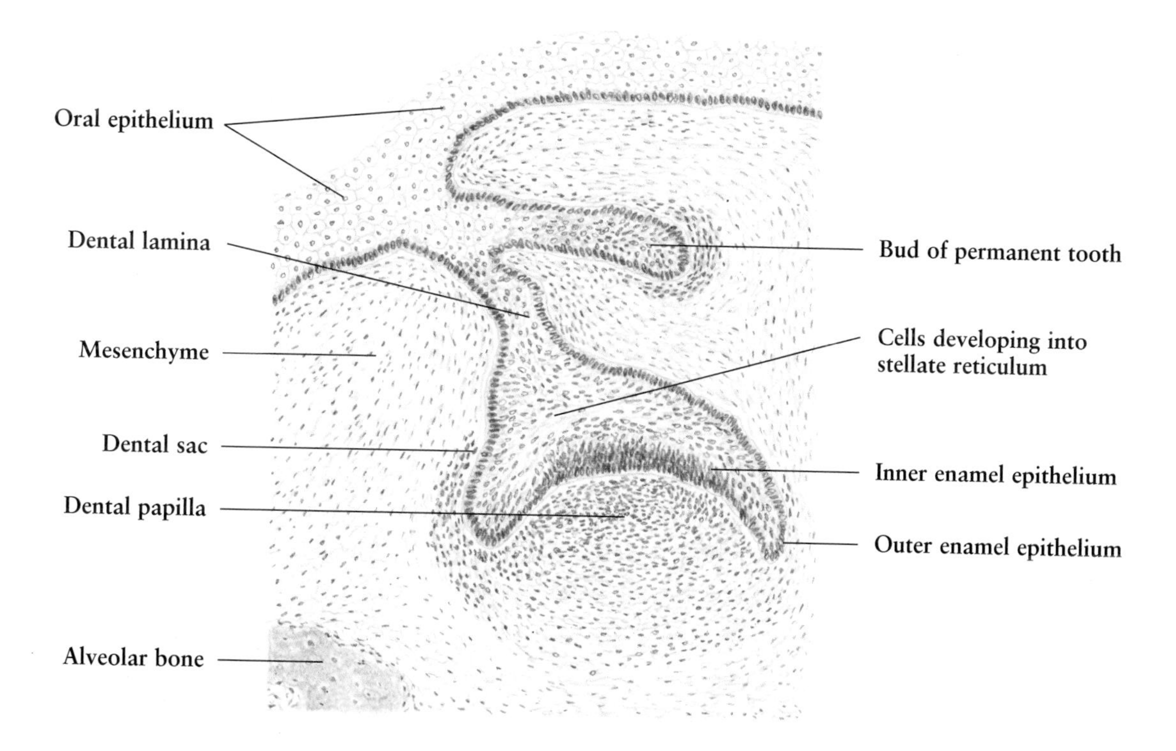

Fig. 10-11.

Tooth Development: Bell Stage

By the **bell stage,** the tooth primordium has a complex structure. The enamel organ becomes bell-shaped in appearance and is characterized by four layers of cells: **ameloblasts** (inner enamel epithelium), **stratum intermedium, stellate reticulum,** and **outer enamel epithelium.**

The **dental pulp** is enveloped nearly completely by **inner enamel epithelium.** The peripheral-most mesodermal cells of the dental pulp differentiate to form **odontoblasts,** which secrete dentin. At the edge of the bell (the future neck of the tooth), the inner and outer enamel epithelia come together to form a fold of epithelial cells that grows toward the root. This is **Hertwig's root sheath.** It is associated with the formation of root dentin by odontoblasts. The root begins to form shortly before the tooth's eruption. When the root dentin is formed completely, Hertwig's root sheath disappears.

The **dental sac** consists of spindle-shaped cells. It forms a special dense connective tissue, the periodontal membrane, from which the cementum develops. By the 16th fetal week of human development, the connection (**dental lamina**) between the tooth bud and the oral epithelium is discontinuous.

Fig. 10-12.

Tooth Development: Enamel and Dentin

This figure is an enlargement of the boxed area in **Fig. 10-11,** demonstrating details of the formation of **enamel** and **dentin,** which commences at about the 20th fetal week.

The **ameloblasts** are differentiated into a single layer of cylindrical epithelial cells. They are separated, by a basement membrane that later breaks down, from the highly polarized **odontoblasts** arranged in a regular manner at the peripheral surface of **dental pulp.** The odontoblasts secrete collagen and matrix as **predentin,** and then calcification transforms it to dentin. The **processes** of the odontoblasts appear to remain in the dentin, housed by the **dentinal tubules.** Once dentin formation has been initiated, the ameloblasts begin to form **enamel** on the dentinal surface. The ameloblasts secrete the organic matrix, which is rich in protein, mineral salts, and sulfur. This matrix becomes mineralized immediately to form enamel prisms made up of short, needle-like crystals.

The **stratum intermedium** consists of one or two layers of fusiform cells, adjacent to the ameloblast layer. The **stellate reticulum** is composed of cells with ovoid nuclei and several processes, which form a network filling the enamel pulp. These cells secrete a soft gel-like matrix that protects the critical shape of the inner enamel epithelium.

Figure 10-11. Tooth Development: Bell Stage
Human • H.E. stain • Low magnification

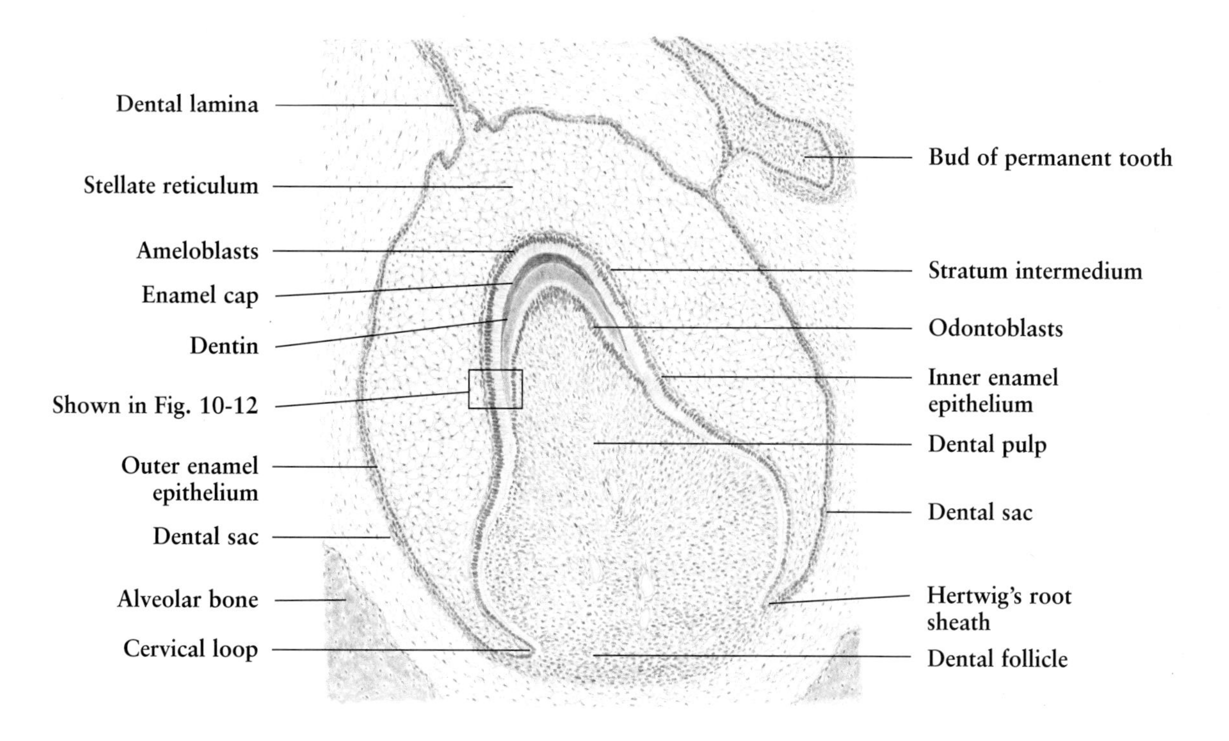

Figure 10-12. Tooth Development: Enamel and Dentin
Human • H.E. stain • High magnification

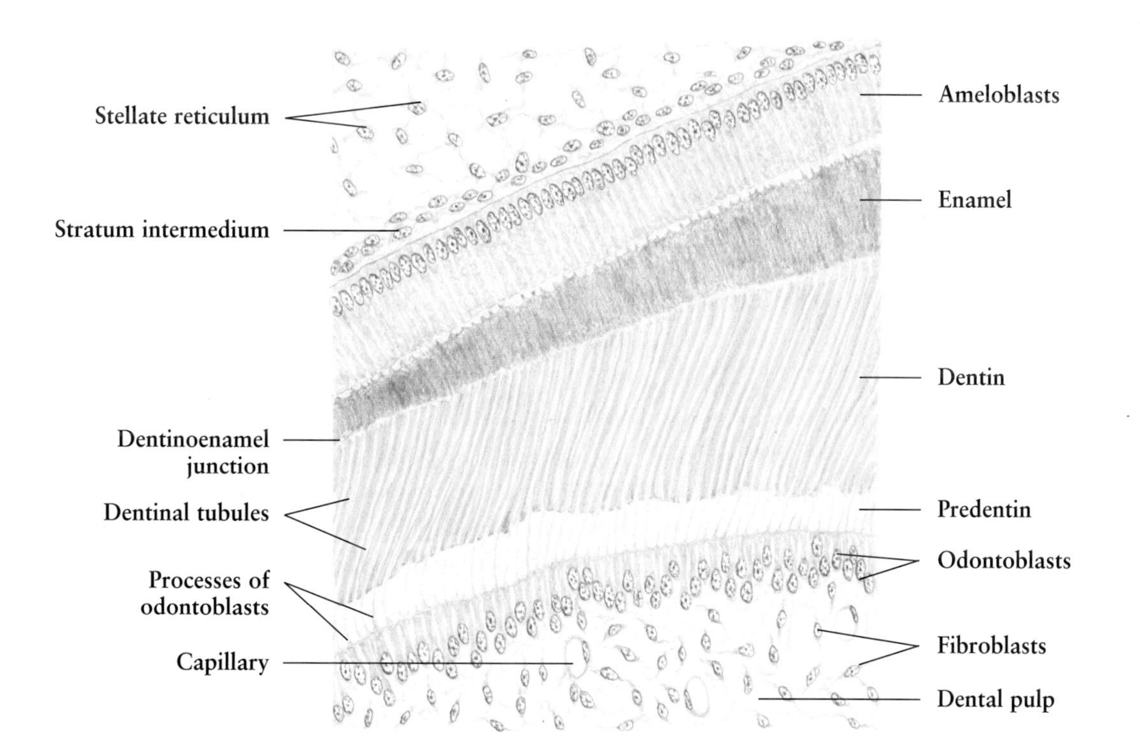

Figs. 10-13 and 10-14.

Tongue

The **tongue** is a freely movable muscular organ covered by a mucous membrane, parts of which are modified to conform to its function as an organ of mastication and taste.

The bulk of the tongue consists of striated muscle fibers and glands. The muscle fibers are both intrinsic and extrinsic. Most are confined to the tongue, but others originate outside, such as the **genioglossus muscle.** The muscles are separated into left and right pairs by the **lingual septum.** These muscle fibers run in three main directions: *longitudinal, transverse,* and *sagittal.* They are arranged in interlacing groups and embedded in loose connective tissue and adipose tissue.

Glands are located between the muscle fibers, scattered throughout the tongue. *Serous glands* are found in the body of the tongue, their ducts opening anterior to the sulcus terminalis; *mucous glands* in the base of the tongue, their ducts opening behind the molars, and *mixed glands* near the tip, their ducts opening on the inferior surface of the tongue.

The *mucosa* is composed of thick **stratified squamous epithelium** and its underlying **lamina propria,** which contains many blood vessels, lymphatics, and nerve fibers. The ventral portion of the mucosa is smooth; its epithelium is nonkeratinized. The dorsal mucosa, however, is provided with numerous variably shaped lingual **papillae** including *filiform, fungiform, foliate,* and *circumvallate.* They are covered by a stratified squamous epithelium that is keratinized only in relation to the filiform papillae. The presence of lingual papillae give the tongue a furred or roughened appearance.

Figure 10-13 is a frontal section through the body of the tongue from a monkey; it demonstrates the alignment of muscle fibers and the differences between dorsal and ventral mucosa. Note the **aponeurosis linguae** in the deep zone of the **lamina propria** of the dorsal mucosa. It serves as an insertion for lingual muscle fibers. This figure also shows the **lingual arteries, veins,** and the bundles of **nerve fibers.**

In **Fig. 10-14,** a longitudinal section through the tip of the tongue, the **fungiform** and **filiform papillae** can be identified on the dorsal surface. The ventral mucosa appears smooth. **Mixed glands** are seen at the bottom of this figure.

Figure 10-13. Tongue: Body

Monkey • Frontal section • H.E. stain • Low magnification

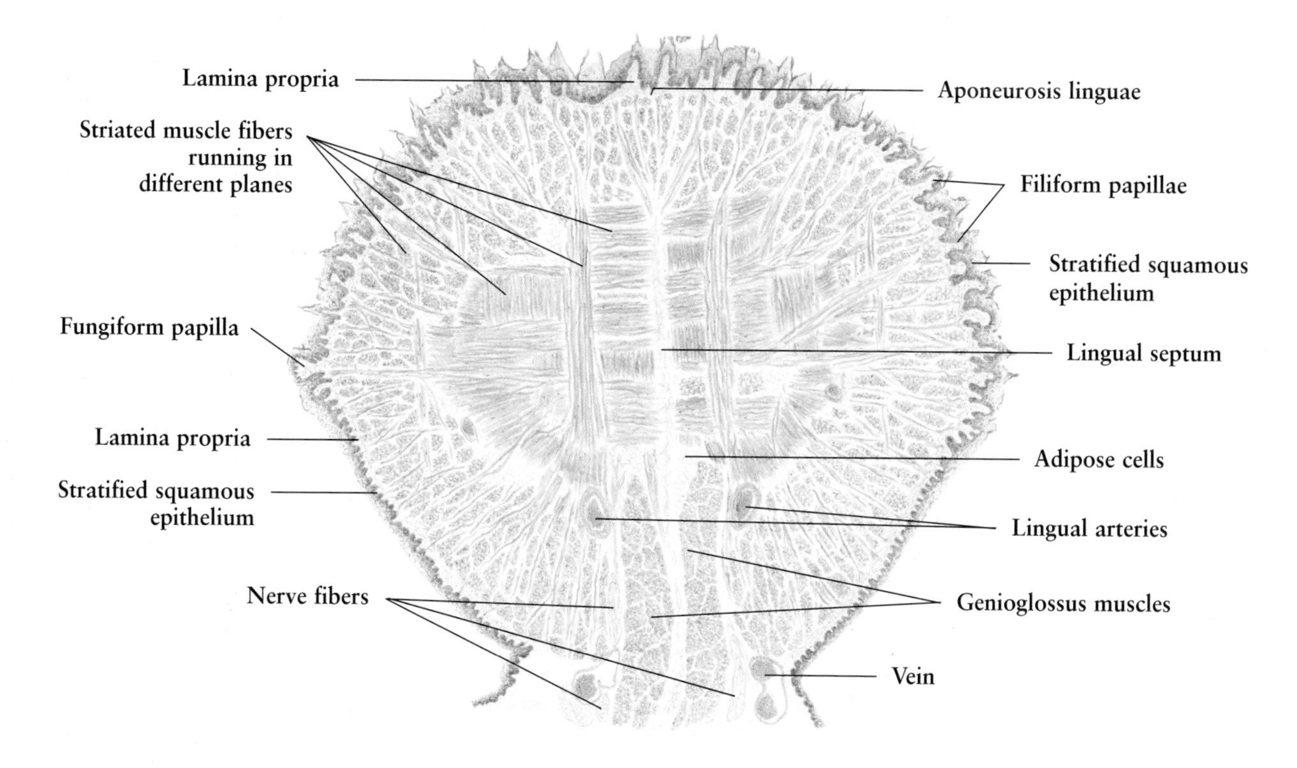

Figure 10-14. Tongue: Apex

Human • Longitudinal section • H.E. stain • Low magnification

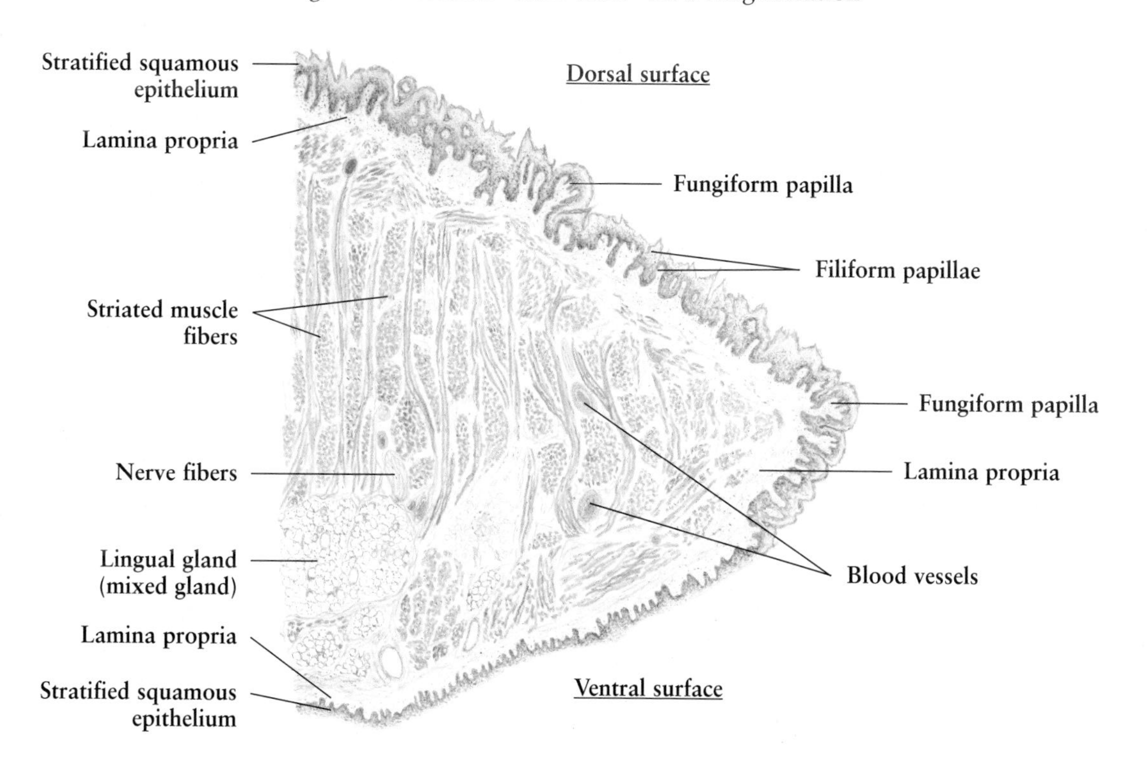

Fig. 10-15. —

Tongue: Filiform and Fungiform Papillae

Filiform papillae are the most numerous type. They are distributed over the entire tongue surface, mainly in rows parallel to the V-shaped sulcus. Each papilla has a primary core of dense connective tissue of the **lamina propria** with **secondary papillae.** The covering epithelium is the heavily **keratinized** and quite hard stratified squamous epithelium, with a pointed **tip** oriented toward the pharynx.

Fungiform papillae are mushroom-(fungus-) shaped. They are larger than the filiform papillae, fewer in number and scattered irregularly over the tongue surface. Fungiform papillae have a thin, nonkeratinized **stratified squamous epithelium** and a richly vascularized (blood vessels) connective tissue of the **lamina propria** with **secondary papillae,** giving them a red appearance. The epithelium of fungiform papillae may contain occasional taste buds.

Fig. 10-16. —

Tongue: Foliate Papilla

Foliate papillae are leaflike, parallel folds of the mucosa on the posterolateral margins of the tongue. The connective tissue of the **lamina propria** forms three **secondary papillae** with **capillaries** and nerve fibers. The papillae are covered by nonkeratinized **stratified squamous epithelium** with numerous **taste buds** (for details, see **Fig. 6-30**) in the **furrows** between papillae. Small **serous glands of von Ebner** open into the bottom of the furrows.

Foliate papillae occur only in infants and some animals such as the rabbit; they are rudimentary in adult humans.

Figure 10-15. Tongue: Filiform and Fungiform Papillae
Human • H.E. stain • Low magnification

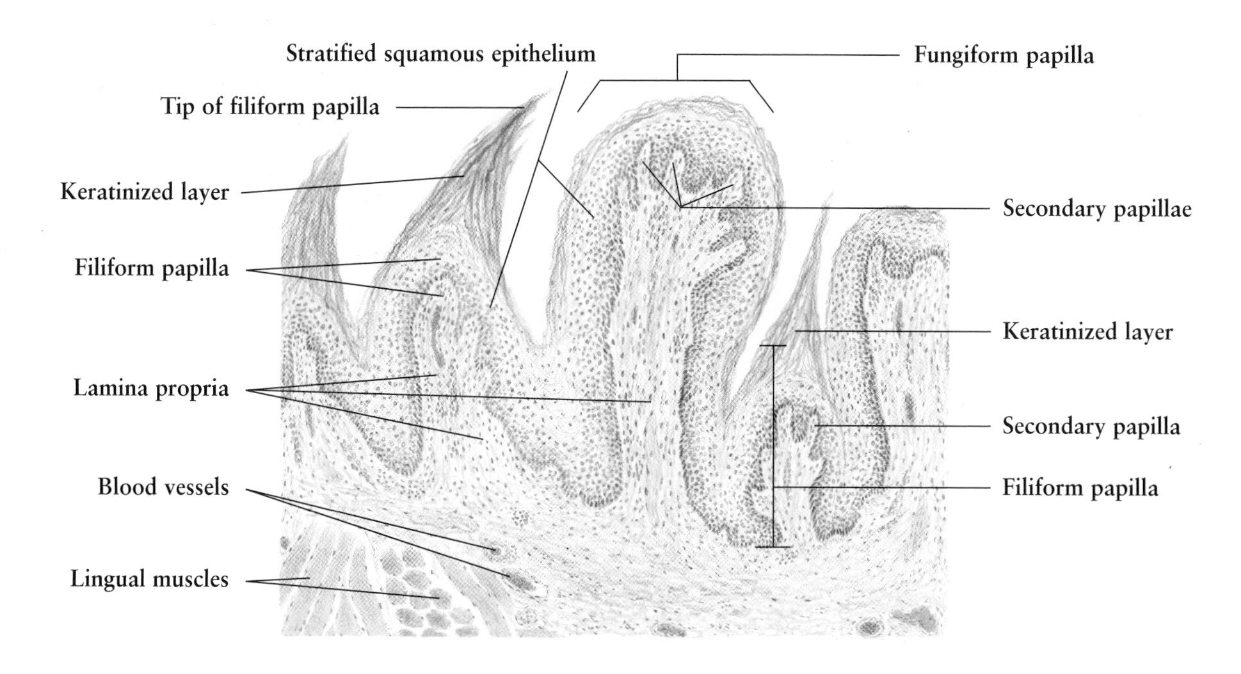

Figure 10-16. Tongue: Foliate Papilla
Rabbit • H.E. stain • Low magnification

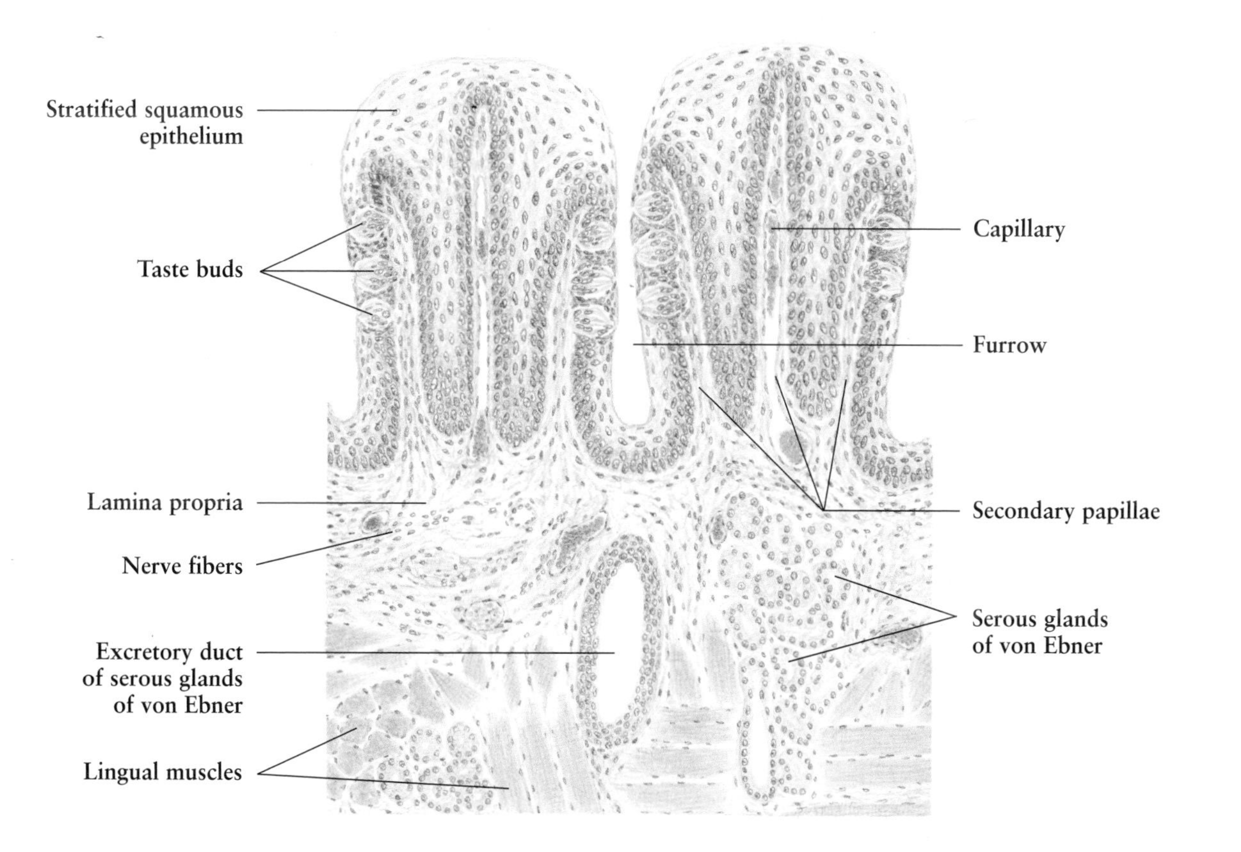

Fig. 10-17.

Tongue: Circumvallate Papilla

Circumvallate papillae are the largest type of tongue papillae, 1–3 mm in width and 1–3 mm in height. The 10–12 circumvallate papillae of a human tongue are located along the V-shaped sulcus terminalis. Each papilla protrudes slightly from the surface and is encircled by a deep moatlike, **circular furrow** with numerous **taste buds** (for details, see **Fig. 6-30**) in the epithelium of the lateral wall. Like other types of papillae, the connective tissue of the **lamina propria** contains **blood vessels** and **nerve fibers,** and forms numerous **secondary papillae.** The **stratified squamous epithelium** covering circumvallate papillae is of a nonkeratinized type.

The **serous glands of von Ebner** located under the papilla open into the bottom of the furrow. These glands secrete a watery fluid that washes away food constituents from the furrow, allowing reception of new gustatory stimuli by the taste buds.

Fig. 10-18.

Tongue: Lingual Tonsil

The **lingual tonsil** is a lymphatic organ, located at the root of the tongue behind the circumvallate papillae. It is composed of an aggregation of wide-mouthed epithelial **crypts,** each surrounded by **lymphatic tissue** and separated from one another by a **connective tissue capsule.** Each simple crypt is lined by a continuation of the nonkeratinized surface **stratified squamous epithelium,** but infiltrated by numerous lymphocytes from the underlying lymphatic tissue, which consists of a layer of lymphatic nodules with germinal centers. **Excretory ducts** of underlying **mucous lingual glands** open into the crypts or onto the tongue surface.

Unlike lymph nodes, the lingual tonsil has no lymphatic sinuses. Hence, lymph is not filtered through it. The only established function of the lingual tonsil is formation of lymphocytes, which migrate through the epithelium of the crypt wall into the oral cavity.

Figure 10-17. Tongue: Circumvallate Papilla
Human • H.E. stain • Low magnification

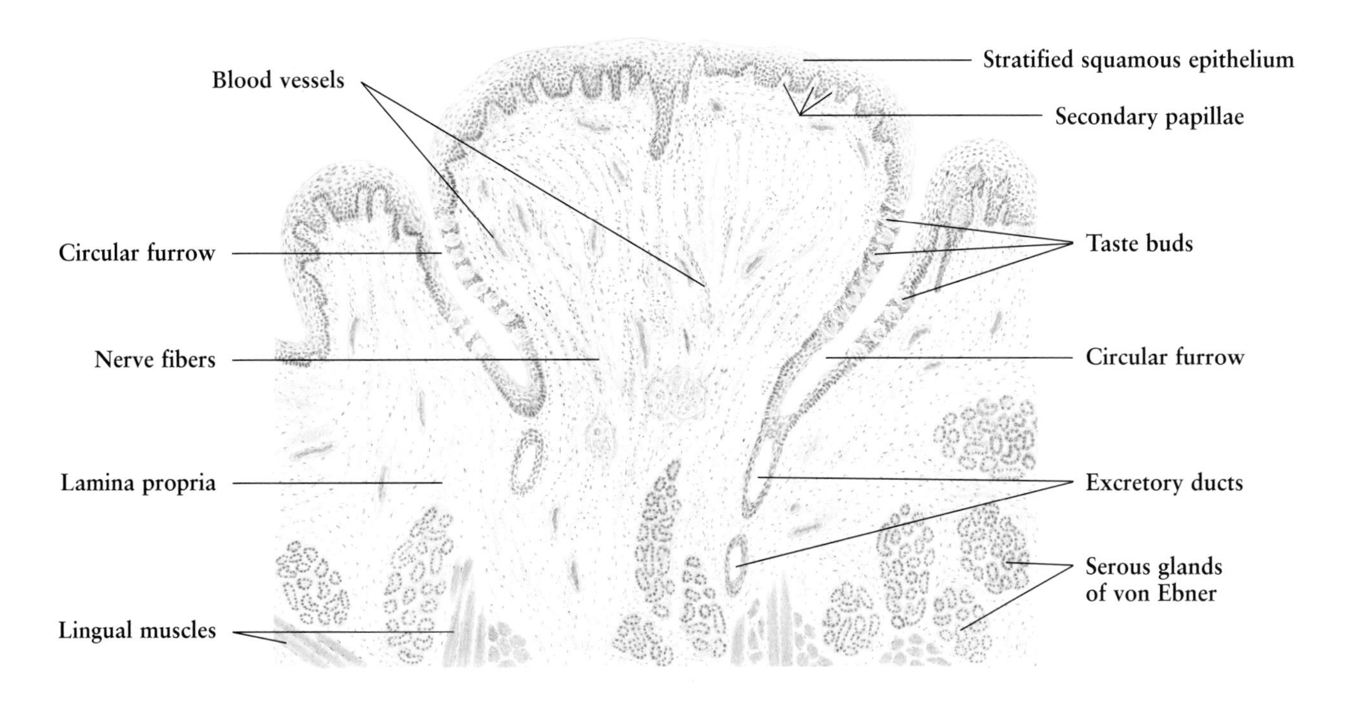

Figure 10-18. Tongue: Lingual Tonsil
Human • H.E. stain • Low magnification

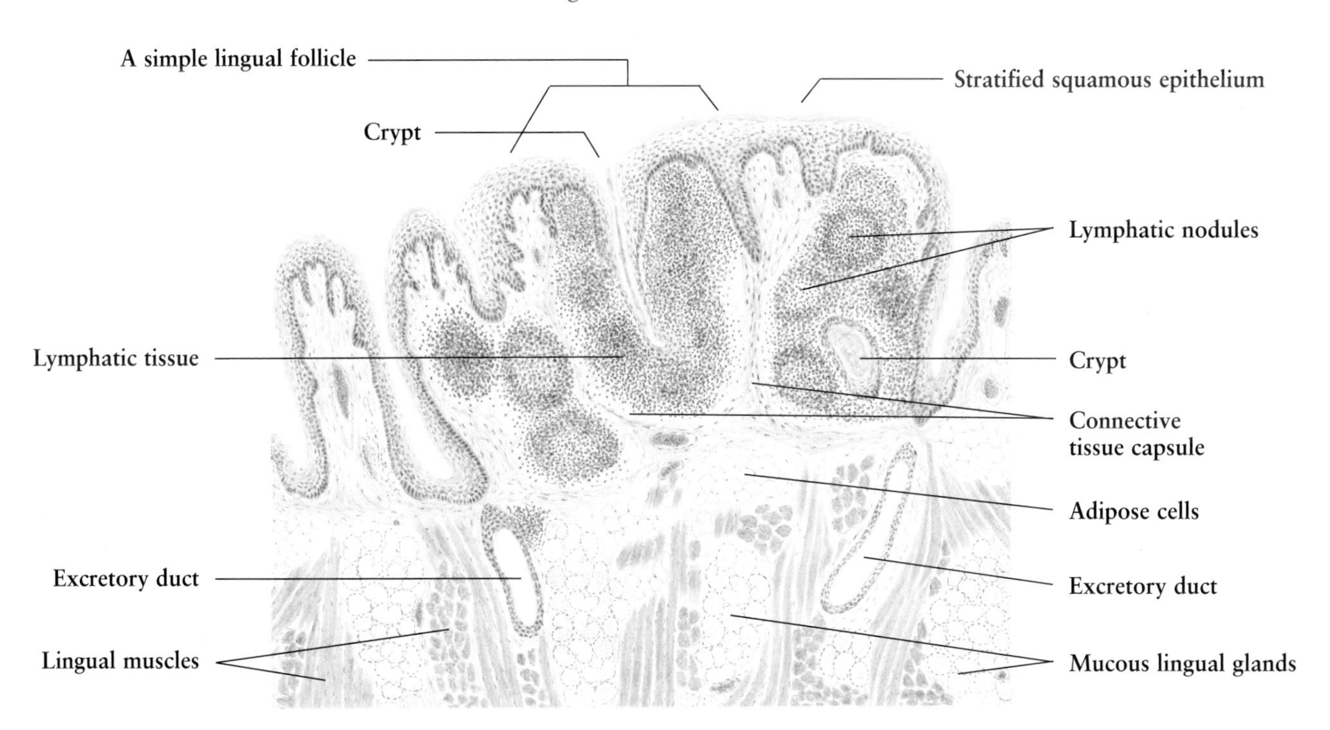

Fig. 10-19.

Parotid Gland

The *salivary glands* include many small glands and three paired major glands. The small glands are scattered in the lip, cheek, palate, and tongue, and their secretion forms a portion of the saliva. The three paired major glands are *parotid, submandibular,* and *sublingual,* which together secrete a large portion of the saliva. *Saliva,* amounting to 1000–1500 ml/day, is a hypotonic, watery mixture of secretions from the small and major salivary glands; it contains variable amounts of inorganic ions, mucins, enzymes (mainly amylase, maltase, and lysozyme), antibodies, desquamated epithelial cells, and degenerating granulocytes and lymphocytes (salivary corpuscles). Saliva has lubricative, digestive, cleaning, and immunological functions.

The **parotid glands** are the largest pair of salivary glands, located inferior and anterior to the ears. They are compound tubuloacinar, purely serous glands. The connective tissue from the fibrous capsule penetrates the parenchyma and separates it into *lobules.* Their serous secretion is delivered from the *acini,* through (in sequence) the *intercalated duct, striated duct (intralobular duct), interlobular duct,* and the *main (Stensen's) duct* to the oral cavity at the inner surface of the cheeks opposite the second upper molar teeth. The main duct is lined by a stratified columnar or stratified squamous epithelium.

Figure 10-19 shows the details of a lobule. The **acini** are composed of pyramidal-shaped acinar cells with basally located round nuclei, basophilic cytoplasm, and apical secretory droplets. They are enclosed in a basement membrane with myoepithelial cells. The long **intercalated duct** is lined by a layer of flattened or low cuboidal epithelial cells with **myoepithelial cells.** The **striated duct** consists of simple columnar epithelial cells with basal striations, which are identified under the electron microscope as deep infoldings of the basal plasma membrane and vertically arranged mitochondria. The **interlobular duct,** surrounded by connective tissue, is lined by columnar, then pseudostratified epithelium with occasional goblet cells. As the diameter of the duct increases, the lining epithelium becomes stratified columnar or cuboidal type, and finally stratified squamous epithelium, continuous with the lining epithelium of the oral cavity.

Among the acini are scattered **adipose cells.** Note that an individual adipose cell is larger in size than a single acinus. The **interlobular connective tissue** contains blood vessels, **nerve fibers,** and interlobular excretory ducts, in addition to collagen fibers and fibroblasts.

Fig. 10-20.

Submandibular Gland

The **submandibular glands** lie bilaterally along the body of the mandible; their excretory ducts open at the floor of the mouth on either side of the frenulum of the tongue. They are compound tubuloacinar glands of mixed type. **Serous acini** usually predominate; some **mucous acini** are also present, but most are capped with **serous demilunes,** forming the **mixed acini. Adipose cells** are scattered among the acini.

The duct system consists of *intercalated ducts, striated ducts, interlobular ducts,* and *excretory (Wharton's) ducts.* In comparison with parotid glands, the **intercalated duct** in a submandibular gland is shorter. The **striated duct,** however, is much longer, indicating a more active transport of various ions both out of and into the secretion passing along the duct.

Like the parotids, the submandibular glands possess a strong fibrous capsule whose connective tissue penetrates the mesenchyma and separates it into lobules. The **interlobular duct** is invested by **interlobular connective tissue,** which contains blood vessels and **nerve fibers.**

Figure 10-19. Parotid Gland
Human • H.E. stain • High magnification

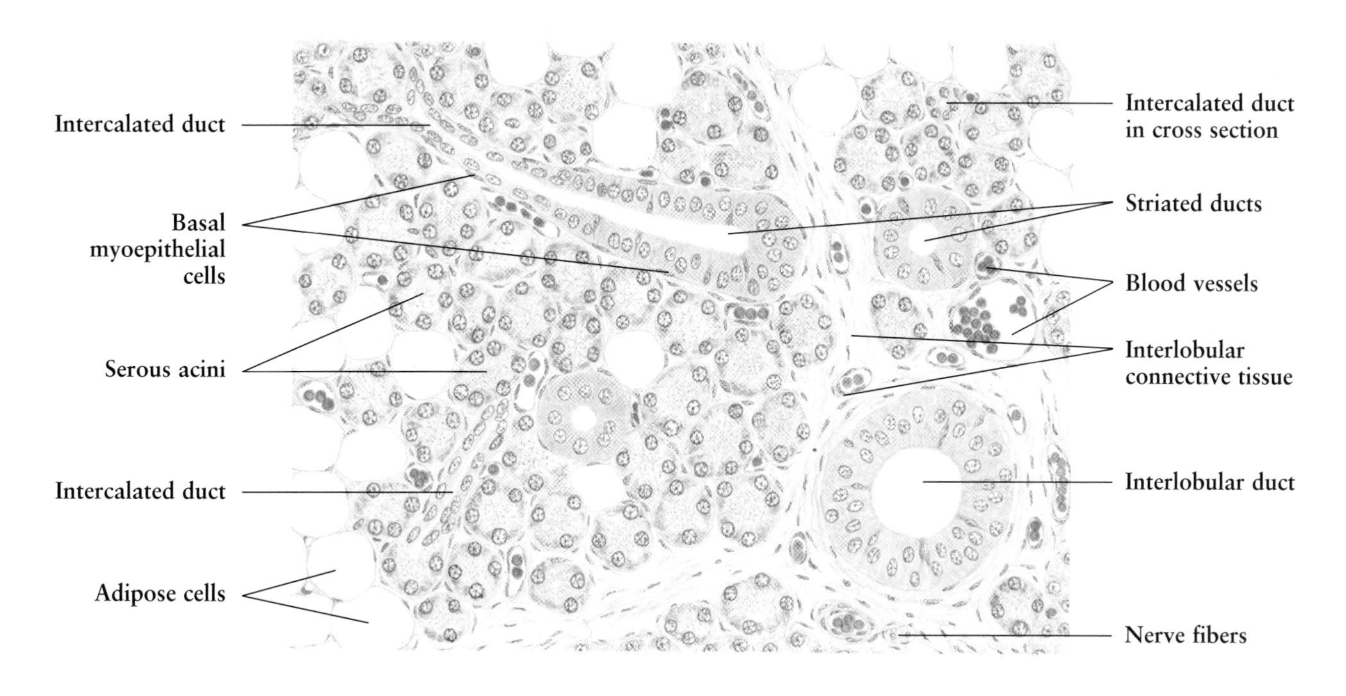

Figure 10-20. Submandibular Gland
Human • H.E. stain • High magnification

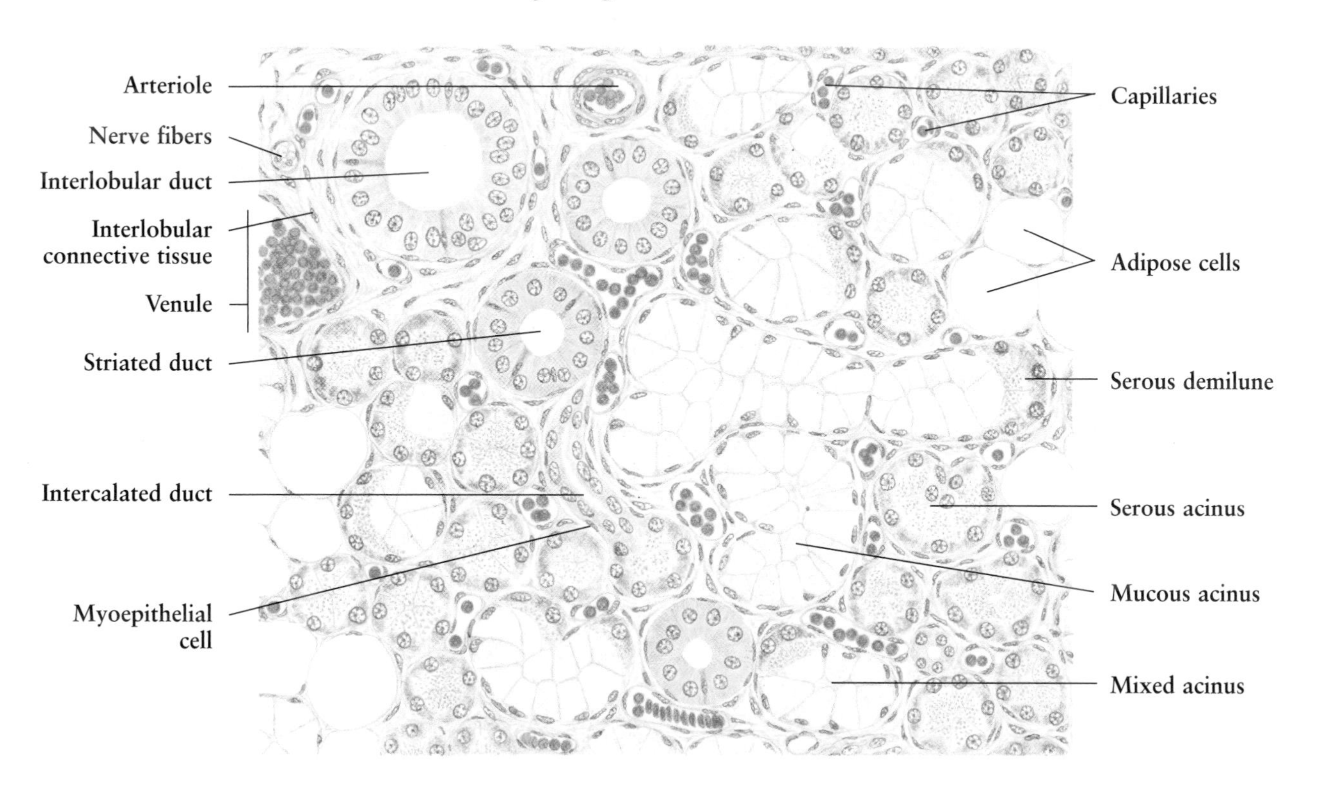

Fig. 10-21. Sublingual Gland

The **sublingual glands** are compound tubuloacinar glands of mixed type. They are formed in aggregates that lie beneath the mucous membrane of the floor of the mouth. Each group of glands has a separate duct that opens into the mouth. There are some 15 minor ducts emptying along the sublingual plica, and 1 major duct opening near the orifices of the submandibular ducts. The capsule of a sublingual gland is thinner and has fewer septa.

The majority of the acini are **mucous,** some with **serous demilunes (mixed acini).** Purely **serous acini** are infrequently present. **Myoepithelial cells** are conspicuous, associated with secretory units. The duct system of the sublingual glands differs from those of the parotid and submandibular glands in that the intercalated ducts are usually absent and the **striated ducts** are short. The **intralobular ducts** are lined by simple cuboidal or columnar epithelium; this is replaced by pseudostratified epithelium in the **interlobular ducts.** The major duct is structurally similar to that of the parotid gland.

Fig. 10-22. Esophagus

The **esophagus** is a tubular, muscular organ, roughly 25 cm long. Its function is to convey food from the oropharynx to the stomach. Similar to most components of the digestive tract, its wall consists of four basic layers: *mucosa, submucosa, muscularis externa,* and *adventitia.*

The **mucosa** is in turn composed of three layers: an epithelial lining formed by **stratified squamous epithelium,** a supporting connective tissue, the **lamina propria,** and a thin smooth muscle layer, the **muscularis mucosae.** The **submucosa** is a layer of loose connective tissue containing larger **blood vessels, esophageal glands,** and nerves. The **muscularis externa** is thick and comprises two layers of muscle fibers: **inner circular muscle layer** and **outer longitudinal muscle layer.** Because the first part of swallowing is under voluntary control, the musculature in the upper third is skeletal. In the inferior third, the musculature is smooth, and in the middle segment, it contains both skeletal and smooth muscle fibers. The **adventitia** is a fibroelastic connective tissue that binds the esophagus to adjacent structures. The short segment of the esophagus below the diaphragm is invested by a serosa.

Figure 10-21. Sublingual Gland
Human • H.E. stain • High magnification

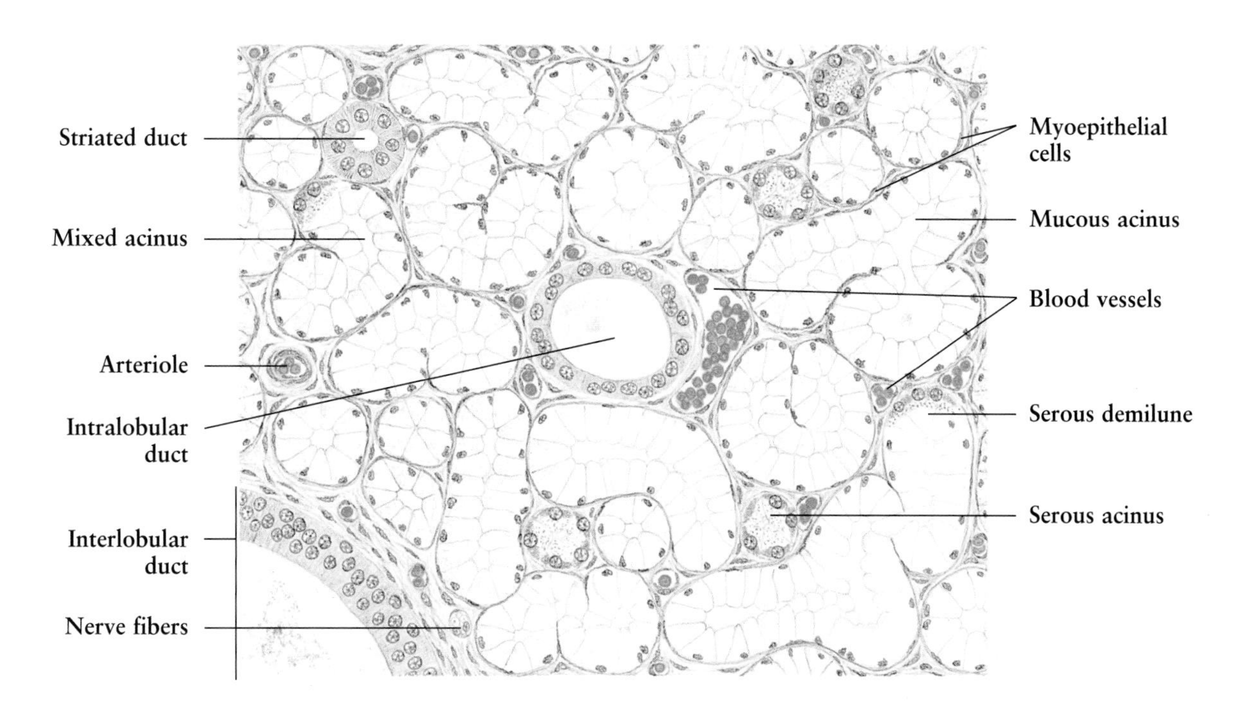

Figure 10-22. Esophagus
Human • Cross section • H.E. stain • Very low magnification

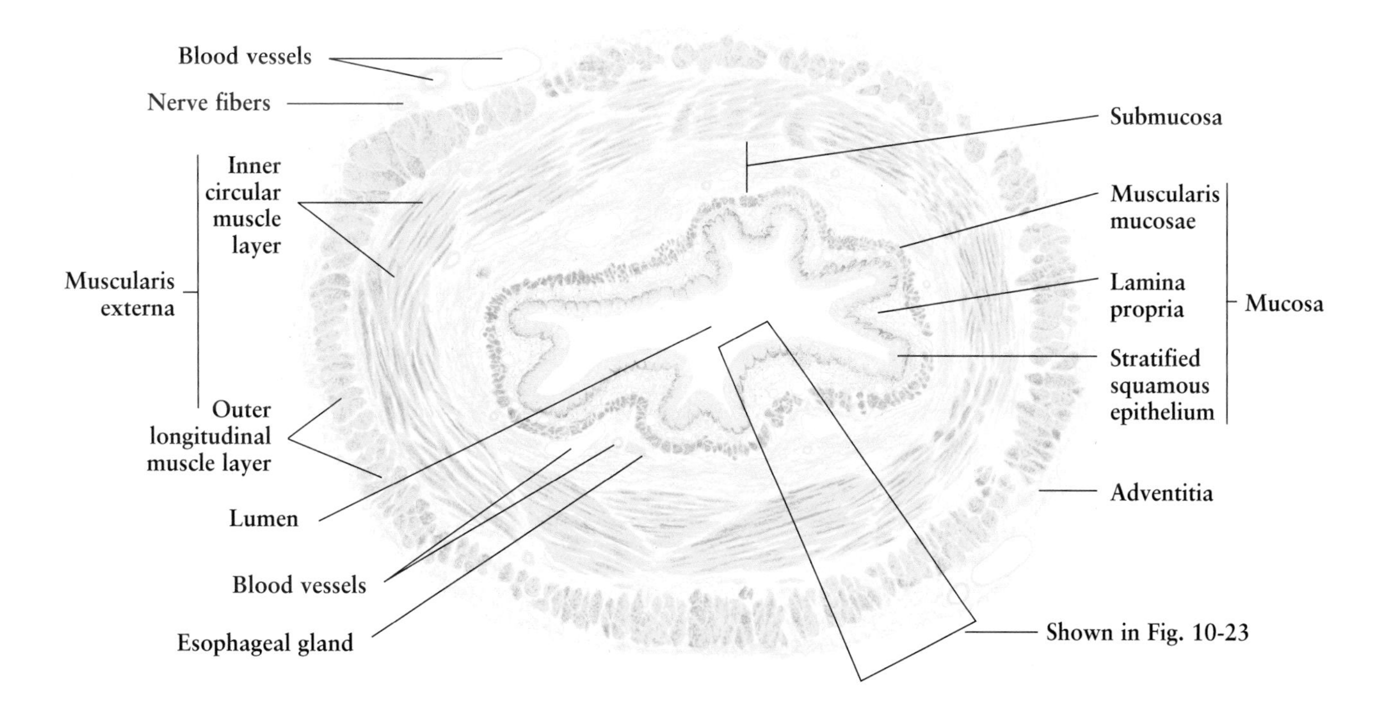

Fig. 10-23. Esophagus

This enlargement of the boxed area in **Fig. 10-22** demonstrates the microscopic mural structure of the middle third of the esophagus.

The **mucosa** consists of stratified squamous epithelium, lamina propria, and muscularis mucosae. The **stratified squamous epithelium** is thick and not normally keratinized in humans. The fine structure of this epithelium resembles that in the oral cavity. The **lamina propria** is a fibroelastic connective tissue containing interwoven fine collagen and elastic fibers, as well as fibroblasts and macrophages. The connective tissue invaginates the epithelium as **papillae,** and is separated from the epithelium by a thin basement membrane. The lamina propria is rich in **blood vessels** and **lymphatic vessels. Diffuse lymphatic tissue** is often seen in this layer (see **Fig. 8-1**). Esophageal cardiac glands are small, branched tubular mucous glands located in the lamina propria at the level of the cricoid cartilage, near the junction with the stomach. They are structurally identical to the cardiac glands in the stomach (see **Fig. 10-24**). The **muscularis mucosae** consists of bundles of longitudinal smooth fibers. Through the contraction of these muscle fibers, the mucosa is thrown into longitudinal folds.

The **submucosa** shows relatively coarse collagen and elastic fibers; these give great flexibility to the esophageal wall. The submucosa contains larger **blood vessels, lymphatic vessels,** and scattered parasympathetic ganglia. This layer also possesses **esophageal glands,** which are tubular glands of mucous type secreting an acid mucus to lubricate the epithelial lining of the esophagus.

The **muscularis externa** is characterized by the arrangement of the muscle fibers in the middle third of the esophagus. Both **inner circular** and **outer longitudinal muscle layers** contain skeletal and smooth muscle fibers. Between the inner circular and outer longitudinal muscle layers are the **myenteric plexuses** (**Auerbach's plexus;** see **Fig. 10-38**).

The **adventitia** is composed of a loose layer of fibroelastic connective tissue; it blends with the surrounding structures of the esophagus.

Figure 10-23. Esophagus
Human • Cross section • H.E. stain • Low magnification

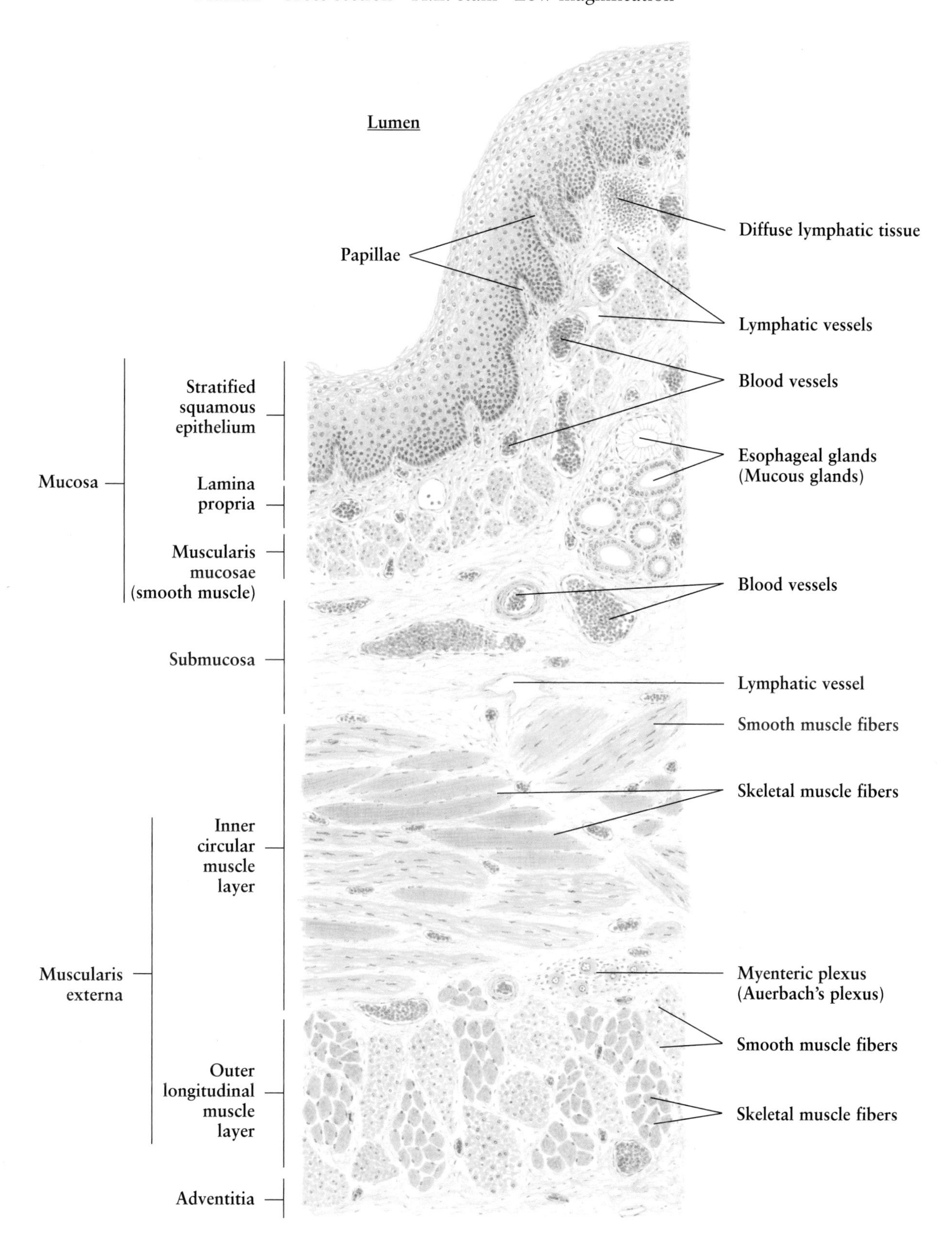

Figs. 10-24 and 10-25.

Cardia and Esophagocardiac Junction

Anatomically, the *stomach* can be divided into five parts: *cardia, fundus, body, pyloric antrum,* and *pyloric channel.* The **cardiac area** is a ring-shaped zone, 5–30 mm wide, surrounding the cardia. The cardia is the opening of the esophagus into the stomach and is characterized by the presence of **cardiac glands** in the **lamina propria.**

The **cardiac glands** are ramified, slightly coiled, tubular glands of mucous type, opening directly into the **gastric pits.** The glands are lined by cuboidal or columnar, mucus-producing cells with a flattened or ovoid nucleus in their base. The supranuclear portion of these cells is occupied by a large quantity of mucous droplets. The cardiac glands are involved in the formation of gastric juice by the secretion of mucus, which contains sodium and potassium chloride, sodium and potassium bicarbonate, and calcium phosphate.

The **esophagocardiac junction** shows an abrupt transition from the **stratified squamous epithelium** of the esophagus into **simple columnar epithelium** of the cardiac area of the stomach. The **lamina propria, muscularis mucosae,** and **submucosa** in the esophagus are continuous with those in the cardia. The lamina propria of the esophagus contains **lymphatic tissue,** whereas the lamina propria of the cardia shows numerous **blood vessels, plasma cells,** and **macrophages,** in addition to the cardiac glands.

Figure 10-24. Cardia
Monkey • H.E. stain • Low magnification

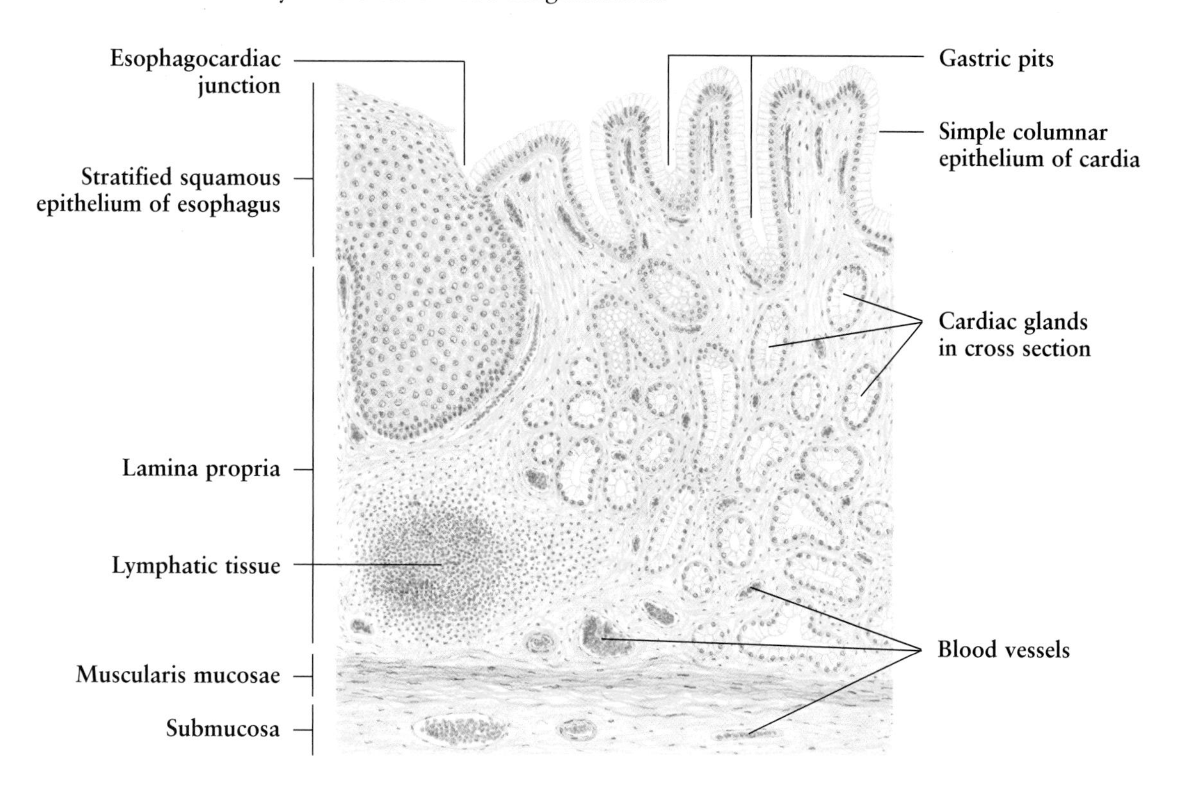

Figure 10-25. Esophagocardiac
Monkey • H.E. stain • High magnification

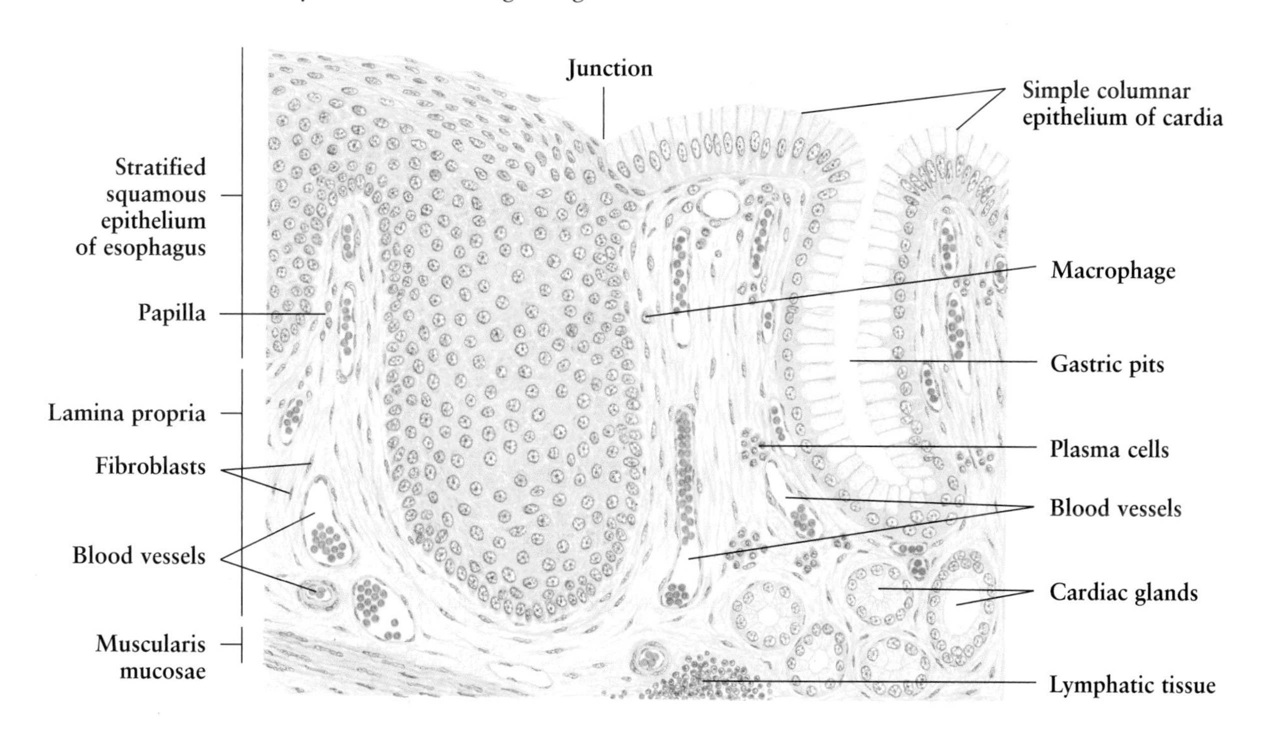

Fig. 10-26. Stomach

The **stomach** is a saclike dilatation of the tubular digestive tract in which ingested food is retained for 2 hours or longer so as to undergo mechanical reduction and primary chemical digestion to form chyme. Mechanical reduction is produced by a strong churning action of smooth muscle, whereas primary chemical digestion is produced by gastric juice secreted by the glands of the stomach mucosa. The *gastric juice* contains *hydrochloric acid, mucus,* and *hydrolytic enzymes* (principally *pepsin, lipase,* and *renin*). In addition, the gastric mucosa produces an *intrinsic factor* for the absorption of vitamin B_{12}, and some hormones including *gastrin.* Some absorption occurs in the stomach, although this is limited to *water, salts, glucose,* and some *drugs; alcohol* is a notable exception.

The stomach wall consists of four layers: *mucosa, submucosa, muscularis externa,* and *serosa.* The **mucosa** contains a surface epithelium of simple columnar type, lamina propria, and muscularis mucosae. The **simple columnar epithelium** lines the luminal surface and the pits. The **lamina propria** is a fine connective tissue, housing **gastric glands** that open into the pits. The lamina propria also contains some **lymphatic tissue.** The **muscularis mucosae** is a layer of smooth muscle fibers. The **submucosa** is a loose connective tissue with collagen and elastic fibers, some larger **blood vessels,** and lymphatic vessels, as well as the submucous plexus. The stomach has a very thick **muscularis externa,** which is formed by three layers of smooth muscle fibers: **inner oblique layer, middle circular layer,** and **outer longitudinal layer.** The middle circular and outer longitudinal layers are continuous with the two muscle layers of the esophagus. At the pylorus, the middle circular layer thickens as the *pyloric sphincter* (see **Fig. 10-29**). Interposed between muscle layers are the **myenteric plexuses** of autonomic nerves and their ganglia. The **serosa (visceral peritoneum)** is composed of a thin layer of loose connective tissue containing **adipose cells;** it is covered by **mesothelium.**

On the basis of the differences in the glands and pits, the mucosa of the stomach is divided into three zones: *cardia, body and fundus,* and *pylorus.* The cardia is illustrated in **Figs. 10-24** and **10-25,** and the pylorus is demonstrated in **Figs. 10-28** and **10-29.** This illustration shows structural features of the body and fundus, and the details of the gastric glands are shown in **Fig. 10-27.**

Figure 10-26. Stomach
Human • H.E. stain • Low magnification

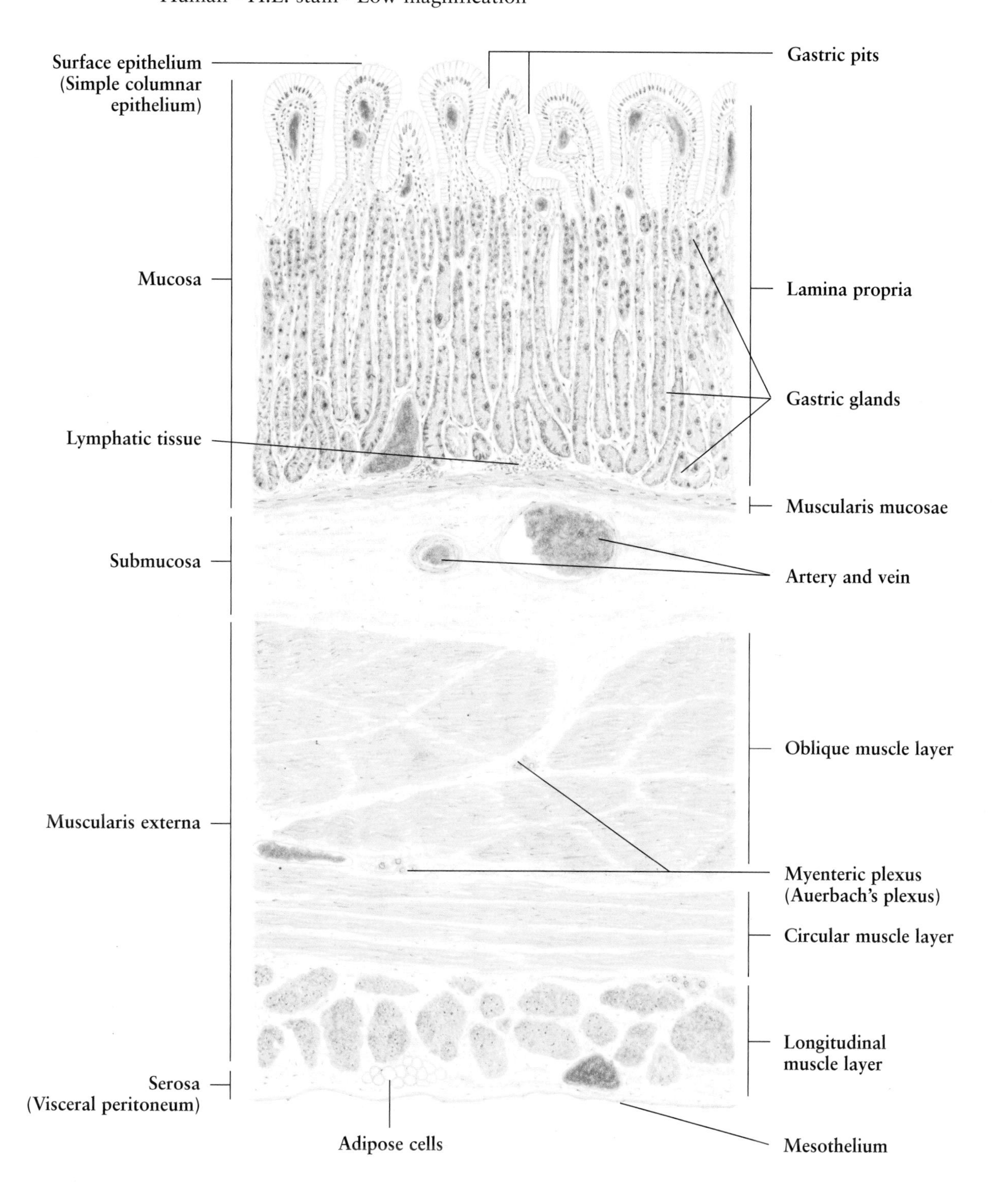

Fig. 10-27.

Stomach: Gastric Glands

This figure shows the mucosa of the *fundic area* of the stomach. The **surface epithelium** of the mucosa, composed of columnar mucous cells, covers the entire luminal surface and lines the **gastric pits.** Their nuclei are ovoid, and located in the basal region of the cells. The basal cytoplasm is basophilic while the apical cytoplasm stains faintly, with a foamy appearance in routine preparations. These mucous cells secrete neutral glycoprotein mucus, which forms a film to protect the mucosa from the high concentration of hydrochloric acid and pepsin in the stomach.

The **lamina propria** is crowded with **gastric glands** that open into the bottom of the gastric pit. Two or three glands may commonly open into a single gastric pit. Gastric glands are straight, tubular glands, with some branches, usually packed closely side by side. A gland generally can be divided into three regions: **isthmus, neck,** and **fundus.** Gastric glands contain a mixed population of cells of five types: *chief cells, parietal cells, mucous neck cells, undifferentiated cells,* and *endocrine cells.*

1. **Chief cells** predominate in the fundus of the gland and intermingle with parietal and mucous neck cells toward the neck region. They are low columnar cells and have the typical appearance of protein-producing cells, with a spherical nucleus located toward the base, basophilic basal cytoplasm, and apical ectoplasm filled with acidophilic zymogen granules. Chief cells secrete inactive *pepsinogen* which, in the acid medium of the stomach, is converted into the highly active proteolytic enzyme pepsin that is able to hydrolyze protein into smaller peptides.
2. **Parietal cells** are located mainly in the isthmus and neck regions, but are also scattered in the fundic region. Characteristically they are large, rounded, or pyramidal cells with central spherical nuclei and extensive eosinophilic cytoplasm. Electron microscopy has shown that the cytoplasm contains numerous mitochondria and *intracellular canaliculi,* deep invaginations of the luminal surface with associated microvilli. In this figure, the canalicular system appears as an irregular, pale area. Parietal cells are involved in the formation of *hydrochloric acid* and *intrinsic factor.*
3. **Mucous neck cells** are present in the neck of gastric glands, typically as aggregates or single cells between parietal cells. These cells usually have an irregular shape, a slender base, and expanded apex, caused by the pressure of adjacent cells. The nucleus is basally located. The basal cytoplasm is basophilic, and the apical cytoplasm contains secretory granules. Mucous neck cells produce an *acid mucus.*
4. **Undifferentiated cells** are very few in number. They are located in the isthmus region, bordering on the mucous neck cells or scattered between the parietal cells. Undifferentiated cells have a large, spherical nucleus and pale cytoplasm. They are thought to be precursors of surface epithelial cells in the gastric pits and mucous neck cells.
5. **Endocrine cells** are scattered throughout the stomach and small and large intestines. They are usually small pyramidal cells with dense granules in the cytoplasm, located between the basement membrane and the epithelial cells of the gastric glands. These cells have been well demonstrated by silver precipitation, immunocytochemistry, and electron microscopy. In this figure, the endocrine cells are shown as small pyramidal cells with orange granules. Most of these cells have the characteristic features of so-called APUD (amine precursor uptake and decarboxylation) cells. Endocrine cells produce some true peptide hormones, including *secretin, gastrin,* and *cholecystokinin.*

Figure 10-27. Stomach: Gastric Glands
Human • H.E. stain • High magnification

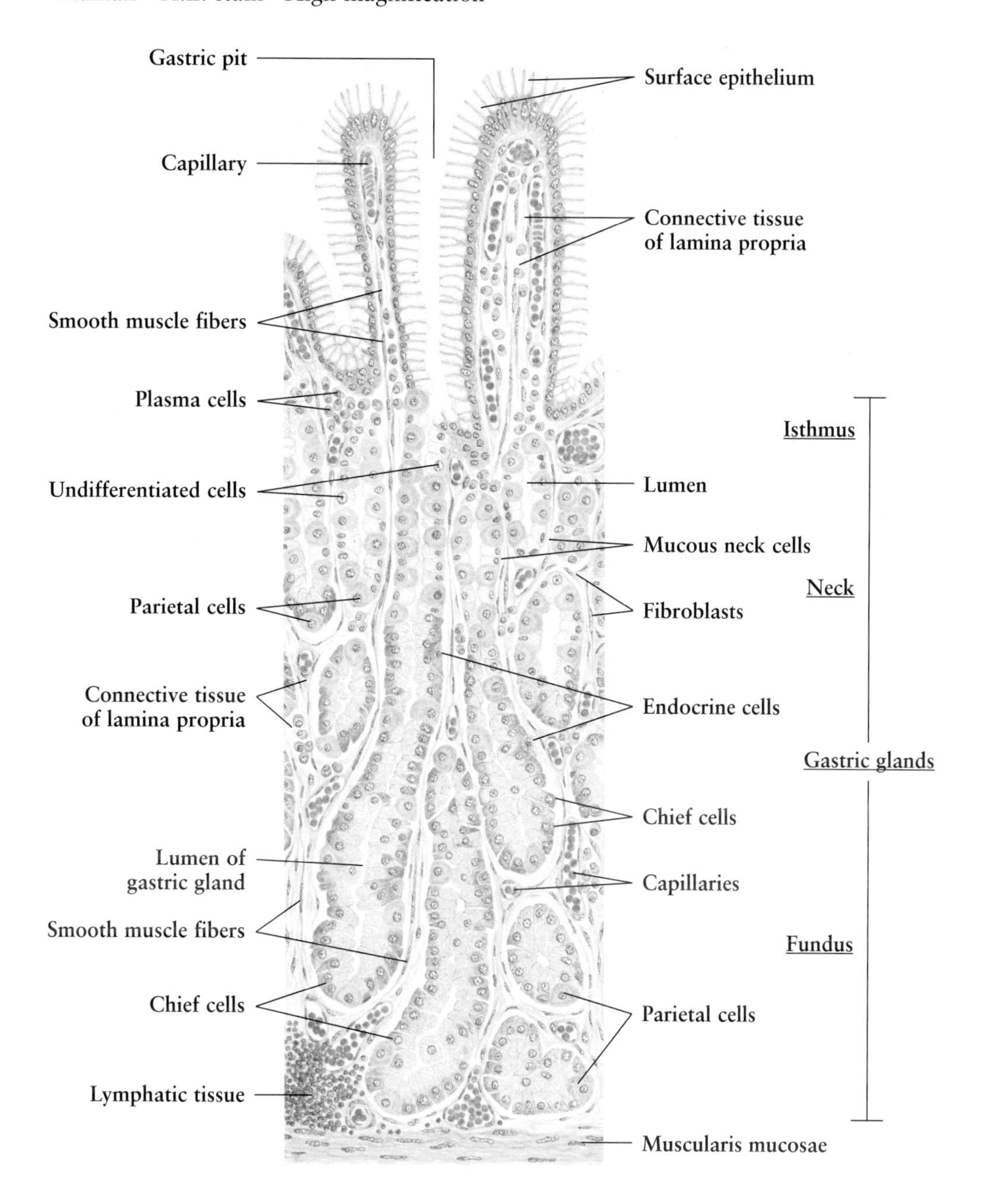

In the **lamina propria,** scanty **connective tissue,** consisting of a delicate meshwork of collagen and reticular fibers with a few **fibroblasts,** fills the spaces between gastric glands. It is more obvious toward the mucosal surface, where the spaces between pits are more extensive. A small amount of **smooth muscle fibers** runs parallel to the long axis of the gastric glands. The **capillary** network surrounds the gastric glands. A large number of **plasma cells** are present at the upper portion of the lamina propria, while some **lymphatic tissue** is usually seen in the lower portions.

The **muscularis mucosae** is not thick; it usually is formed by an inner circular and an outer longitudinal layer of smooth muscle fibers. In some regions, there is an external layer of oblique fibers.

Fig. 10-28.

Stomach: Pyloric Mucosa

In the **pyloric stomach,** the mucosa shows deep, irregularly shaped **gastric pits,** extending at least midway to the muscularis mucosae. The epithelium is simple columnar, consisting mainly of **surface epithelial cells.**

The **pyloric glands** are short, coiled, and branched. They have a relatively wide lumen that opens into the bottom of the gastric pits. They are composed almost exclusively of mucus-secreting cells; these cells are similar to the mucous neck cells of the gastric glands, with clear cytoplasm, indistinct granulation, and a flattened basal nucleus. The pyloric glands produce *mucus,* which lubricates and protects the entrance to the duodenum. *Endocrine cells* are scattered in the pyloric glands, secreting the peptide hormone *gastrin.* They are not shown in this figure.

The **connective tissue** of the **lamina propria** contains **blood vessels, fibroblasts,** and **smooth muscle fibers.** The **muscularis mucosae** is similar to and continuous with that in other areas of the stomach.

Fig. 10-29.

Gastroduodenal Junction

In the **gastroduodenal** (pyloric-duodenal) **junction,** the four layers of the pylorus of the stomach are continuous respectively with those of the duodenum, but with dramatic transitional changes.

In the **mucosa,** the *glandular arrangement* of the stomach is transformed into the *villous arrangement* of the duodenum, accompanied by an abrupt change of the surface epithelium into cells showing a striated **border** (for details, see **Fig. 10-30**).

The **submucosa** contains no glands on the pyloric side, but is filled on the intestinal side with **duodenal (Brunner's) glands.** They are characteristic of the duodenum.

The most dramatic transition in the gastroduodenal junction is present in the **muscularis externa,** where the circular muscle layer is extremely thickened to form the **pyloric sphincter.**

The **serosa** shows no conspicuous differences between the pyloric and duodenal sides.

Figure 10-28. Stomach: Pyloric Mucosa
Monkey • H.E. stain • High magnification

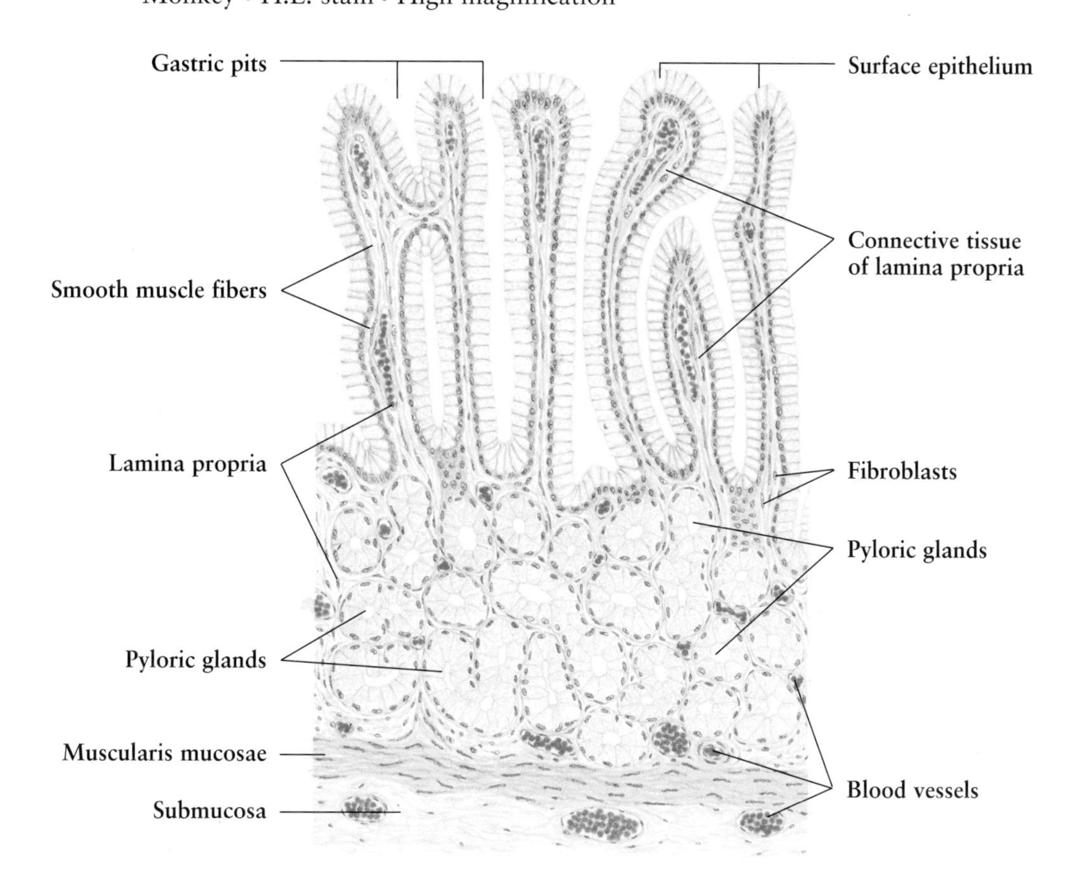

Figure 10-29. Gastroduodenal Junction
Monkey • H.E. stain • Low magnification

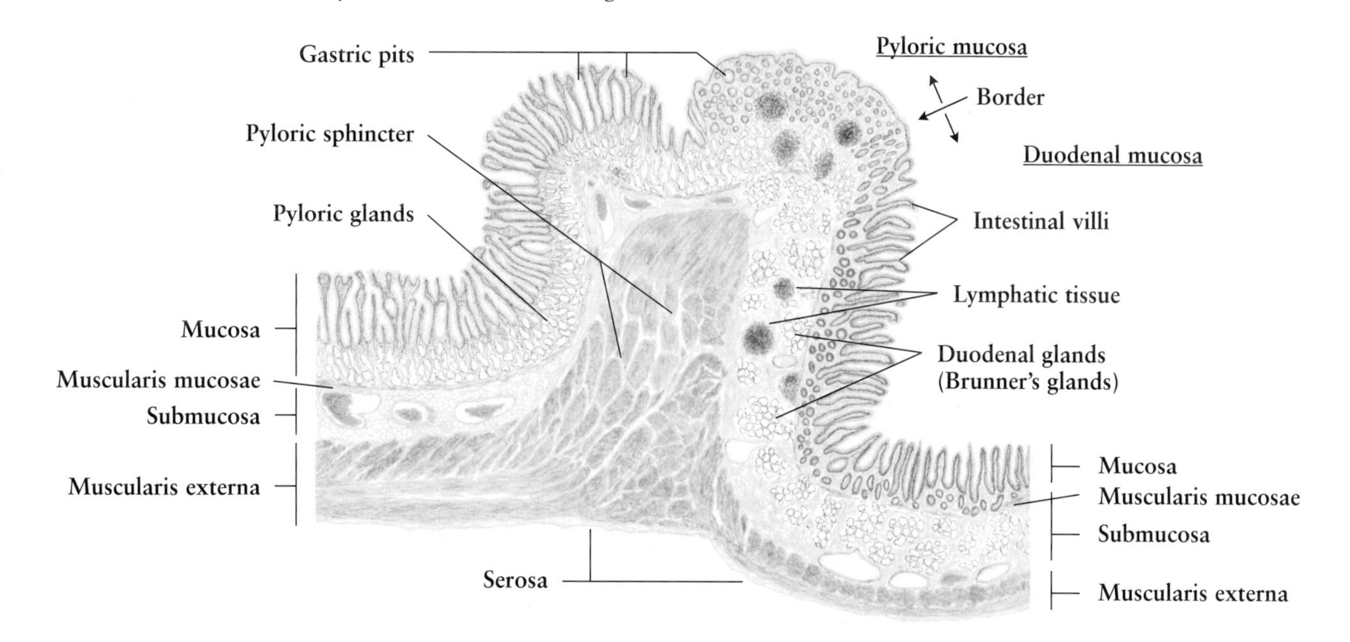

Fig. 10-30.

Gastroduodenal Junction: Epithelium

This illustration demonstrates an abrupt transition from **pyloric surface epithelial cells** to **intestinal epithelial cells** at the gastroduodenal junction. Both epithelial cell types are columnar, and have a basally oriented ovoid nucleus. The pyloric surface epithelial cells are mucus-secreting, with a pale cytoplasm and foamy appearance. The absorptive epithelial cells of the duodenum have an eosinophilic cytoplasm, with **striated border** at the upper free surface of the cell. Scattered among the epithelial cells of the duodenum are **goblet cells.**

Note that the lamina propria underlying the epithelium contains a number of **plasma cells, macrophages, fibroblasts,** and **smooth muscle fibers,** as well as **capillaries.**

Fig. 10-31.

Duodenum

The **duodenum,** about 25 cm long, is the initial segment of the small intestine, connecting the pylorus and jejunum. The main function of the duodenum is to neutralize gastric acid and to continue further the digestive processes.

The **mucosa** of the duodenum contains numerous finger-like **villi,** which have as their core the connective tissue of the **lamina propria,** covered by **intestinal epithelium. Intestinal glands** (crypts of Lieberkühn) are housed in the lamina propria. The **muscularis mucosae** is formed of a layer of smooth muscle fibers. The entire coat of mucosa is thrown into plicae circulares.

The **submucosa** is a richly vascularized loose connective tissue, filled with characteristic **duodenal (Brunner's) glands.** The larger **blood vessels** and **submucosal (Meissner's) plexuses** of autonomic nerves are found in this coat.

The **muscularis externa** is of smooth muscle, composed of an **inner circular layer** and an **outer longitudinal layer.** Between these two layers of muscle are **myenteric (Auerbach's) plexuses.**

The **serosa** is formed of a thin layer of connective tissue covered by mesothelial cells. In some areas, the serosa is composed of fibrosa.

Figure 10-30. Gastroduodenal Junction: Epithelium
Monkey • H.E. stain • High magnification

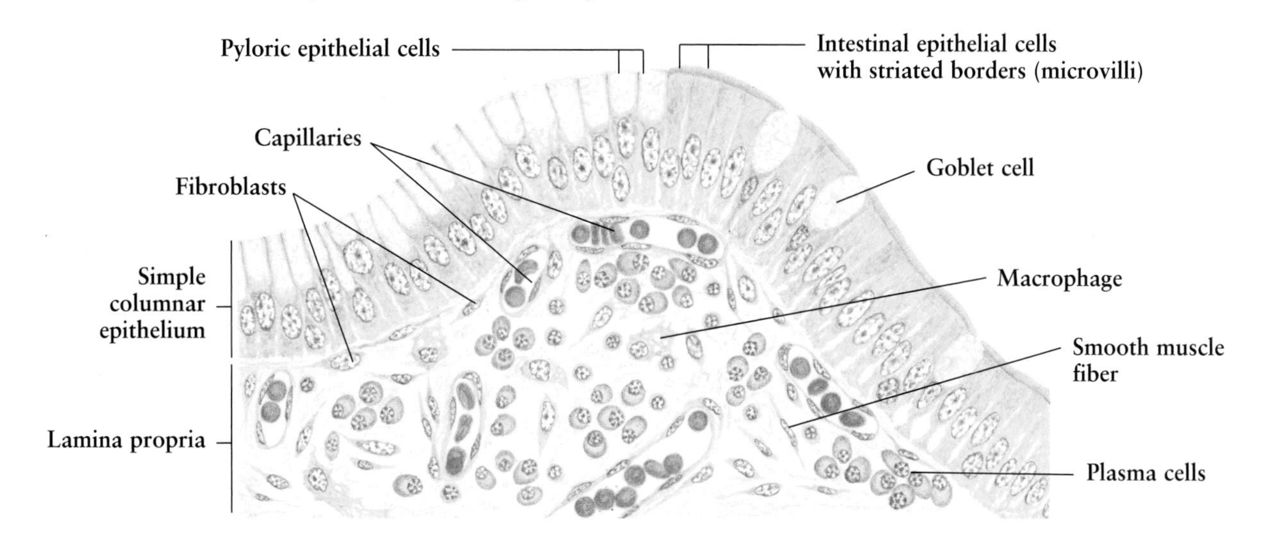

Figure 10-31. Duodenum
Human • Cross section • H.E. stain • Low magnification

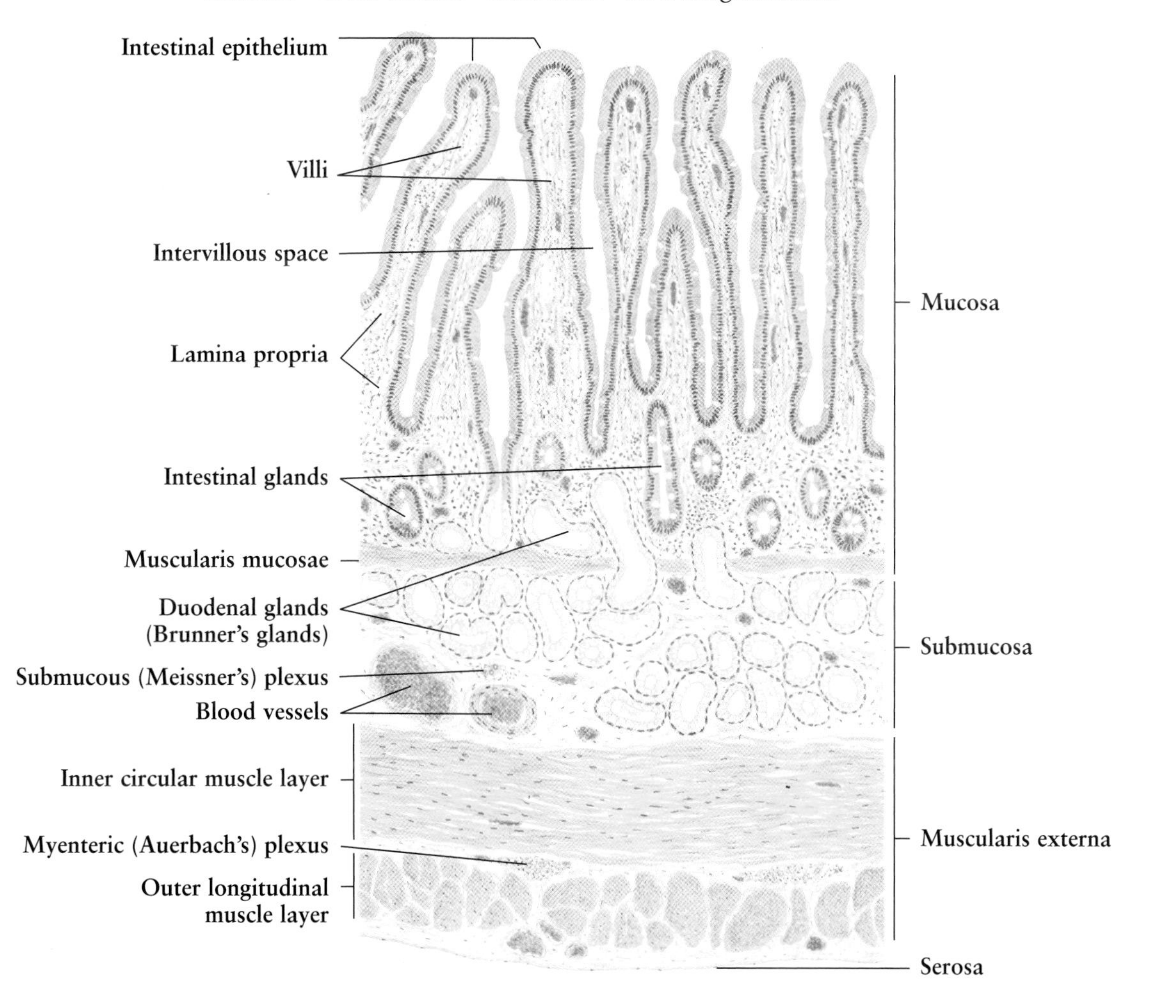

Fig. 10-32. Duodenum: Mucosa

This figure is an enlargement of the **duodenal mucosa** of the human in **Fig. 10-31.**

The **villi** of the duodenal mucosa are finger-like processes of **lamina propria,** covered by **intestinal epithelium.** As the core of a villus, the lamina propria is a loose connective tissue, containing numerous **plasma cells,** a small number of **fibroblasts,** and **smooth muscle fibers.** The villus also possesses a **blood capillary** network and **central lacteal** that receives the absorbed fat and conducts it to the larger lymphatic vessels. The **intestinal epithelium** is composed of simple columnar cells with a **striated border** on the free surface of the cell. Under the microscope, do not mistake the **intervillous space** for the lumen of intestinal glands. **Goblet cells** are scattered among the intestinal epithelium. The portion of the lamina propria under the villus base is filled with **intestinal glands** lined by secreting **epithelial cells, Paneth cells,** and **endocrine cells.** The connective tissue, surrounding the intestinal glands, contains a large number of **plasma cells,** some **fibroblasts,** and **smooth muscle fibers** as well as **capillaries.** The **muscularis mucosae** is formed of a single layer of circular smooth muscle.

The **submucosa** is occupied by **duodenal (Brunner's) glands.** They are tubular, coiled, and branched glands that open into the intestinal glands via short **ducts** which penetrate the muscularis mucosae. Brunner's glands are lined with cuboidal or columnar cells that have a structure similar to that of pyloric glands with clear, vacuolated cytoplasm and flattened, basal nuclei. The duodenal glands secrete an *alkaline mucus,* which protects the intestinal epithelium against pancreatic enzyme lysis and erosion by the gastric juice through the buffering capacity of the high bicarbonate content of the mucus. These glands are also believed to secrete a *proteolytic enzyme* that is activated by gastric hydrochloric acid, *enterokinase,* which activates the pancreatic trypsinogen to trypsin, as well as an *urogastrone,* which is a peptide that inhibits the secretion of hydrochloric acid in the stomach.

The connective tissue surrounding the duodenal glands also contains **capillaries** as well as a **submucosal (Meissner's) plexus.**

Figure 10-32. Duodenum: Mucosa
Human • H.E. stain • Medium magnification

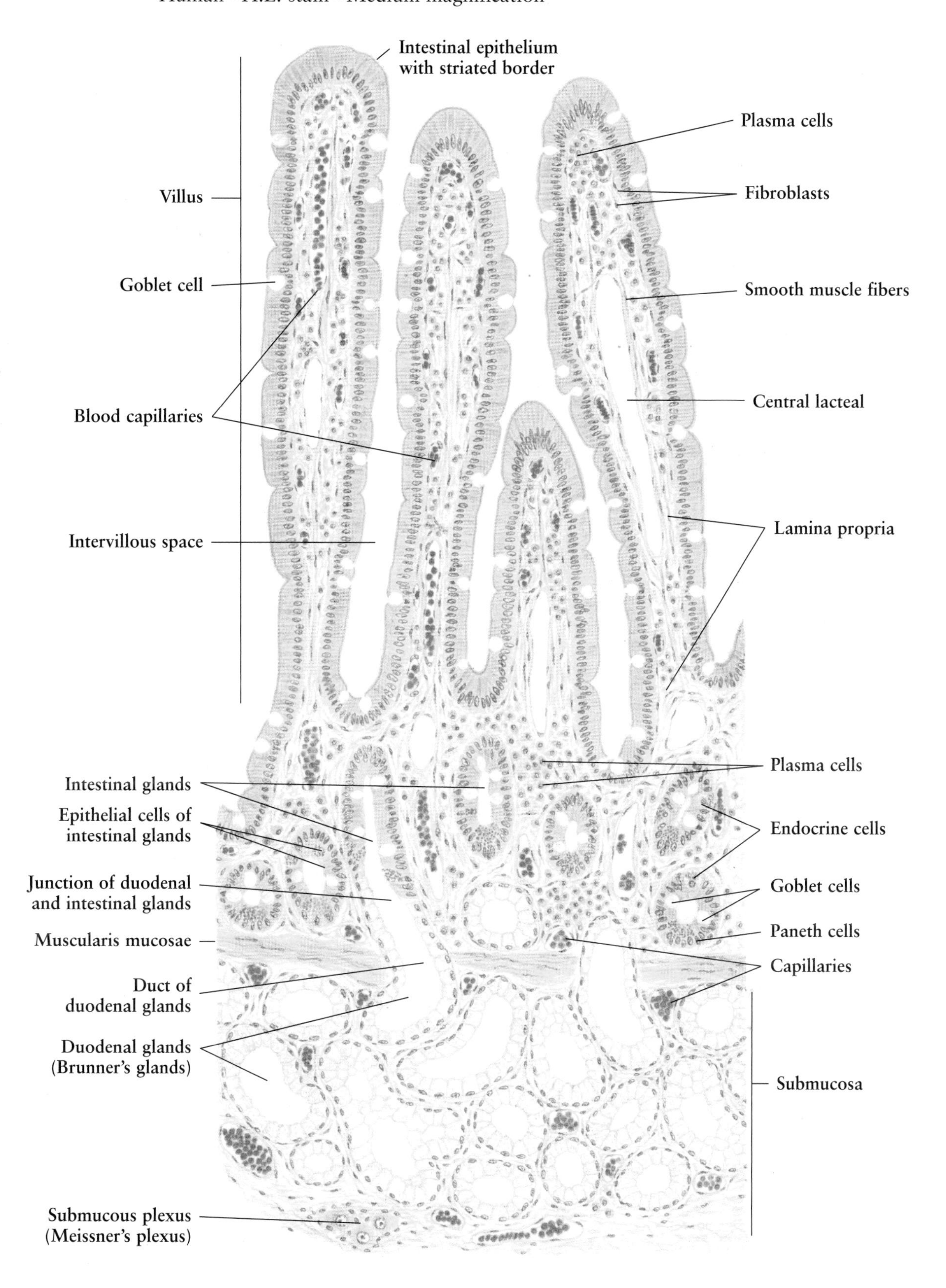

Figs. 10-33 and 10-34.

Jejunum and Ileum

The small intestine includes the duodenum, jejunum, and ileum. **Figures 10-33** and **10-34** display the principal features of the **jejunum** and the **ileum,** respectively.

Like the duodenum, the wall of these two segments of the small intestine consists of four layers: *mucosa, submucosa, muscularis externa,* and *serosa.* The **mucosa** is composed of the intestinal epithelium, lamina propria, and muscularis mucosae. The **villi** (see **Fig. 10-35**) are finger-like or leaflike projections of the **lamina propria,** lined by a simple columnar epithelium of absorptive and goblet cells. The villi of the ileum are less numerous and shorter than those of the jejunum. The **intestinal glands** (Lieberkühn's crypts, see **Fig. 10-36**) open into the bottom of the intervillous space and are also lined with simple columnar epithelium containing goblet cells. The **muscularis mucosae** is a layer of circular smooth muscle fibers that separate the lamina propria from the submucosa. The **submucosa** is a vascularized loose connective tissue containing larger **blood vessels, lymphatic vessels,** and **adipose cells** as well as submucosal (Meissner's) plexus (see **Fig. 10-37**). The **muscularis externa** is composed of an inner circular and an outer longitudinal layer of smooth muscle; their contractions produce the continuous peristaltic activity of the small intestine. Between these two layers of smooth muscle is the myenteric (Auerbach's) plexus (see **Fig. 10-38**). The **serosa** is a loose connective tissue lined on its peritoneal surface by mesothelium.

A characteristic feature of the small intestine is the presence of **plicae circulares** (valves of Kerckring). These are permanent circular or spiral folds of the entire thickness of the mucosa with a core of submucosa, particularly numerous in the jejunum. The plicae circulares, together with the villi intestinal glands and striated border (microvilli), significantly enhance the surface area for digestion and absorption.

Another prominent feature of the small intestine is the presence of lymphoid aggregations of various sizes within the lamina propria. Especially in the ileum these large aggregations are known as **aggregated lymphatic nodules** or **Peyer's patches** (see **Fig. 8-2**), which are associated with the function of the immune system.

Figure 10-33. Jejunum
Human • Longitudinal section • H.E. stain • Low magnification

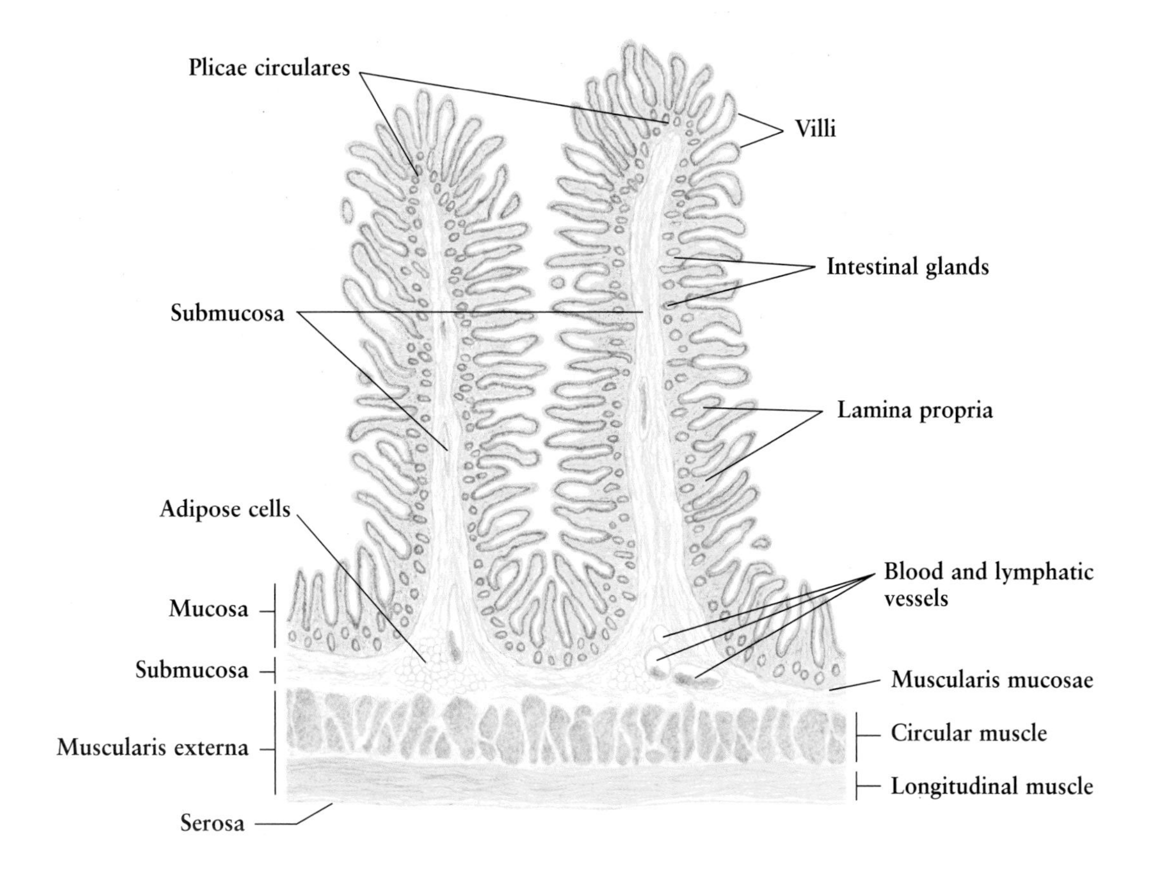

Figure 10-34. Ileum
Monkey • Longitudinal section • H.E. stain • Low magnification

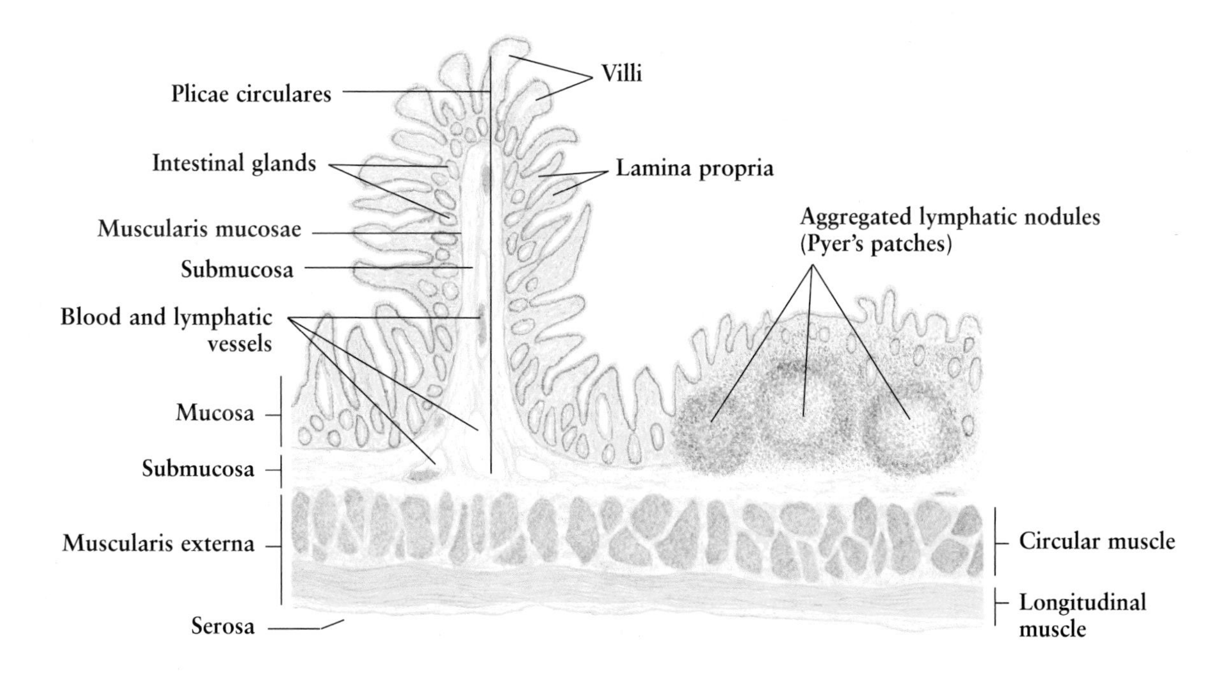

Fig. 10-35.

Intestinal Villus

Intestinal villi are finger-like or leaflike projections of the mucosa, 0.5–1.5 mm in length, found only in the small intestine. They vary in form and length in different regions. Those of the duodenum are broad, spatulate structures, but they become finger-like in the ileum. Each has a *core* of lamina propria covered with *simple columnar epithelium.*

The **simple columnar epithelium** is composed of *absorptive cells* and *goblet cells* resting on a very thin basement membrane. The **absorptive cells** are tall columnar cells, with a clear cytoplasm and an ovoid nucleus located toward the base. Each cell has a **striated border** formed by closely packed, parallel microvilli covered by a layer of the goblet cell's derived mucus. The striated border is strongly PAS (periodic acid-Schiff's reaction) positive as the result of the presence of a particularly thick *glycocalyx* and the surface *mucus;* both surface features protect against autodigestion. **Goblet cells** are scattered among the absorptive cells; they increase in number from the duodenum to the terminal ileum. These cells have a slender base that contains a darkly stained nucleus and an expanding apex with mucous secretory granules. The mucus is an *acidic glycoprotein,* which forms a protective film lying on the glycocalyx of the microvilli of the absorptive cell. In addition, **endocrine cells** are found in the epithelium. Those pyramidal cells have a pale-stained cytoplasm and a small nucleus, and are located between absorptive cells and basement membrane. Migrating **lymphocytes** are also found between epithelial cells.

The core of the villus is a highly cellular loose connective tissue containing a **central lacteal** and a network of **blood capillaries.** The blood capillaries, into which digestive products are absorbed, lie close to the basal surface of the epithelium. The central lacteal receives the absorbed lipid and transports it into the circulatory system via the thoracic duct. The cell types in the core include **lymphocytes, plasma cells, macrophages,** and **fibroblasts.** Also seen within the connective tissue are **smooth muscle fibers,** which are derived from the muscularis mucosae separating lamina propria from the underlying submucosa. Contraction of the smooth muscle fibers empties the contents of the lacteal into the intestinal lymphatic vessels.

Fig. 10-36.

Intestinal Gland

Intestinal glands (crypts of Lieberkühn) are tubular structures within the lamina propria, 0.3–0.5 mm in depth, which open between the bases of the intestinal villi. They are lined with simple columnar epithelium composed of *absorptive cells, goblet cells, endocrine cells,* and *undifferentiated cells.*

The **absorptive cells, goblet cells,** and **endocrine cells (basal granular cells)** resemble those of the intestinal villus in both structure and function (see **Fig. 10-35**). The **Paneth cells,** located only in the bases of intestinal glands, are pyramidal in shape with a broad base and a narrow apex. Their round or ovoid nucleus lies toward the base of the cell, and the cytoplasm is filled with eosinophilic secretory granules in routine preparation. Paneth cells secrete a *lysozyme* that digests some bacterial cell walls; it may also phagocytose some bacteria. The **undifferentiated cells** are the source of replacement for other cell types, both in the intestinal glands and on the villi. They are columnar cells, with a large, basally located nucleus and poorly stained cytoplasm. With frequent mitosis, the undifferentiated cells form the cells that differentiate into absorptive cells, goblet cells, Paneth cells, and endocrine cells.

Intestinal glands are not packed as closely as gastric glands. The spaces between them are filled with loose connective tissue of the lamina propria that contains **blood capillaries, lymphocytes, plasma cells,** and **fibroblasts,** as well as **smooth muscle cells.**

Figure 10-35. Intestinal Villus
Human • Longitudinal section • H.E. stain • High magnification

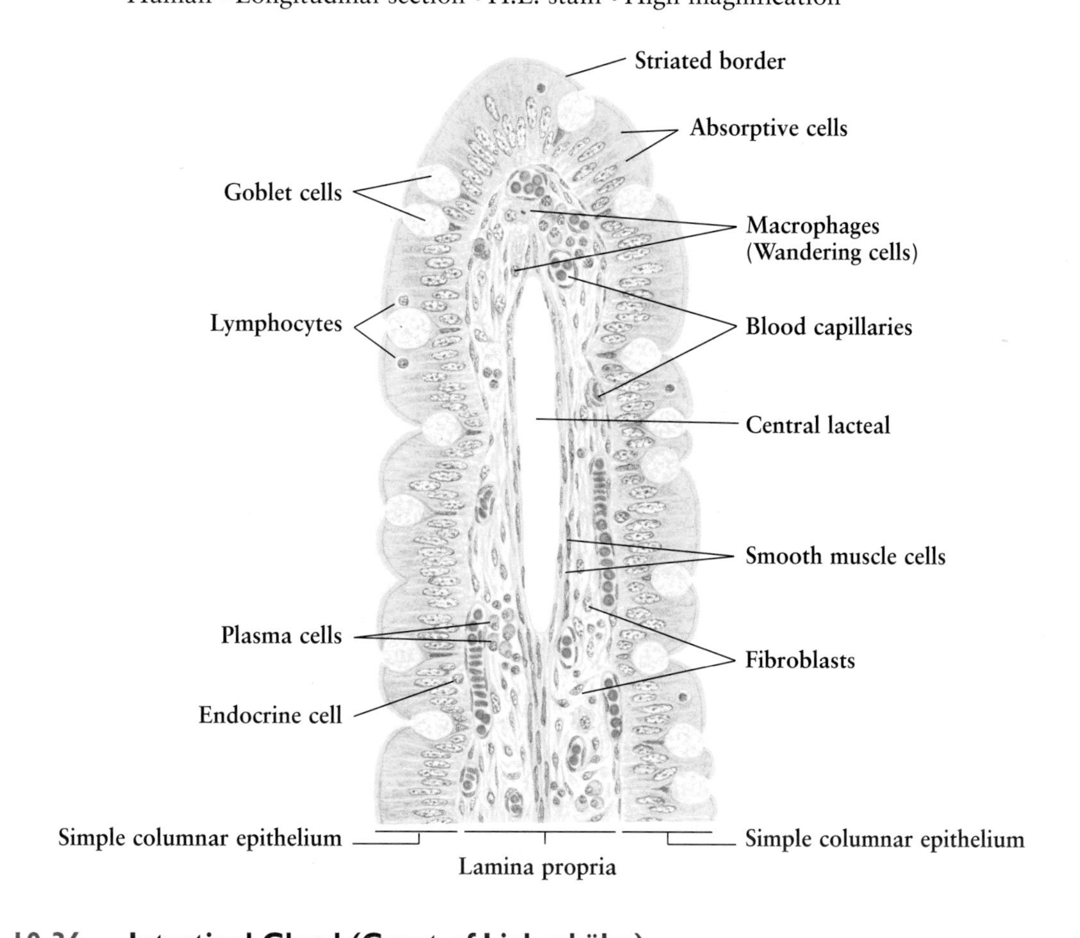

Figure 10-36. Intestinal Gland (Crypt of Lieberkühn)
Human • Longitudinal section • H.E. stain • High magnification

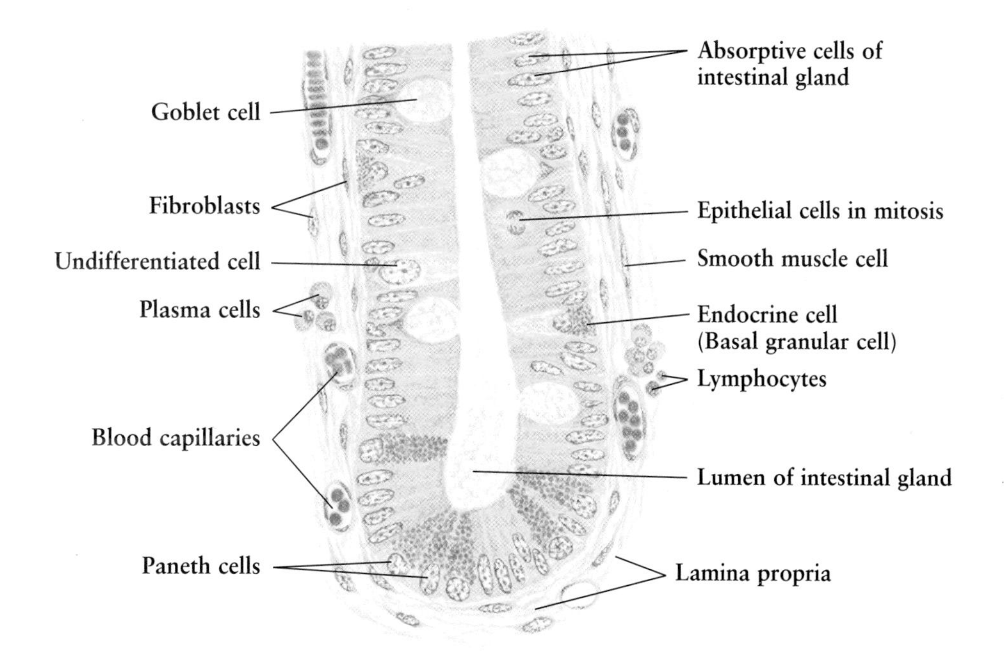

Figs. 10-37 and 10-38.

Submucosal and Myenteric Plexuses

Within the intestinal wall, nervous tissue forms two interconnected plexuses. This tissue includes sympathetic and parasympathetic fibers, and the ganglia that innervate the smooth muscle of the wall, blood vessels, villi, and glands. These are known as the **submucosal (Meissner's) plexus** in the submucosa and the **myenteric (Auerbach's) plexus** between the circular and longitudinal layers of the muscularis externa.

These two plexuses have similar structural features but differ in their location. The whole ganglion is invested by a thin connective tissue **capsule,** which contains collagen, reticular fibers, and fibroblasts. The **ganglion cells** are large multipolar neurons, and have a basophilic cytoplasm and a large, round nucleus with a conspicuous nucleolus. They are surrounded by small **satellite cells** that act as neuroglia in the central nervous system. The **nerve fibers** are unmyelinated, which can be seen with **Schwann cells** both inside and outside the ganglion. Multiple connections are present between the two plexuses.

The ganglion cells contain motor neurons and association neurons, which belong to the parasympathetic system. Because the myenteric and submucosal plexuses contains both sympathetic and parasympathetic fibers, they *control* the contraction of musculature and the secretion of glands, as well as the dilation and constriction of blood vessels.

Figure 10-37. Submucosal (Meissner's) Plexus
Human • Duodenum • H.E. stain • High magnification

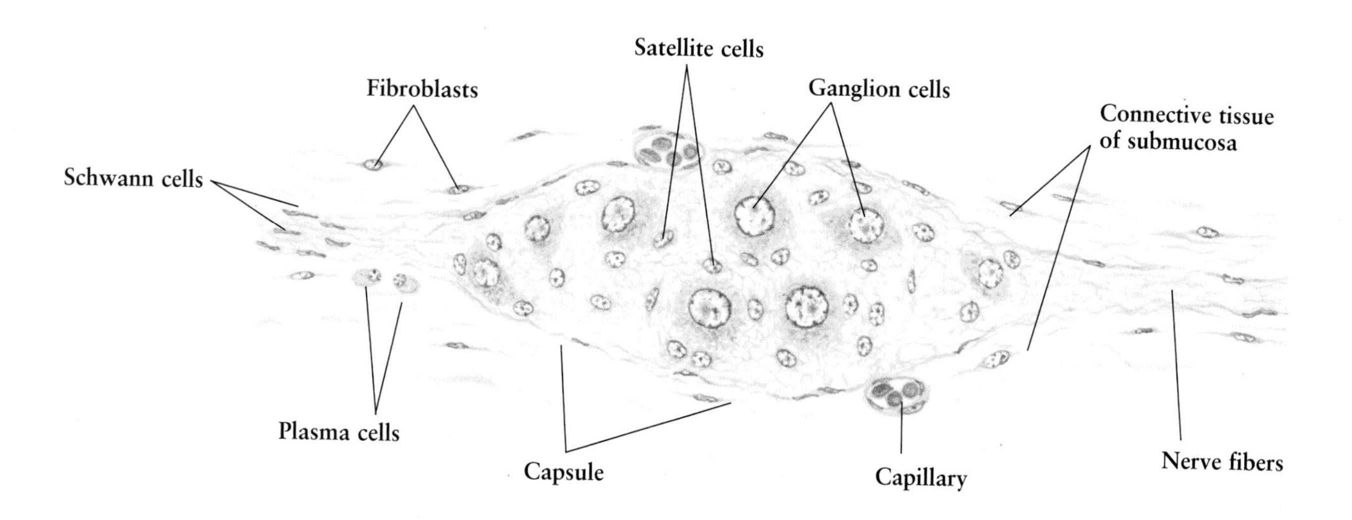

Figure 10-38. Myenteric (Auerbach's) Plexus
Human • Large intestine • H.E. stain • High magnification

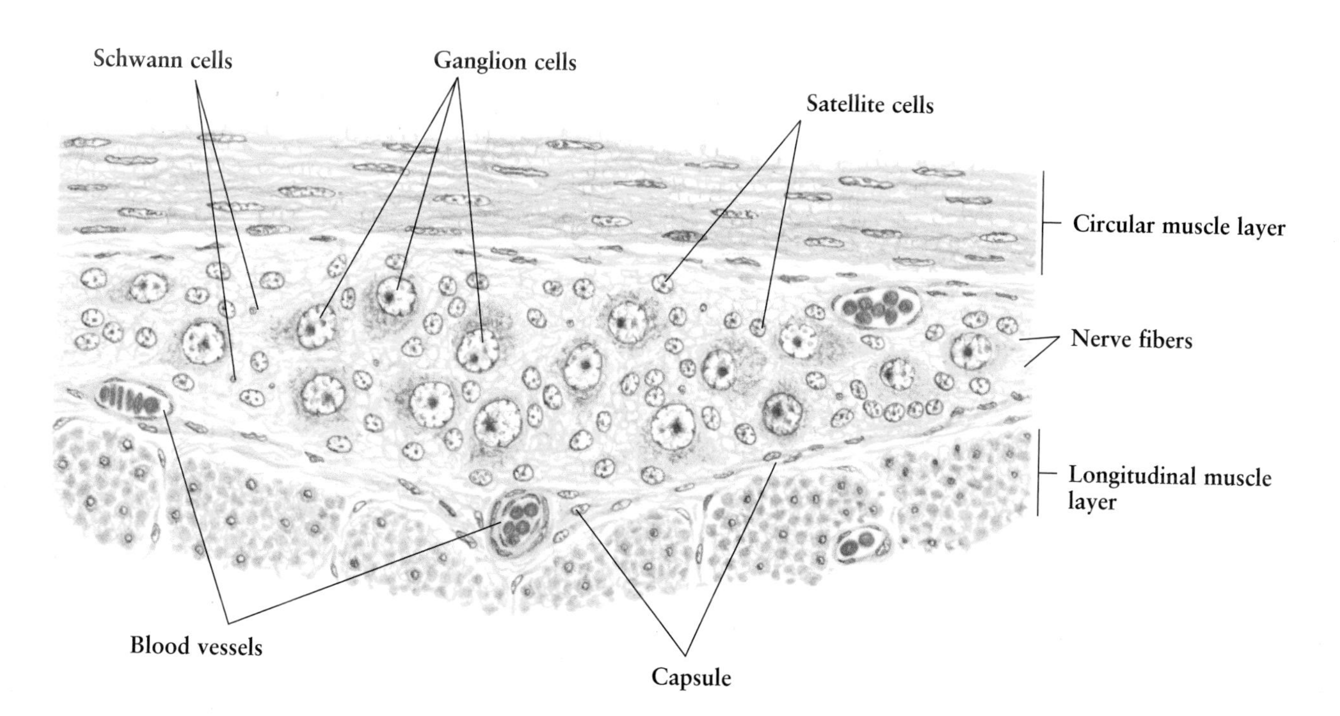

Fig. 10-39.

Ileocecal Valve

The **ileocecal junction** is located in the lower-right quadrant of the abdomen and is fixed to the posterior abdominal wall. At the ileocecal junction the **mucosa** and **submucosa** are thrown into anterior and posterior folds to form two valves, called the **ileocecal valves,** which are supported by a thickened **extension of circular smooth muscle** derived from the **muscularis externa** of the ileum and the cecum. The mucosa of the valve shows an abrupt transition from a villiform pattern in the small intestine to the glandular form in the large intestine. Note the difference between the **villi** of the ileum and the **mucosal glands** of the cecum. **Lymphatic tissue** is found in the mucosa. The submucosa is a loose connective tissue rich in **blood vessels** and lymphatic vessels.

Fig. 10-40.

Vermiform Appendix

The **vermiform appendix** is a wormlike, blind diverticulum of the cecum arising about 2.5 cm below the ileocecal valve. It is 10–15 cm long and as much as 8 mm wide. The wall of the appendix contains the four major layers of the digestive tract, but it lacks plicae circulares and villi. Its lumen is narrow, and often contains cellular debris.

In the **mucosa,** the **intestinal glands (crypts of Leiberkühn)** embedded in the **lamina propria** are few, with irregular shapes and varying lengths. The luminal surface epithelium and the intestinal glands are lined mostly by goblet cells and some absorptive cells; at the bottom of the glands, a few Paneth cells and numerous endocrine cells can be seen. The vermiform appendix is characterized by the mass of **lymphatic tissue** located in the lamina propria. The **submucosa** is thick, with blood and lymphatic vessels, nerves, and adipose cells housed in loose connective tissue. The **muscularis externa** consists of two complete muscle layers: inner circular and outer longitudinal fibers. The **serosa** is identical to that covering the remainder of the intestine. The vermiform appendix is suspended by the **mesentery,** which contains a large **artery, vein,** and **nerve fibers.**

Because the appendix is rich in lymphatic tissue and intestinal endocrine cells, it is probably not correct to classify it as a rudimentary or functionless organ. The appendix is a common site of acute and chronic inflammation.

Figure 10-39. Ileocecal Valve
Monkey • H.E. stain • Low magnification

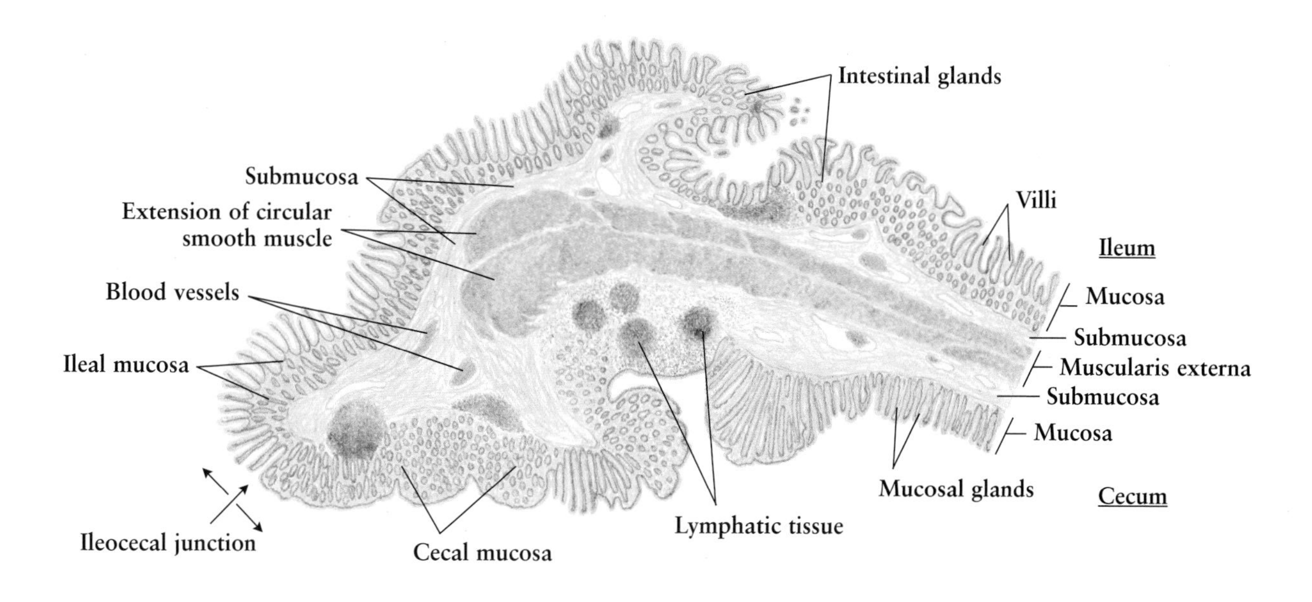

Figure 10-40. Vermiform Appendix
Human • Cross section • H.E. stain • Low magnification

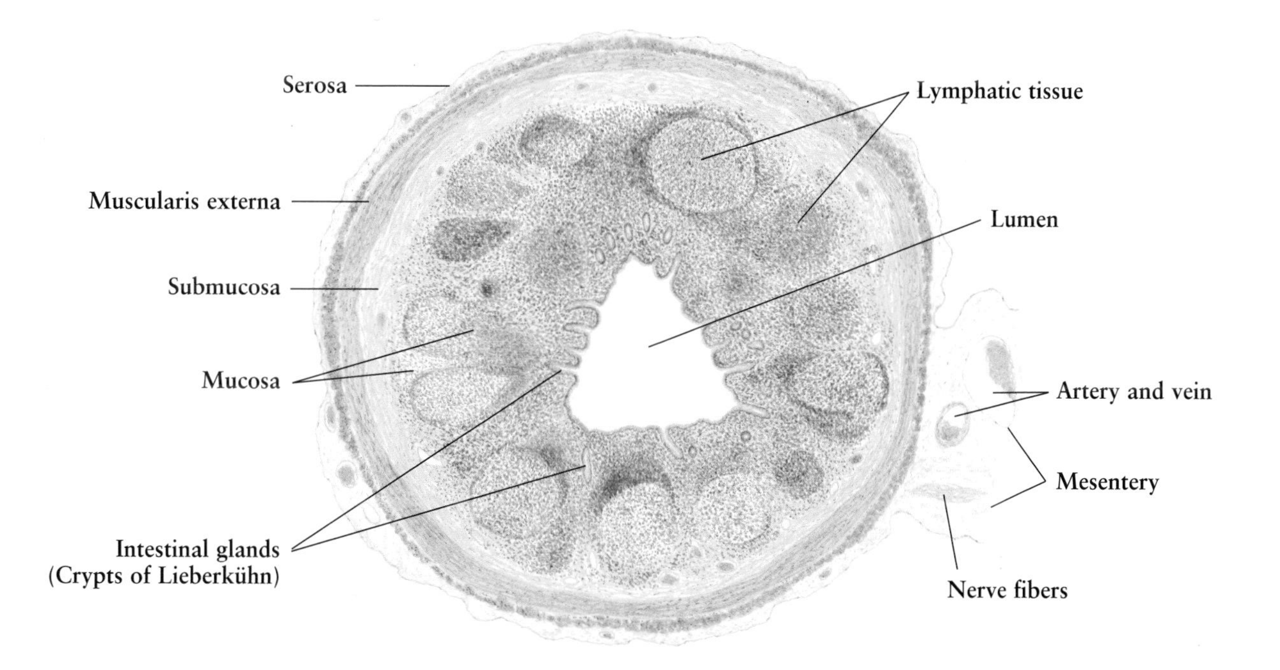

Fig. 10-41. Colon

The *large intestine,* about 180 cm long, consists of *cecum, appendix, colon,* and *rectum.* The **colon,** which constitutes most of the length of the large intestine, is divided into three parts: ascending, transverse, and descending colon. The principal functions of the colon are absorption of water from the semisolid feces, propulsion of increasingly solid feces to the rectum before defecation, and secretion of mucus (which lubricates and protects the mucosa of the colon from trauma).

The wall of the colon has the typical four layers of the gastrointestinal tract: *mucosa, submucosa, muscularis externa,* and *serosa.* However, there are no permanent plicae circulares or villi as seen in the mucosa of the small intestine. The **mucosa** is characterized by the presence of the **mucosal glands.** The **submucosa** is a loose connective tissue with large **blood vessels,** lymphatic vessels, **lymphatic tissue,** and submucosal plexus. The **muscularis externa** consists of **inner circular** and **outer longitudinal layers** of smooth muscles. The outer longitudinal muscle thickens to form three longitudinal bands or the **tenia coli,** one of which lies within the mesentery. The **serosa** is similar to that of the small intestine, except for the appendices epiploicae, which are large accumulations of adipose tissue.

Fig. 10-42. Colon: Mucosa

The **lamina propria** of the **mucosa** is occupied by the **intestinal glands,** which are deep and closely packed. The **epithelium** covering the luminal surface and the intestinal glands is composed of numerous **goblet cells,** some **absorptive cells,** undifferentiated cells, and **endocrine cells (basal granular cells).** The connective tissue between the glands is vascularized and cellular, containing a network of **capillaries** and several cell types such as **lymphocytes, plasma cells, macrophages, fibroblasts,** and **smooth muscle cells.** The **muscularis mucosae** is well developed; however, it may be irregular or deficient at the site of lymphatic nodules.

Figure 10-41. Colon
Human • Transverse section • H.E. stain • Low magnification

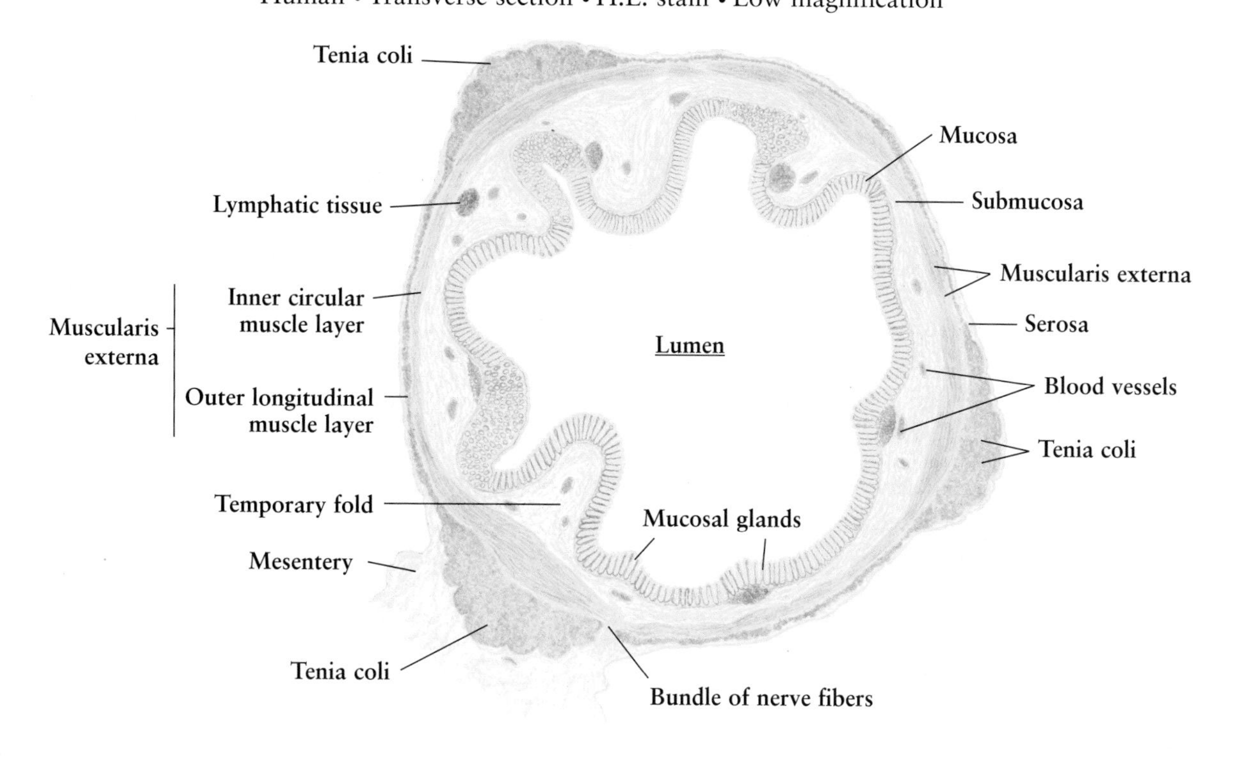

Figure 10-42. Colon: Mucosa
Human • H.E. stain • High magnification

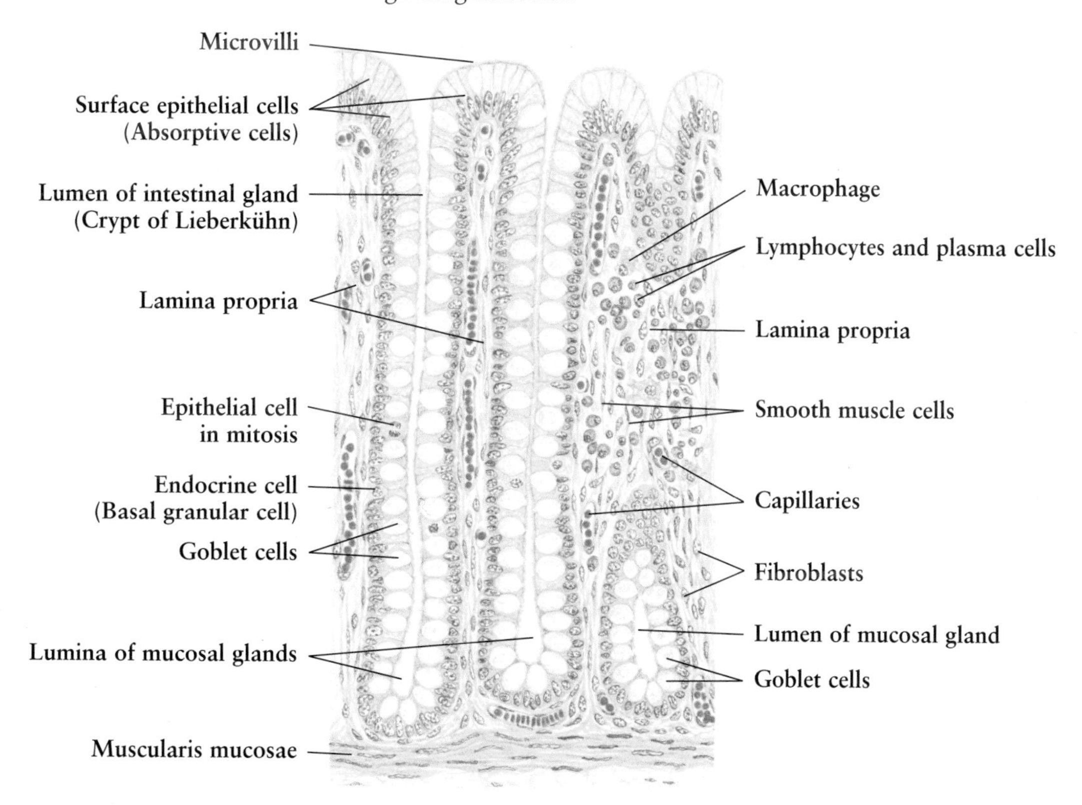

Fig. 10-43.

Rectoanal Junction

The **rectum** is divided into an upper and lower part. The mucosa of the upper part has a structure similar to that of the colon. In contrast, the mucosa of the lower part is thrown into several longitudinal folds known as the *rectal columns of Morgagni;* at their lower termination these columns unite with one another to form the *anal valves.* In this region there is a plexus of small, thin-walled veins, which may cause *internal hemorrhoids* or *piles* if chronic congestion, dilation, and convolution occur.

At the level of the anal valves, the covering epithelium exhibits an abrupt change from the **simple columnar epithelium** in the **rectum** to the nonkeratinized **stratified squamous epithelium** in the **anal canal.** In the underlying **lamina propria,** the **intestinal glands** become shorter and fewer and the **lymphatic tissue** becomes more conspicuous. The **muscularis mucosae** becomes irregular.

At the lower rectum and anal canal, the inner circular layer of the smooth muscle of the muscularis externa thickens, forming the *internal sphincter.* The outer layer is composed of skeletal muscle that forms the *external sphincter,* providing voluntary control of defecation. At the anus the stratified squamous epithelium becomes keratinized, and underneath are large apocrine sweat glands and sebaceous glands.

Fig. 10-44.

Liver: Hepatic Lobules

The **hepatic lobule** (liver lobule) is the structural unit of the liver; it is composed of parenchymal cells called the *hepatocytes* (hepatic cells), arranged in *plates* or *cords* that are interconnected with one another. The hepatic lobule is a polygonal prism in shape measuring 1–2 mm, and usually appears hexagonal in cross section, although this is difficult to see in human liver. Each hepatic lobule has a **central vein** in the middle from which the **plates of hepatic cells** radiate to the periphery like the spokes of a wheel from a central hub. In the pig liver, as shown in this figure, each lobule is outlined completely by **interlobular connective tissue** which, at the corners of the lobule, constitutes the **portal triads** that contain *blood vessels, bile ducts,* and *lymphatic vessels.* The organization of pig liver makes it ideal for understanding the concept of the hepatic lobule.

Figure 10-43. Rectoanal Junction
Human • Longitudinal section • H.E. stain • Low magnification

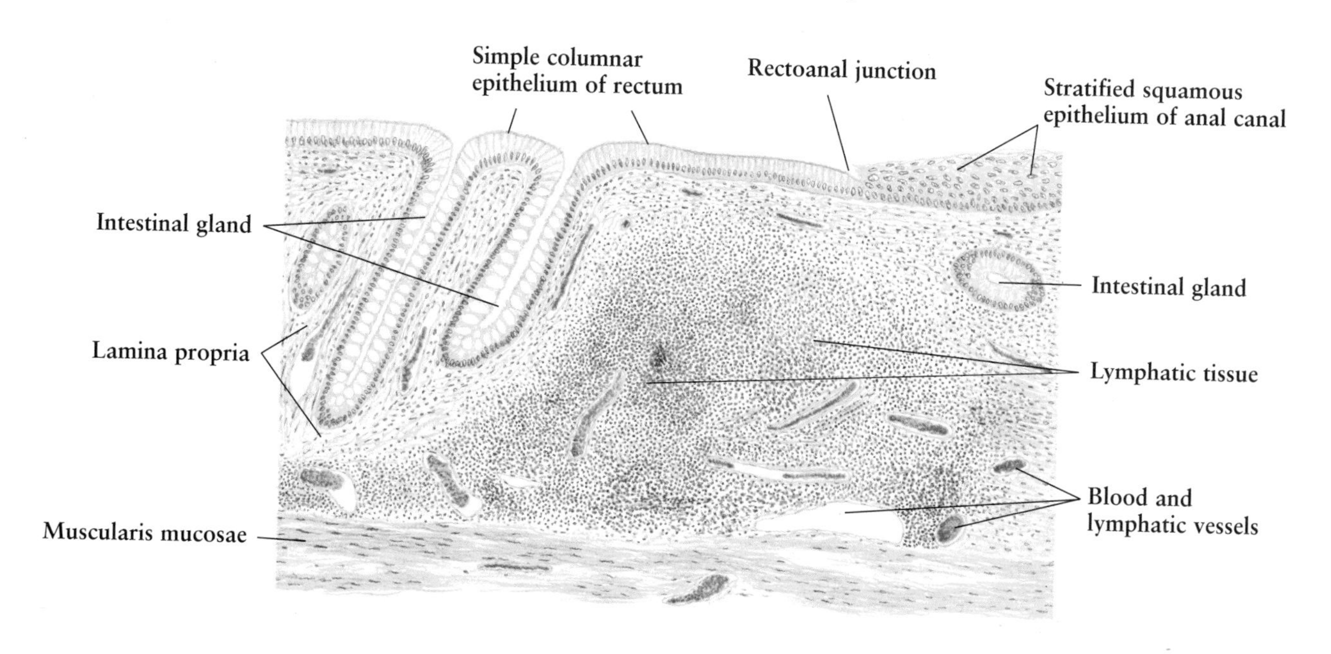

Figure 10-44. Liver: Hepatic Lobules
Pig • H.E. stain • Low magnification

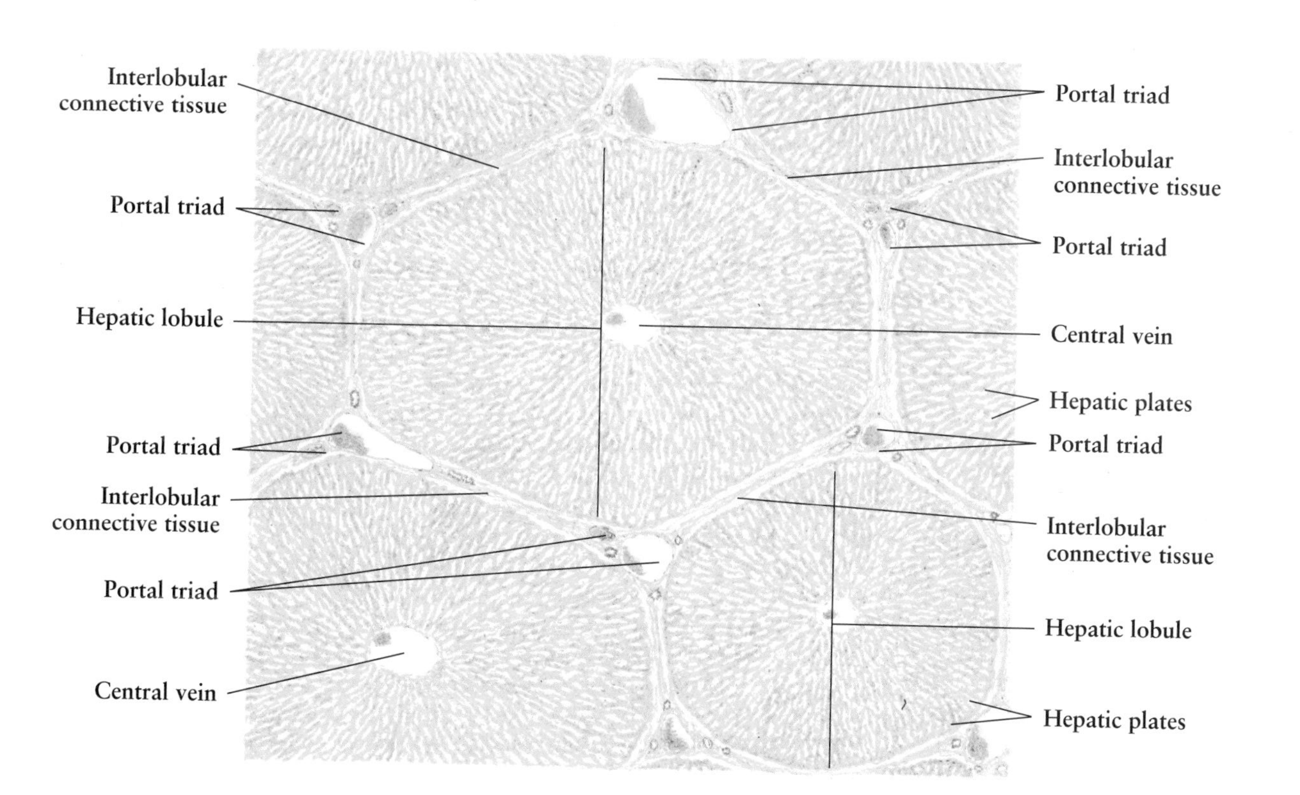

Fig. 10-45.

Liver: Hepatic Lobules

The **liver** is the largest compound gland (about 1500 g in the adult human), located beneath the diaphragm in the upper-right quadrant of the abdominal cavity. It is divided into right and left lobes by the line of attachment of the gastrohepatic and falciform ligaments. The liver is enclosed by a thin capsule of dense connective tissue and mostly covered by a visceral layer of peritoneal mesothelium. On the inferior surface there is a transverse fissure known as the *porta hepatis,* where the *portal vein,* the *hepatic artery,* and the *hepatic (bile) ducts* enter or leave the liver. The major functions of the liver include *detoxification* of metabolic waste products, *synthesis* of plasma proteins and plasma lipoprotein, *storage* of carbohydrates, and *secretion* of bile salts, bilirubin, and IgA.

As a structural unit of the liver, each **hepatic lobule** has a **central vein** in the middle, with **hepatic plates** radiating from it to the periphery. The hepatic plate consists of a single layer of hepatocytes, anastomosing with one another to form a network in three dimensions. Between the hepatic plates are **sinusoids** filled with blood and phagocytic cells. The human liver has a structural pattern of lobules similar to that of pig liver, but human hepatic lobules are not compartmentalized by distinct connective tissue of the interlobular septa. A **portal triad** is present in the connective tissue at the junction of three neighboring lobules. Each human hepatic lobule is surrounded by several portal triads. Although the lobules vary in orientation, their hexagonal outline can be identified by drawing a line connecting the portal triads around the lobule.

Each **portal triad** contains vessels of four types: *interlobular vein, interlobular artery, interlobular bile duct,* and *lymphatic vessel.* The **interlobular veins,** preterminal branches of the portal vein, transport nutrients and some poisonous substances from the intestine to the liver, where the hepatocytes absorb, process, and detoxify them. The arterial blood carried by the **interlobular arteries,** preterminal branches of hepatic arteries, is rich in oxygen. It is utilized by the hepatocytes for metabolism. The blood from both portal vein and hepatic artery mixes after entering the lobule. It flows between the hepatic plates in the sinusoid to the central vein. The blood from the central vein is collected by the **sublobular vein,** then passes via the hepatic vein to the inferior vena cava. The sublobular vein drains two or three central veins, running without any accompanying vessel. It has a larger lumen than the central vein, and its endothelium is enclosed by a layer of **connective tissue.** The **interlobular bile duct** collects bile from the bile canaliculi within the lobule, then drains into the left and right hepatic ducts, and finally to the common bile duct. Lymphatic capillaries are not found in intralobular structures, but **lymphatic vessels,** which collect hepatic lymph and drain to the thoracic duct, are present in the portal triad.

Figure 10-45. Liver: Hepatic Lobules

Human • H.E. stain • Low magnification

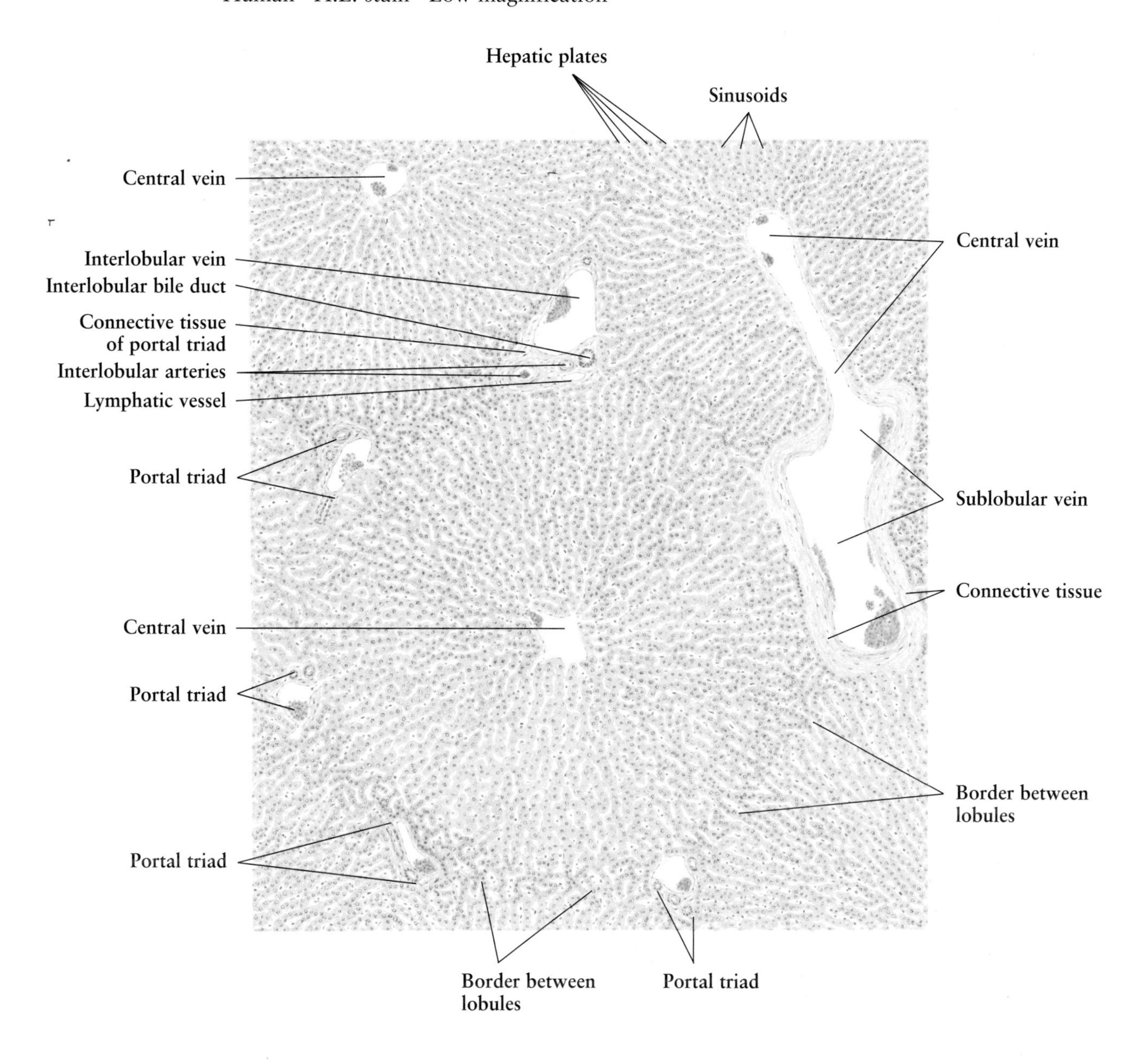

Fig. 10-46.

Liver: Intralobular Structure

This illustration demonstrates a high magnification of the intralobular structure. A **hepatic plate** is composed of **hepatocytes,** usually only one cell thick. The spaces between hepatic plates are occupied by the **sinusoids.** Each hepatocyte has two sides bathed with blood, giving a huge surface area for exchange of metabolites. The sinusoids are lined with a very thin discontinuous **endothelium,** supported by a network of reticular fibers (see **Fig. 2-10**). There is a narrow space between the endothelial lining and the hepatocytes, known as the **space of Disse,** which is filled with plasma. Therefore, the hepatocytes actually do not directly contact the blood. Within the sinusoids there are numerous phagocytic cells, the **Kupffer cells.**

The **central vein** links up the sinusoids. It is lined with a layer of **endothelial cells** that is continuous with that of the sinusoids. The wall of the central vein does not show the typical three-layer structure. Surrounding the endothelial lining is a small amount of **collagen fibers, fibroblasts,** and reticular fibers (see **Fig. 2-10**).

Under higher magnification, very thin holes or canals between hepatocytes can be observed. These are the **bile canaliculi.** Also found between hepatocytes are **fat-storing cells** with a pale cytoplasm and a small dense nucleus.

Fig. 10-47.

Liver: Hepatocytes and Sinusoid

This figure is based on a plastic epon section of rabbit liver stained with toluidine blue. It shows the fine structure of *hepatocytes* and *sinusoids.*

Large, polygonal **hepatocytes** form the **hepatic plate.** Their cytoplasm in H.E.-stained sections is eosinophilic; in this figure, however, it is filled with granular structures which at the level of electron microscopy are identified as the mitochondria and lysosomes. The pale areas correspond with granular or agranular endoplasmic reticulum. Each hepatocyte has one or two **nuclei,** each with a conspicuous nucleolus. The hepatocytes are responsible for almost all the functions of the liver. **Bile canaliculi** can be seen as tiny cavities between adjacent hepatocytes; they have no independent walls. Instead, the canalicular "walls" are formed by the cytoplasmic membrane of hepatocytes. The bile made by the hepatocytes is first secreted into the lumen of the bile canaliculi.

Between the hepatic plates are the **sinusoids.** They differ from capillaries in that they are of greater diameter (9–12 μm), and their endothelial lining is atypical, with an incomplete basement membrane. The sinusoidal **endothelial cell** has a flattened nucleus and greatly attenuated cytoplasm. The endothelial lining appears incomplete, with gaps between adjacent cells and fenestrations in the attenuated cytoplasm. **Kupffer cells** are located among the endothelial cells, but they are often found within the lumen of the sinusoids. These cells are irregular, large, stellate in shape, and contain a large ovoid nucleus. The cytoplasm has many processes or pseudopodia and contains **lysosomes** and engulfed **phagosomes.** Kupffer cells are actively phagocytic, involved with the spleen in the removal of aged erythrocytes and other debris from the circulation.

The **space of Disse** is the narrow plasma-filled space between the hepatocyte surface and the endothelial lining. The space of Disse is continuous with the sinusoid through the gaps between the endothelial cells of the sinusoid walls, and through the fenestrations in the attenuated cytoplasm of the endothelial cells. Numerous irregular microvilli are projected from the luminal surface of the hepatocyte into the space of Disse, thus greatly increasing the plasma membrane surface available for bidirectional exchange of metabolites between the liver and the blood.

Figure 10-46. Liver: Intralobular Structure
Human • H.E. stain • High magnification

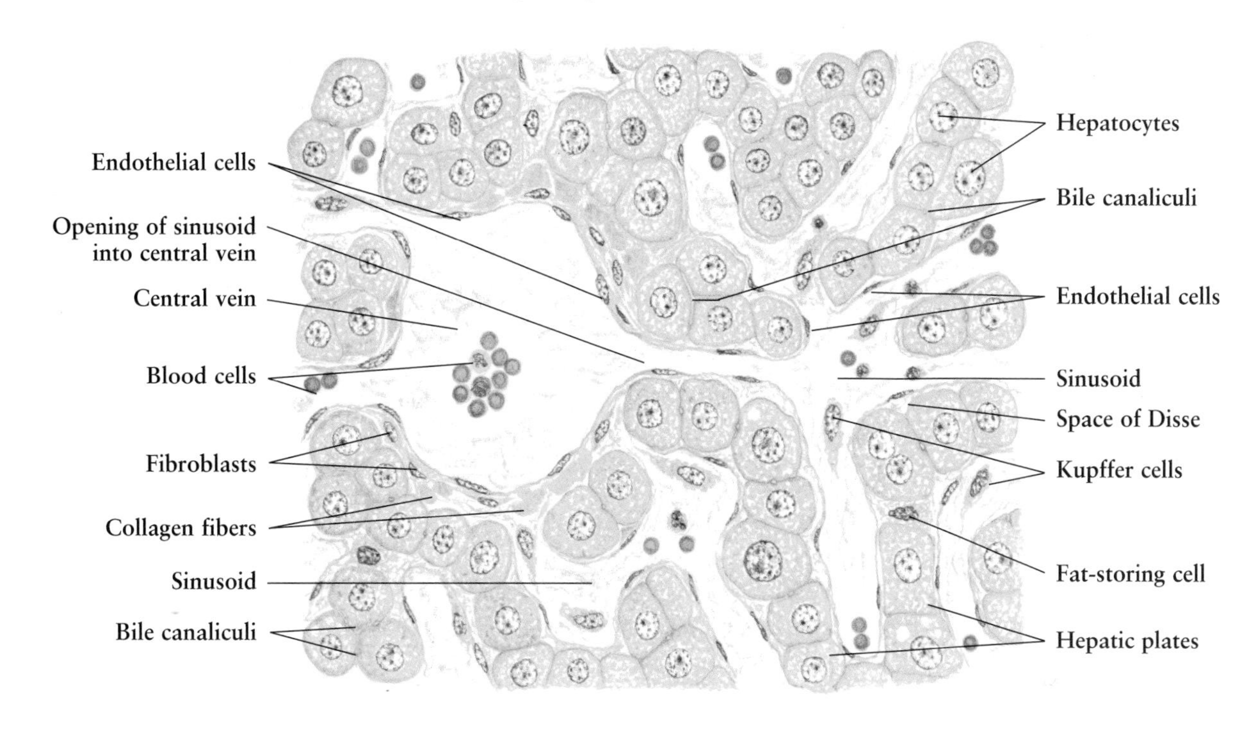

Figure 10-47. Liver: Hepatocytes and Sinusoid
Rabbit • Epon section • Toluidine blue stain • High magnification

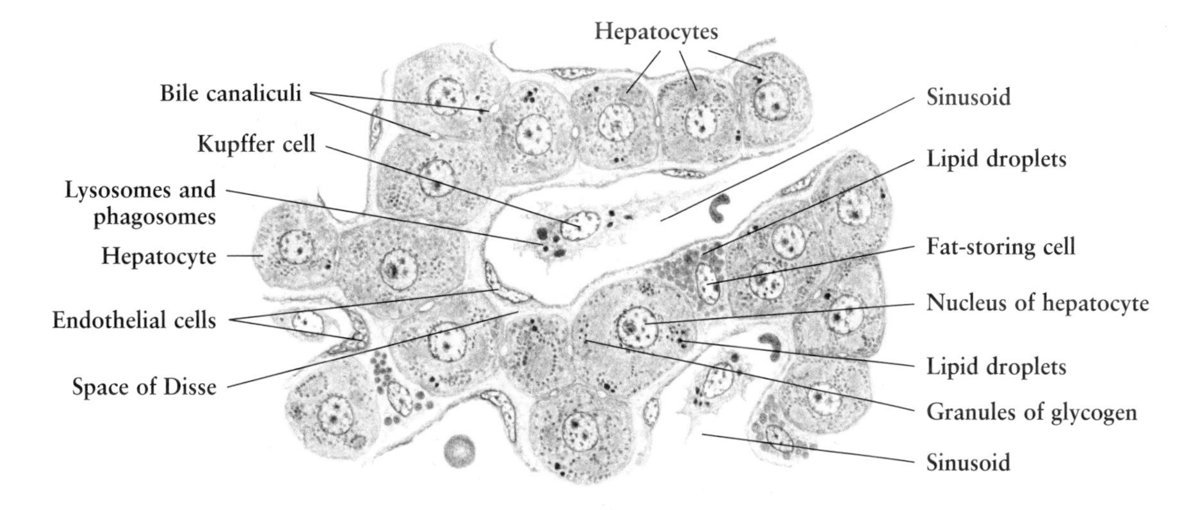

Fat-storing cells are located between hepatocytes or within the space of Disse. They are smaller than the hepatocytes and have an ovoid nucleus. The cytoplasm appears pale but always contains lipid droplets. The cell has at least two sides bathed in the blood plasma in the space of Disse. The fat-storing cells are involved in the accumulation of fat and vitamin A, as well as the production of intralobular collagen and reticular fibers. Under certain pathological conditions, they may be transformed into adipose cells or fibroblasts.

Fig. 10-48.

Hepatocytes: Glycogen

Hepatocytes contain a high quantity of glycogen, which is polymerized from glucose absorbed from the small intestine. Glycogen is water soluble and thus removed in ordinary histological preparation, leaving a characteristic appearance of irregular, unstained areas within the cytoplasm. It can however be demonstrated by histochemical methods such as the periodic acid-Schiff (PAS) reaction. In this figure, the PAS agent imparts a red color to the **granules of glycogen** within the hepatocytes. Note that the nuclei of the hepatocytes are not stained in this illustration.

Fig. 10-49.

Liver: Kupffer Cell

This is a figure of rabbit liver after intravital injection of trypan blue, demonstrating **Kupffer cells** with phagocytosed particles. These cells are present in the **sinusoid** and have a large, ovoid nucleus with a conspicuous nucleolus. The cytoplasm is filled with phagocytosed dye particles, so that the contour of Kupffer cell is outlined by the blue-stained cytoplasm. The phagocytic vacuoles are also found within the cytoplasm. No other cell types take up the dye particles.

Also shown in this drawing are **hepatocytes, bile canaliculi, fat-storing cells, endothelial cells** of the **sinusoid,** the **space of Disse,** and **erythrocytes.**

Fig. 10-50.

Liver: Bile Canaliculi

As shown in **Figs. 10-46, 10-47,** and **10-49, bile canaliculi,** about 0.5–1.0 μm in diameter, are formed by the cytoplasmic membranes of adjacent **hepatocytes.** They therefore have no wall of their own. The bile canaliculi are not easily seen in conventional preparations, but they are revealed by special staining methods, such as Golgi's silver impregnation used in this figure. In the plane of the hepatic plate, each hepatocyte is surrounded by a hexagonal network of bile canaliculi. The canaliculi anastomose with one another to form a three-dimensional network. In the periphery of the lobule, they connect with the canal of Hering (see **Fig. 10-52**), which drains into the bile duct in the portal triad.

Bile contains *bile salts, bilirubin,* and *steroids;* bile salts are associated with the digestion and absorption of fats. When bile canaliculi are broken because of the necrosis of hepatocytes or obstruction of the bile duct, bile enters the blood stream through the space of Disse, resulting in the accumulation of bilirubin. Bilirubin is a metabolic product of hemoglobin and imparts a yellow appearance to the skin and sclera, a condition known as *jaundice.*

Figure 10-48. Hepatocytes: Glycogen
Rabbit liver • PAS reaction • High magnification

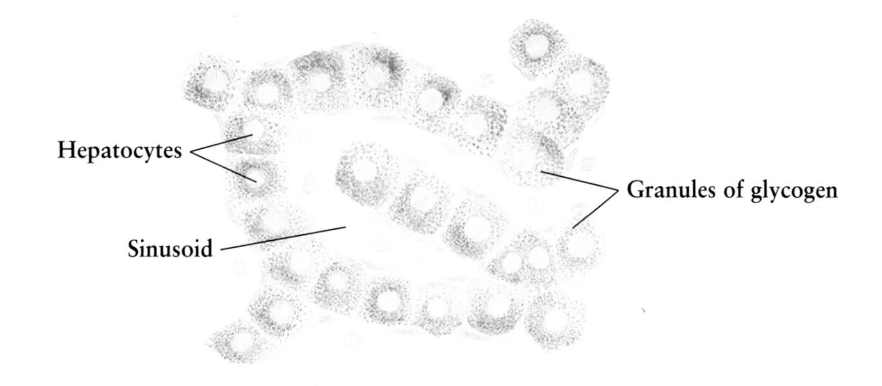

Figure 10-49. Liver: Kupffer Cell
Rabbit • Intravital injection of dye • Eosin stain • High magnification

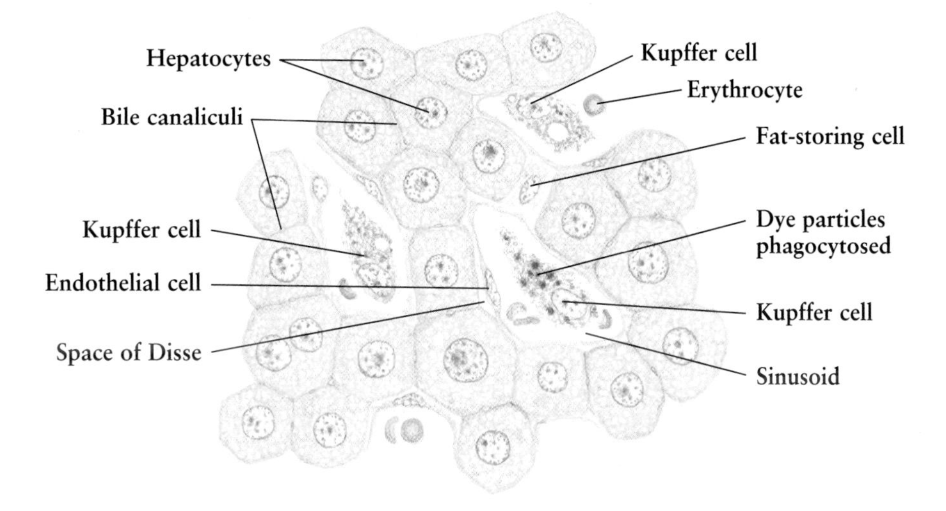

Figure 10-50. Liver: Bile Canaliculi
Human • Silver impregnation • High magnification

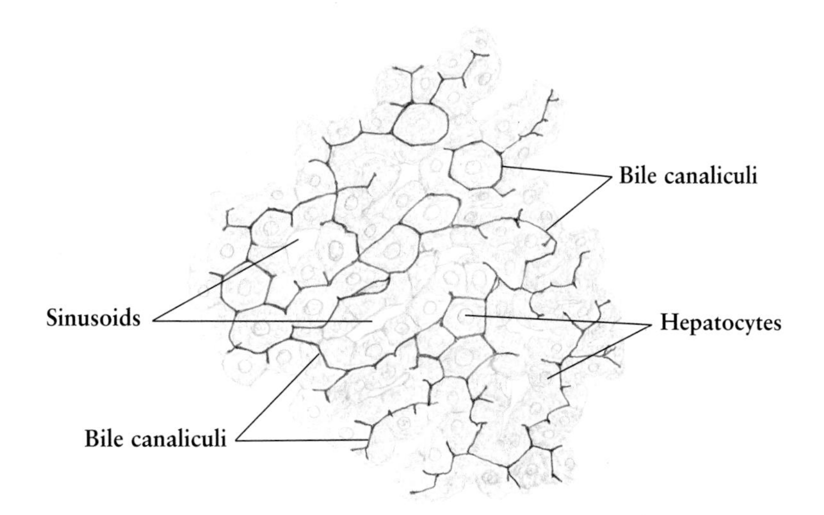

Figs. 10-51 and 10-52.

Liver: Portal Triad

The **portal triad** is a triangular or round area between three adjacent lobules. It consists of fibroconnective tissue, and contains *interlobular veins, interlobular arteries, interlobular bile duct,* and *lymphatic vessels.*

The **interlobular veins** are the largest in diameter in the portal triad. They are preterminal branches of the portal vein and carry nutrient-enriched blood to the sinusoids, which are separated from the hepatocytes by the **space of Disse.** The interlobular veins are lined by a layer of flattened endothelial cells and surrounded by **connective tissue** in the portal triad. The **interlobular arteries** are pretermined branches from the hepatic artery, responsible for the transportation of oxygen to the sinusoids for hepatocytes. Structurally the interlobular arteries resemble the arterioles, whose walls consist of a layer of endothelium and one or two layers of smooth muscle fibers. The **interlobular bile ducts** are formed of cuboidal or columnar epithelial cells. They connect intralobular bile canaliculi with the **canal of Hering.** The interlobular bile ducts ultimately empty bile into the common hepatic bile duct. The **lymphatic vessels** often appear as slits and have a very thin wall lined by endothelium.

With special staining methods, the portal triad shows many **collagen fibers** (Fig. 10-52). This fibroconnective tissue also contains **fibroblasts** and a few fine unmyelinated nerve fibers of the autonomic nervous system, which are not readily seen in conventional preparations.

Figure 10-51. Liver: Portal Triad
Human • H.E. stain • High magnification

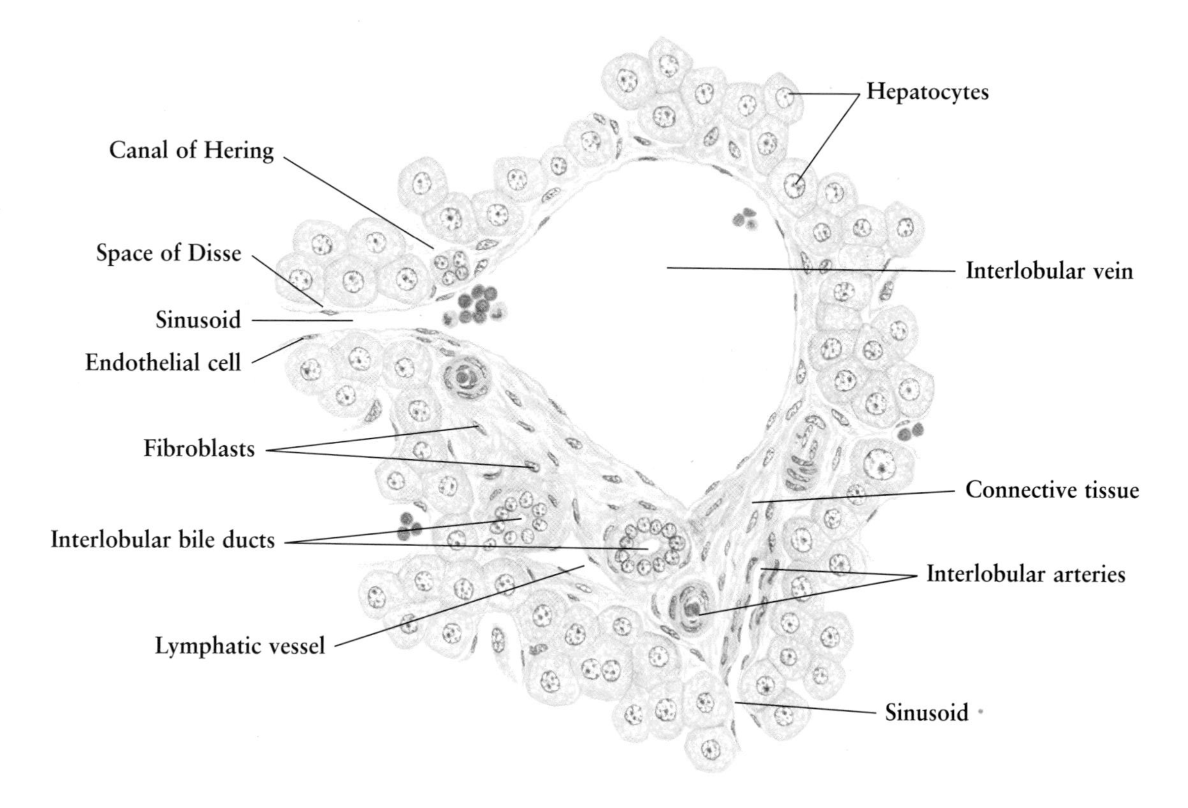

Figure 10-52. Liver: Portal Triad
Pig • Azan stain • High magnification

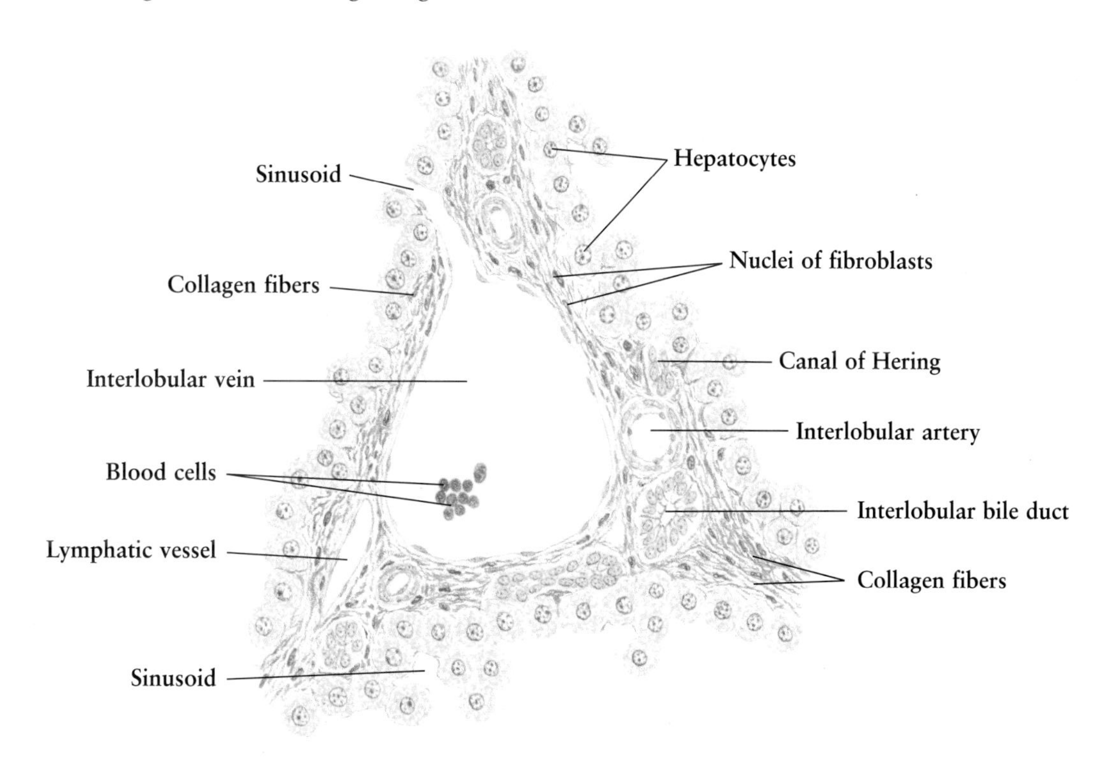

Fig. 10-53.

Gallbladder

The **gallbladder** is a small, pear-shaped sac, attached to the lower surface of the liver. It is only 7–10 cm in length and about 4 cm in diameter, with a capacity of 30–60 ml. The gallbladder connects with the common bile duct through the cystic duct. The wall of the gallbladder is thin, and composed of three layers: *mucosa, muscularis,* and *adventitia.*

The luminal surface of the gallbladder is lined with **simple columnar epithelium,** which rests on a basement membrane (see **Fig. 1-7**). The epithelial cells are tall and distinct, with an ovoid nucleus located toward the base. The cytoplasm is eosinophilic, with a few short microvilli at the free surface. The **lamina propria** that supports the epithelium consists of loose **connective tissue** containing collagen and elastic fibers, fibroblasts, and a **vascular plexus.** In the neck region of the gallbladder, the lamina propria contains numerous tubuloacinar glands. The entire **mucosa** is thrown into numerous folds, and the deep epithelial invaginations can be seen in sections as crypts, called **Rokitansky-Aschoff sinuses.** There is a single layer of muscle and no submucosa in the wall of the gallbladder. The **muscularis** is irregularly arranged **smooth muscle** bundles intermingled with a considerable amount of collagen and elastic fibers. The **adventitia** is a layer of loose connective tissue containing many **blood vessels,** lymphatic vessels, and nerve fibers. The part of the gallbladder that does not face the liver is covered by a serous membrane.

The main functions of the gallbladder are storage and concentration of bile (by absorbing water from it by the epithelium). Bile is discharged from the gallbladder into the intestine following contraction of the muscularis, upon stimulation by cholecystokinin.

Fig. 10-54.

Pancreas

The **pancreas** is a large, elongated (12–15 cm) organ, located in the concavity of the duodenum behind the peritoneum of the posterior abdominal wall. It is enclosed by a loose connective tissue that penetrates the parenchyma as connective septa, dividing the parenchyma into lobules. The pancreas is composed of an *exocrine portion* and an *endocrine portion.*

The **exocrine portion** consists of compound, tubuloacinar glands divided into **lobules.** Each is surrounded with **interlobular connective tissue,** which contains excretory ducts, blood and lymphatic vessels, and nerves. A lobule contains numerous secretory units, the **acini,** which connect to the **intercalated ducts.** The intercalated ducts drain into the **intralobular duct,** that, in turn, drains into the **interlobular duct,** which ultimately converges upon the main pancreatic duct. The **endocrine portion** consists of numerous isolated groups of endocrine cells, known as **islets of Langerhans.** They are scattered among the exocrine acini and are characterized by a pale-staining appearance in the H.E.-stained preparations. The islets, which produce hormone, are shown as endocrine organs in **Figs. 14-14** and **14-15.** Also found in the lobule are **adipose cells.**

The exocrine portion secretes an alkaline pancreatic juice that is released through the main pancreatic duct into the duodenum. This juice contains *trypsin, chymotrypsin, amylase, lipase, carboxyl peptidase, ribonuclease,* and *deoxyribonuclease.* The secretion of pancreatic juice is mainly regulated by two hormones, *secretin* and *cholecystokinin,* which are secreted by the duodenal mucosa. Stimulation of vagus parasympathetic activity also induces some pancreatic secretion.

Figure 10-53. Gallbladder

Monkey • H.E. stain • Medium magnification

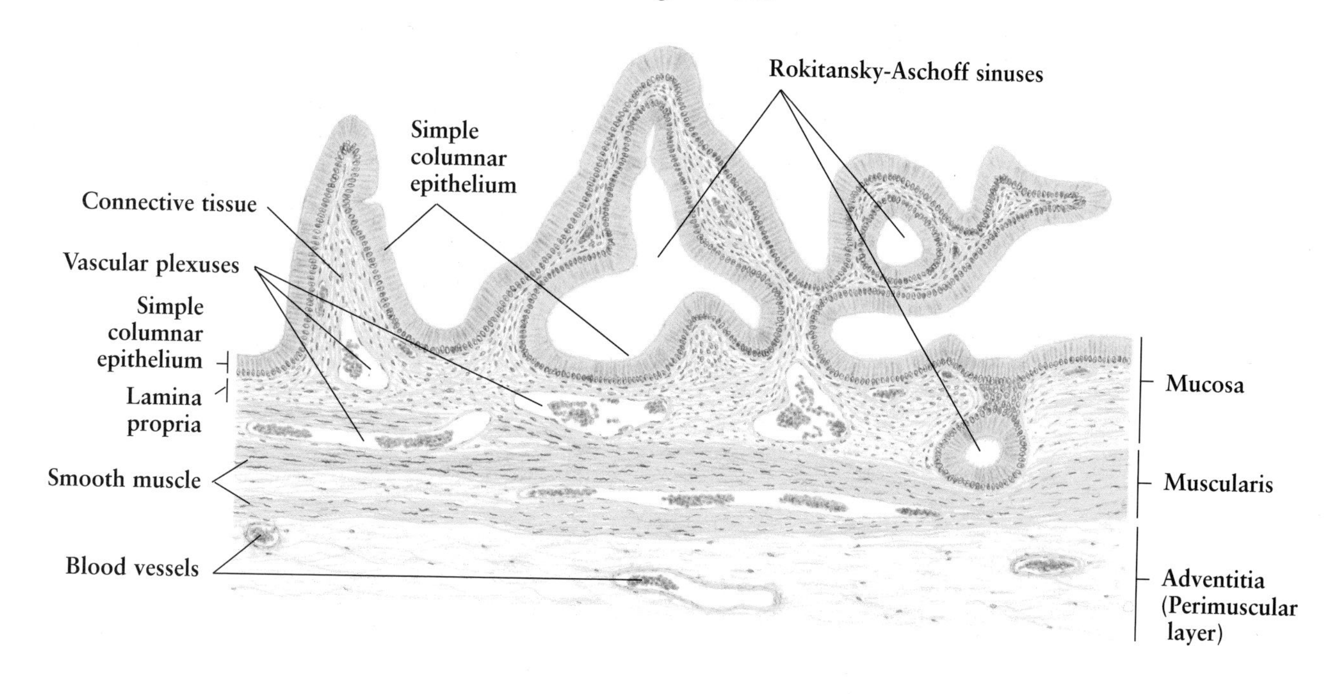

Figure 10-54. Pancreas

Human • H.E. stain • Low magnification

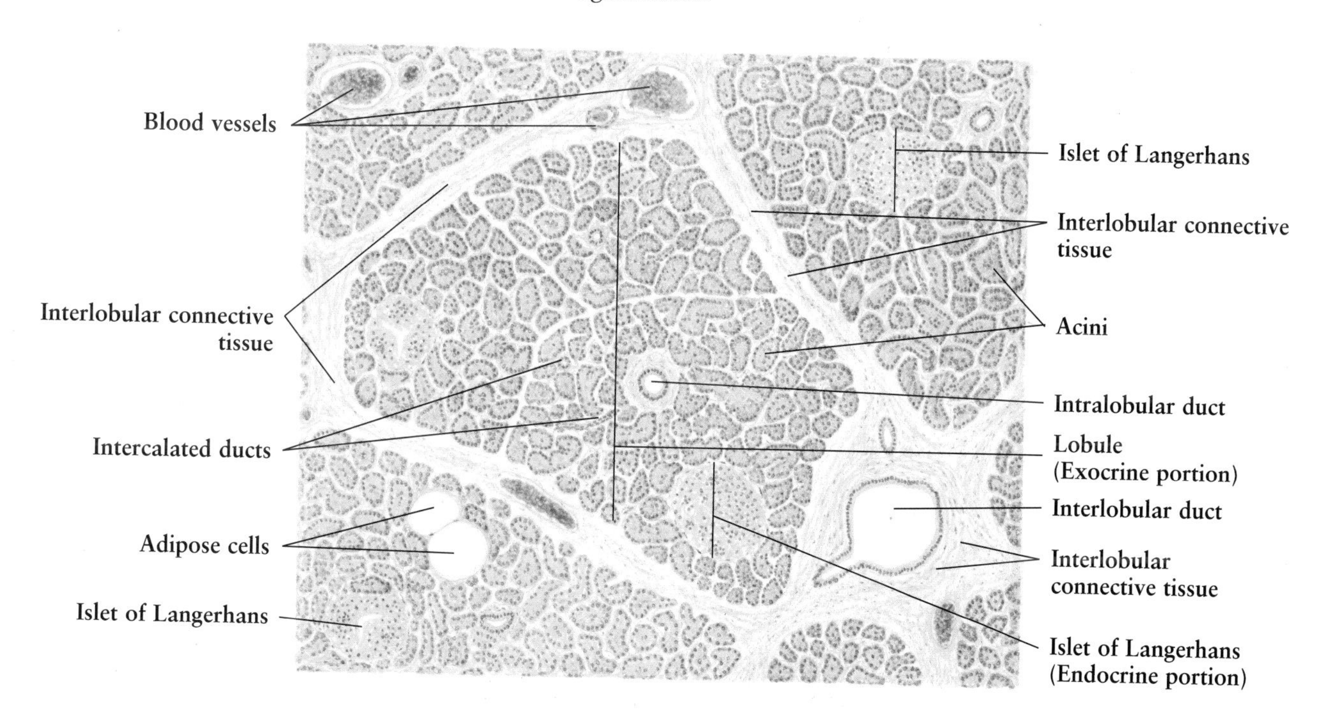

Fig. 10-55.

Pancreas: Exocrine Portion

The **exocrine portion** of the pancreas is a typical serous gland, consisting of tubular, round, or pear-shaped acini surrounded by a basement membrane. Myoepithelial cells are not present. Each **acinus** is composed of several pyramidal **acinar cells** arranged around a small central **lumen.** The acinar epithelial cell has a spherical nucleus with abundant chromatin and one to three nucleoli lying toward the base. The basal cytoplasm is basophilic **ergastoplasm** rich in rough endoplasmic reticulum; the apical cytoplasm contains numerous eosinophilic secretion granules, the **zymogen granules.** Present in the acinar lumen are so-called **centroacinar cells.** These are flattened epithelial cells with ovoid nuclei and pale-stained cytoplasm, which represent the beginning of the duct system.

With the centroacinar cells as the beginning, the intercalated duct connects with the intralobular duct. The **intercalated duct** is formed of flattened or low cuboidal epithelial cells, which possess a spherical or oval nucleus and scanty cytoplasm. In longitudinal section, the intercalated duct can be seen as two rows of nuclei. The **intralobular duct,** with a large lumen, is composed of simple columnar epithelium surrounded by a thin layer of loose **connective tissue.** Between acini are small amounts of delicate connective tissue with **capillaries.** Also shown in this figure are the **islets of Langerhans** and **adipose cells.**

Fig. 10-56.

Pancreas: Glandular Acinus

This figure demonstrates the fine structure of a glandular acinus of the exocrine portion of the pancreas. The acinus consists of pyramidal glandular epithelial cells, the **acinar cells,** which have large and spherical **nuclei** located in the base of the cells. The **secretory granules** are always found in the apical cytoplasm. The **intercalated duct,** continuous with the acinus, consists of a single layer of epithelial cells that penetrate to the lumen of the acinus and are evident as **centroacinar cells.** All acinar cells rest on the **basement membrane,** and the acinus is surrounded by a loose connective tissue that contains **capillaries** and **fibroblasts.**

Figure 10-55. Pancreas: Exocrine Portion
Human • H.E. stain • High magnification

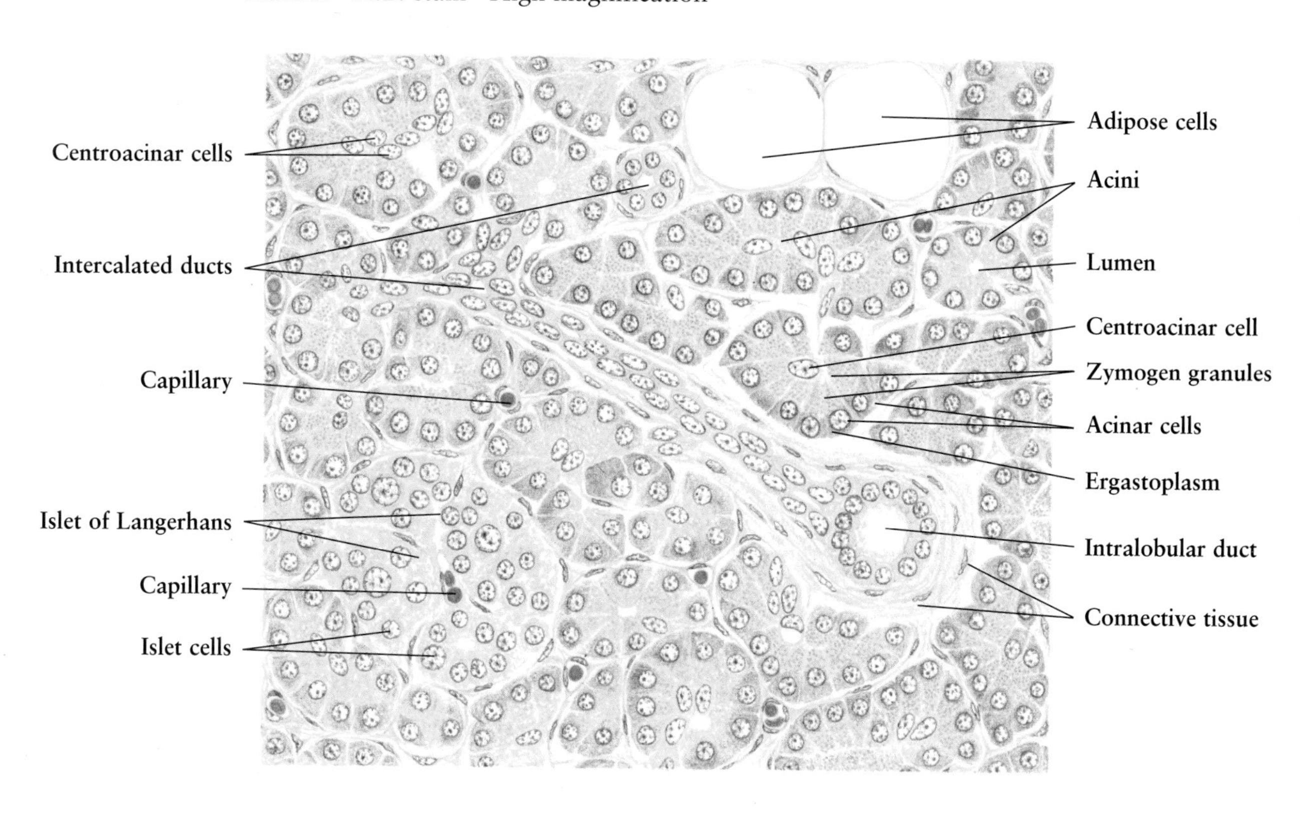

Figure 10-56. Pancreas: Glandular Acinus
Human • Epon section • Toluidine blue stain • High magnification

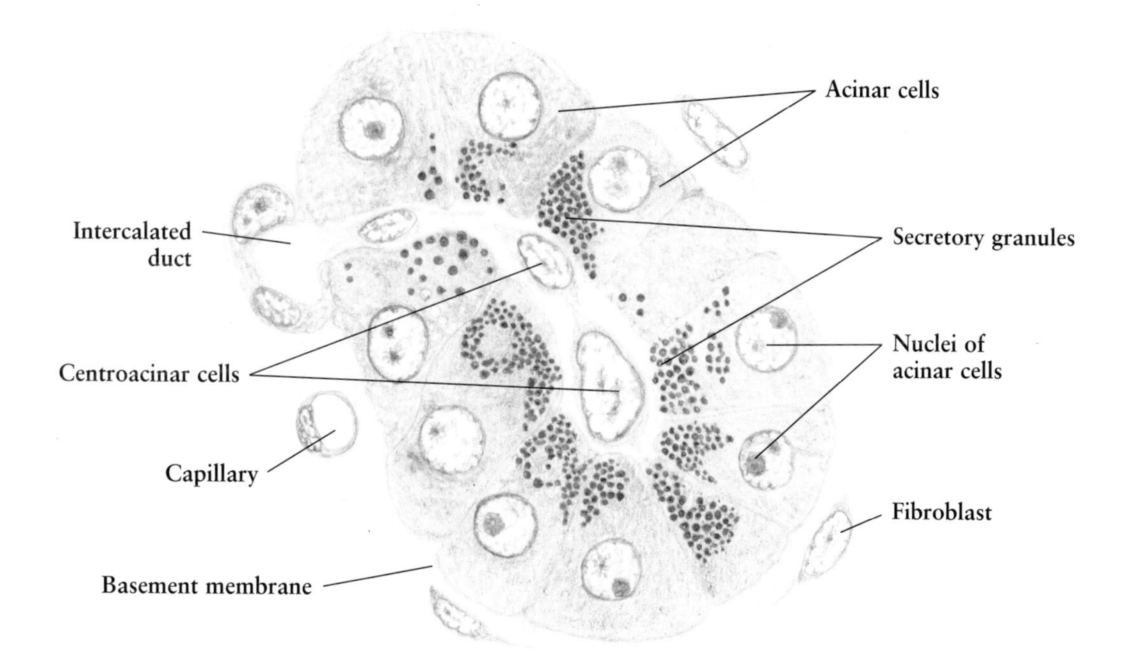

11. Urinary System

The **urinary system** consists of the two *kidneys,* the two *ureters,* the *urinary bladder,* and the *urethra.* This system regulates *water* and *electrolyte homeostasis* by the production of *urine,* the medium in which various metabolic waste products (particularly nitrogenous compounds such as urea and creatine) are eliminated. Urine, produced in the kidneys, passes down the ureters to the urinary bladder where it is temporarily stored and then eventually evacuated to the exterior through the urethra.

Each **kidney** is a large compound tubular gland and contains numerous **nephrons.** The nephron is the basic structural and functional unit; it consists of a **renal corpuscle** and a long, convoluted **renal tubule.** Nephrons are involved in (1) the *formation* of filtrate by ultrafiltration of blood plasma, (2) the *selective resorption* of most of the filtered water and other small molecules (such as amino acids, glucose, and sodium ions), and (3) the *secretion* of some excretory products (such as hydrogen and potassium ions). In addition, nephrons produce the hormones *renin* and *erythropoietin.* Renin is associated with the regulation of blood pressure, and erythropoietin stimulates production of erythrocytes in bone marrow.

The **ureter** is an excretory duct that drains urine away from the renal pelvis to the urinary bladder. Each ureter is lined by transitional epithelium. The **urinary bladder** is a hollow muscular organ, and is characterized by a thick lining of transitional epithelium. The **urethra** delivers urine from the urinary bladder to the outside. The male urethra is a duct common to the urinary and reproductive systems; the female urethra belongs only to the urinary system.

Fig. 11-1.

Kidney: Renal Pyramid

The **kidneys** are bean-shaped organs, 10–12 cm long and 3.5–5 cm thick. They are bilaterally positioned in the upper-posterior abdominal wall with one on either side of the vertebral column. Each kidney is invested by a tough fibroconnective tissue **capsule** that does not penetrate the parenchyma. The *hilum* of the kidney is a depressed region with a great deal of adipose tissue. Through this area the renal artery and vein, ureters, lymphatic vessels, and nerves enter or leave the kidney.

The basic structural and functional unit of the kidney is the *nephron.* It is responsible for filtration, excretion, and resorption. There are more than a million nephrons densely packed in each kidney. Each nephron consists of a *renal corpuscle* and a long, *convoluted renal tubule,* which is divided into a *proximal convoluted tubule, loop of Henle,* and a *distal convoluted tubule* that joins to the *collecting tubule.*

The substance of the kidney is made up of an outer **cortex** and an inner **medulla.** A portion of each nephron may be located in both cortex and medulla, but the renal corpuscles are only present in the cortex. The cortex can be divided into an **outer zone** and **inner zone** based on the distribution of the corpuscles. The **medullary rays** are groups of straight collecting tubules and loops of Henle in the cortex. Between the medullary rays are **cortical labyrinths,** composed of renal corpuscles and convoluted tubules. According to the combination of renal tubules at different areas, the medulla may also be divided into an **outer zone** and **inner zone.** The outer zone is further subdivided into an **outer stripe** and **inner stripe.**

The medulla is arranged into 10–18 pyramidal structures, known as **renal pyramids.** One is shown in this figure. The base and sides of the renal pyramids are within the cortex, and their vertices, the **renal papilla,** protrude into the **minor calyx.** The renal pyramids are separated by a layer of connective tissue and the **renal column,** which possesses cortical structure. One renal pyramid is also called a *lobe.* A *renal lobule* is composed of a single medullary ray and the cortical tissue surrounding it.

After entering the hilum, the renal artery ramifies to form **interlobar arteries,** which run between the medullary pyramids. At the corticomedullary junction, the interlobar arteries give rise to **arcuate arteries,** which run parallel to the surface of the kidney. The interlobular arteries, derived from the arcuate arteries, run between the medullary rays toward the capsule. The interlobular arteries branch into several afferent glomerular arterioles, which supply the capillaries of the **glomeruli.** From there blood passes into the efferent glomerular arterioles, which branch again to form a second capillary network, supplying the majority of other portions of the same nephrons. The blood is collected into the interlobular veins, which join the **arcuate veins.** Then blood flows into the **interlobar veins,** which collectively form the renal vein through which blood leaves the kidney.

Figure 11-1. Kidney: Renal Pyramid
Human • M.G. stain • Low magnification

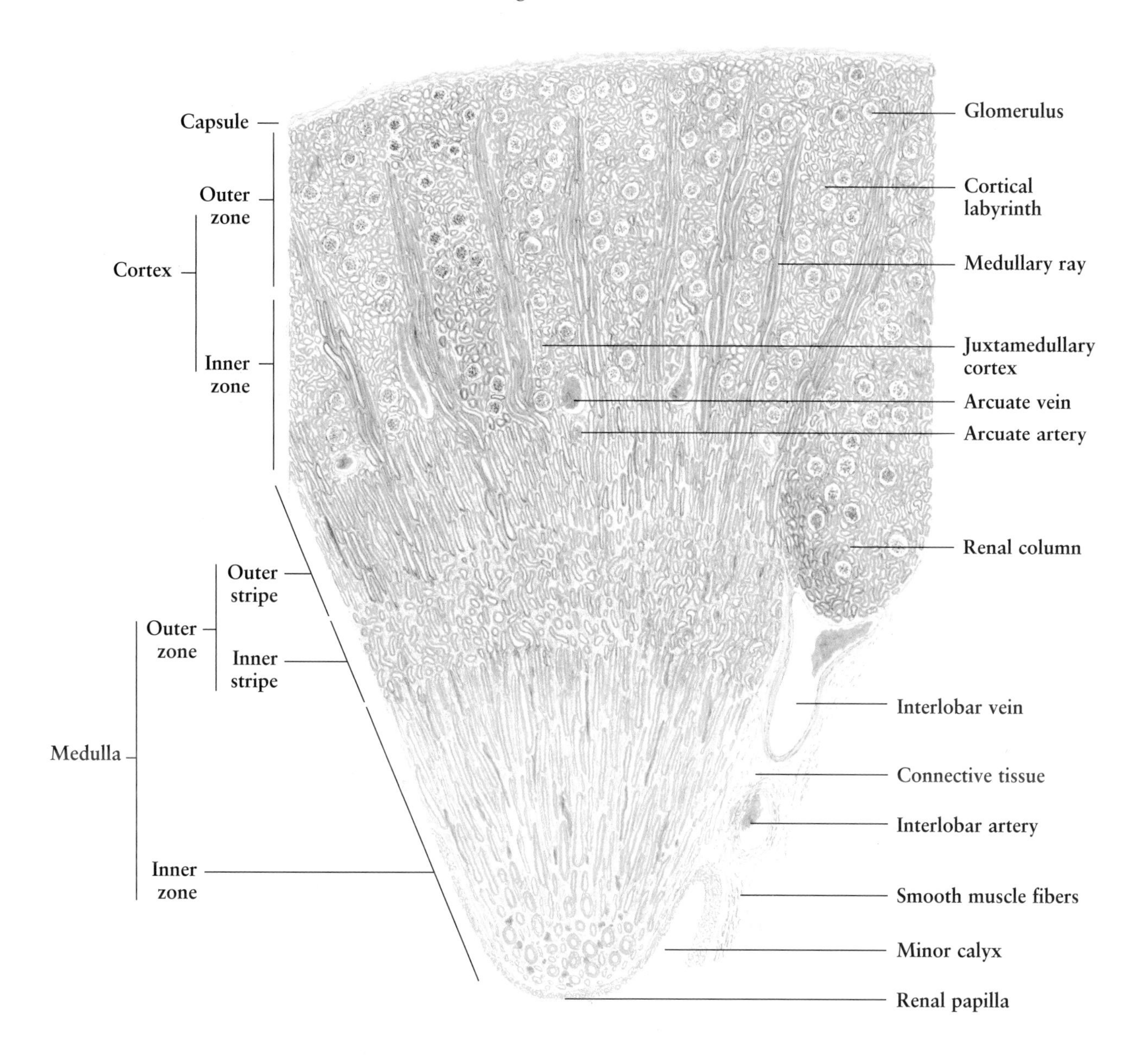

Fig. 11-2.

Cortex: Renal Corpuscle

The **renal corpuscle,** also known as a *Malpighian corpuscle,* is the essential portion of the nephron. It consists of a **renal glomerulus** and **Bowman's capsule.** The corpuscle has two poles: the *vascular pole* and *urinary pole.* At the **vascular pole,** the **afferent arteriole** (which supplies the capillary network of the glomerulus) enters, and the **efferent arteriole** (which collects the blood from the capillaries) leaves the corpuscle. The **urinary pole** is the beginning of the proximal convoluted tubule; it is continuous with Bowman's space.

The **renal glomerulus** is composed of a cluster of **capillaries.** After entering the corpuscle from the vascular pole, the afferent arteriole immediately branches into numerous capillaries that anastomose with one another and form a complicated capillary network. All capillaries eventually converge to form the **efferent arteriole.** The glomerulus is suspended in Bowman's capsule by means of the vascular pole. The endothelial cells of the capillary are fenestrated, rest on a thin basement membrane, and are externally surrounded by a layer of pedicels of the **podocytes** that constitute the visceral lining of Bowman's capsule. These three layers are called the glomerular *filtration barrier,* through which the filtrate first enters Bowman's space from the blood circulation. The capillary loops are supported by stalks of special connective tissue containing **extraglomerular mesangial cells** at the vascular pole. This cell type (**mesangial cells**) may also be found between the capillaries within the glomerulus.

Bowman's capsule is a double-walled envelope that holds the glomerulus. The **parietal layer** of the capsule is composed of simple squamous epithelial cells that are continuous with the cuboidal cells of the proximal tubules at the urinary pole. The visceral layer is made up of **podocytes** whose feet surround the glomerular capillaries, forming one of the layers of the filtration barrier.

The walls of **afferent** and **efferent arterioles** are composed of three layers, similar to those of arterioles elsewhere. At the vascular pole, however, the smooth muscle cells in the tunica media of the afferent arteriole are modified into epithelioid cells with a large nucleus and pale-staining cytoplasm. These are the **juxtaglomerular cells,** responsible for the synthesis of the hormone *renin,* which is associated with enhancing blood pressure.

The **proximal convoluted tubule,** starting at the urinary pole, is about 14 mm long and 50–60 μm wide. It follows a tortuous course through the cortical labyrinth, medullary ray, and the outer stripe of outer medulla, and is continuous with the thin segment of the loop of Henle. The proximal convoluted tubule is lined by a layer of cuboidal cells, which are characterized by the **brush border** at the free surface, striations at the basal surface, and a large, round nucleus in the center of the cytoplasm. The striations are recognized at the electron microscopic level as numerous mitochondria and an infolded basal cytoplasmic membrane. The function of these epithelial cells is to reabsorb the glucose, amino acids, ascorbic acids, chloride, and sodium from the filtrate within the tubule.

The **distal convoluted tubule** is situated between the straight part of the proximal tubule and the collecting tubule, and is composed of a straight part and a distal convoluted part. The epithelial cells lining the distal convoluted tubule are cuboidal, with a round nucleus and basal striations at the base of the cells. The distal convoluted tubule is responsible for the resorption of sodium, an activity that is promoted by the hormone *aldosterone.* At the portion of the distal convoluted tubule contacting the parent renal corpuscle between afferent and efferent arterioles, the epithelial cells become densely packed, with the nuclei close together. Thus the region appears

Figure 11-2. Cortex: Renal Corpuscle
Human kidney • H.E. stain • High magnification

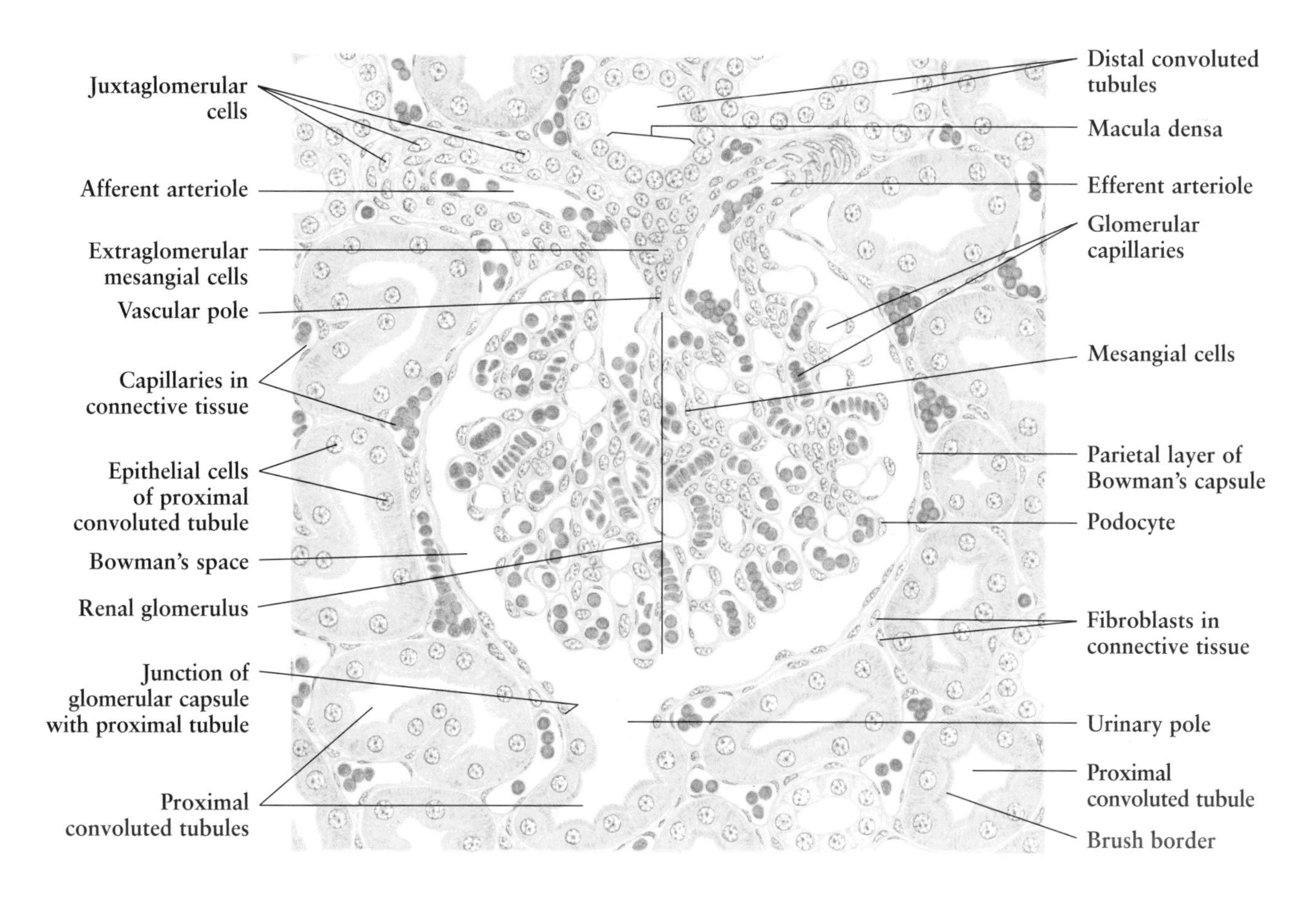

darker and is known as **macula densa.** The macula cells are immediately adjacent to the afferent and efferent arterioles, to the juxtaglomerular cells, and to the extraglomerular mesangial cells. These three cell groups together are called the *juxtaglomerular apparatus.*

In the cortex, the connective tissue is sparse, with some collagen fibers and a small number of **fibroblasts.** However, there is an abundance of **capillaries,** which are supplied by the efferent arterioles.

Fig. 11-3. Cortex: Medullary Ray

The **medullary rays** are finely striated extensions of the medullary tissue that project up into the cortex. They run parallel to the axis of the renal pyramid and represent the central axis of a renal lobule. A medullary ray consists of the straight portion of **proximal tubules,** the thick ascending limb of Henle's loops (**distal tubules**), and straight **collecting tubules.** The space between the tubules is filled with a very thin layer of **loose connective tissue,** containing some **fibroblasts** and a large number of **capillaries.**

Fig. 11-4. Medulla: Outer Stripe of Outer Zone

This drawing is made from a section cut at the **outer stripe** of the **outer medulla.** It illustrates a *thick descending limb, thin descending limb, thick ascending limb,* and *collecting tubule.*

The structure of the **thick descending limb** is similar to that of the proximal convoluted tubule seen in **Fig. 11-2.** The **thin descending limb** may occasionally be found at the level of the medulla through which this section was cut. It is lined by a simple squamous epithelium like a capillary. It differs from the capillary in several ways. For instance, the thin descending limb has a regular, rounded shape, its epithelial cells have thicker cytoplasm, and it lacks blood cells within the lumen. The **thick ascending limb** is lined by a layer of low cuboidal epithelial cells that resemble the distal convoluted tubule (see **Fig. 11-2**). These epithelial cells, however, are shorter and their round nuclei tend to bulge into the lumen. Additionally, they have a few microvilli at the free apical surface and striations at the base of the cell. These two parts of Henle's loop are involved in the resorption of water and sodium. The **collecting tubule** is lined by low cuboidal to tall columnar epithelial cells. The cells are generally pale-staining with a clear, regular cell border. Their nuclei are round or ovoid and situated at the center of the cell. The collecting tubules conduct urine from the nephron to the minor calyx, with some resorption of water.

The **capillary** network embedded in the sparse **loose connective tissue** containing a few **fibroblasts** can also be seen.

Figure 11-3. Cortex: Medullary Ray
Human kidney • H.E. stain • High magnification

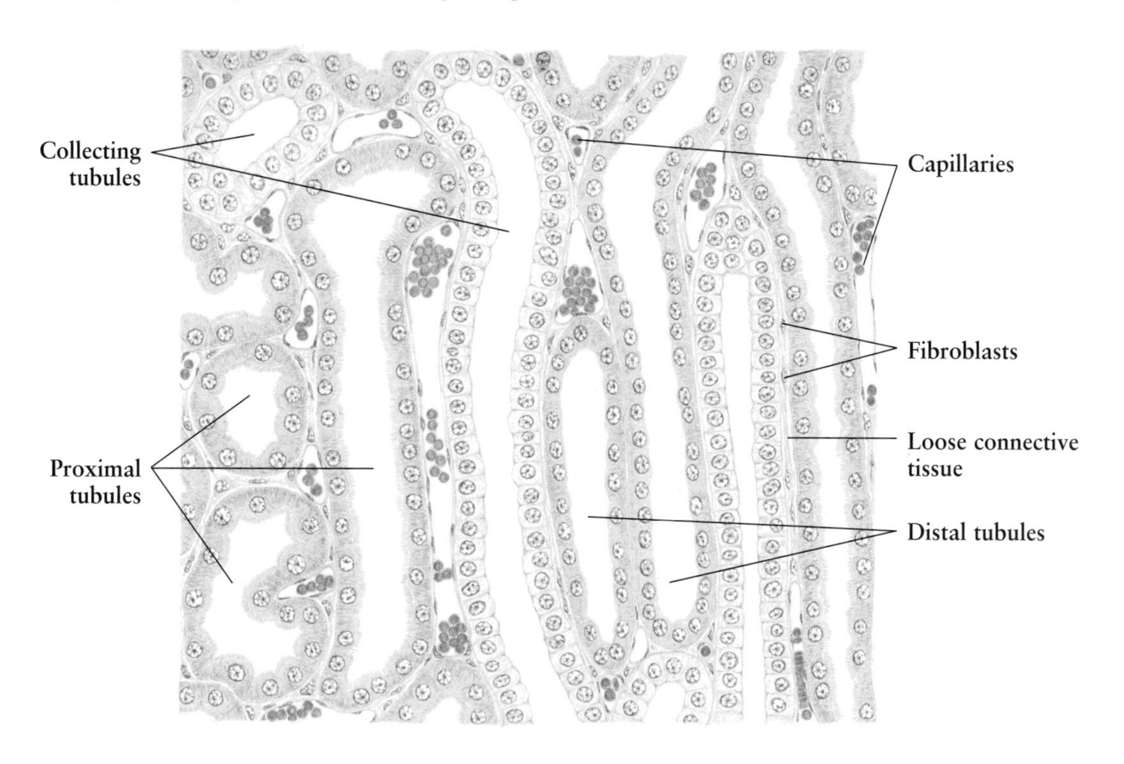

Figure 11-4. Medulla: Outer Stripe of Outer Zone
Human kidney • H.E. stain • High magnification

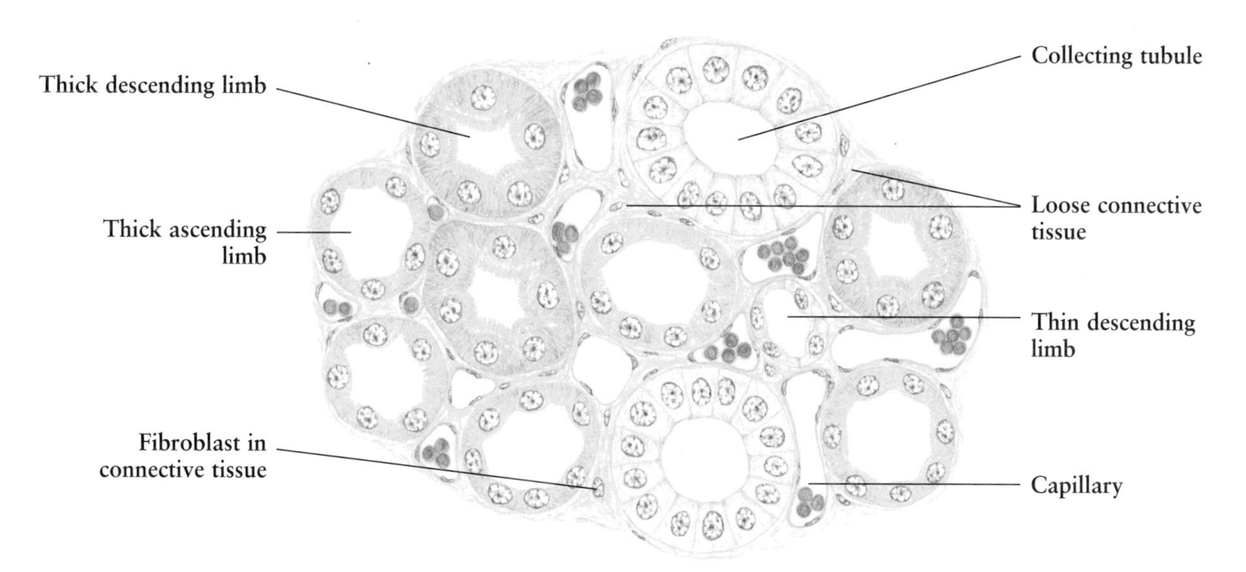

Fig. 11-5.

Medulla: Inner Stripe of Outer Zone

This is a transverse section through the **inner stripe** of the **outer medulla** of a human kidney, illustrating the presence of a **thin descending limb, thick ascending limb,** and **collecting tubule.** At this level, the thick descending limb (proximal tubule) is typically not present, but the thin descending limbs are quite abundant. Note that the **connective tissue,** containing **fibroblasts** and **capillaries,** gradually increases in volume.

Fig. 11-6.

Medulla: Inner Zone

At the **inner medulla** near the renal papilla, the collecting tubules converge to form **papillary ducts,** which convey urine into the pelvicalyceal space. The papillary ducts are lined by a layer of columnar or high columnar epithelial cells; they have a structure similar to that lining the collecting tubules. In addition to the papillary ducts, only **thin segments of Henle's loop** approach this inner zone. They run down to this zone and return to the outer medulla. There are no histological differences between the descending and ascending thin segment of Henle's loop.

Among these ducts and tubules there is a great amount of connective tissue. It contains **capillaries, collagen fibers** that are stained green by a special method used in this sample, and **interstitial cells.** These are fibroblast-like cells, whose function, apart from their collagen production, remains unknown.

The peripheral margin of the inner zone close to the papilla is covered by **transitional epithelium,** which is composed of only two or three layers of epithelial cells. These cells have a large nucleus, pale-staining cytoplasm, and a clear intercellular border.

Figure 11-5. Medulla: Inner Stripe of Outer Zone
Human kidney • H.E. stain • High magnification

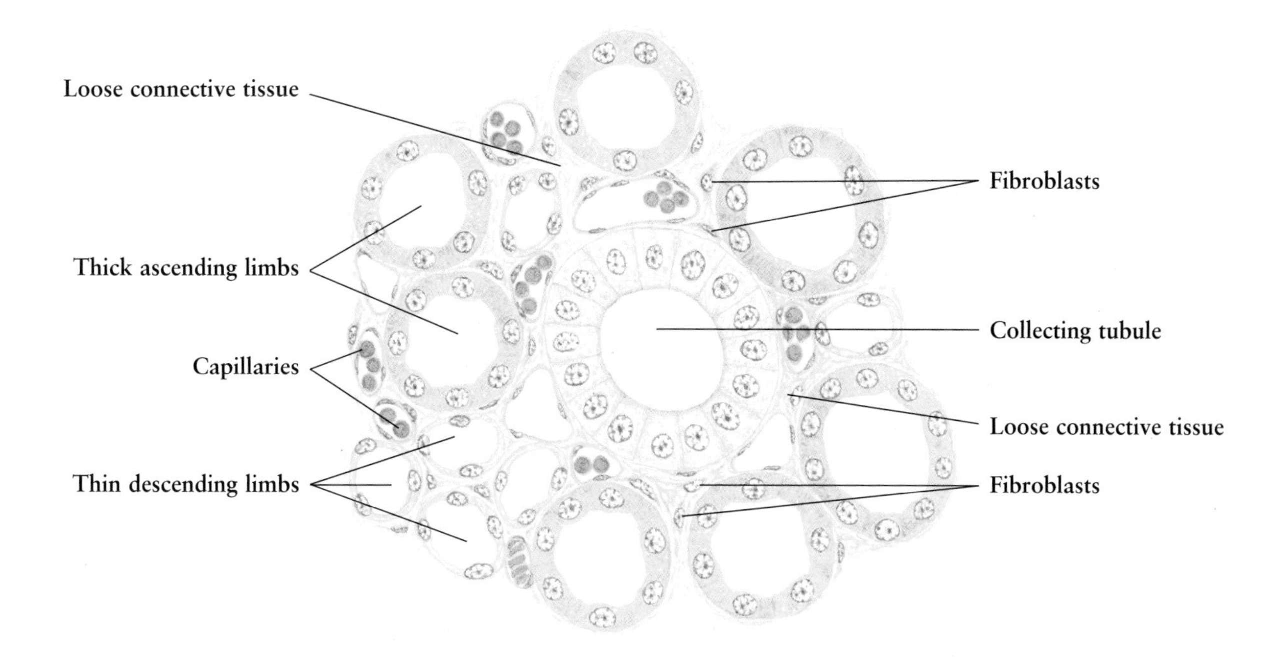

Figure 11-6. Medulla: Inner Zone
Human kidney • M.G. stain • High magnification

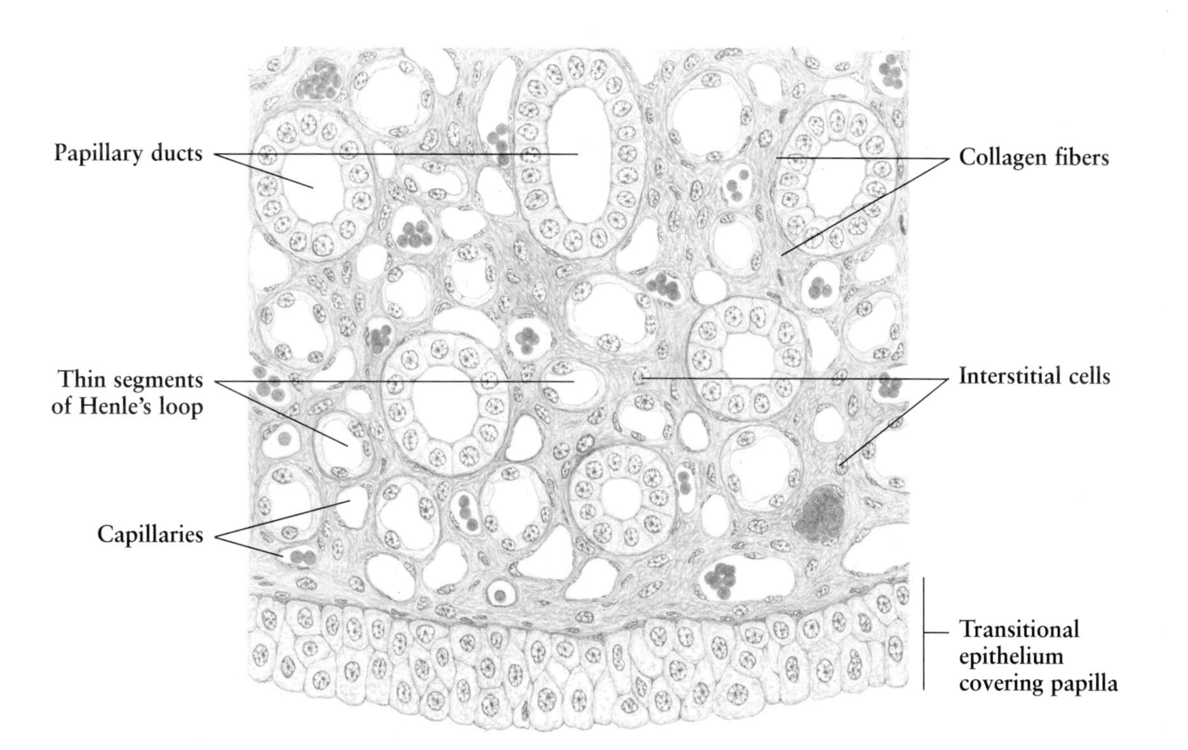

Fig. 11-7. Ureter

The **ureters** are muscular tubes, 25–30 cm long and 5–10 mm in diameter, leading from the renal pelvis to the urinary bladder. The wall of the ureter consists of three layers: *mucosa, muscularis*, and *adventitia* of connective tissue.

The **mucosa** is composed of the transitional epithelium and lamina propria. The **transitional epithelium,** which rests upon a basement membrane, consists of several layers of epithelial cells. The underlying **lamina propria** is a loose connective tissue containing many capillaries. The mucosa is usually thrown into longitudinal folds.

The smooth muscle fibers in the **muscularis** are loosely arranged into three layers: **internal longitudinal layer, intermediate circular layer,** and **external longitudinal layer.** Unlike the intestine, the muscle fibers in the ureter do not form a dense, regular layer, but run in loose strands separated by connective tissue and an elastic network. It is by the peristaltic action of the smooth muscle in the wall that the urine is conducted from the pelvis to the urinary bladder.

The **adventitia** is composed of loose connective tissue that contains many large **blood vessels,** lymphatics, and nerves. It is continuous with the surrounding connective tissue and **adipose tissue.**

Fig. 11-8. Ureter: Mucosa

This picture demonstrates the fine structure of the **mucosa** of the ureter under high magnification. The lining of the mucosa is **transitional epithelium,** which is composed of several layers of epithelial cells. These light-staining cells have clearly distinct boundaries, and the **superficial cells** are large, polyploid, and may contain two nuclei. The connective tissue of the **lamina propria** is abundant and rich in **blood vessels** and a network of elastic fibers.

Figure 11-7. Ureter

Human • Cross section • H.E. stain • Low magnification

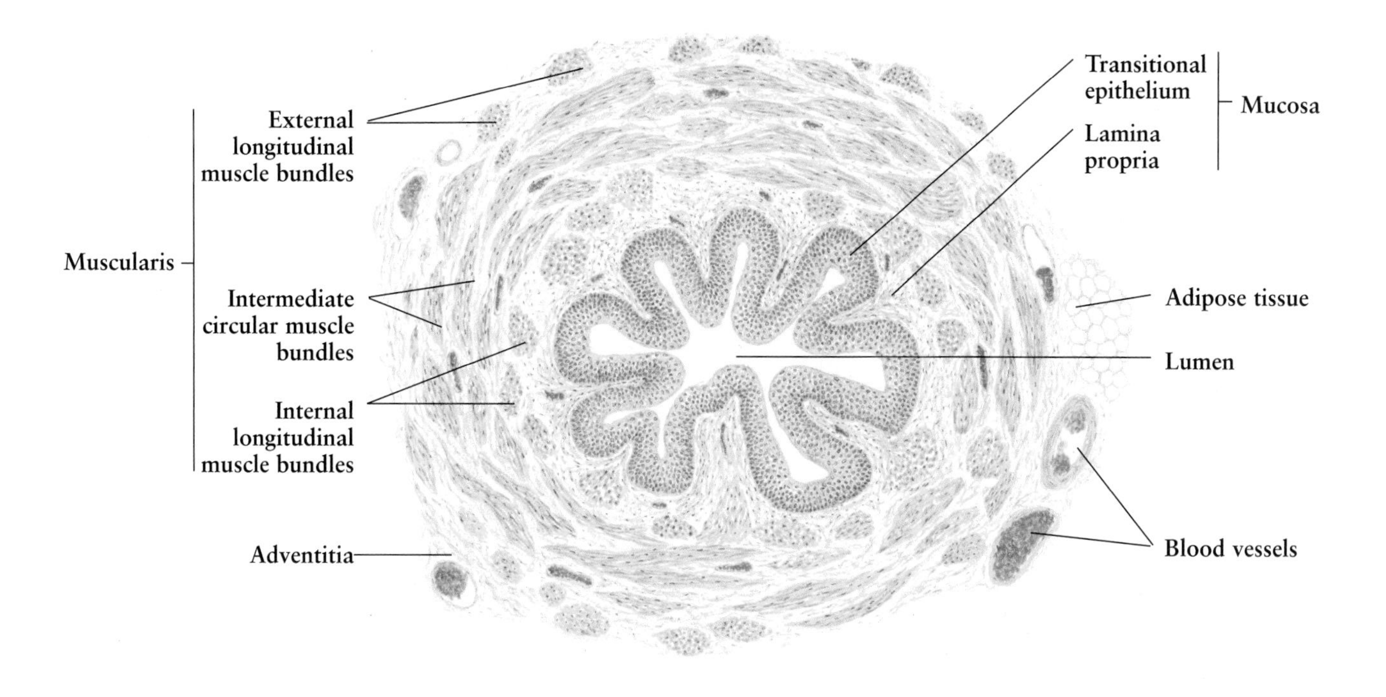

Figure 11-8. Ureter: Mucosa

Human • H.E. stain • High magnification

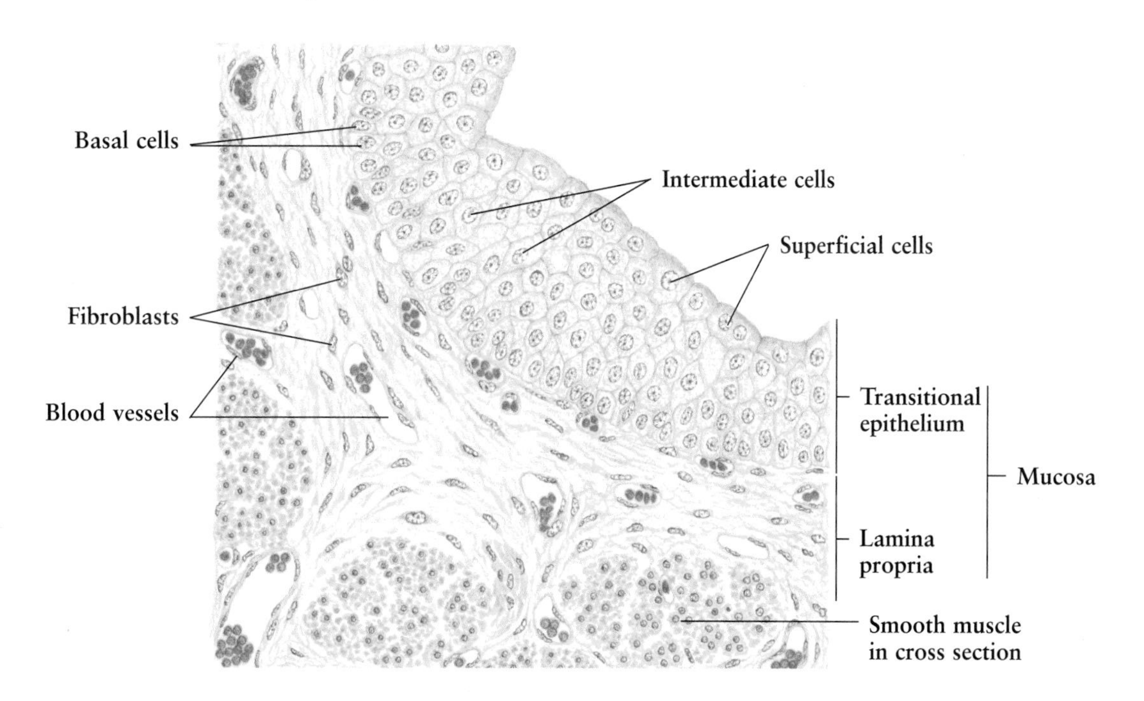

Fig. 11-9.

Urinary Bladder

The **urinary bladder** has a similar structure to the ureter. The bladder wall, however, is composed of four layers: *mucosa, submucosa, muscularis,* and *adventitia.*

The **mucosa** consists of **transitional epithelium** resting on a basement membrane and the underlying fibroelastic connective tissue of the **lamina propria,** which is rich in elastic fibers, capillaries, and fibroblasts.

The **submucosa** is an abundant, loose connective tissue, containing large blood vessels, lymphatic vessels, and nerve fibers. There is no muscularis mucosae separating the submucosa from the lamina propria.

The **muscularis** consists of three layers: **internal longitudinal layer, intermediate circular layer,** and **external longitudinal layer.** In the trigone, the muscularis thickens and forms the internal sphincter, surrounding the internal opening of the urethra.

The **adventitia** is a loose connective tissue. This layer contains large **blood vessels,** lymphatic vessels, and nerve fibers, as well as **adipose tissue.** Its surface is covered by a **serosa,** a simple squamous epithelium, the mesothelium.

Fig. 11-10.

Urinary Bladder: Mucosa

This illustration shows details of the **mucosa** of the urinary bladder. The **transitional epithelium** (see **Figs. 1-12** and **1-13**) is composed of three layers of cells: the **superficial cells, intermediate cells,** and **basal cells.**

The superficial cells are the largest in size, with a darker-staining area near the luminal surface. At the electron microscopic level, the luminal cytoplasmic membrane is composed of asymmetric unit membranes with a thicker leaflet adjacent to the lumen than the one facing the cytoplasm. It is believed that the thick asymmetric unit membrane is able to resist erosion by the hyperosmotic urine. When the bladder contracts, the excess luminal cell membranes enter the cytoplasm in the form of discoidal vesicles, which serve as a readily available source of cell membrane to allow a rapid increase in luminal surface as the bladder dilates.

The basement membrane cannot be seen easily at the light microscopic level. The **lamina propria** is a connective tissue rich in elastic fibers, and contains densely arranged **fibroblasts.** Also shown in this figure is the **submucosa,** which contains **blood vessels, lymphatic vessels,** and **nerve fibers.**

Figure 11-9. Urinary Bladder
Human • H.E. stain • Low magnification

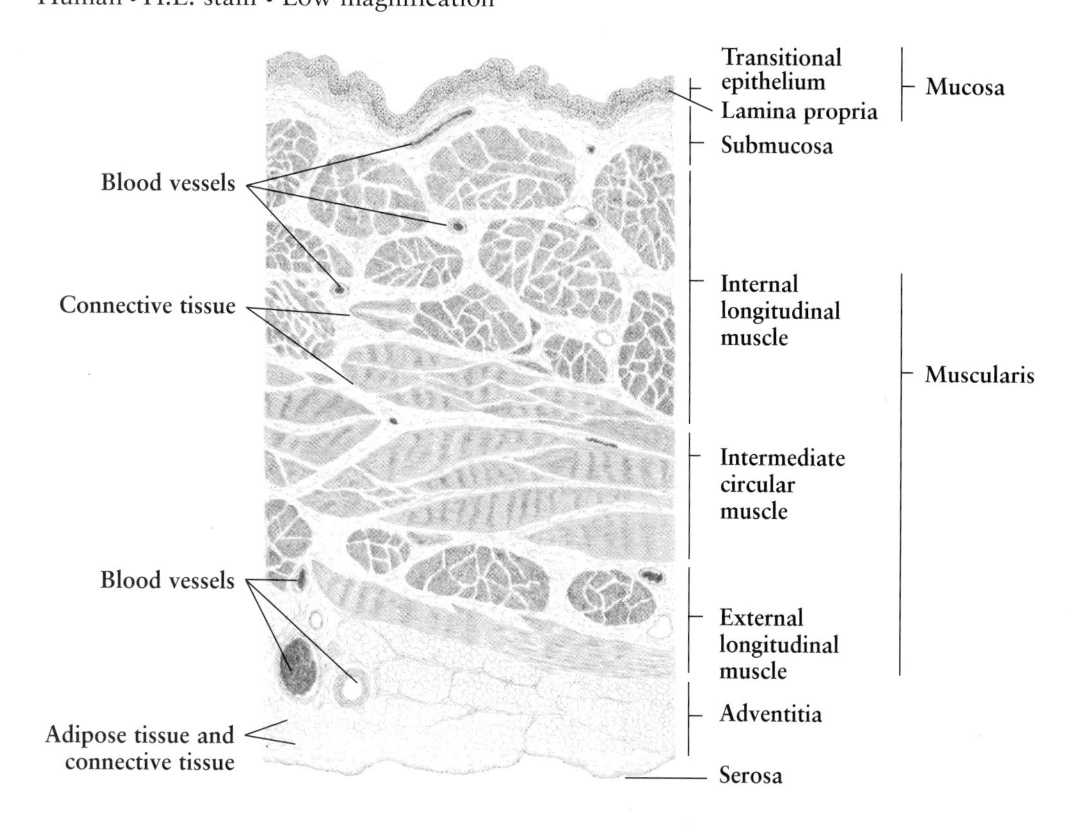

Figure 11-10. Urinary Bladder: Mucosa
Human • H.E. stain • High magnification

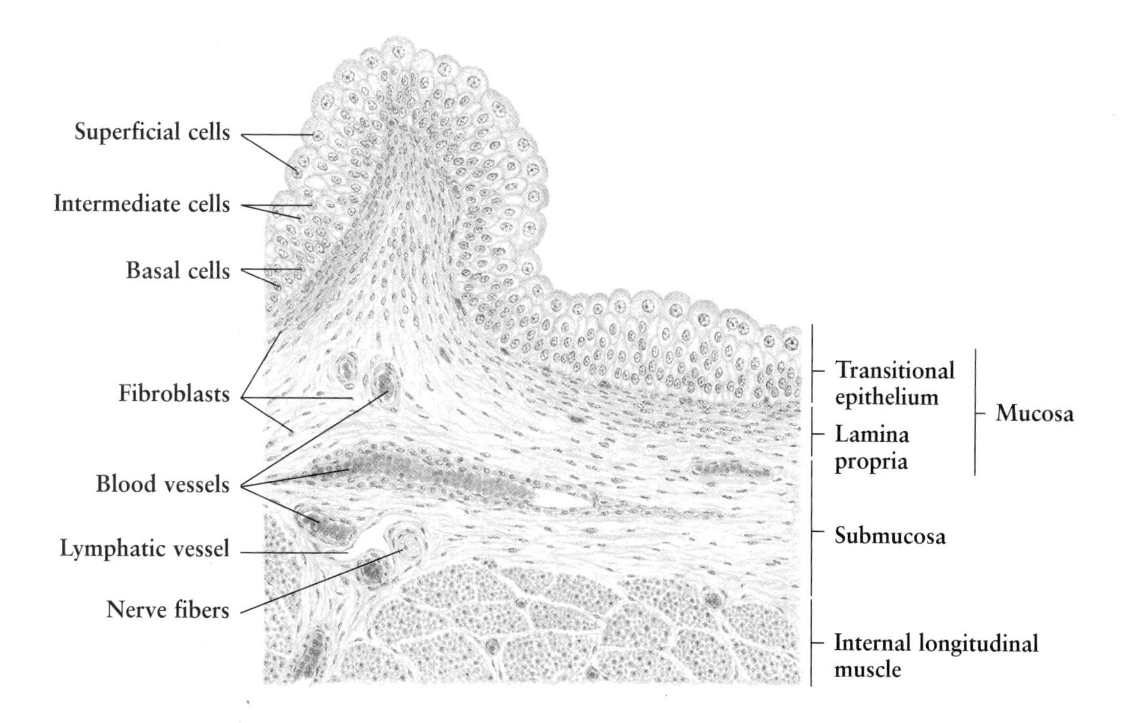

Fig. 11-11. Male Urethra: Spongy Part

The **male urethra** is 18–20 cm in length and serves as the terminal portion of both the urinary tract and the reproductive tract. It has three distinct, successive portions: *prostatic part, membranous part,* and *spongy (cavernous) part.* The epithelium of the prostatic part is transitional, while that of the membranous part is stratified cuboidal or columnar.

The **spongy part** of the male urethra is located within the **corpus spongiosum,** and opens at the glans penis. The lining epithelium is a **stratified columnar,** but it changes to stratified squamous epithelium at the terminal dilatation of the penile urethra, the fossa navicularis. The epithelium is surrounded by a thin layer of connective tissue of the **lamina propria** and the corpus spongiosum. Also found in this figure are the **tunica albuginea (corporis spongiosi), tunica albuginea,** and **corpora cavernosa.**

Fig. 11-12. Male Urethra: Spongy Part

The **spongy part** of the male urethra is lined by **stratified columnar epithelium** with clear cells. The underlying **lamina propria** is a thin layer of connective tissue rich in elastic fibers, **fibroblasts,** and **blood vessels.** Along the entire length of the spongy part, there are many shallow recesses in the mucosa, known as **urethral lacuna (of Morgagni).** Some of these continue into the **excretory duct** of branching mucous glands, the **urethral glands (of Littré)** located in the lamina propria or erectile tissue. **Intraepithelial glands** (see **Fig. 1-15D**) can be found within the stratified epithelium. These two kinds of glands secrete mucus to lubricate the mucosa of the urethra. The structure of the **corpus spongiosum** is composed of **cavernous spaces** and **trabeculae,** which contain **smooth muscle fibers** and **helicine arteries.**

Figure 11-11. Male Urethra: Spongy Part

Human • Cross section • H.E. stain • Low magnification

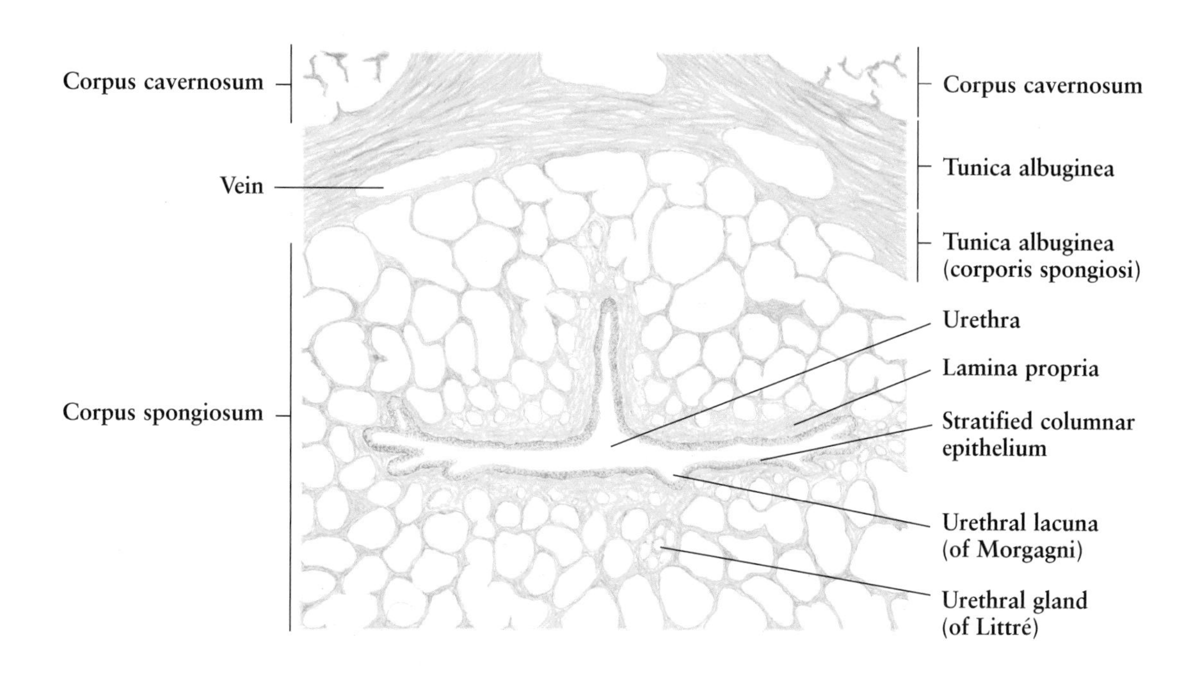

Figure 11-12. Male Urethra: Spongy Part

Human • Cross section • H.E. stain • High magnification

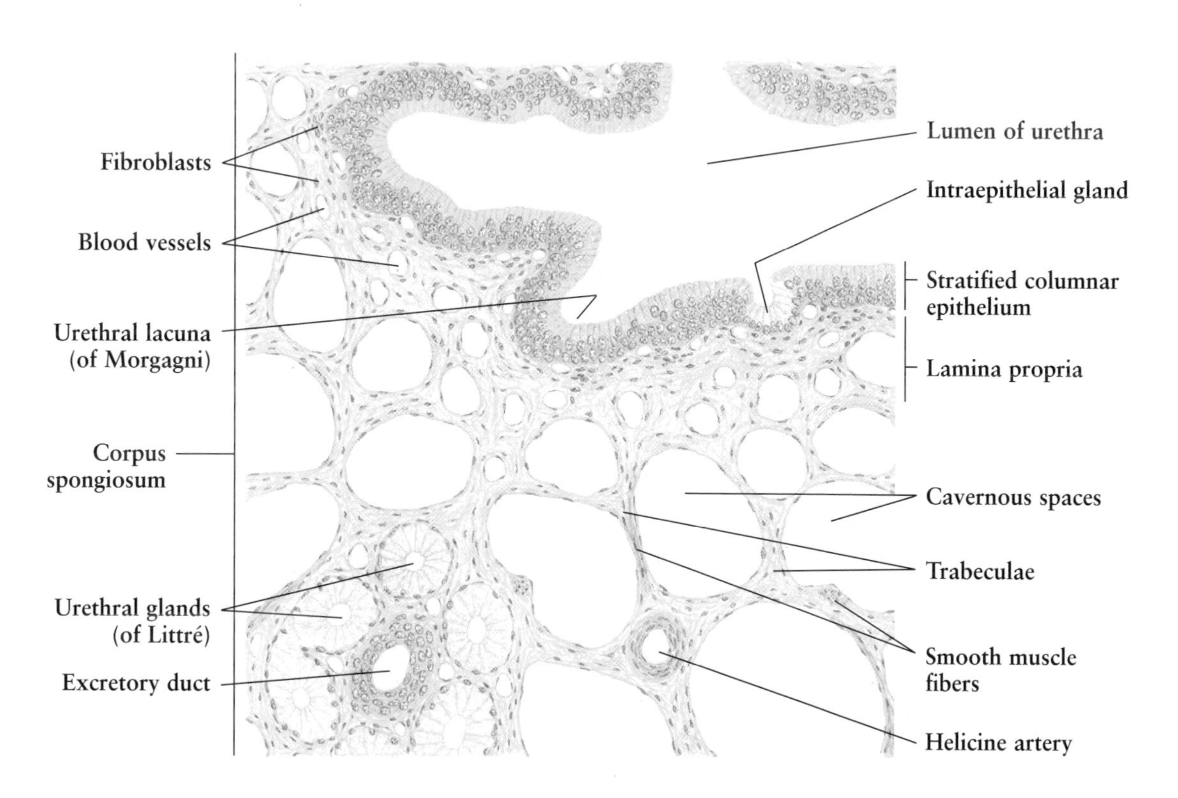

Fig. 11-13.

Female Urethra

The **female urethra** is about 4 cm long, and conducts urine from the urinary bladder to the exterior. Its wall consists of three layers: *mucosa, muscularis,* and *adventitia.*

The **mucosa** is lined by **stratified squamous epithelium,** and is plicated to form longitudinal folds. Hence the **lumen** appears irregular. Many **gland-like lacunae,** the shallow invaginations of the epithelium, are present. The underlying **lamina propria** is a loose connective tissue with abundant elastic fibers and numerous **venous plexuses** similar to the spongy body in the male urethra.

The **muscularis** is composed of two layers of smooth muscle: **inner longitudinal layer** and **outer circular layer.** The circular smooth muscle fibers unite with those of the urinary bladder to form a sphincter muscle around the internal opening of the urethra. Some **striated muscle fibers** can be found among the circular smooth muscle fibers. The urethra has an external sphincter of striated muscle at its orifice.

The **adventitia** is a **loose connective tissue** continuous with surrounding tissues such as the **smooth muscle surrounding the vagina.**

Fig. 11-14.

Female Urethra: Mucosa

In contrast to the male urethra, the lining of the **female urethra** is a nonkeratinized **stratified squamous epithelium.** The basal cells are densely arranged and stain dark, while the intermediate and superficial cells appear large and clear, similar to those of the vaginal mucosa. The **gland-like lacunae** are simple invaginations of the epithelium and do not form true glands.

The underlying connective tissue of the **lamina propria** projects up into the epithelium, forming **papillae.** The **loose connective tissue** contains abundant elastic fibers and numerous fibroblasts. It is characterized by the presence of numerous thin-walled **venous plexuses,** which give it a structural feature of erectile tissue resembling that of the corpus cavernosum of the male.

The female urethra lacks a submucosa between the lamina propria and muscularis. The connective tissue of the lamina propria is continuous with the inner **longitudinal smooth muscle.**

Figure 11-13. Female Urethra
Human • Cross section • H.E. stain • Low magnification

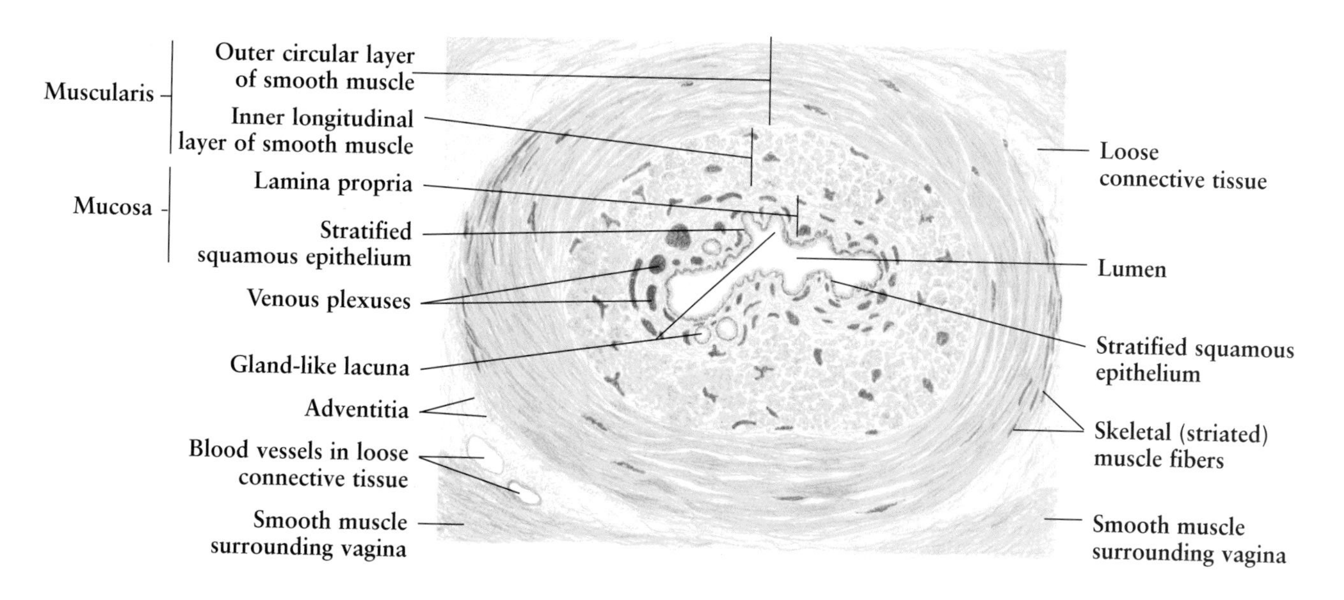

Figure 11-14. Female Urethra: Mucosa
Human • Cross section • H.E. stain • High magnification

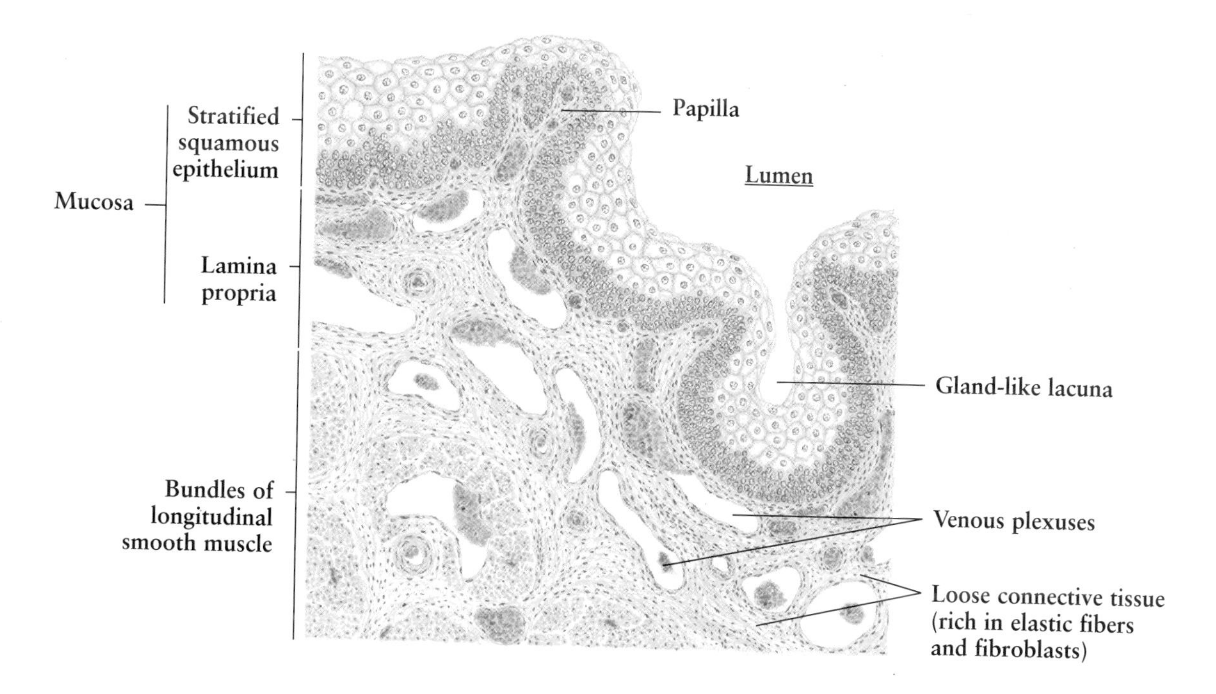

12. MALE REPRODUCTIVE SYSTEM

The **male reproductive system** consists of the *testes, genital ducts, accessory glands,* and *penis.* It has three major functions: (1) to produce spermatozoa; (2) to secrete male hormones (androgens, principally testosterone); and (3) to copulate and facilitate fertilization through the deposition of spermatozoa into the female genital tract.

The **testes** are responsible for the production of spermatozoa and male hormones. The parenchyma of the testes is composed of numerous tortuous looped **seminiferous tubules** in which the **spermatozoa** are produced. Between these tubules are small groups of Leydig cells. **Leydig cells** produce *testosterone,* which promotes the production of spermatozoa and acquisition of the male secondary sexual characteristics. The testes are contained in the scrotum (external to the abdominal cavity), where they are maintained below body temperature (35°C). This temperature is critical for the development of spermatozoa.

The **genital ducts** consist of **ductuli efferentes, epididymis, ductus deferens,** and **ejaculatory duct.** The spermatozoa are collected from the testes by the ductuli efferentes and temporarily stored in the epididymis. During copulation, the spermatozoa are conducted by the ductus deferens to the ejaculatory duct, which expels them by strong contraction of the smooth muscle through the penis into the female genital tract.

The **accessory glands** include the paired **seminal vesicles,** the single **prostate gland,** and the small paired **bulbourethral glands.** The seminal vesicles produce a thick nutritive fluid, whereas the prostate gland secretes a thin milky secretion. These two secretions in addition to the spermatozoa constitute *semen.* The bulbourethral gland provides a mucous fluid that lubricates the urethra for the passage of semen during ejaculation.

The **penis** is the copulatory organ and is composed of three erectile bodies: the paired **corpora cavernosa** (corpora cavernosa penis), and a single **corpus spongiosum** (corpus cavernosum urethrae). An erection occurs when the erectile tissues are engorged with blood and distended as the result of parasympathetic stimulation by tactile or erotic stimulation.

Fig. 12-1.

Testis and Epididymis

The **testes** are paired organs lying within the scrotum, and are responsible for the *production* of sperm and male hormones (androgens). Each testis is covered by a dense, fibrous, connective tissue, the **testicular capsule.** The parenchyma of the testis is divided into about 250 pyramidal lobules by connective tissue septa from the capsule at the posterior aspect.

The capsule is composed of three layers: an outer layer or **tunica vaginalis,** a single layer of attenuated mesothelial cells derived from the peritoneum; a middle layer or **tunica albuginea,** a thick layer of dense collagenous fibrous tissue containing some smooth muscle fibers; and an innermost layer or **tunica vasculosa,** a delicate loose connective tissue rich in blood vessels.

The parenchyma of the testis is composed of closely packed coils of **seminiferous tubules** within a stroma of **interstitial connective tissue** that contains blood vessels and groups of interstitial cells (Leydig cells). These are endocrine cells, which produce the male hormone testosterone. The convoluted seminiferous tubules are lined by a complex germinal epithelium of spermatogenic or sperm-forming cells and supporting cells.

The mature spermatozoa pass from the convoluted seminiferous tubules via the short **tubuli recti** into the **rete testis** at the mediastinum testis. The rete testis are continuous with the **ductuli efferentes** in the **epididymis,** which is also covered by the tunica vaginalis. About 8–20 ductuli efferentes grouped in coni vasculosi join the **ductuli epididymidis** to form the head of the epididymis at the posterior and superior surface of the testis. The highly convoluted ductulus epididymidis forms the body of the epididymis. At the caudal portion, the ductulus epididymidis becomes continuous with the relatively straight ductus deferens, which is not shown in this illustration. The epididymis is the major storage site of of newly formed spermatozoa.

This illustration shows the positional relationship among these tubules, which are described in the following figures.

Figure 12-1. Testis and Epididymis
Human • H.E. stain • Low magnification

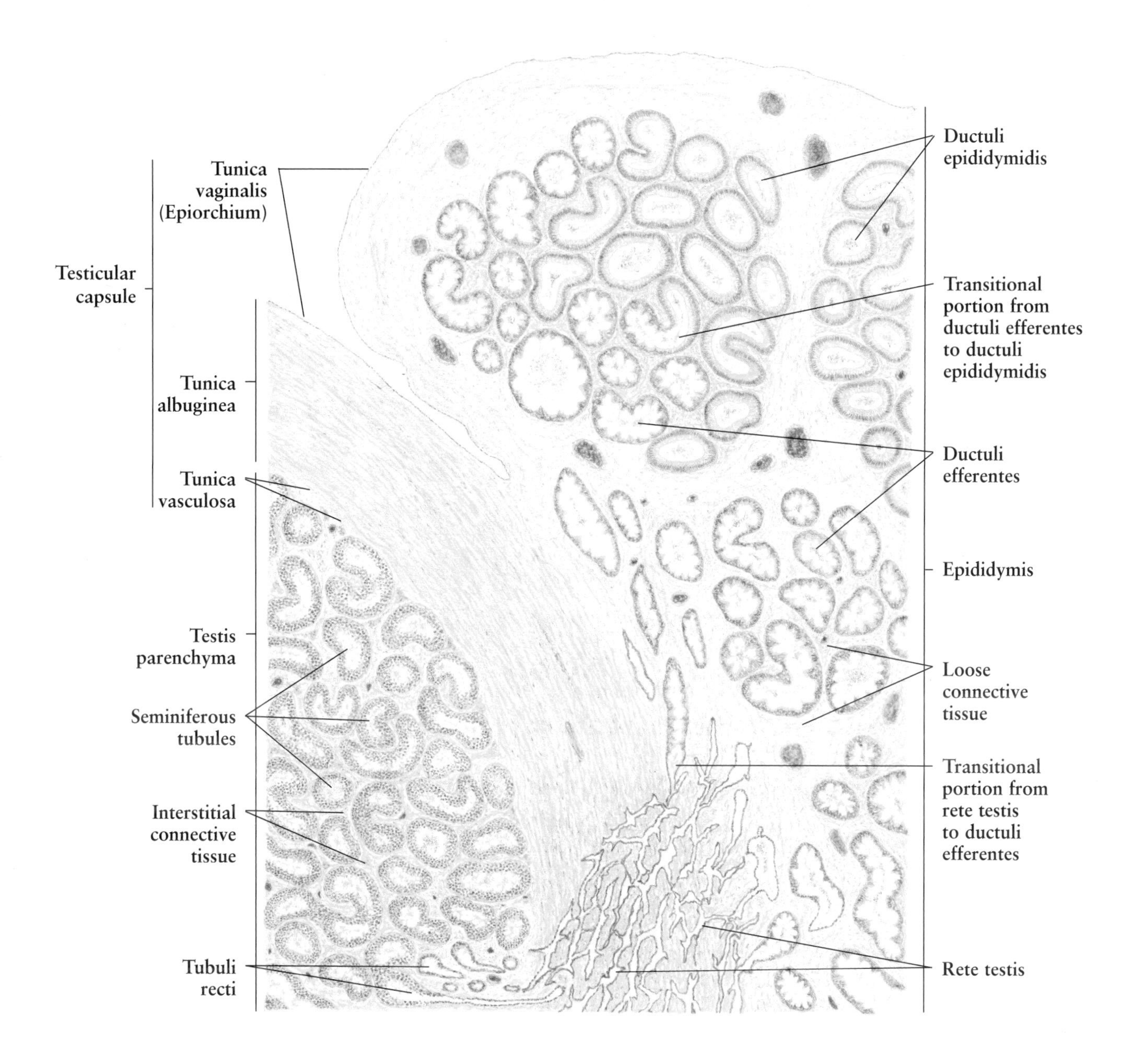

Fig. 12-2. Seminiferous Tubule

The **seminiferous tubules** are highly convoluted tubules, 30–70 cm in length and 0.2 mm in diameter. They are lined by a stratified epithelium, composed of two distinct populations of cells: *spermatogenic series* and *supporting cells.* The basal layer of the germinal cells lie on a **basement membrane,** which is surrounded by one to three layers of **myoid cells.** The myoid cells are structurally similar to smooth muscle fibers, but their function remains to be known.

The **spermatogenic series** contains *spermatogonia, primary spermatocytes, secondary spermatocytes, spermatids,* and *spermatozoa.* These cell types are arranged in layers that show their differentiation sequentially from the basal layers of the epithelium to the **lumen.**

Spermatogonia are stem cells; they are found in the basal layer of the germinal epithelium, resting upon the basement membrane. According to their nuclear characteristics, three types can be recognized: type A dark spermatogonia, type A pale spermatogonia, and type B spermatogonia. **Type A dark spermatogonia** are characterized by their ovoid, darkly staining nucleus, which contains one or two nucleoli associated with the nuclear envelope. Following mitotic division of type A dark spermatogonia, about half the daughter cells maintain type A dark cells, and the other half form type A pale spermatogonia. These cells possess an ovoid, pale-staining nucleus. **Type A pale spermatogonia** in turn divide by mitosis to give rise to type B spermatogonia, and more type A pale cells. **Type B spermatogonia** have a spherical nucleus with peripherally massed chromatin and a centrally located nucleolus. After mitotic division of type B spermatogonia, daughter cells differentiate into primary spermatocytes that remain connected by intercellular bridges.

Primary spermatocytes are the largest germ cells found within the seminiferous tubules, present in the middle layers of the germinal epithelium. These cells are readily recognized by their spherical outline and large nucleus with either thin threads or coarse clumps of chromatin. **Primary spermatocytes in mitosis** are also observed. Primary spermatocytes, which have 46 chromosomes (44 autosomes plus one X chromosome and one Y chromosome) undergo meiotic division to give rise to secondary spermatocytes, which contain 23 chromosomes (22 autosomes plus either one X or one Y chromosome).

Secondary spermatocytes are smaller than primary spermatocytes. They rapidly undergo the second meiotic division (a regular mitosis) to produce spermatids that contain 23 single chromosomes. Therefore they are rarely seen in routine preparation.

Spermatids undergo a long metamorphic phase known as *spermiogenesis* to become **spermatozoa,** which have a head and a long tail. The spermiogenesis is a cytological transformation consisting of several steps, which include *condensation* of the nucleus to form the head of the spermatozoon, *formation* of the head cap from the Golgi complex, *formation* of the tail from the centrioles and *aggregation* of mitochondria around the middle piece of the tail, and *shedding* of the superfluous part of the spermatid cytoplasm.

The supporting cells, **Sertoli cells** or sustentacular cells, are relatively few in number, resting at intervals on a basement membrane. A Sertoli cell has a huge, irregular outline and is characterized by its ovoid nucleus with very prominent nucleolus and deep invaginations of the nuclear envelope. The cytoplasm is pale-staining and contains small fibrils, lipid droplets, and small granules. The differentiating germ cells attach to the Sertoli cells and are nourished by them. The Sertoli cells are believed to provide structural and metabolic support for the developing spermatogenic cells and form a blood–testis barrier to large molecular weight compounds.

Figure 12-2. Seminiferous Tubule
Human • H.E. stain • High magnification

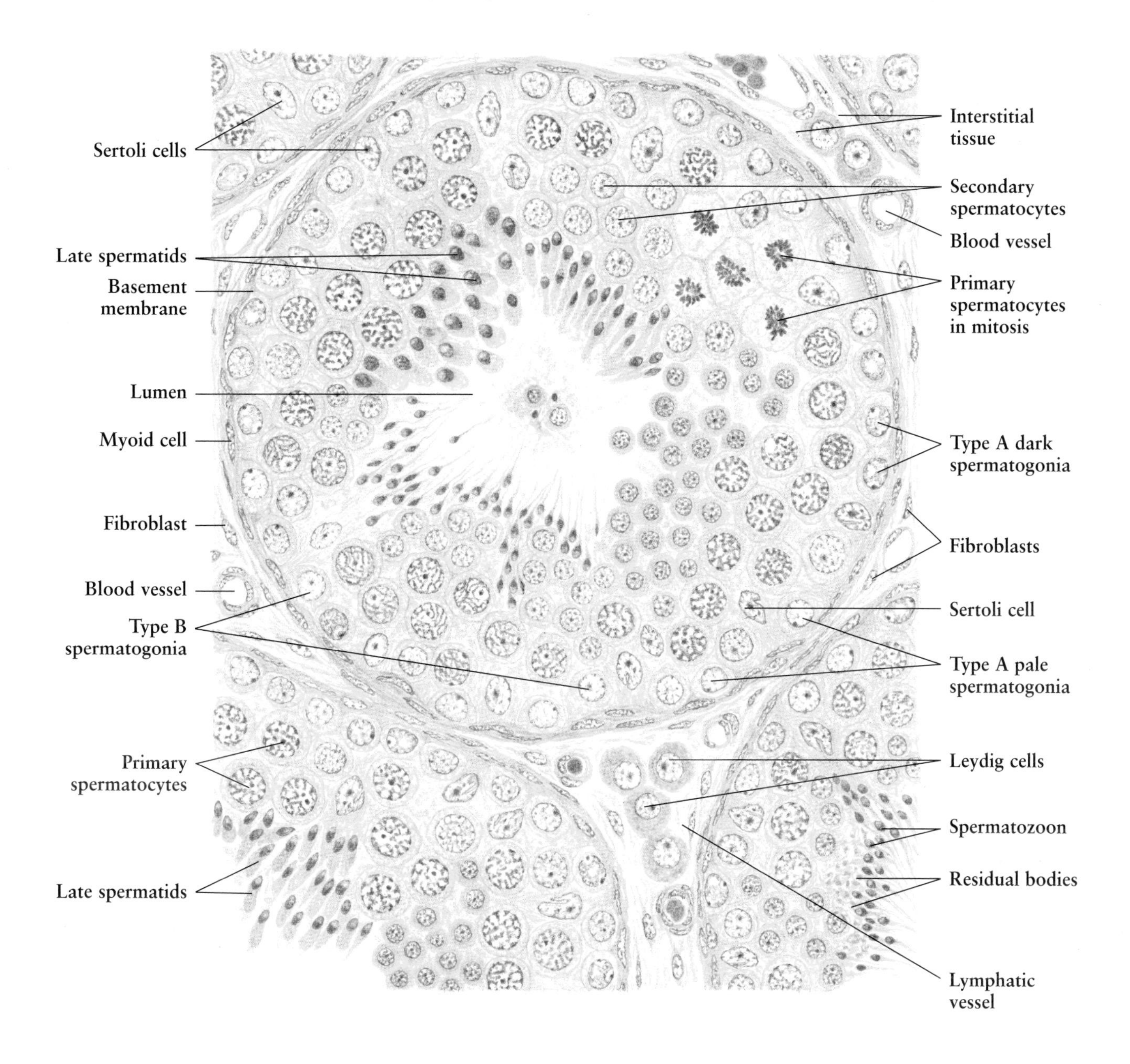

Between the seminiferous tubules is the loose connective tissue of **interstitial tissue** rich in **blood vessels** and **lymphatic vessels.** A marked feature of this tissue is the presence of special interstitial cells, called **Leydig cells.** They are found either singly or in a small group close to the blood vessels. Leydig cells are large in size, and have a round nucleus with a big nucleolus. The eosinophilic cytoplasm contains abundant smooth endoplasmic reticulum. Leydig cells are responsible for the production of the male hormone, ***testosterone.***

Fig. 12-3.

Cycle of Human Seminiferous Epithelium

In the human, it takes about 64 days to produce mature sperm cells after an initial type A spermatogonium enters a cell cycle. Six stages or six cell associations can be recognized, consisting of germinal cells in various states of differentiation. Unlike that in rodents in which each cell association occupies an entire cross section of the tubule, the human germinal epithelium is a mosaic of such stages. Three or four stages of the cell cycle can be seen in a single section of a tubule (also see **Fig. 12-2**). **Figure 12-3** demonstrates the germinal cell cycle of human seminiferous epithelium, represented by six stages. The following table (**Table 12-1**) will help in interpreting the cell associations of the cycle.

		Stage 1	Stage II	Stage III	Stage IV	Stage V	Stage VI
Type A dark spermatogonia		Present	Present	Present	Present	Present	Present
Type A pale spermatogonia		Present	Present	Present	Present	Present	Present
Type B spermatogonia		Present	Present				
Primary sperma-tocytes	Leptotene			Present	Present	Present	
	Zygotene						Present
	Pachytene	Present	Present	Present	Present	Present	
	Diplotene						Present
	Division						Present
Secondary spermatocytes							Present
Spermitids		Present; A	Present; B	Present; C	Present; D	Present; E	Present; F
Spermatozoa		Present	Present				

The designation of A-F in Table 12-1 represents the different states of the spermatids during the cytological transformation into the mature sperm cells.

Figure 12-3. Cycle of Human Seminiferous Epithelium
According to H.E.-stained section • High magnification

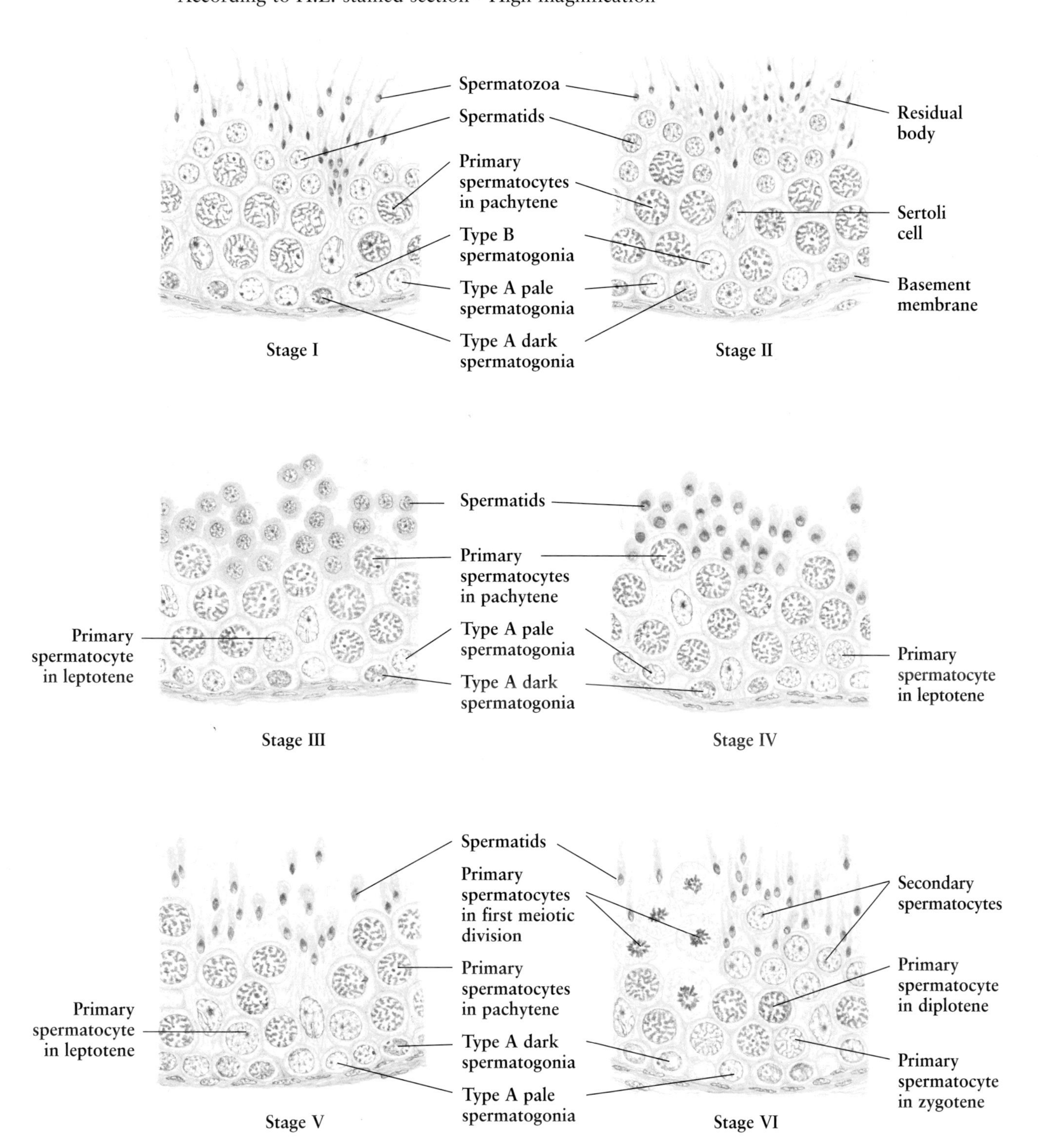

Fig. 12-4.

Tubuli Recti and Rete Testis

The spermatozoa from the seminiferous tubule gain access to the rete testis through a short, straight tubule, the **tubulus rectus,** which is located at the apex of each lobule, close to the **mediastinum testis.** Tubuli recti, 1 mm in length and 0.1–0.25 mm in diameter, connect **seminiferous tubules** with the rete testis. At the point directly following the seminiferous tubule, the germinal cells disappear, and only Sertoli cells remain, forming a **simple columnar** (or cuboidal) **epithelium** lying on a thin basement membrane. The tubuli recti are surrounded by a thin layer of **loose connective tissue.**

The **rete testis** is composed of a network of irregular labyrinthine, anastomosing spaces in the mediastinum testis. They are lined by a **simple cuboidal epithelium** that is nonsecretory. The epithelium is supported by a delicate basement membrane. The **dense connective tissue** in the mediastinum testis contains abundant collagen fibers, **fibroblasts,** and **blood vessels.** Spermatozoa are typically not seen within the lumina of rete testis, indicating the rapid passage of spermatozoa through the rete testis to the ductuli efferentes. This figure also shows the **transitional portion** from the rete testis to the **ductuli efferentes.**

Fig. 12-5.

Ductuli Efferentes

The rete testis empties into the head of the epididymis via 8–20 convoluted ducts, the **ductuli efferentes,** grouped in coni vasculosa with the tip toward the rete testis and the base toward the ductulus epididymidis. The ductuli efferentes are lined by a single layer of **epithelium** that has a distinctive scalloped outline in section. This appearance results from the composition of the epithelium, groups of tall columnar **ciliated cells** alternating with groups of much shorter **nonciliated cells** with numerous microvilli on their free surface. The cilia beat toward the ductulus epididymidis to assist in transporting **spermatozoa** through the ductuli efferentes. The shorter cells are responsible for the absorption of the fluid produced within the seminiferous tubules.

The epithelium lies upon a thin **basement membrane,** and is surrounded by a layer of **smooth muscle fibers.** Between the ductuli efferentes is a **loose connective tissue** that contains a network of **blood vessels** and **lymphatic vessels** in addition to **fibroblasts** and collagen fibers.

Figure 12-4. Tubuli Recti and Rete Testis

Human • H.E. stain • Low magnification

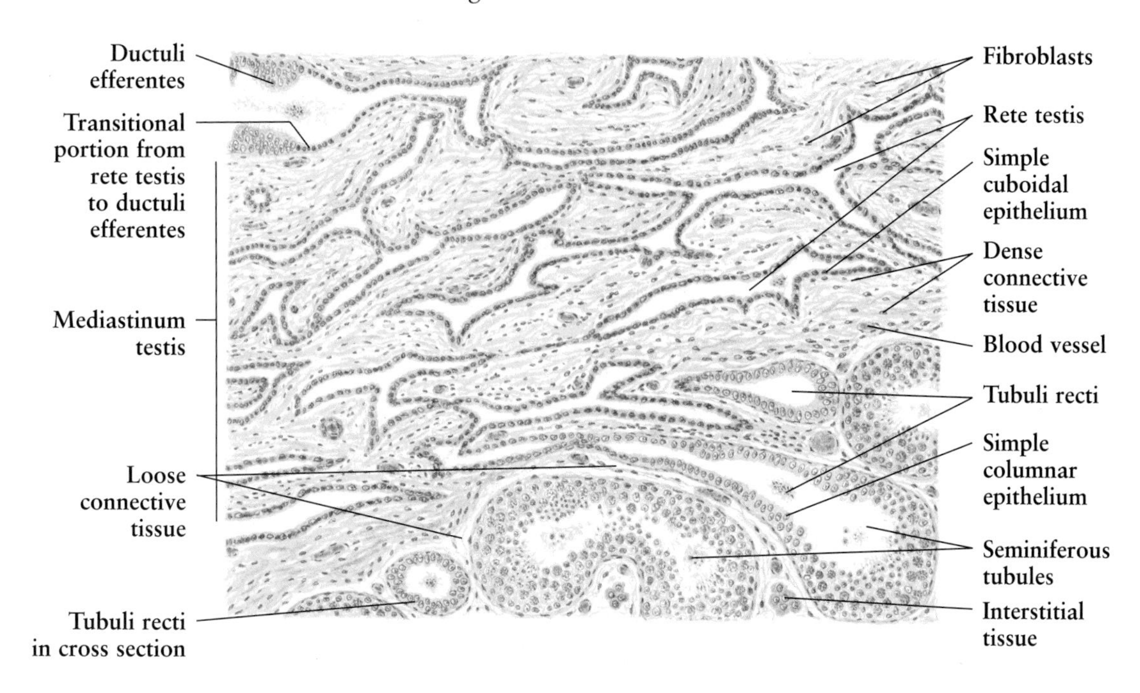

Figure 12-5. Ductuli Efferentes

Human • H.E. stain • High magnification

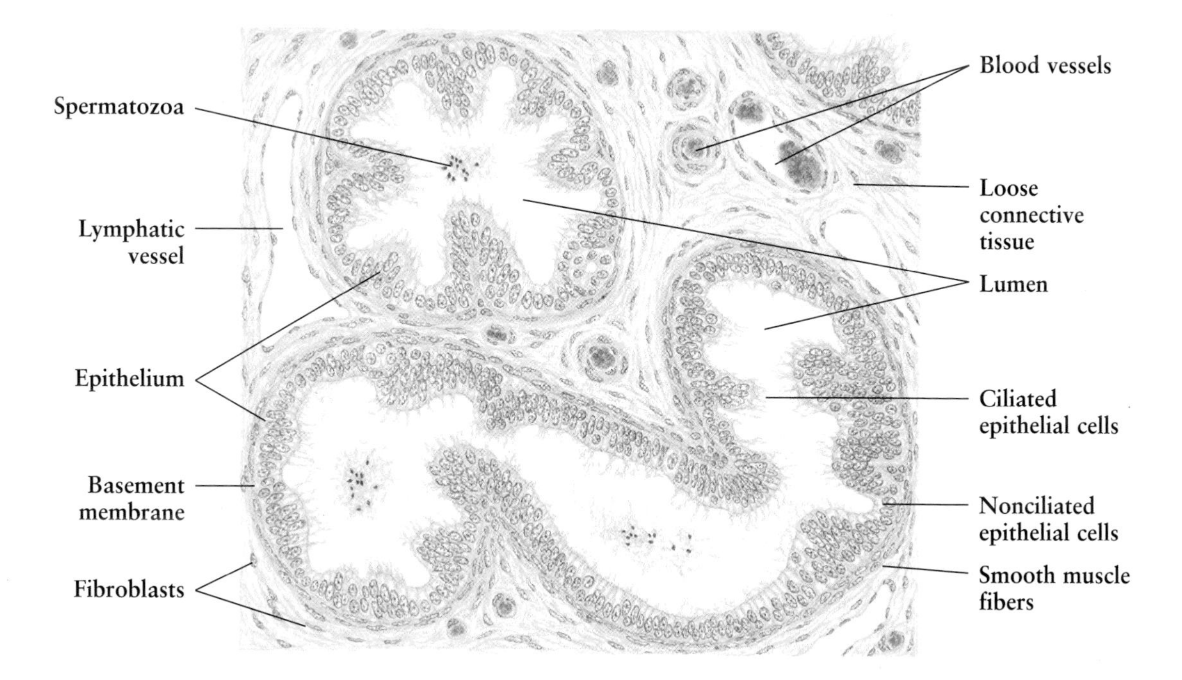

Fig. 12-6.

Ductulus Epididymidis

The **ductulus epididymidis,** 5–6 meters long, is a highly convoluted single duct that forms the body and tail of the epididymis. The epithelium lining the duct is **pseudo-stratified,** composed of tall columnar cells and basal cells. The **columnar cells** have an ovoid nucleus and numerous **nonmotile stereocilia** at the free surface. They function both in resorption of fluid and in synthesis of a secretion necessary for sperm maturation. The **basal cells** are small round or pyramidal cells, but their function remains obscure. The **epithelium** rests on a definite **basement membrane,** surrounded by a layer of circularly oriented **smooth muscle fibers.** The muscle layer gradually thickens toward the tail of the epididymis. The **connective tissue** between the ducts is rich in **fibroblasts** and collagen fibers, as well as **blood** and **lymphatic vessels.**

Numerous **spermatozoa** are present within the lumen of the ductulus epididymidis, which is continuous with the ductus deferens. The ductulus epididymidis acts as a long storage site, as spermatozoa pass through it slowly they acquire motility and the capacity for fertilization. Also shown in the figure is the **transitional portion** from the **ductulus efferentes** to the ductulus epididymidis.

Fig. 12-7.

Spermatic Cord

The **spermatic cord** is a long vascular strand extending from the deep inguinal ring through the inguinal canal down into the scrotum. It consists of **ductus deferens, spermatic artery, pampiniform vein plexus, lymphatic vessels,** and nerves, all embedded in a **loose connective tissue** containing numerous **adipose cells.** The spermatic cord is surrounded by a layer of striated muscle, the **cremaster muscle,** which arises from the internal oblique muscle of the abdominal wall. The testis is suspended within the scrotum by the cremaster muscle.

Figure 12-6. Ductulus Epididymidis

Human • H.E. stain • High magnification

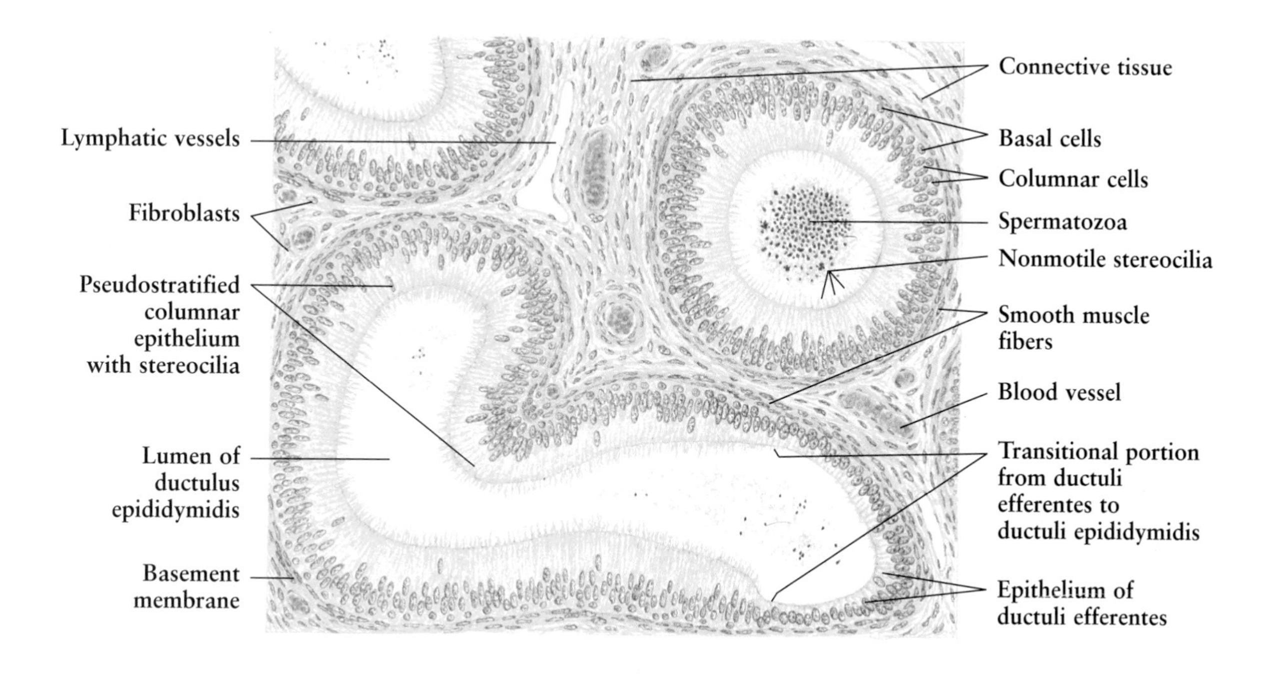

Figure 12-7. Spermatic Cord

Human • Cross section • H.E. stain • Low magnification

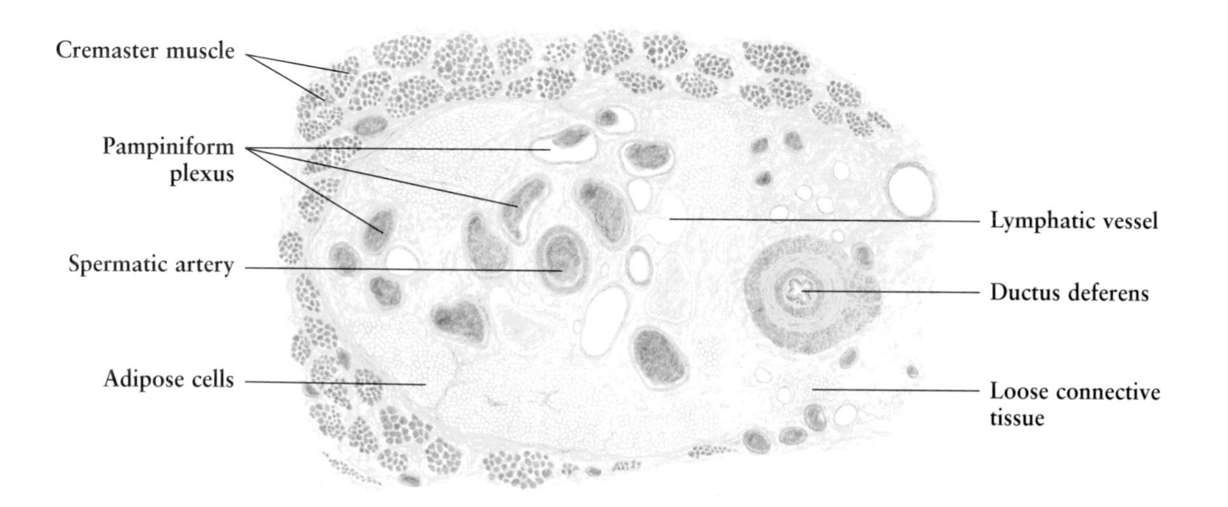

Fig. 12-8.

Ductus Deferens

The **ductus deferens** is a thick-walled muscular tube, about 30 cm long, which conducts spermatozoa from the tail of the epididymis to the ejaculatory ducts. Its wall consists of three coats: *mucosa, muscularis,* and *adventitia.*

The **mucosa** is composed of a **pseudostratified columnar epithelium** and the underlying **lamina propria.** The mucosa is thrown into longitudinal folds that allow expansion of the duct during ejaculation. The **muscularis** is very thick. It consists of three layers: **internal longitudinal, intermediate circular,** and **external longitudinal.** Electron microscopic studies have found that almost every smooth muscle cell has an autonomic motor nerve ending. This finding explains the fast contraction of the ductus deferens during ejaculation. The **adventitia** is a loose connective tissue containing **nerve fibers, blood vessels,** and **lymphatic vessels.**

Fig. 12-9.

Ductus Deferens: Mucosa

The **ductus deferens** is lined by a **pseudostratified columnar epithelium** of tall **columnar cells** and small **basal cells.** The columnar cells are characterized by **stereocilia** at their free surface, and a basal ovoid nucleus with prominent nucleolus. There is a delicate **basement membrane,** which separates the epithelium from the underlying lamina propria. The **lamina propria** is connective tissue with a network of elastic fibers. It contains numerous **fibroblasts** and abundant **blood vessels** in addition to the elastic fibers.

Figure 12-8. Ductus Deferens
Human • Cross section • H.E. stain • Low magnification

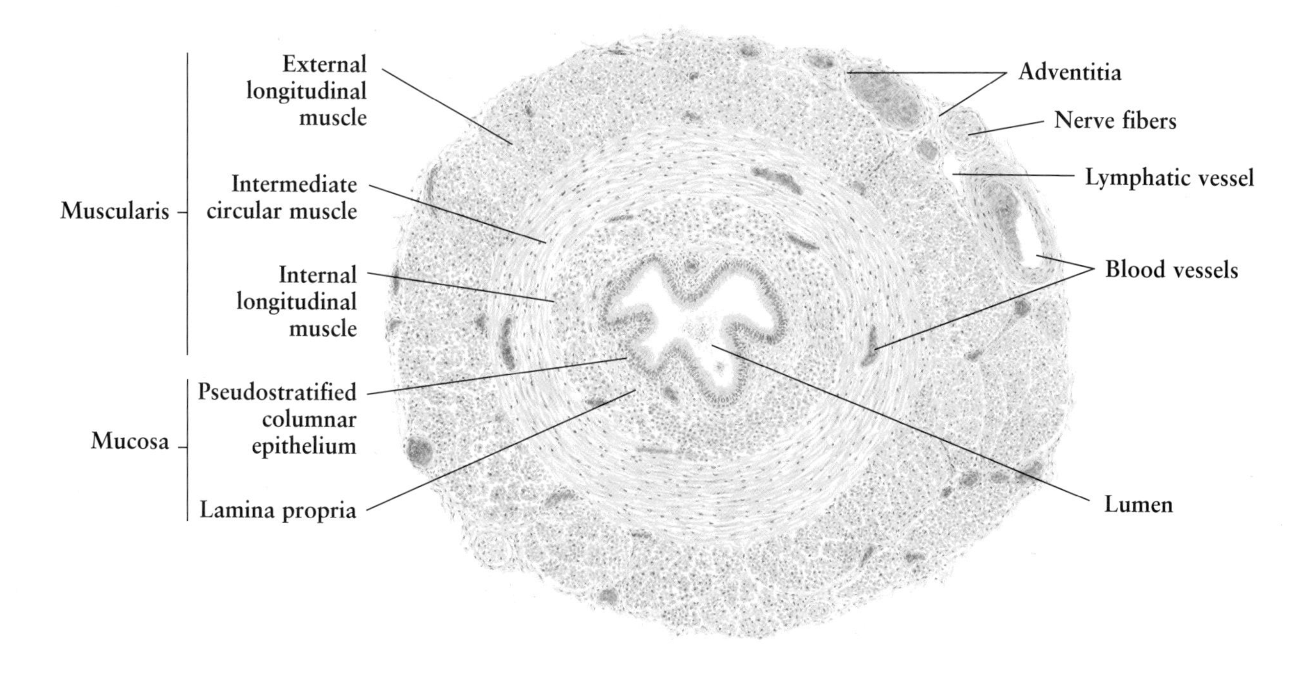

Figure 12-9. Ductus Deferens: Mucosa
Human • H.E. stain • High magnification

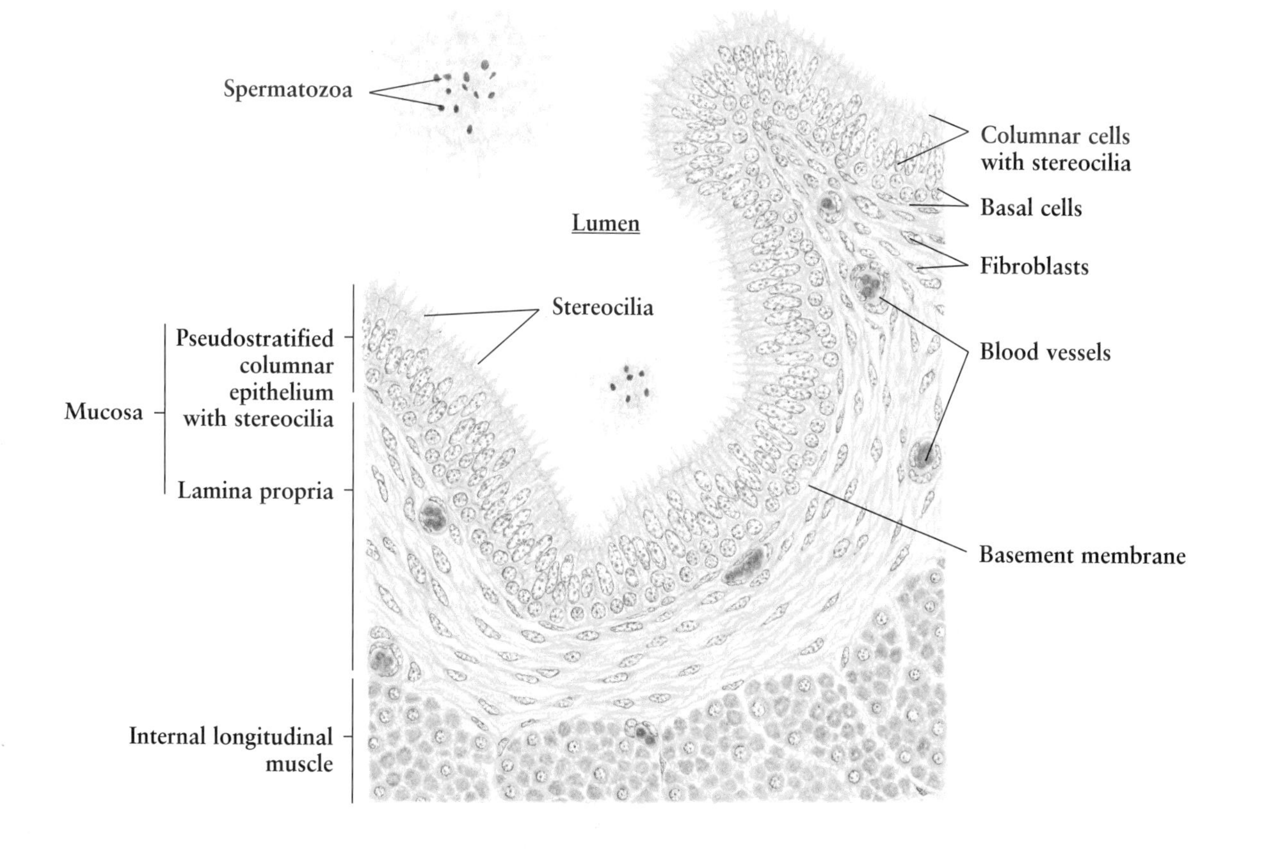

Fig. 12-10.

Seminal Vesicle

The **seminal vesicles** are paired saccular glands, situated posterior to the prostate gland and lateral to each ampulla of the ductus deferens. Each seminal vesicle is a single highly convoluted, elongated tube that develops as an outpocketing from the ductus deferens at the termination of the ampullary portion.

The wall of the seminal vesicle is composed of an irregular secretory **mucosa, muscularis** consisting of **internal circular** and **external longitudinal smooth muscle layer,** and **loose connective tissue** as its **adventitia.** The loose connective tissue contains numerous **blood vessels** and an elastic fiber network. During ejaculation, the contraction of the smooth muscle supplied by the sympathetic nervous system forces secretion from the seminal vesicle into the ejaculatory duct.

Fig. 12-11.

Seminal Vesicle: Mucosa

The **mucosa** of the seminal vesicle is composed of a **simple columnar epithelium** and a **lamina propria,** a network of vascularized loose **connective tissue.** The architecture of the mucosa is relatively complicated. Primary folds of the mucosa branch to form secondary and tertiary folds that merge with one another. As a result, the epithelial surface area increases considerably.

The secretion of the epithelium is a yellowish viscid, alkaline fluid. It enables spermatozoa to start moving and protects them against the acid secretion in the female reproductive tract. Fructose, as well as other products of the seminal vesicle, are believed to nourish spermatozoa. The secretion also contains certain concentrations of prostaglandin. The activity of the seminal vesicle is maintained by the stimulation of testosterone.

Figure 12-10. Seminal Vesicle
Human • H.E. stain • Low magnification

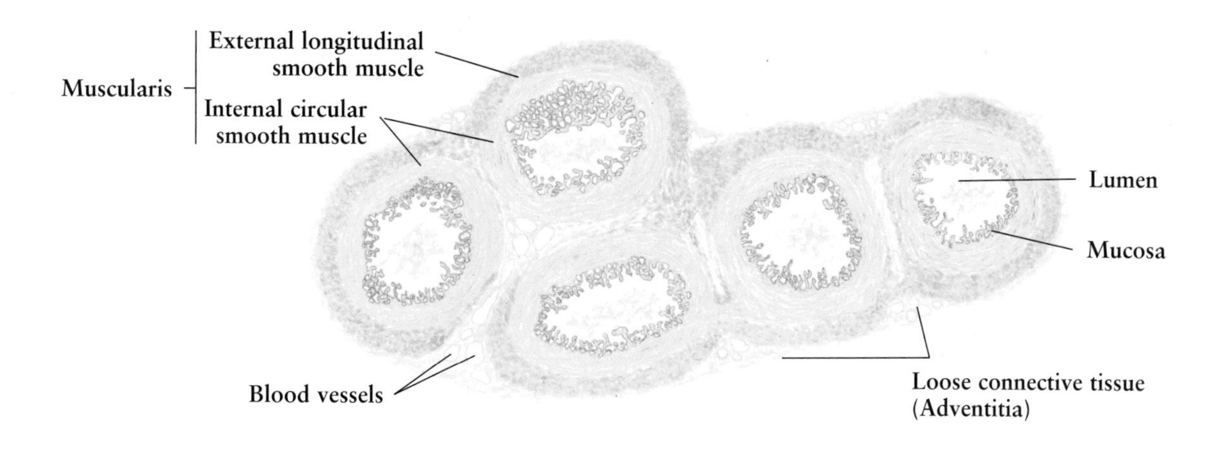

Figure 12-11. Seminal Vesicle: Mucosa
Human • H.E. stain • High magnification

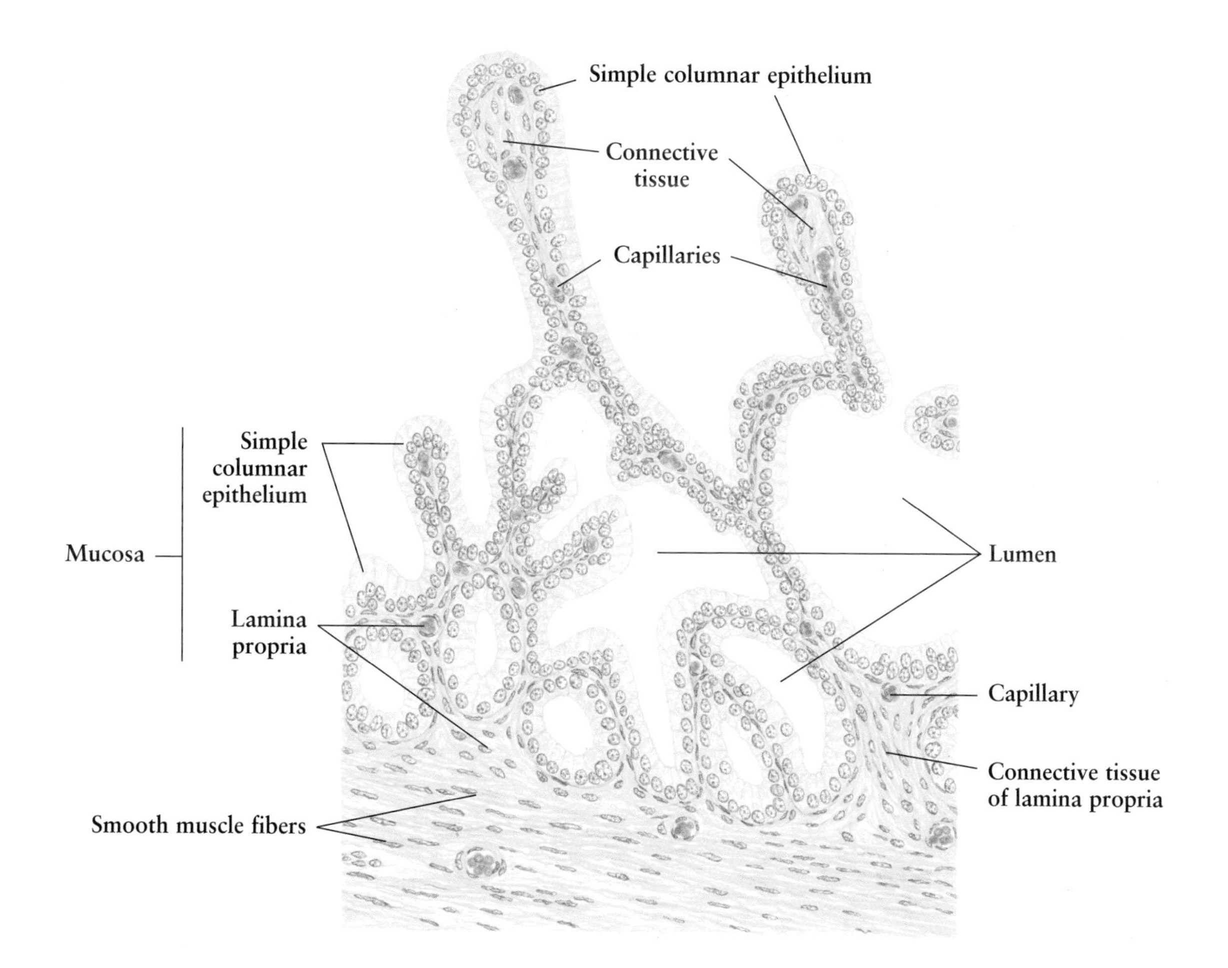

Fig. 12-12.

Prostate Gland

The **prostate gland,** about the shape and size of a chestnut, is the largest accessory gland of the male reproductive tract. It is located below the urinary bladder and is traversed by the prostatic urethra (prostatic part). The organ is enclosed by a dense connective tissue capsule, rich in **smooth muscle fibers,** which penetrate the parenchyma as **septa** to divide it into about 40 indistinct lobules.

The glandular elements are distributed in three layers, more or less centrally arranged around the **prostatic urethra** into which they open. They include small **mucosal glands,** lying in the mucosa of the prostatic urethra, intermediate-sized **submucosal glands,** arranged in a ring surrounding the mucosal glands, and large **main prostatic glands,** constituting the bulk of the prostate, located toward the periphery of the gland. The main prostatic glands drain into the prostatic uretha through 15-30 **prostatic ducts.** For unknown reasons, the mucosal and submucosal glands undergo hyperplasia, in older men, producing *nodular prostatic hyperplasia,* which narrows the urethra.

Figure. 12-12 shows the central part of the prostate gland under low magnification. The lumen of the prostatic urethra, where the **colliculus seminalis** is located, has a crescent appearance in cross section. Two **ejaculatory ducts,** which are continuous with the ductus deferens, open into the slitlike lumen of the urethra. Directly below the urethral lumen is the **utriculus of prostate,** a small, pocket-like indentation of the colliculus between the orifices of the ejaculatory ducts. Around these ducts is a half ring of smooth muscle fiber-rich stroma, from which the septa radiate to the peripheral capsule. Between septa are the acini of **main glands.**

Fig. 12-13.

Prostate Gland: Acini

The secretory **acini** have large, irregular **lumina;** they anastomose with each other to form a labyrinth. Acini are lined by a **simple (glandular) epithelium** resting on a very thin basement membrane that is hard to see even under the electron microscope. The epithelial cells vary from columnar to low cuboidal, depending on endocrine status and glandular activity. The cytoplasm of the epithelial cells is pale-staining, and the nucleus is round or ovoid, located at the base of the cells. Between epithelia is fibroelastic **connective tissue** containing numerous **smooth muscle fibers** and **capillaries.** The epithelium together with the underlying connective tissue often rises into folds, projecting into the lumen of the glands. **Concretions** are calcified prostatic secretions that have formed concentric condensations. They are often found within acini of older men.

The secretion of the prostate gland, which makes up about 70% of the semen, is a thin, milky fluid with a characteristic odor. It is slightly alkaline and is rich in citric acid and proteolytic enzymes, chiefly fibrinolysin, which aids in liquefaction of the coagulated semen after it has been deposited within the vagina.

Figure 12-12. Prostate Gland

Human • Cross section through colliculus seminalis • H.E. stain • Low magnification

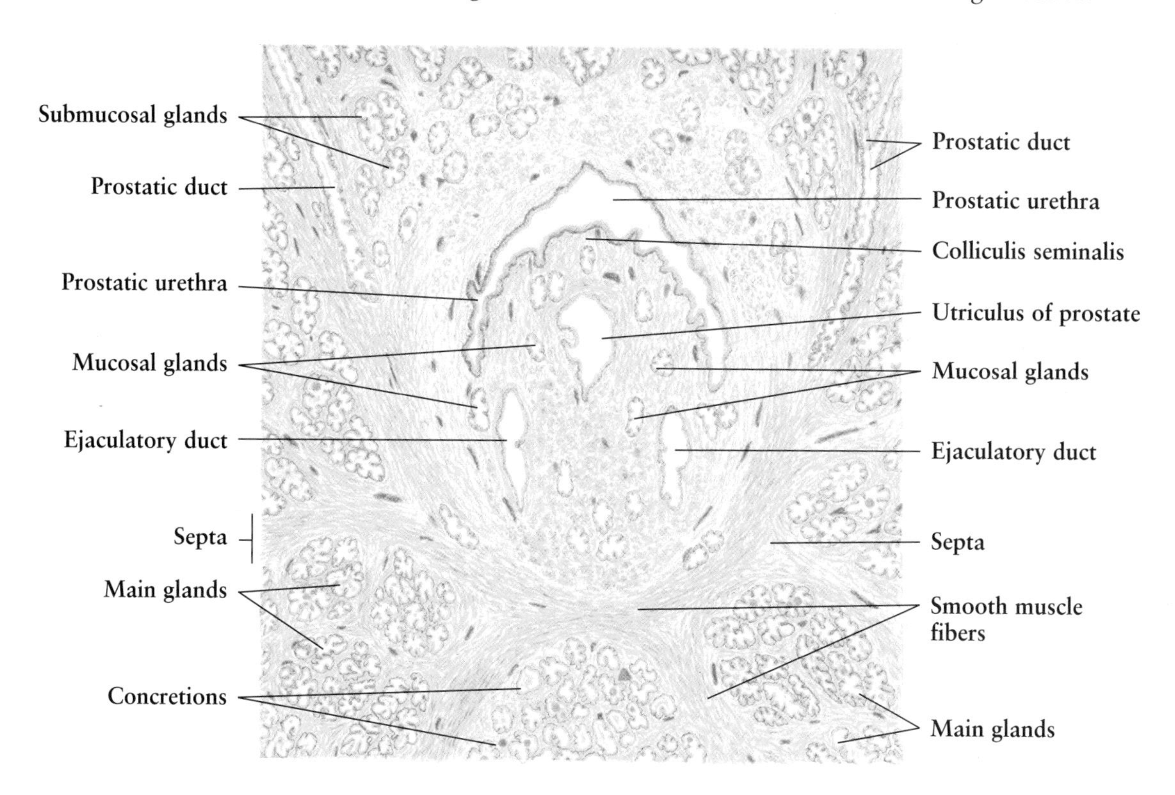

Figure 12-13. Prostate Gland: Acini

Human • H.E. stain • High magnification

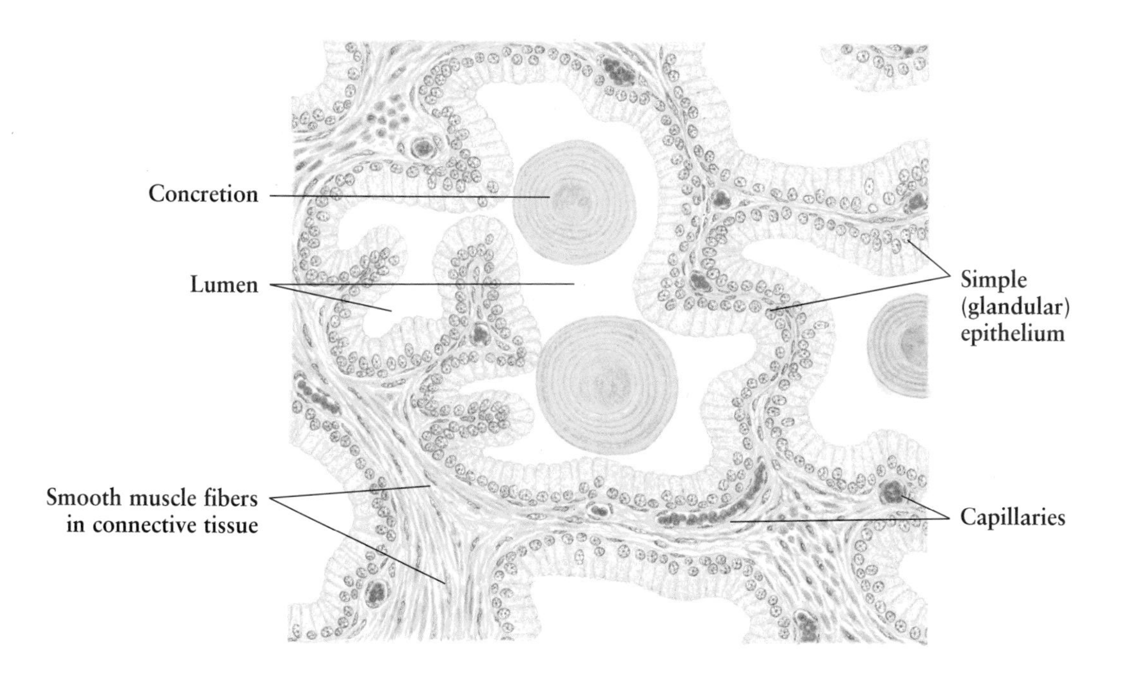

Fig. 12-14.

Bulbourethral Gland

The **bulbourethral glands** (of Cowper) are paired, pea-sized tubuloacinar glands, lying on either side of the urethral bulb, which open into the floor of the cavernous urethra by way of its long duct. The glands are invested by a dense **connective tissue capsule** containing some **skeletal muscle fibers** from the urogenital diaphragm. The **connective tissue septa** with elastic and smooth and skeletal muscle fibers pass into the gland to divide it into lobules.

The secretory units with myoepithelial cells around them are either tubular, acinar, or saccular. The **glandular epithelium** is composed of simple columnar cells with small, round, and basally located nuclei and clear supranuclear cytoplasm filled with secretory granules. The **excretory ducts** are lined by a pseudostratified epithelium similar to that of the urethra. Numerous **capillaries** are found in the connective tissue surrounding the acini.

The secretion of the bulbourethral glands is a clear, viscous, alkaline mucous fluid. It is discharged first during sexual arousal to lubricate or neutralize the acidic lumen of the urethra.

Figure 12-14. Bulbourethral Gland
Human • H.E. stain • High magnification

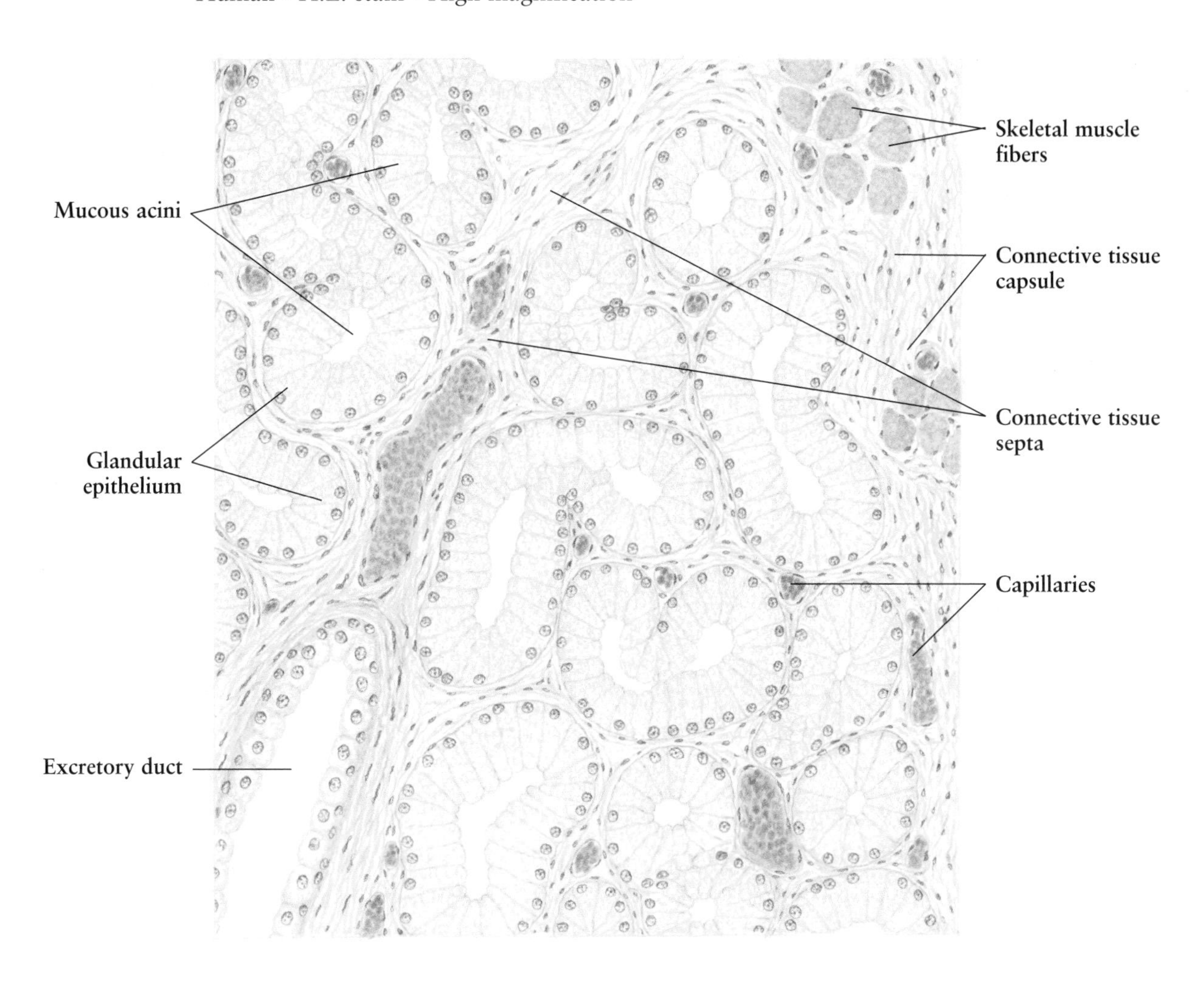

Fig. 12-15.

Penis

The **penis** is a copulatory organ that serves as the common outlet for urine and seminal fluid. **Figure 12-15** is drawn from a transverse section of an adult human penis, demonstrating the general arrangement of the penile tissue. The erectile tissue of the penis consists of three cylindrical bodies: the paired *corpora cavernosa* (corpora cavernosa penis) dorsally and the single *corpus spongiosum* (corpus cavernosum urethrae) ventrally.

Each **corpus cavernosum** is enveloped by a thick fibrous sheath, the **tunica albuginea.** The paired corpora cavernosa are incompletely separated by a middle septum, the **septum penis,** which is continuous with the tunica albuginea. Each corpus cavernosum is made up of a network of **trabeculae,** which create a pattern of **cavernous sinuses,** and contains a **deep artery** (arteria profunda penis) and its branches, the **helicine arteries.** The **corpus spongiosum** is invested by **tunica albuginea corporis spongiosi.** This sheath is much thinner than that of corpora cavernosa and contains many elastic and smooth muscle fibers. The corpus spongiosum houses the cavernous portion (spongy part) of the **urethra** (see **Figs. 11-11** and **11-12**), and the structure resembles that of the corpora cavernosa.

The three cylindrical masses of erectile tissue are surrounded by subcutaneous tissue, the **fascia penis,** which is devoid of adipose tissue but contains many smooth muscle fibers. The **superficial dorsal vein** (venae dorsalis penis subcutanea), **deep dorsal vein** (venae dorsalis penis), **dorsal artery** (arteriae dorsalis penis), and **dorsal nerve** (nervus dorsalis penis) run in the fascia penis. The **skin** of the penis is thin, delicate, and hairless. It is loosely attached to the fascia penis, allowing the penile skin mobility over the underlying structure.

The median **penile raphe** is a dark line on the ventral surface. It is continuous with the scrotal raphe, indicating where the urogenital folds fused during the fetal period.

Fig. 12-16.

Penis: Corpus Cavernosum

The erectile tissue, the **corpus cavernosum,** is composed of connective tissue septa, the **trabeculae,** and large irregular blood spaces, the **cavernous sinuses.** The core of the trabeculae consists of a mixture of collagen and elastic fibers, fibroblasts, and **smooth muscle fibers,** which occur singly or in small bundles. The cavernous sinuses are lined by a thin layer of **endothelial cells** resting upon a very thin basement membrane.

The **deep artery (arteria profunda penis)** runs in the trabecula, parallel to the long axis of the penis. There is no distinct internal elastic lamina. The **muscularis** is composed of inner longitudinal and outer circular layers of smooth muscle fibers. The deep artery gives rise to coiled arteries, the **helicine arteries,** which open directly into the cavernous sinuses. The smooth muscle of the arteries and trabeculae is supplied by both sympathetic and parasympathetic fibers.

Erection is initiated by parasympathetic stimulation, resulting in relaxation of the arterial smooth muscle. The arteries dilate to allow more blood to enter the cavernous sinuses, which become large and hard. The engorged cavernous sinuses, compress and restrict the venous outflow directly under the tunica albuginea; the result is an erection.

Figure 12-15. Penis
Human • Cross section • H.E. stain • Very low magnification

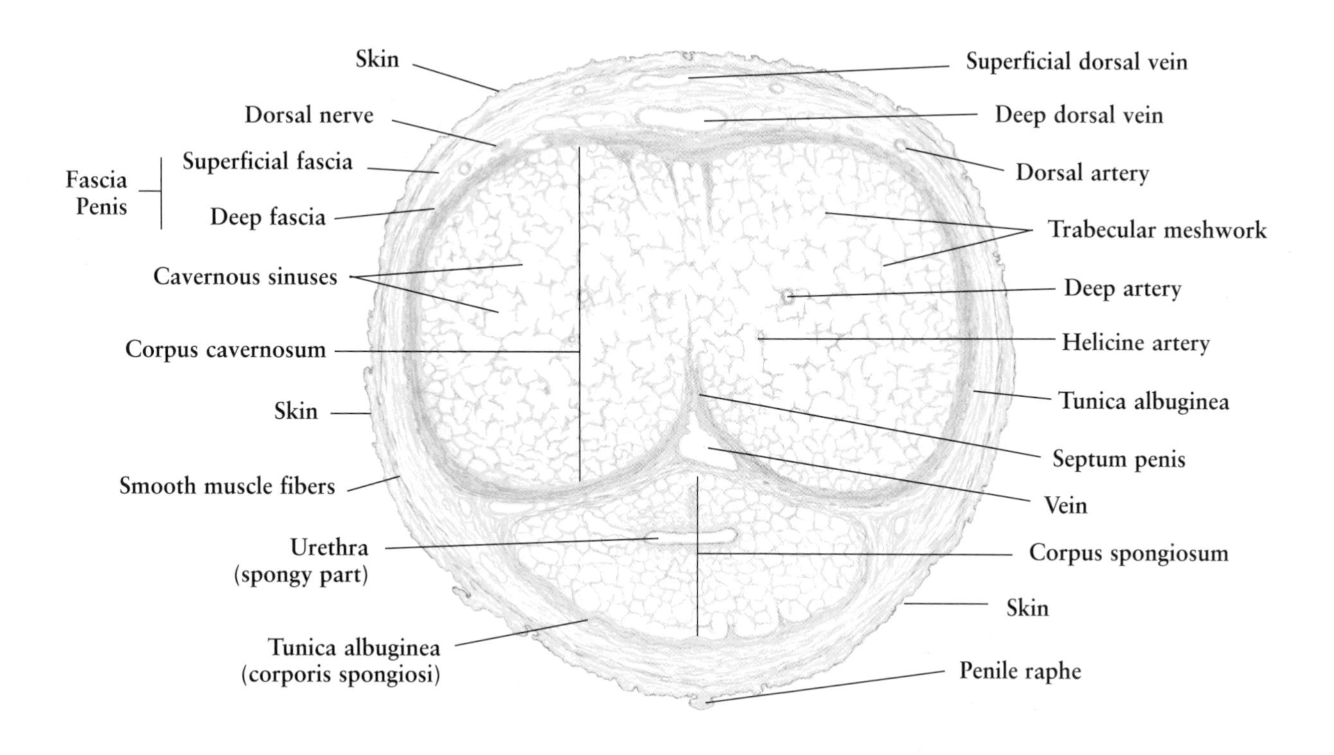

Figure 12-16. Penis: Corpus Cavernosum
Human • H.E. stain • Medium magnification

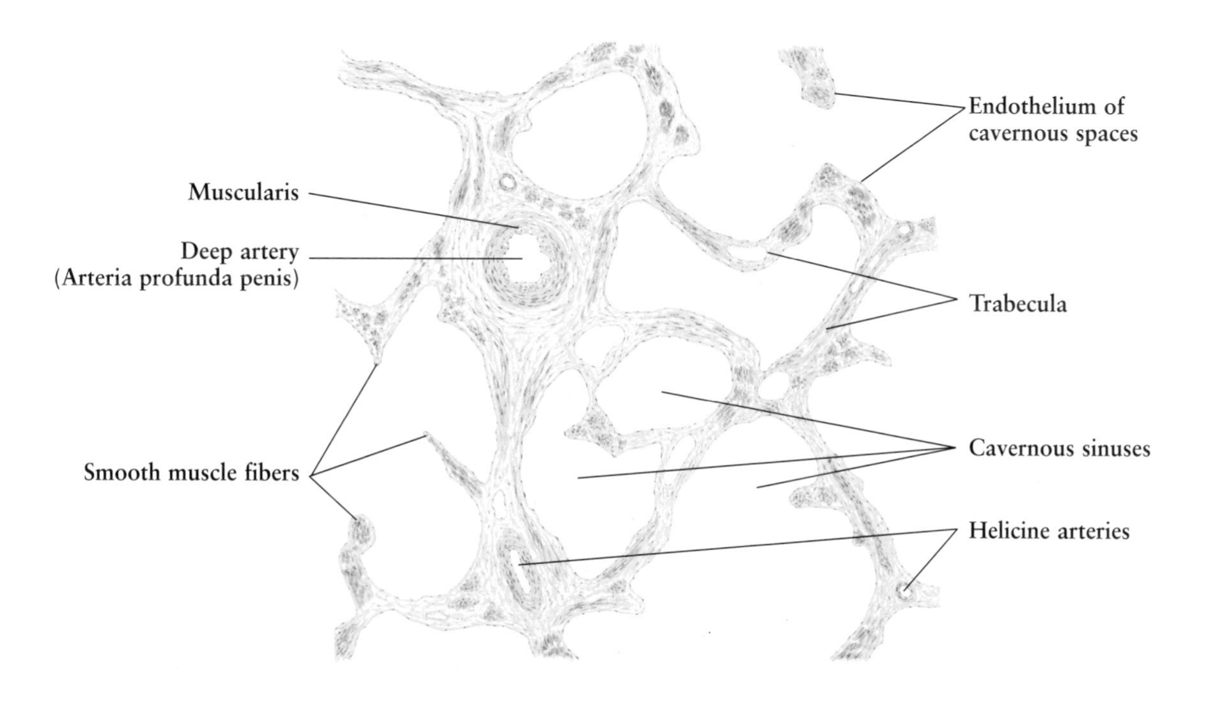

Fig. 12-17. Penis: Skin and Tunica Albuginea

The **skin** of the penis is thin and hairless in type, composed of epidermis and dermis. The **epidermis** consists of keratinized **stratified squamous epithelium** with characteristic **melanin granules** in the **basal cells.** The **dermis** is a dense fibroelastic connective tissue that is rich in blood vessels. It also contains some small **sebaceous glands** that are not associated with hair follicles.

The subcutaneous stratum (hypodermis) is a loose connective tissue, consisting of **superficial fascia (fascia penis superficialis)** and **deep fascia (fascia penis profunda).** The former shows numerous **smooth muscle fibers** grouped in bundles, collagen and elastic fibers, fibroblasts, and blood vessels. The latter contains large **arteries** and **veins,** lymphatic vessels, and **nerves.**

The **tunica albuginea** is a very thick and strong dense connective tissue composed of bundles of regularly arranged **collagen fibers.** These bundles of collagen fibers are arranged mainly **longitudinally** in its outer layer and **circularly** in its inner layer. The inner circular layer is much thicker than the outer longitudinal one. In the flaccid state, the collagen fiber bundles have a wavy course; during erection, however, they straighten out to accommodate the considerable increase in volume of the corpora cavernosa.

Figure 12-17. Penis: Skin and Tunica Albuginea
Human • Cross section • H.E. stain • Low magnification

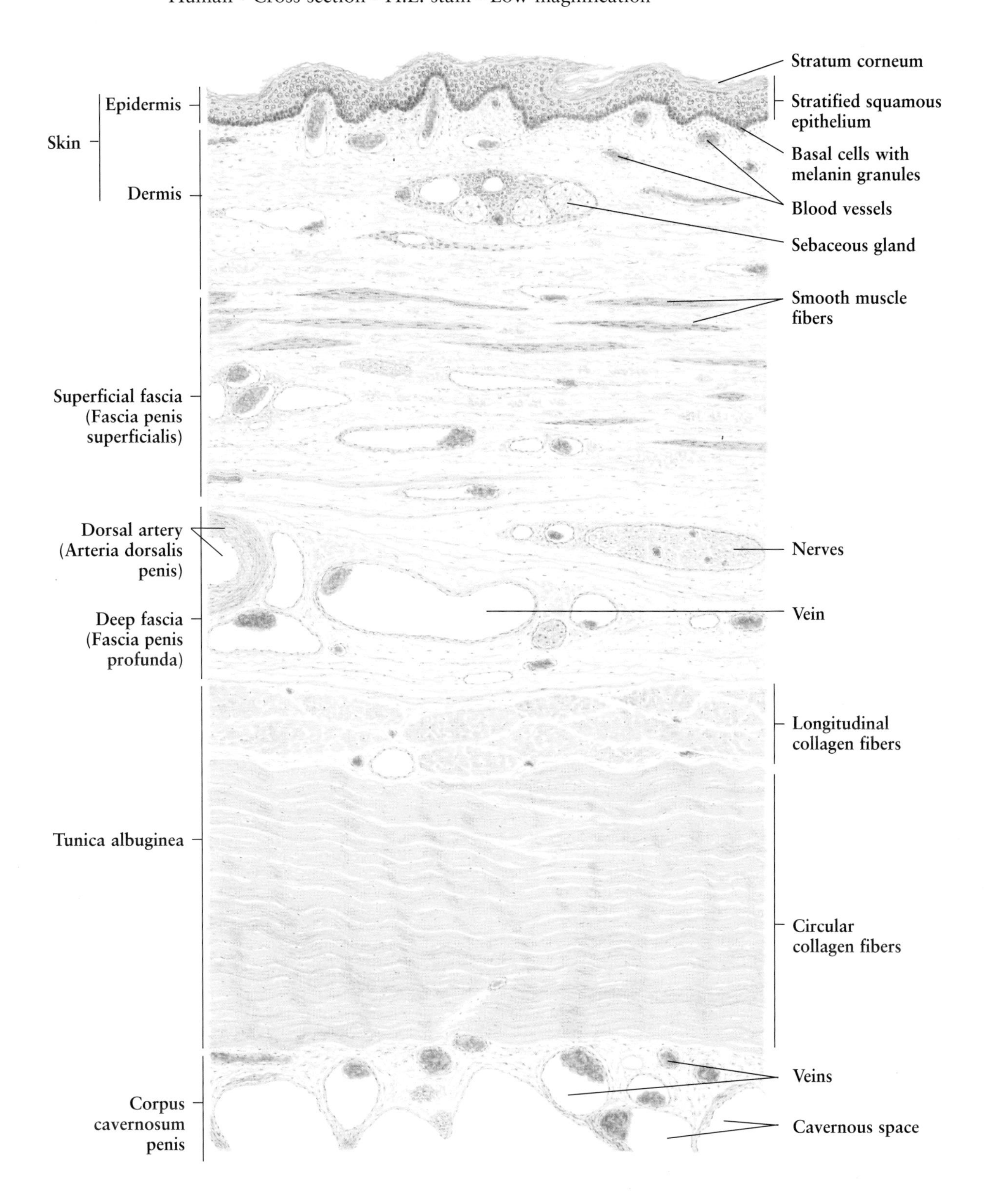

Fig. 12-18.

Scrotum

The **scrotum** is a saclike organ that holds the paired testes. It and its contents are located beneath the root of the penis. It is characterized by the presence of tunica dartos and pigmented epithelium.

The **epidermis** of the **skin** consists of **stratified squamous epithelium** with a thick **layer of keratinization** on the surface. The **basal cells** contain large amounts of **melanin granules,** giving the skin a dark brown color. Beneath the epidermis is a very thick loose connective tissue, the **tunica dartos,** which contains numerous **smooth muscle fibers** in bundles, in addition to numerous collagen and elastic fibers. The smooth muscle bundles are mainly arranged into two layers: inner longitudinal and outer circular. The muscle bundles are packed loosely. Unlike the skin of the penis, that of the scrotum has hair. The **hair follicle** can be found in the tunica dartos. It also contains **sebaceous glands, eccrine sweat glands, nerve fibers,** and **blood vessels.**

Internal to the tunica dartos is a thin layer of **fascia spermatica externa** with densely packed collagen fibers. Next to the fascia spermatica externa are fascia spermatica interna and musculus cremaster internus. The **fascia spermatica interna,** which is separated from fascia spermatica externa by a layer of very **loose connective tissue,** possesses numerous small veins. The **musculus cremaster internus** is a smooth muscle layer arranged longitudinally, continuous with that of the spermatic cord. At the innermost side, the musculus cremaster internus is adjacent to the **periorchium** of lamina parietalis, which is lined by a simple squamous epithelium, the mesothelium, that used to be a part of the peritoneum.

The main function of the scrotum is to regulate the temperature within the scrotum so as to provide a conducive environment for the testis. This regulation is essential for the development of germinal cells in the seminiferous tubules. A temperature below that of the body is required for development of germinal cells in the seminiferous tubules, because high temperature may hinder the maturity of germinal cells.

Figure 12-18. Scrotum
Human • H.E. stain • Low magnification

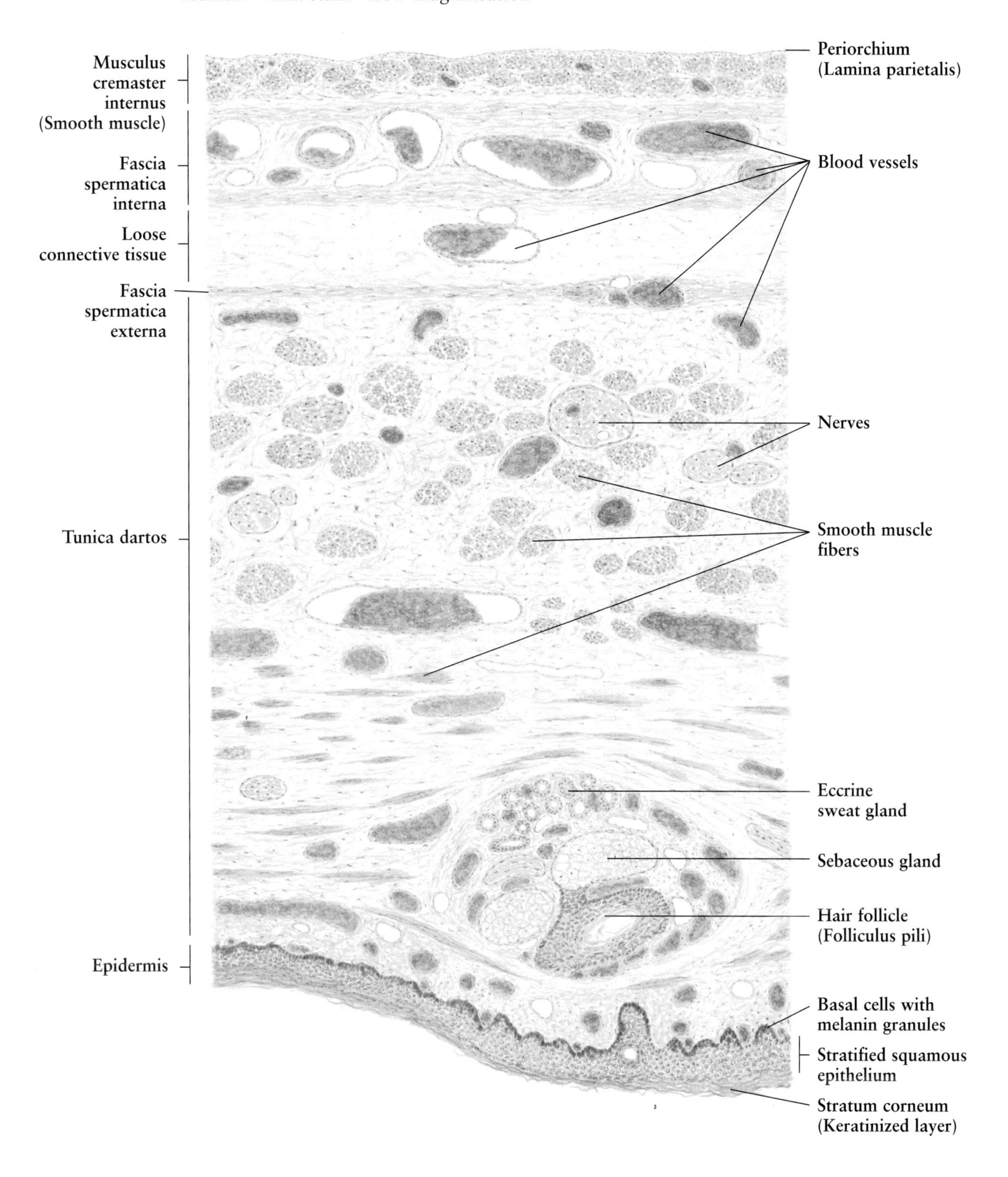

13. Female Reproductive System

The **female reproductive** system consists of the *ovaries, oviducts, uterus, vagina,* and *external genitalia.* The *placenta* and *mammary glands* are also included in this chapter. The principal functions of the system include (1) *production* of female gametes, the ova, by a process of oogenesis; (2) *reception* of male gametes, the spermatozoa; (3) *provision* of a suitable environment for fertilization of ova by spermatozoa and for development of the fetus; (4) a *mechanism* for the expulsion of the developed fetus to the external environment; and (5) *nutrition* of the newborn.

The **ovaries** are small, paired, almond-shaped organs, located in the pelvic cavity. The surface of the ovary is covered by a modification of the peritoneum, which is a simple squamous or cuboidal epithelium, also called the **germinal epithelium.** The ovary is subdivided into the **cortex** and **medulla.** The cortex contains numerous **ovarian follicles** at various stages of development. The medulla is a vascular loose connective tissue, containing large blood vessels that provide a blood supply to the cortex. The ovaries are responsible for the production of the oocytes and the synthesis of the hormones *estrogen* and *progesterone.* These hormones promote development of the ovarian follicles and development of the uterine endometrium, respectively.

The **oviducts** are paired, muscular tubes, which convey ova from the ovaries to the uterus. Each is subdivided into four regions: the **infundibulum, ampulla, isthmus,** and **pars interstitialis.** The pars interstitialis traverses the wall of the uterus. The ampulla is the usual site of fertilization.

The **uterus** is a pear-shaped organ; it is subdivided into three regions: the **fundus, body,** and **cervix.** The wall of the uterus is composed of the **endometrium, myometrium,** and **adventitia** (or **serosa**). In humans, the endometrium undergoes hormonally modulated alterations, known as the *menstrual cycle,* which is usually of 28 day duration. The uterus provides a suitable environment for implantation of the fertilized ovum and development of the embryo and fetus during pregnancy.

The **vagina** is an expansible fibromuscular tube, and its wall is composed of an inner **mucosa,** a middle **muscularis,** and an external **adventitia.** It is specialized for the reception of the penis during copulation and the passage of the fetus to the external environment.

The **external genitalia** include the **labia majora, labia minora, clitoris,** and the mucus-secreting **vestibular glands.** Only the labia minora is shown in this chapter.

The **placenta** is a transient organ, consisting of a **maternal portion** and **fetal portion.** It permits the exchange of various materials between the fetal and maternal circulatory system. The placenta also secretes the hormones *estrogen, progesterone, human chorionic gonadotropin,* and *chorionic corticotropin.*

The **mammary glands** are not one of the genital organs, but they are important glands of the female reproductive system. They are modified sweat glands, composed of about two dozen compound tubuloacinar glands. During pregnancy, the glandular tissues undergo structural changes in preparation for the production of the milk which nurtures the newborn.

Fig. 13-1. Ovary

The **ovaries** are paired, slightly flattened, ovoid organs, 3–4 cm long, 2 cm wide, and 1 cm thick. Each is suspended by a fold of peritoneum, the **mesovarium,** to the broad ligament of the uterus. At the region of attachment to the mesovarium is the **hilum,** through which blood vessels, autonomic nerves, and lymphatic vessels enter and leave the ovary.

The ovary is covered by a simple squamous or cuboidal epithelium, the **germinal epithelium,** which is a specialization of the peritoneal mesothelium. The germinal epithelium rests on a thin basement membrane. The term *germinal epithelium* is a misnomer here, because it does not give rise to any oogonia as previously thought. Beneath the epithelium, there is a pale-stained thick layer of dense connective tissue, the **tunica albuginea.**

The parenchyma of the ovary consists of two zones: *cortex* in the peripheral zone and *medulla* in the core. There is no distinct boundary between the two zones. The stroma of the **cortex** is composed of networks of reticular fibers and numerous spindle-shaped cells, arranged in irregular whorls. The cortex contains a great number of **follicles** in all stages of development. As shown in this figure, the **primordial follicles, primary follicles, secondary follicles,** and **mature (Graafian) follicle** can be recognized in the cortex. Also shown in this figure are a **corpus luteum** and a **corpus albicans,** as well as **atretic follicles.** The **medulla** is a loose fibroelastic connective tissue containing numerous large **blood vessels, lymphatic vessels,** nerves, and some scattered strands of smooth muscle fibers. The connective tissue of the medulla is continuous with that of the mesovarium at the hilum.

The *function* of the ovary is the storage and development of oocytes before their release, and the production of steroid hormones which stimulate the development and function of secondary sex organs, placenta, and mammary glands.

Figure 13-1. Ovary

Human • H.E. stain • Very low magnification

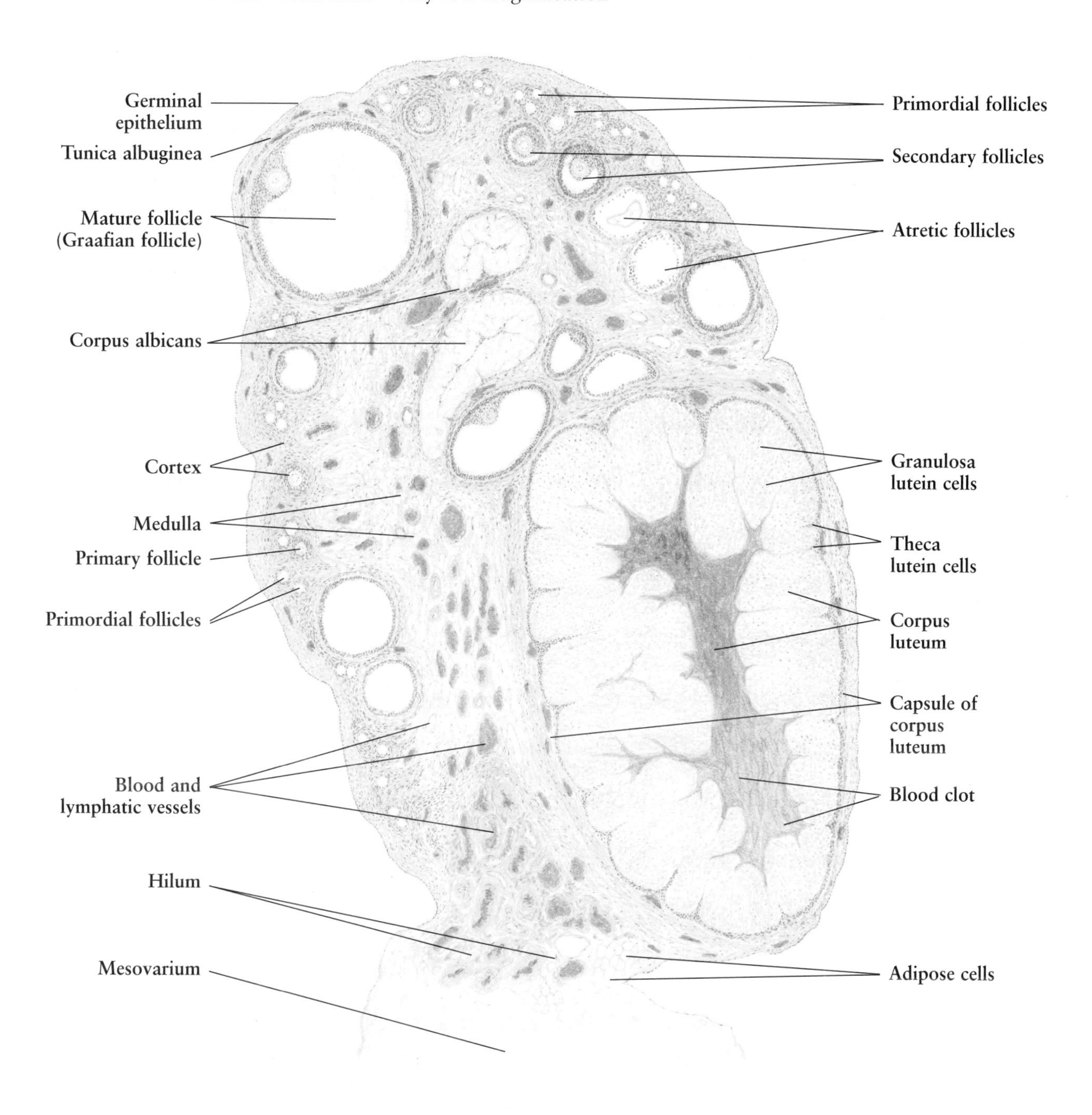

Fig. 13-2. Ovary: Cortex

This figure shows the **cortex** of the ovary, which contains numerous follicles in different stages of development. According to the stage of development, the follicles can be distinguished as: *primordial follicles, primary follicles, secondary follicles,* and *mature* or *Graafian follicles.* The primary and secondary follicles are also called *growing follicles.*

A **primordial follicle** consists of an immature ovum, the **primary oocyte,** surrounded by a single layer of flattened epithelial cells, the **follicular cells,** outside of which is a thin basement membrane. The primary oocyte, 20–40 μm in diameter, has a little cytoplasm and a large nucleus with a prominent nucleolus and finely dispersed chromatin. In the newborn female, the number of primordial follicles is estimated to be 400,000. Only about 400–450 of these follicles complete their maturation and are released as an ovum during the reproductive years; most undergo atresia, which may take place at any stage of development. After menopause, almost no follicles remain.

The progressive development of follicles occurs after the onset of puberty. When stimulated to develop, a primordial follicle enlarges to form a **primary follicle.** Entering the stage of primary follicle, the primary oocyte enlarges because of an increase in the volume of the ooplasm. The flattened follicular cells become **cuboidal** in shape and increase in number (**unilaminar primary follicle**). The primary oocyte continues to enlarge until it reaches a size that is about three times that of the primordial oocyte. The follicular cells, which now are also called the **granulosa cells,** proliferate to form a stratified **follicular epithelium** around the oocyte (**multilaminar primary follicle**). Between the oocyte and the granulosa cells there is a red-stained, homogeneous glycoprotein layer, the **zona pellucida.** It is synthesized by the granulosa cells.

With further development, the primary follicle gradually changes into a **secondary follicle.** Spaces between granulosa cells occur (**early vesicular secondary follicle**), widen, and eventually merge forming a follicular cavity, the **antrum.** The follicle becomes a **vesicular secondary follicle.** The antrum is filled with follicular fluid. As the antrum enlarges, granulosa cells arrange in a stratified manner along the basement membrane, called **membrana granulosa.** The follicular cells surrounding the oocyte form the **corona radiata.** The granulosa cell mass that contains the oocyte and is continuous with the membrana granulosa is known as the **cumulus oophorus.** Early in the stage of the secondary follicle, the adjacent stromal cells organize into a capsule or the **theca folliculi,** which is separated from the membrana granulosa by the basement membrane. Then the theca folliculi differentiates into two layers: an inner layer, the **theca interna,** and an outer layer, the **theca externa.** The theca interna contains enlarged stromal cells and numerous capillaries. The theca externa is composed of closely packed collagen fibers and small fusiform cells.

The **mature** (**Graafian**) **follicle,** about 1 cm in diameter, is a transparent vesicle that protrudes from the surface of the ovary (see **Fig. 13-1**). As a result of the accumulation of follicular fluid, the antrum increases greatly in size, and the follicular epithelium (membrana granulosa) becomes thinner. The oocyte adheres to the follicular epithelium through a pedicle formed by granulosa cells. The primary oocyte, arrested in prophase during fetal development, completes the first meiotic division shortly before ovulation to produce a secondary oocyte and a polar body, the first polar body. The secondary oocyte does not undergo the second meiotic division until after fertilization.

The **Call-Exner body** is a small cavity found among the follicular epithelium. This cavity is filled with PAS-positive materials similar to the basement membrane. Its functional significance is unknown.

Figure 13-2. Ovary: Cortex
Human • H.E. stain • Medium magnification

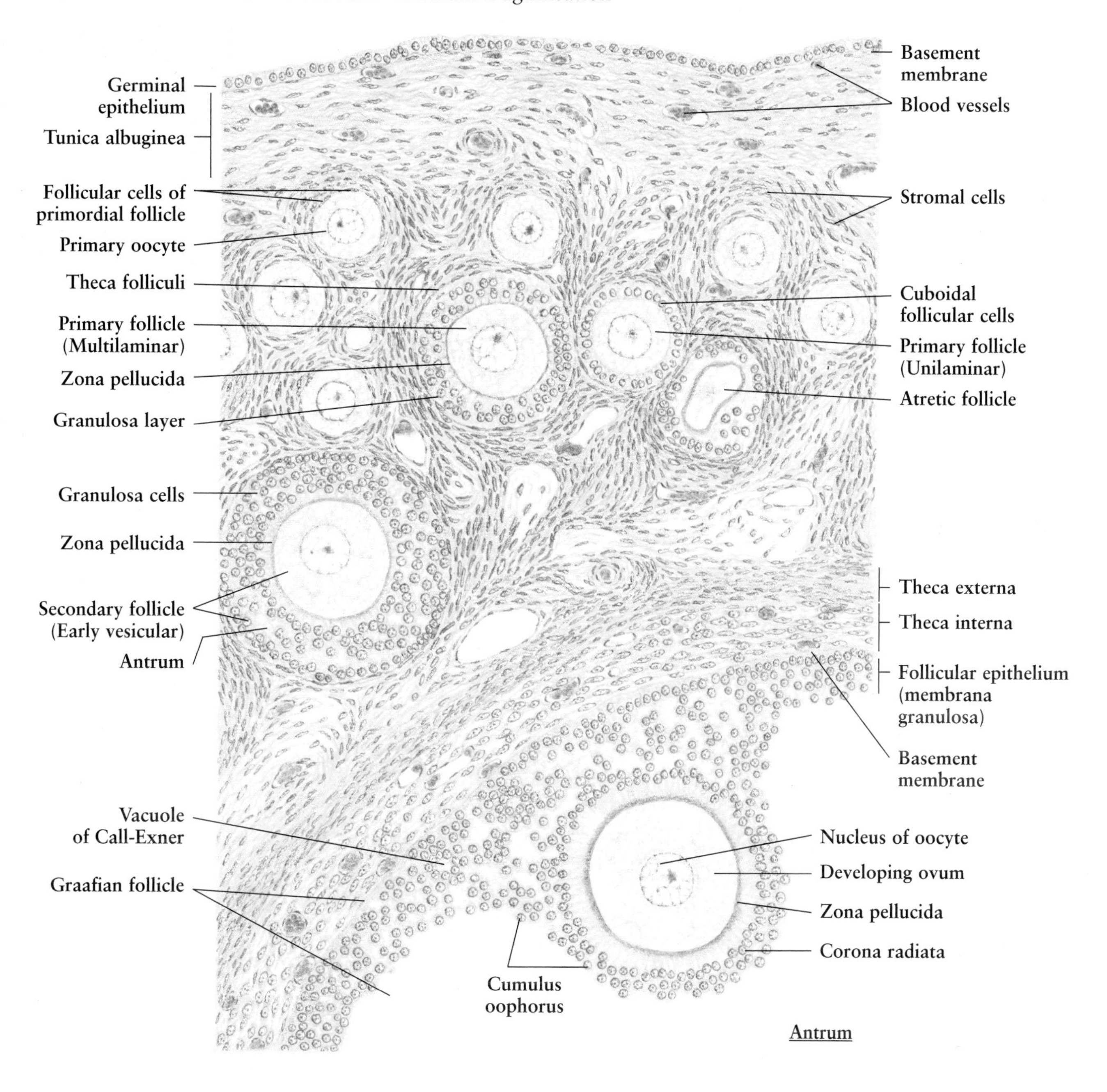

Follicular atresia may occur at any stage in the development of the follicle. The **atretic follicle** shown in this figure is a secondary follicle. The oocyte has died, and the zona pellucida has degenerated. The follicular cells become unorganized; the entire follicle has collapsed.

This figure also demonstrates the **germinal epithelium,** a simple cuboidal epithelium that lies upon a **basement membrane.** The **tunica albuginea** is a dense fibroelastic tissue containing numerous fibroblasts and capillaries. The stromal cells in the cortex are spindle-shaped and densely packed, either surrounding the follicles or arranged in irregular whorls.

Fig. 13-3.

Growth of Ovarian Follicles

The characteristic features of follicles in different stages of development are summarized in the following table (**Table 13-1**). For more information, see **Fig. 13-1.**

	Primordial follicle	Growing follicles				Mature follicle (Graafian follicle)
		Primary follicles		Secondary follicles		
		Unilaminar	Multi-laminar	Early vesicular	Vesicular	
Ovum	Primary oocyte	Primary oocyte	Primary oocyte	Primary oocyte	Primary oocyte	Primary oocyte
Zona pellucida	None	None	Present	Present	Present	Present
Follicular cells	Simple flattened	Simple cuboidal	Stratified	Stratified	Stratified	Stratified
Follicular cavity	None	None	None	Multiple, small	One, large	One, largest
Corona radiata	None	None	None	Forming	Present	Present
Cumulus oophorus	None	None	Forming	Present	Present	Present
Call-Exner bodies	None	None	None	None	Present	Present
Basement membrane	Present	Present	Present	Present	Present	Present
Theca folliculi	None	None	Forming	Theca interna, theca externa	Theca interna, theca externa	Theca interna, theca externa

Figure 13-3. Growth of Ovarian Follicles
According to human H.E.-stained section

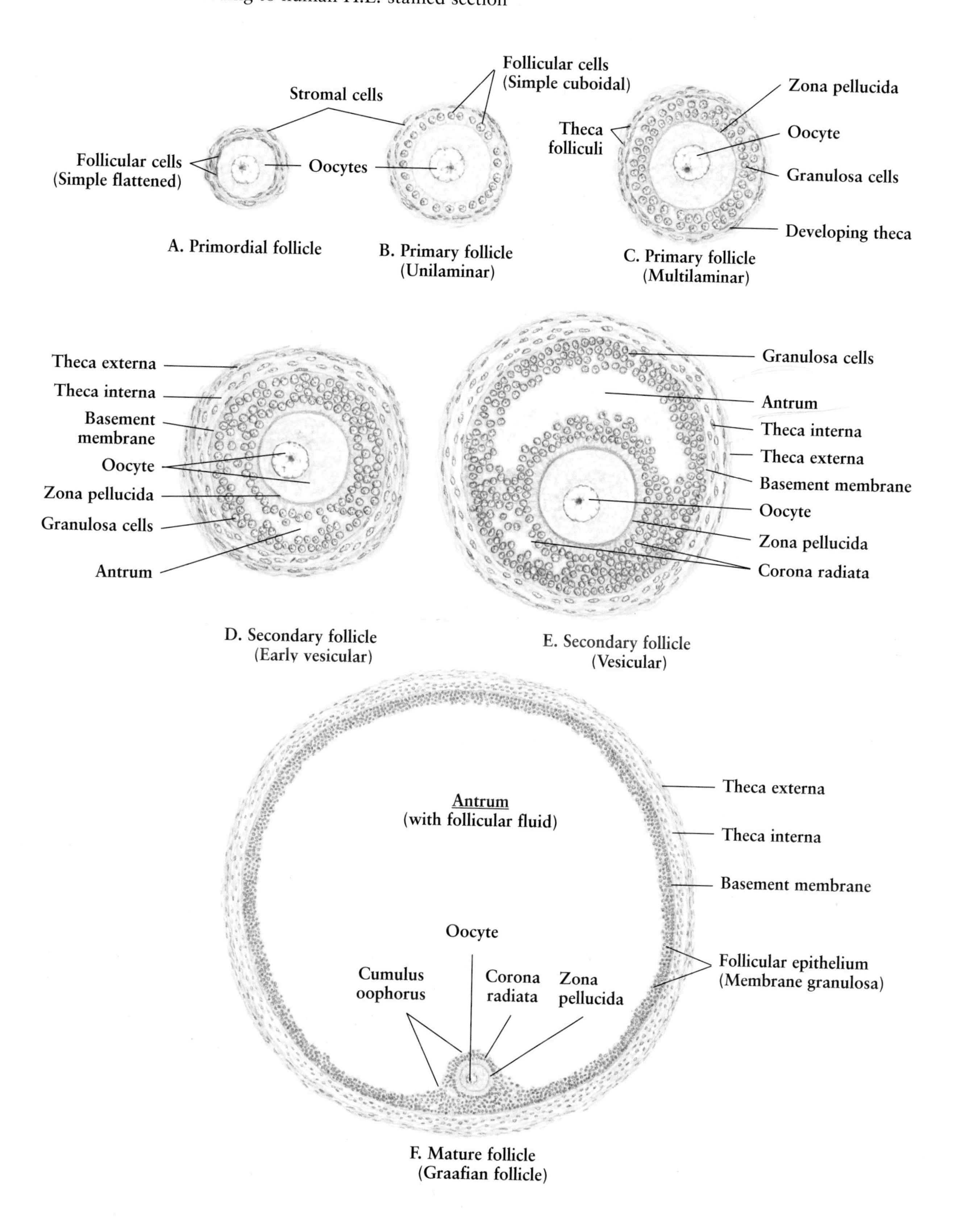

Fig. 13-4.

Corpus Luteum

After ovulation, the ruptured follicle collapses and fills with a blood clot. The follicular wall is thrown into folds and transformed into a temporary endocrine gland, the **corpus luteum,** which is regulated by *luteinizing hormone* (LH) secreted by the anterior pituitary.

The corpus luteum is surrounded by a layer of connective tissue, the **capsule** of the corpus luteum (see **Fig. 13-1**), formed by the theca externa. The parenchyma of the gland consists mainly of two cell types: *granulosa lutein cells* and *theca lutein cells.* The **granulosa lutein cells,** forming the bulk of the gland, are developed from the granulosa cells of the follicular epithelium. They are large, polygonal cells that have a large, round nucleus and pale-staining cytoplasm filled with fine lipid droplets that give the cells a vacuolated appearance. The granulosa lutein cells secrete a steroid hormone, ***progesterone,*** which promotes the exocrine secretion of the uterine glands in the endometrium. The **theca lutein cells,** derived from the cells of the theca interna, are smaller in size than the granulosa lutein cells, and have a dark-staining cytoplasm and a round nucleus. They aggregate peripherally and fill the recesses between the folds of granulosa lutein cells. The theca lutein cells are responsible for the secretion of ***estrogens.*** Among these glandular cells are numerous **capillaries** and a small amount of **connective tissue** that contains some **fibroblasts.**

If the discharged ovum is not fertilized, the corpus luteum begins to degenerate 9 days after ovulation. This is the *corpus luteum of menstruation.* If fertilization and implantation occur, the corpus luteum, known then as the *corpus luteum of pregnancy,* continues to enlarge and functions until the placenta takes over production of these hormones. Eventually both the corpus luteum of menstruation and the corpus luteum of pregnancy involute and become a *corpus albicans* (see **Fig. 13-1**).

Fig. 13-5.

Oviduct: Infundibulum

The **oviducts** (uterine or fallopian tubes) are paired tubular organs, located in the mesosalpinx. One end of an oviduct opens into the peritoneal cavity, the other end opens into the uterine cavity. Oviducts conduct ova from the surface of the ovaries to the uterine cavity, and serve as a site where fertilization takes place. Each is 12–15 cm long and about 1 cm in diameter. Four regions can be recognized: *infundibulum, ampulla, isthmus,* and *interstitial portion.* The **wall** of the oviduct consists of three layers: **mucosa, muscularis,** and **serosa.**

Figure 13-5 displays the **infundibulum** in a longitudinal section. It is a wide, trumpet-shaped **opening** into the peritoneum near the ovary. The margins are thrown out into numerous finger-like processes, the **fimbriae.** These are composed of a loose connective tissue of the lamina propria covered by the simple columnar epithelium present throughout the oviduct. Near the margin of the opening, there is a **border** between the **simple columnar epithelium** of the fimbriae and the simple squamous epithelium, the **mesothelium** of the serosa. When ovulation occurs, the infundibulum moves to overlie the site of rupture of the mature follicle and conducts the ovum into the tube.

Figure 13-4. Corpus Luteum
Human ovary • H.E. stain • High magnification

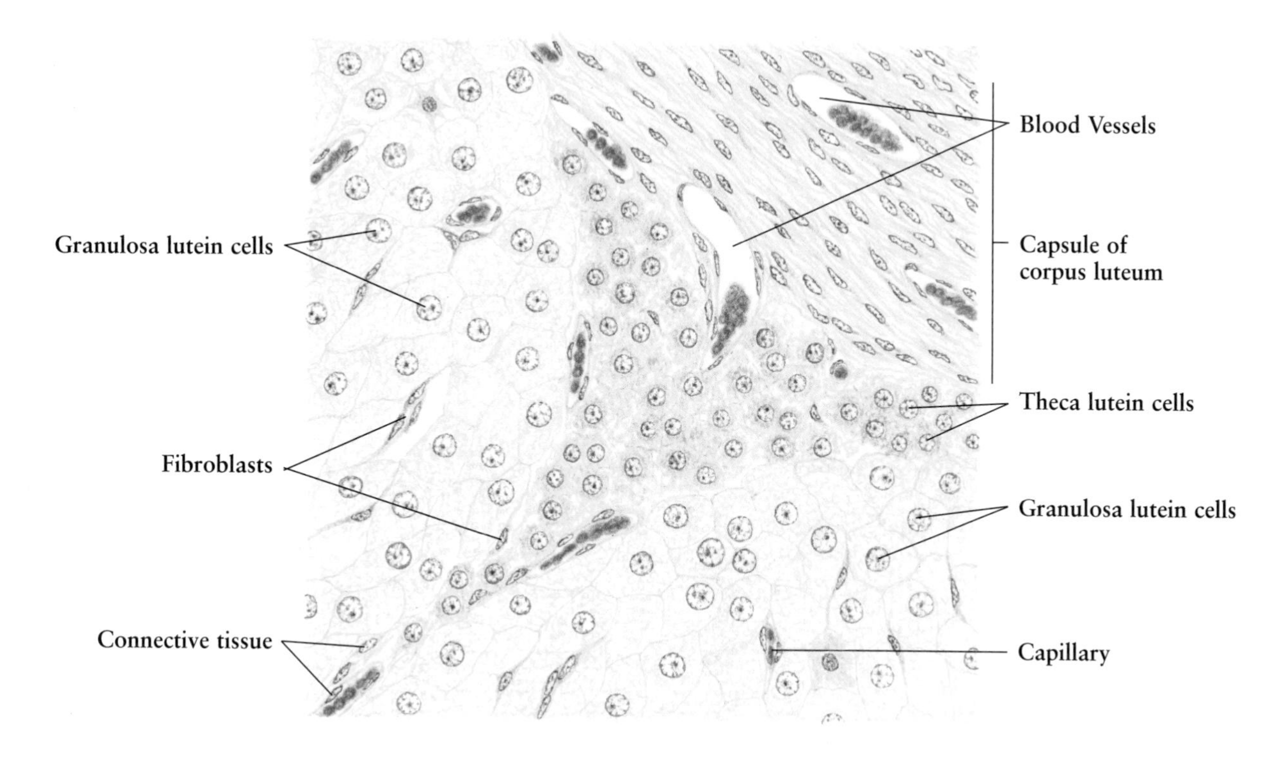

Figure 13-5. Oviduct: Infundibulum
Human • Longitudinal section • H.E. stain • Very low magnification

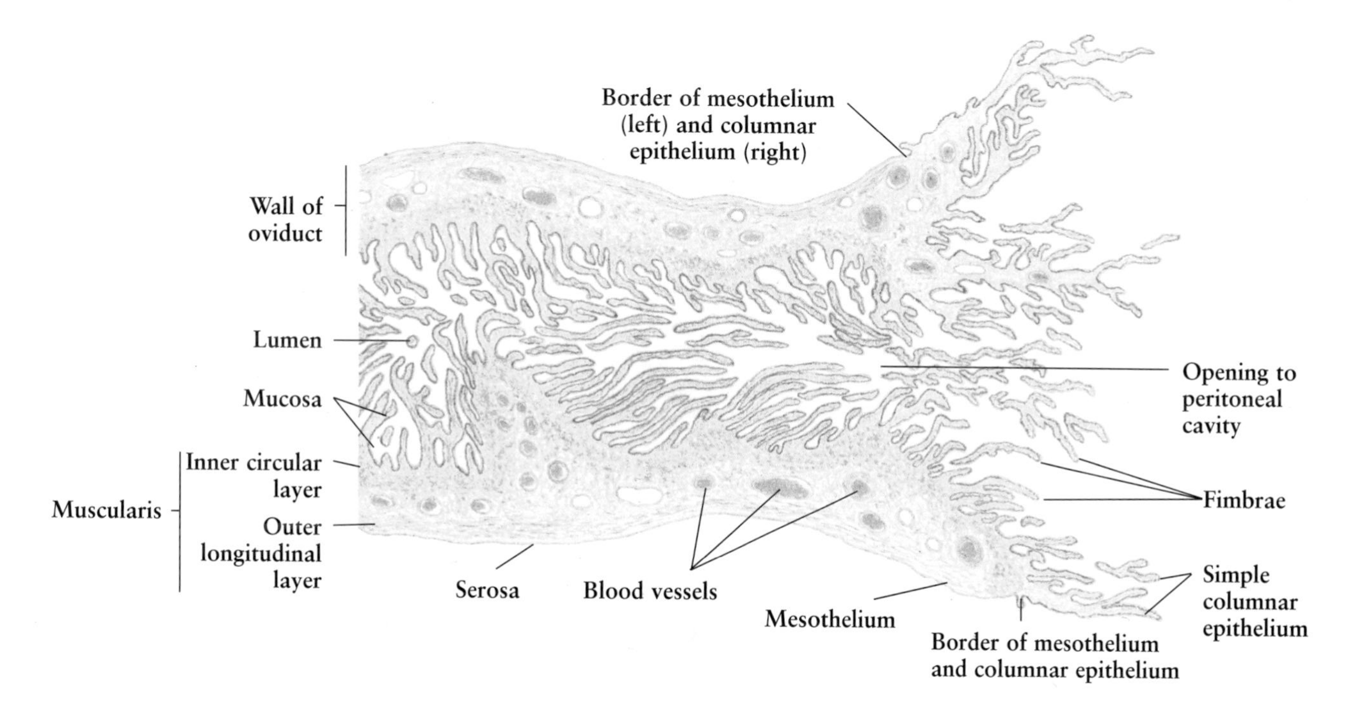

Fig. 13-6.

Oviduct: Ampulla

The **ampulla** is an expanded intermediate segment, comprising two-thirds of the length of the oviduct. Its wall is thin, composed of *mucosa, muscularis,* and *serosa.* The ampulla is characterized by a virtual labyrinth of primary, secondary, and tertiary **longitudinal folds of mucosa** that project into the **lumen.** In the cross section of the ampulla, the branches of the folds divide the lumen into very complicated, irregular spaces. The **muscularis** is composed of **inner circular** and **outer longitudinal layers** of smooth muscle. The smooth muscle fibers in the outer longitudinal layer are loosely arranged in bundles, between which runs connective tissue. In the **mesosalpinx,** large **blood vessels** can be found. The **serosa** consists of loose connective tissue covered by a layer of mesothelium. The ampulla is the normal site where fertilization takes place.

Fig. 13-7.

Oviduct: Isthmus

The **isthmus** is a slender, narrow portion of the oviduct that connects with the uterus. In the cross section shown in this drawing, the **lumen** of the oviduct diminishes in size, and the **longitudinal folds of mucosa** decrease in number and become smaller. The bundles of the **outer longitudinal smooth muscle** are arranged loosely. Among the muscle bundles there are numerous **blood vessels.**

Figure 13-6. Oviduct: Ampulla
Human • Cross section • H.E. stain • Very low magnification

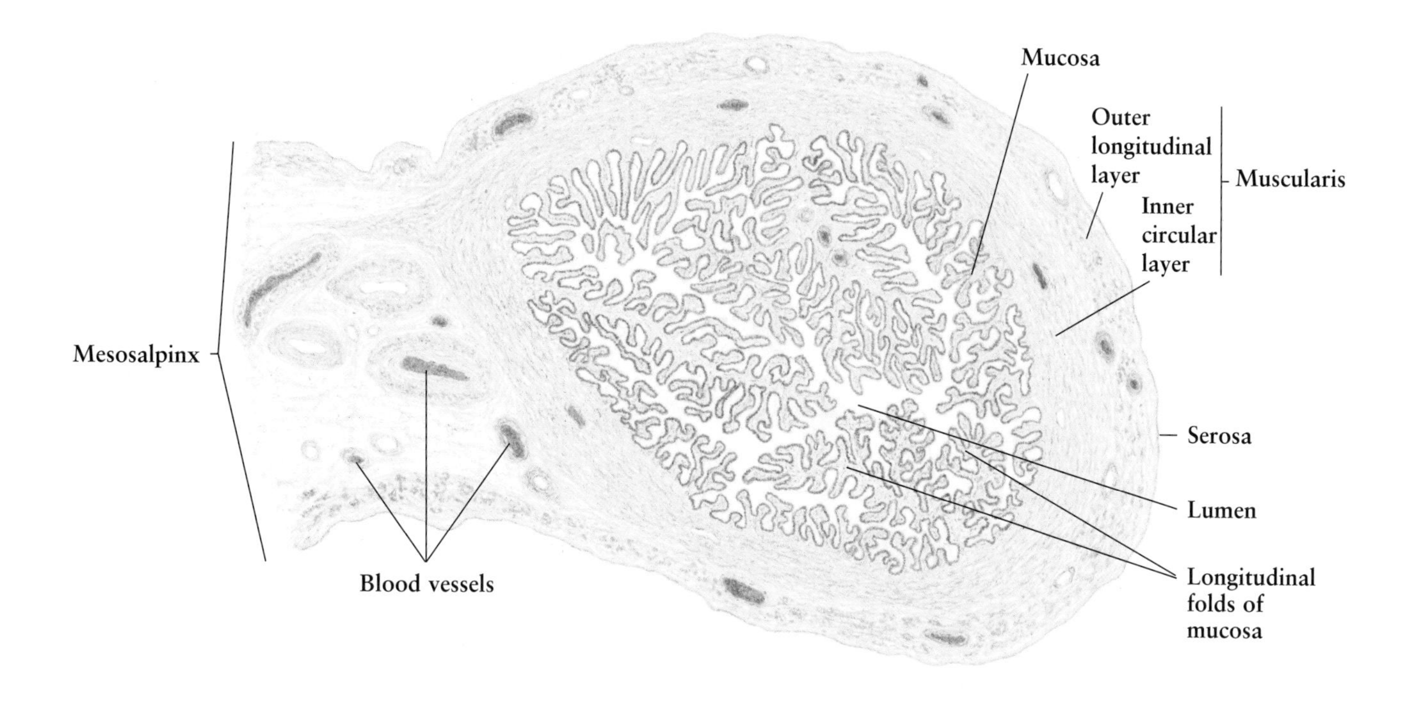

Figure 13-7. Oviduct: Isthmus
Human • Cross section • H.E. stain • Very low magnification

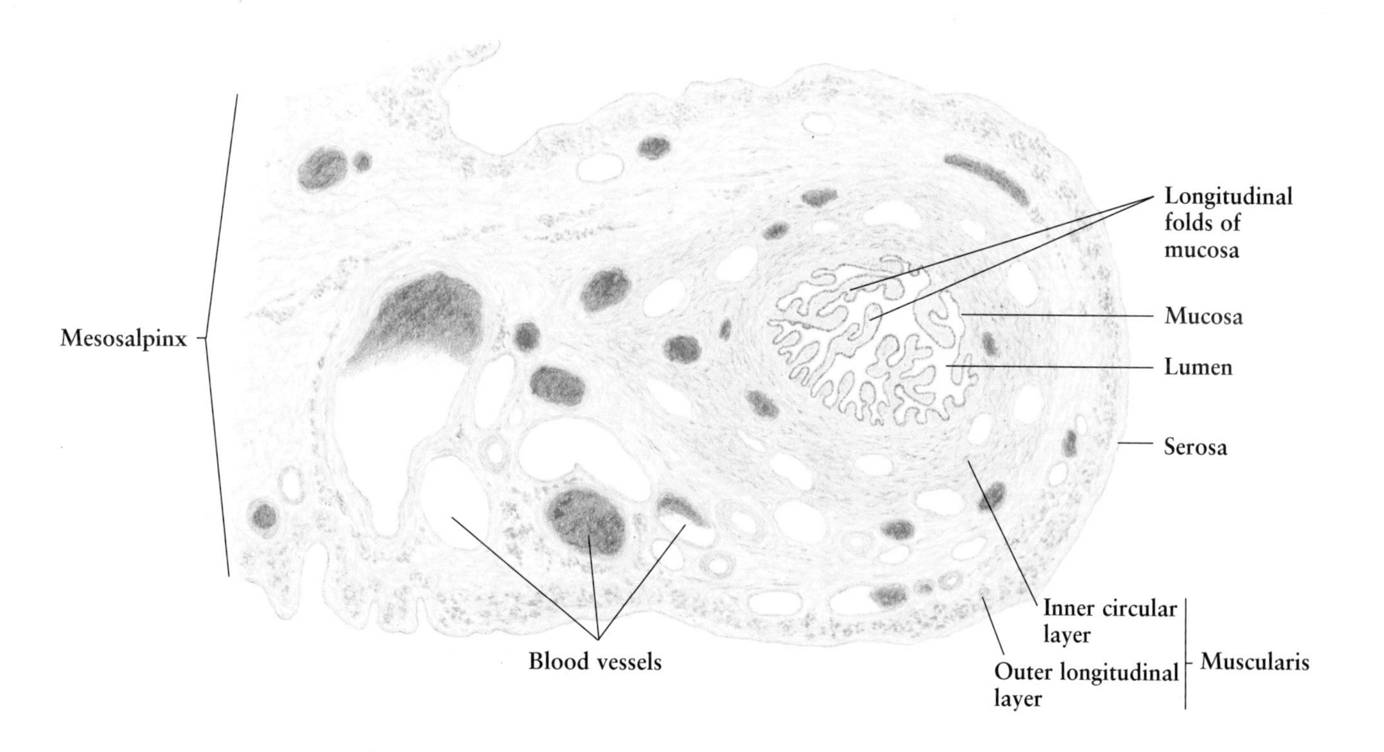

Fig. 13-8.

Oviduct: Interstitial Segment

The **interstitial segment** is the portion of the oviduct that penetrates the uterine wall and opens into the uterus. The **lumen** becomes much smaller, and the mucosal folds disappear. The wall of this segment thickens conspicuously. The smooth muscle fibers in the **outer longitudinal layer** of the **muscularis** are arranged in a looser manner. Numerous **blood vessels** and **lymphatic vessels** are present in the wall of the interstitial segment.

Fig. 13-9.

Oviduct: Mucosa

The **mucosa** is covered by a **simple columnar epithelium** composed of *ciliated* and *nonciliated cells* resting on a very thin **basement membrane.** The **nonciliated cells,** also known as **secretory cells,** are slender and peglike in shape, and have a dark-staining cytoplasm and an elongated, deeply invaginated nucleus. The secretion is associated with the nutrition and protection of the ovum. The **ciliated cells** occur in small groups; between these are scattered secretory cells. The free surfaces of the cells bear numerous cilia, which beat toward the uterus and may have a role in the transportation of the ovum and secretion toward the uterus. The ratio of ciliated to secretory cells and the height of the epithelial cells undergo cyclical variation with the influence of ovarian hormones. The height of the epithelium is greatest during the follicular phase, and lowest during the latter part of the luteal phase as well as during pregnancy.

The **lamina propria** consists of a network of collagen and reticular fibers with numerous fibroblast-like cells, the **fusiform cells.** The fusiform cells may differentiate into typical decidual cells in the case of tubular pregnancy. Many **blood vessels** and **lymphatic vessels** are present in the connective tissue of the lamina propria. There is no muscularis mucosae or submucosa, and the lamina propria is directly adjacent to the **muscularis.**

Figure 13-8. Oviduct: Interstitial Segment
Human • Cross section • H.E. stain • Very low magnification

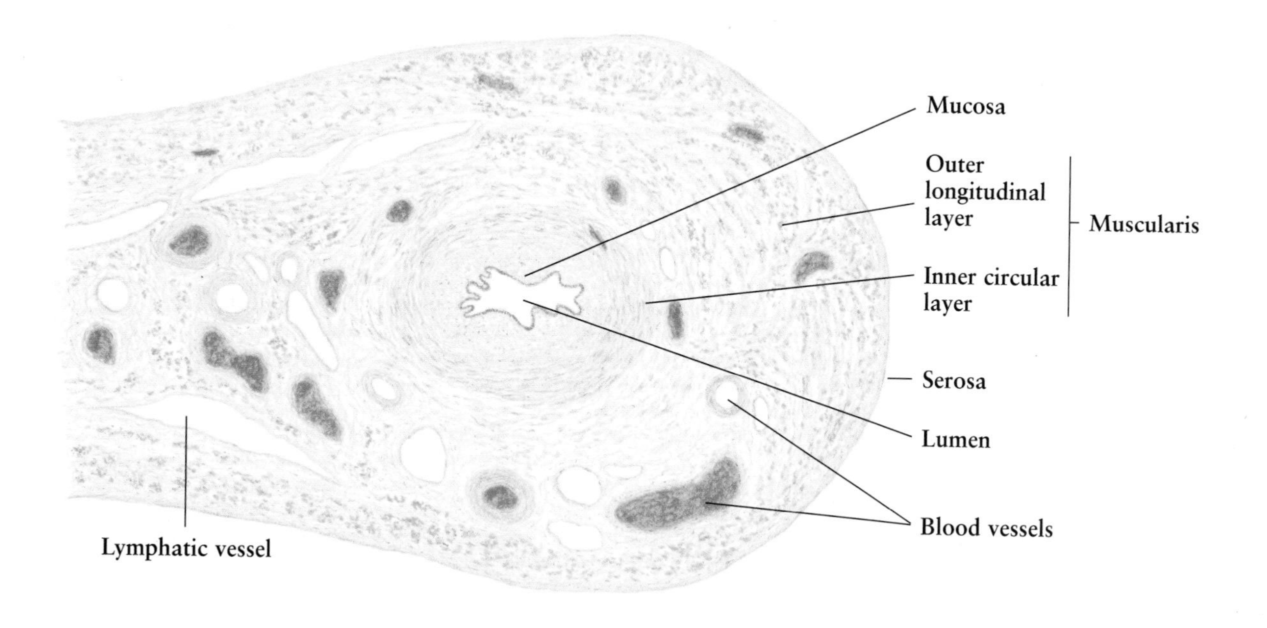

Figure 13-9. Oviduct: Mucosa
Human • H.E. stain • High magnification

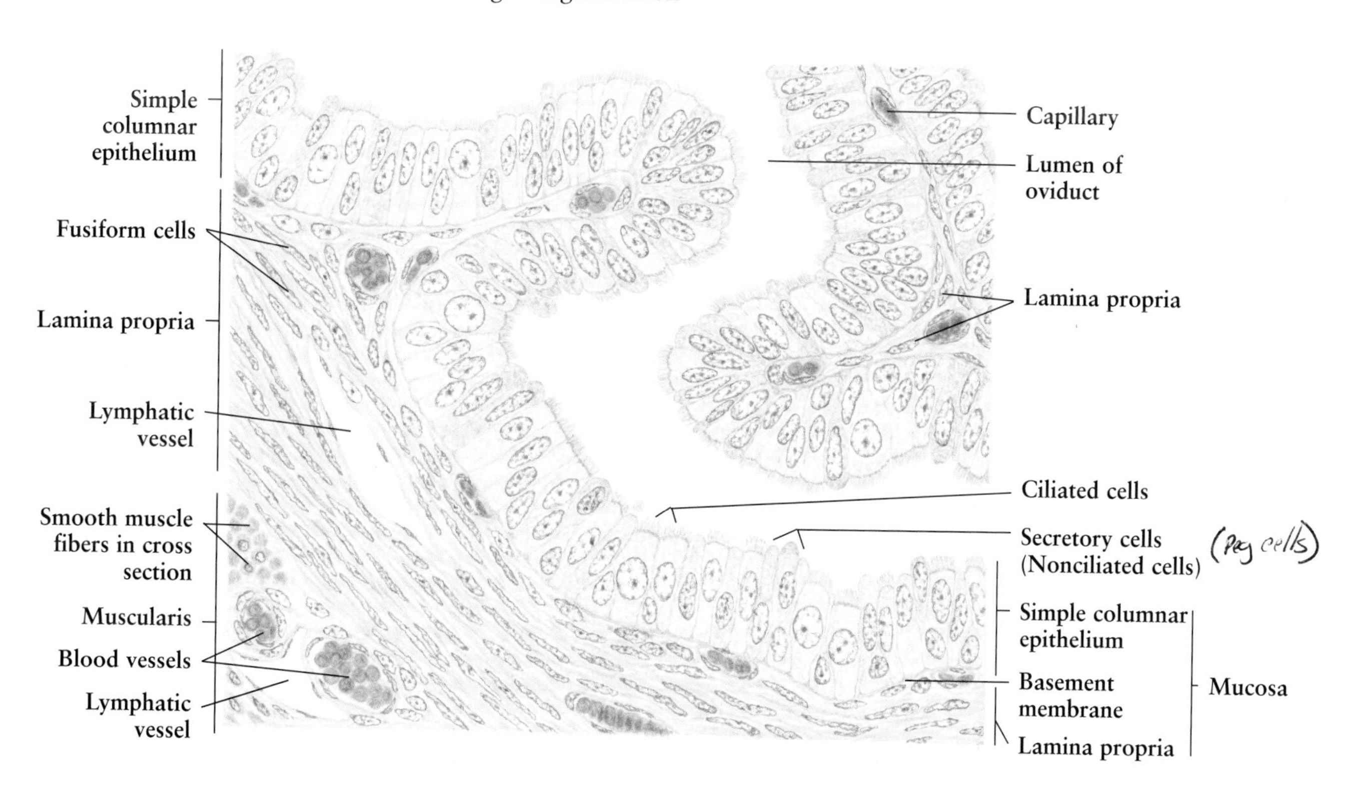

Fig. 13-10.

Uterus

The **human uterus** is a pear-shaped muscular organ, about 7 cm in length, 4 cm in width, and 2–3 cm in thickness. Four portions may be recognized: the *fundus,* the *body* or *corpus uteri,* the *isthmus,* and the *cervix.* The fundus receives on each side the openings of the oviducts. The cervix projects into the vagina as the portio vaginalis. The wall of the uterus consists of three layers: **endometrium,** which corresponds to the mucosa and submucosa, myometrium (muscularis), and perimetrium (serosa).

The **myometrium** is a thick layer, and constitutes the main bulk of the uterine wall. It is composed of flat or cylindrical bundles of smooth muscle that interlace in all directions. Although the smooth muscle bundles are arranged in an ill-defined manner, four layers may be distinguished: *stratum submucosum, stratum vasculare, stratum supravasculare,* and *stratum subserosum.* The **stratum submucosum,** in contact with the endometrium, is made up of muscle bundles, predominantly longitudinal and some oblique. The **stratum vasculare** is composed mainly of longitudinal bundles and contains many blood vessels, including **arcuate arteries.** The **stratum supravasculare** consists chiefly of circular bundles together with some longitudinal fibers. The **stratum subserosum,** just underneath the perimetrium, is a relatively thin layer of muscle fibers arranged longitudinally and circularly.

During pregnancy, under the influence of estrogen, the myometrium increases greatly in size as a result of both hyperplasia and hypertrophy. The length of the smooth muscle fibers may increase to 600 μm from the original length of 40–90 μm. At parturition, strong contractions of the myometrium expel the fetus out of the uterus. After parturition, the uterus returns to its normal size by a reduction in size and number of smooth muscle fibers.

The **perimetrium** is a **serosa,** covering the greater part of the fundus. It is composed of a thin layer of loose connective tissue lined by simple squamous mesothelium.

Figure 13-10. Uterus
Human • H.E. stain • Very low magnification

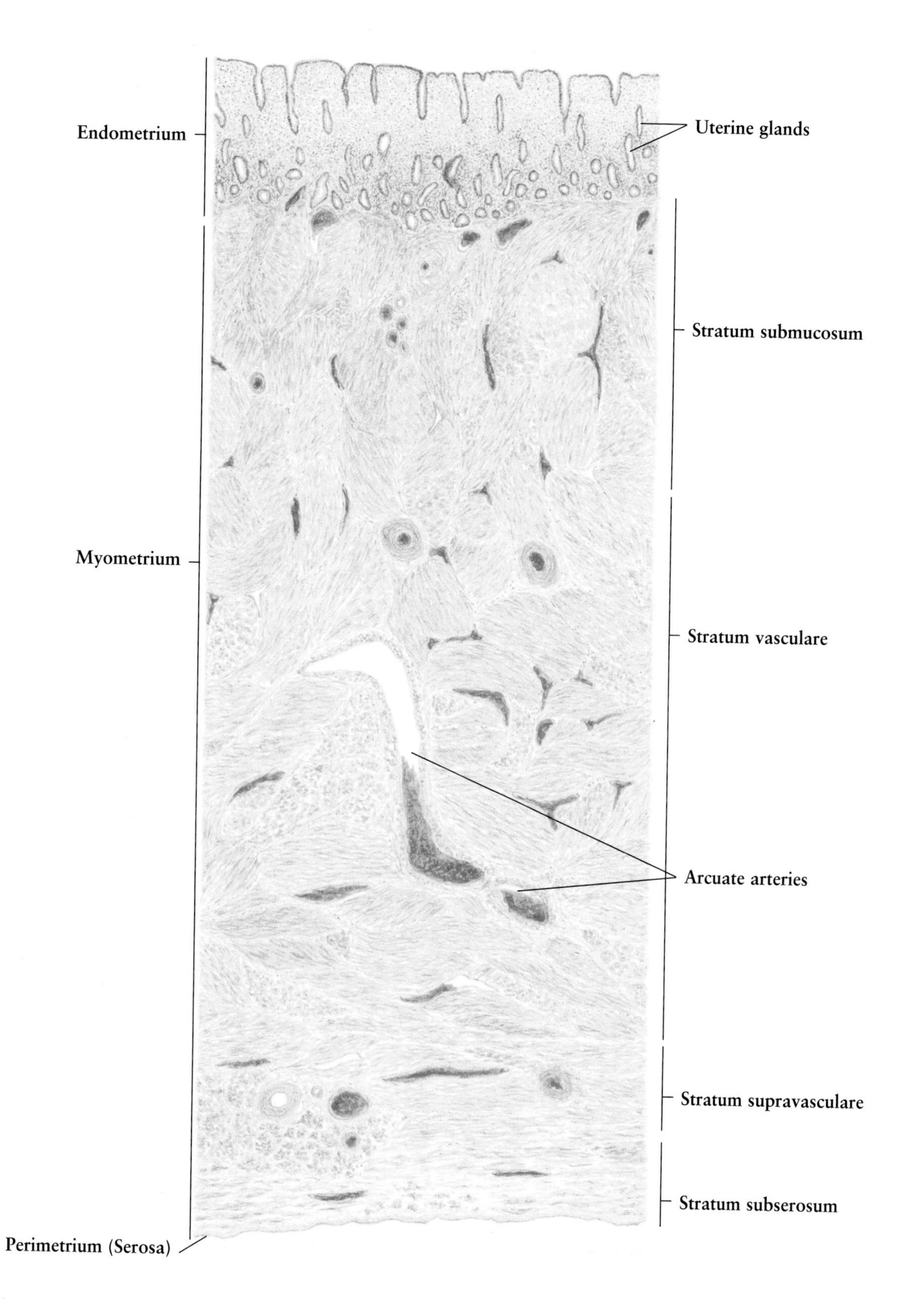

Fig. 13-11.

Uterus: Endometrium

The **endometrium** consists of an *epithelium, uterine glands,* and *endometrial stroma.* Its thickness varies with age and the phase of the menstrual cycle. The superficial **epithelium** is simple columnar in type, composed of a mixture of secretory cells and ciliated cells. The **uterine glands** are simple tubular in type with occasional branchings, extending through the full thickness of the endometrium and opening at the superficial surface. The epithelium lining the glands is similar to the superficial epithelium, but the ciliated cells are rare within the glands.

The **stroma** is a cellular, richly vascularized connective tissue with abundant amorphous ground substance, and delicate reticular, elastic, and collagen fibers.

The endometrium may be divided into two zones on the basis of morphology as well as function: the **stratum functionalis,** the layer sloughed off during every menstrual cycle; and the **stratum basalis,** a relatively thin deep layer adjacent to the **myometrium.** The stratum basalis does not slough off during the menstrual cycle, but regenerates the stratum functionalis during the proliferative phase. The stratum functionalis may be further subdivided into a superficial **stratum compactum** containing numerous stromal cells, and an underlying **stratum spongiosum** rich in blood vessels.

The endometrium is supplied by the branches of the uterine arteries. The **spiral arteries,** which are more or less contorted in shape, spread out superficially into a rich capillary bed. The straight arteries supply the base of the endometrium.

Figure 13-11. Uterus: Endometrium
Human • H.E. stain • Very low magnification

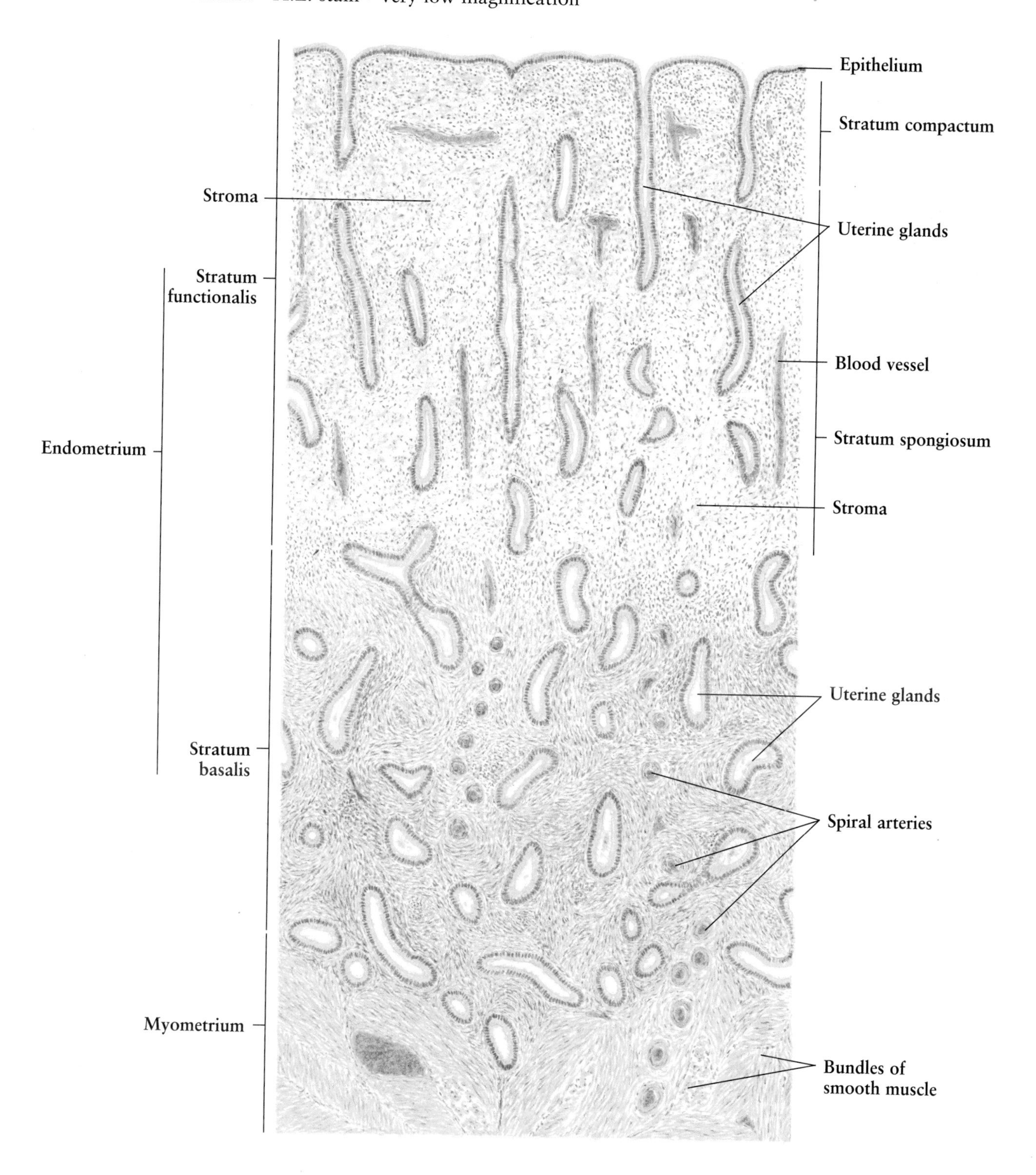

Fig. 13-12.

Endometrium: Stratum Compactum

Figure 13-12 demonstrates the **stratum compactum** of the endometrium under high magnification. The superficial lining **epithelium** is simple columnar in type, resting on a thin **basement membrane.** It consists of a mixture of ciliated and secretory cells. The **ciliated cells** have a centrally located, round or ovoid nucleus, and the free surfaces bear numerous cilia that beat toward the cervix. The **secretory cells** have a nucleus located in the base of the cell, and the cytoplasm is filled with secretory granules.

The **uterine glands** are straight, tubular glands with a narrow lumen. They are lined with simple columnar epithelium, which is similar and continuous to the superficial epithelium, but the ciliated cells become rare.

The **stroma** contains a great number of **stromal cells,** which are irregular stellate cells with large, ovoid nuclei. They lie in a framework of reticular and collagen fibers and amorphous ground substance, which are not visible in the H.E.-stained preparations. The stromal cells resemble the mesenchymal cells, differentiating into decidual cells during pregnancy. The stroma also contains **venules, capillaries,** and **spiral arteries,** which supply the stratum functionalis.

Fig. 13-13.

Endometrium: Stratum Basalis

In the **stratum basalis** there are many **uterine glands,** which may be branched. These are lined by a layer of columnar cells, with an ovoid, basally located nucleus. The stratum basalis is characterized by numerous stromal cells, which are densely packed and form irregular whorls, surrounding the uterine glands or blood vessels. A series of section profiles of the **spiral artery** indicate that the artery extends to the surface from the deep portion of the endometrium.

During the menstrual cycle, the stratum basalis does not slough off. After the menstrual phase, the remaining stromal cells, highly reminiscent of mesenchymal cells, proliferate greatly and form a new stratum functionalis. After frequent mitoses, the epithelial cells of the uterine glands in the stratum basalis form new glandular tubes and regenerate the entire superficial lining of the endometrium.

The bottom of this illustration shows bundles of **smooth muscle fibers** in the myometrium in a variety of profiles depending on the orientation of the bundle. Note some stromal cells invading this layer from the adjacent stratum basalis.

Figure 13-12. Endometrium: Stratum Compactum
Human • H.E. stain • High magnification

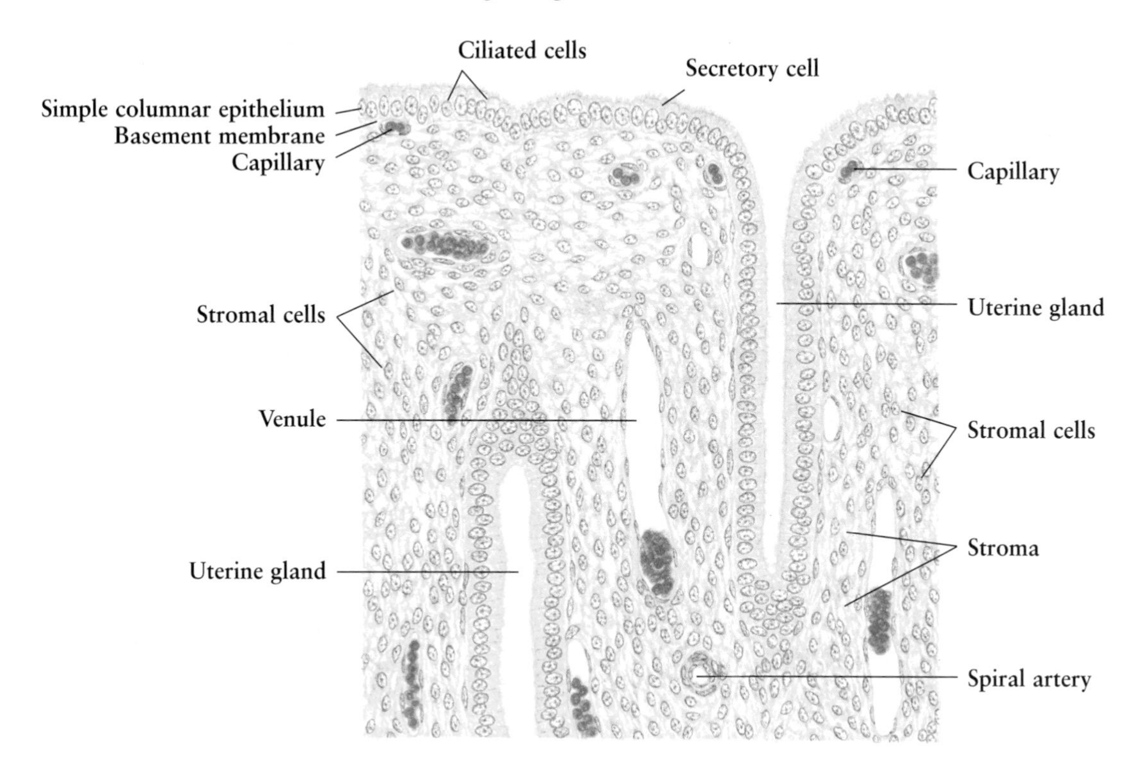

Figure 13-13. Endometrium: Stratum Basalis
Human • H.E. stain • High magnification

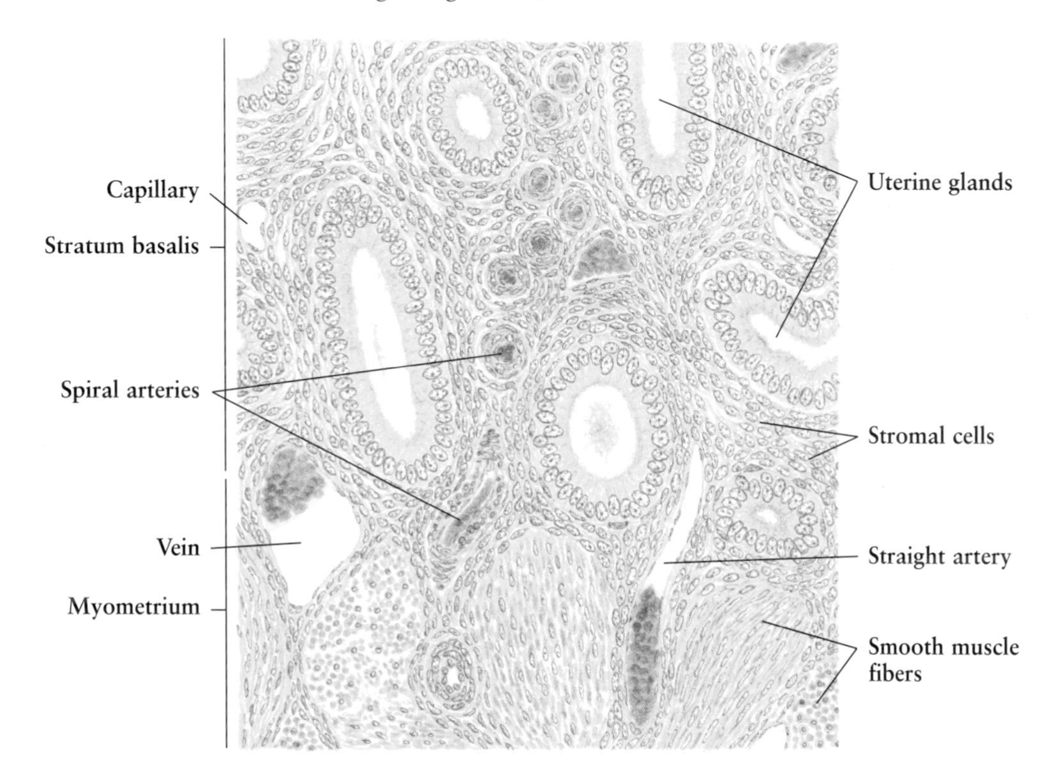

Fig. 13-14.

Menstrual Cycle: Structural Modification of Endometrium

Beginning with puberty and ending at menopause, under the stimulus of ovarian hormones (estrogens and progesterone), which are controlled by the anterior lobe of the pituitary, the **endometrium** periodically undergoes a continuous sequence of histological and physiological changes with an average duration of 28 days. This is the **menstrual cycle,** which is divided into three continuous phases: *menstrual phase* (1st–4th day), *proliferative phase* (5th–14th day), and *secretory phase* (15th–28th day). **Figure 13-14** is a collection of drawings showing the structural characteristics of the endometrium in different phases of the menstrual cycle.

The **proliferative phase** begins at the end of a menstrual flow and is characterized by the rapid regeneration of the uterine glands, lining epithelium, and stroma. After the menstrual phase, the epithelial cells from the remnants of torn glands proliferate rapidly, covering the denuded surface of the endometrium, and forming regrowing glands with straight tubules. At the same time, proliferation of the stromal cells and deposition of the ground substance also occurs, causing the regenerating endometrium to increase in thickness, (**Fig. 13-14B**). By the end of the proliferative phase, the endometrium is 2–3 mm thick; the glands become wavy in outline and their lumina begin to widen. The spiral arteries grow into the regenerating tissue, but are not found in the superficial third of the endometrium (**Fig. 13-14C**). This phase coincides with growth of the ovarian follicles and secretion of estrogen, so it is also known as the **follicular phase.**

The **secretory** (or luteal) **phase** begins after ovulation and depends on the progesterone secreted by the corpus luteum. By the 20th day (**Fig. 13-14D**), the endometrium has almost reached its maximum thickness (4–5 mm), and the glands become tortuous. Toward the end of this phase (**Fig. 13-14E**), the glands become highly coiled and their lumina widen, take on a serrated profile, and are filled with secretion that is thick and rich in glycogen. The spiral arteries continue to elongate and convolute, and extend into the superficial portion of the endometrium. The edema of the stroma increases, and stromal cells enlarge to become decidual cells.

The **menstrual phase** occurs in the absence of fertilization and implantation of the ovum, which results in the cessation of the function of the corpus luteum. The rapid dropping of the level of progesterone and estrogens induces the contraction of spiral arteries, which leads to ischemia and necrosis of the functional layer (**stratum functionalis**) of the endometrium. Then the vessel walls rupture and the blood escapes into the stroma and breaks out into the uterine lumen with the detached debris of the endometrium. Thus the menstrual discharge contains blood, stromal tissue, and glandular secretion. The menstrual flow lasts 3–4 days and has a volume of about 35 ml. At the end of the menstrual phase, the whole functional layer sloughs off, leaving only the basal layer (**stratum basalis**) (**Fig. 13-14A**).

Figure 13-14. Menstrual Cycle: Structural Modification of Endometrium
Human • H.E. stain • Low magnification

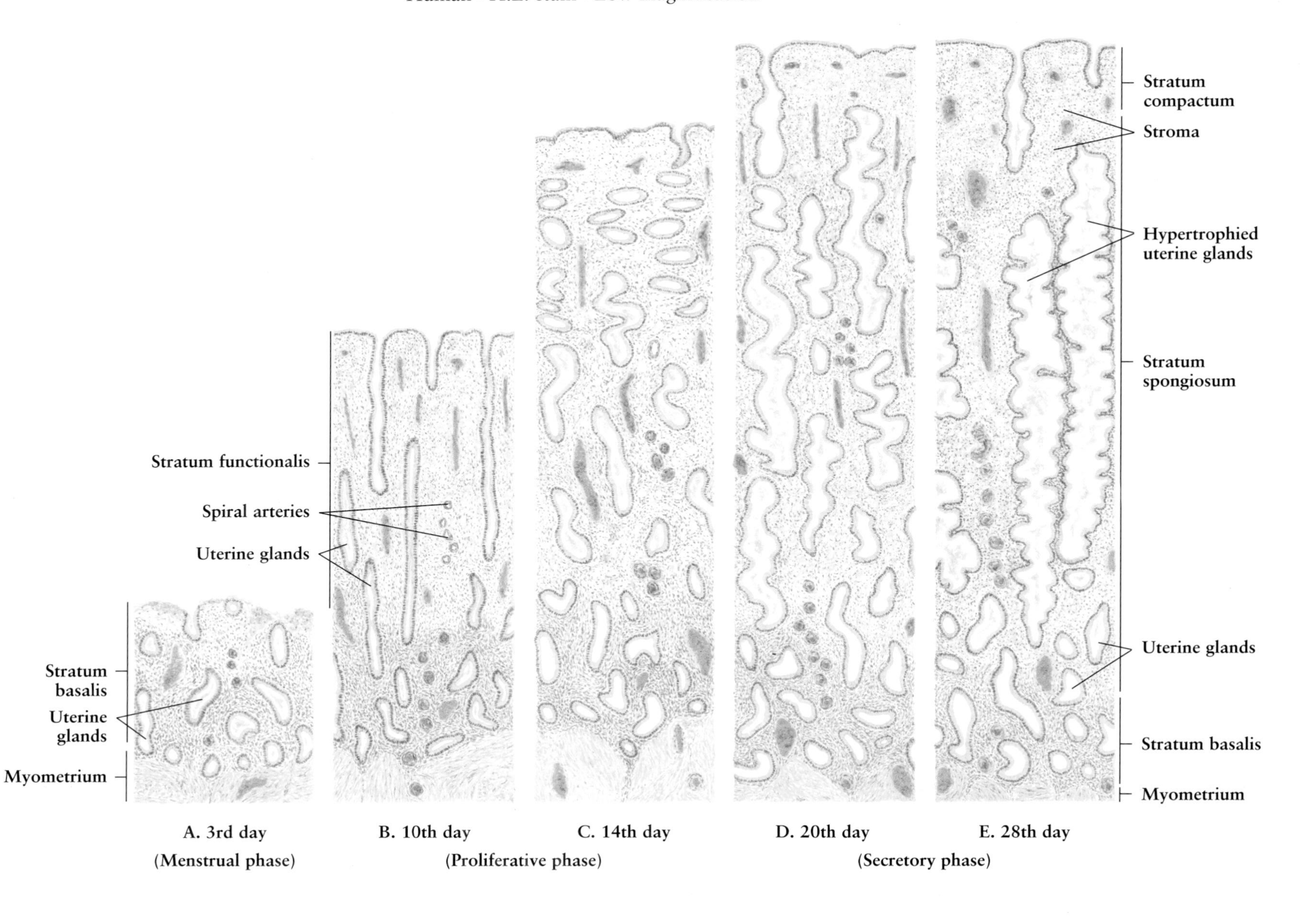

Fig. 13-15. Endometrium: Secretory Phase

During the **secretory phase** of the menstrual cycle, under the stimulation of progesterone, the **endometrium** approaches its maximum thickness as the result of the great proliferation and edema of the stroma, and the dilation and secretion of the uterine glands. **Figure 13-15** provides details of these changes.

The hypertrophied **uterine glands** are dilated and tortuous, and characterized by a saw-tooth appearance. The **epithelial cells** of the uterine glands contain copious secretion in the apical cytoplasm; this pushes their nuclei basally. The **lumina** of the uterine glands become wider and is filled with **secretory material.** This material is thick and rich in glycogen, and provides nutrition for the fertilized ovum, which arrives in the uterine cavity about 3 days after ovulation.

The **stroma** is filled with edematous fluid, and the **stromal cells** become decidual cells with a epithelioid appearance. These cells are large, polyhedral, and have a pale-stained cytoplasm and a large nucleus with prominent nucleolus and loose chromatin. Some stromal cells surround the spiral arteries to form a sheet around the vessels. At this phase the stroma is infiltrated by a great number of **lymphocytes** and **macrophages.** Also demonstrated in this figure are the **capillaries** and **spiral arteries,** which have grown nearly to the surface of the endometrium.

Figure 13-15. Endometrium: Secretory Phase
Human • H.E. stain • High magnification

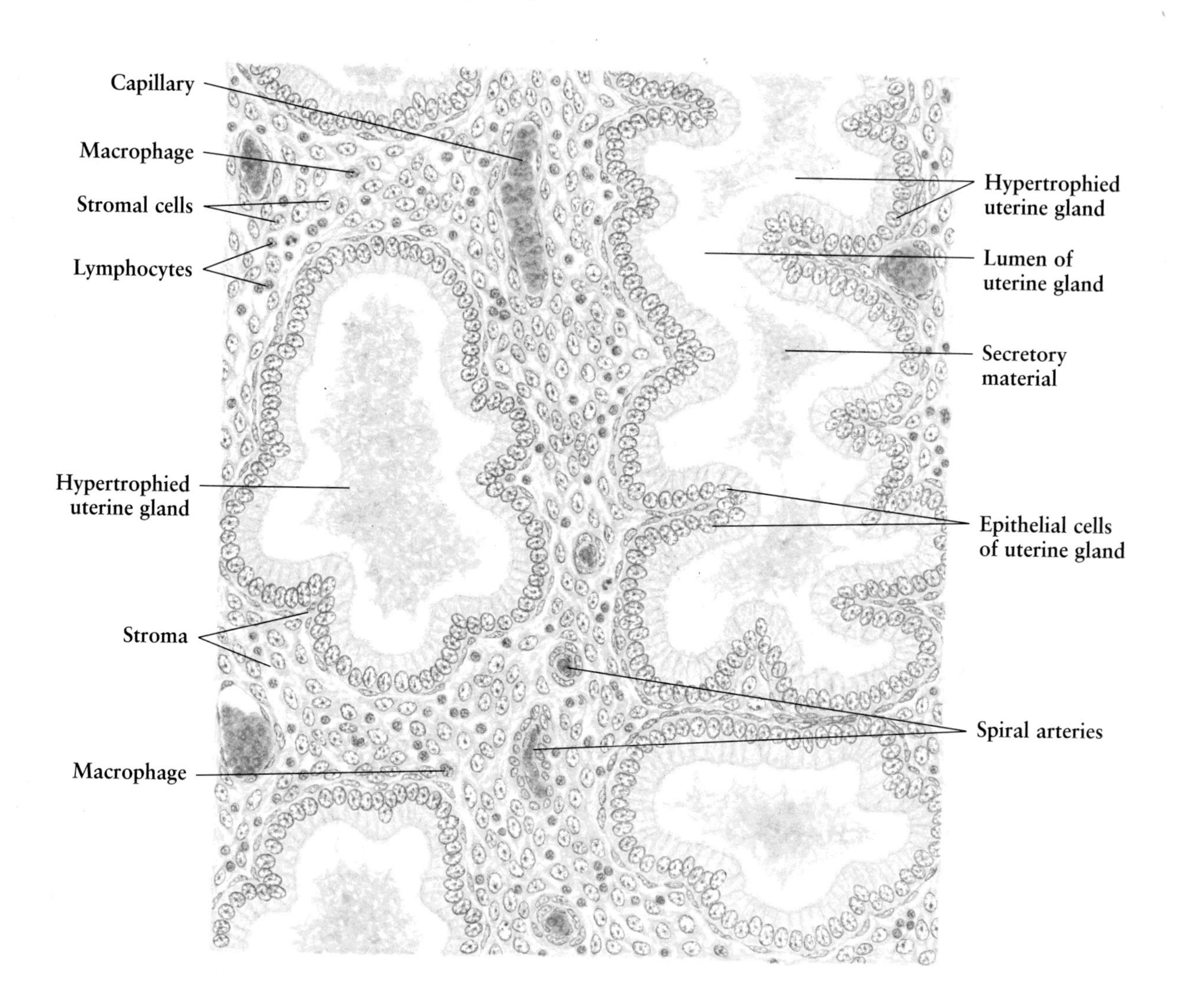

Fig. 13-16.

Uterine Cervix

The **uterine cervix** is the most inferior, tubelike segment of the uterus. Its central cavity, the **cervical canal,** about 3 cm in length, is continuous, above, with the uterine cavity through a constriction known as the **internal os,** and opens, below, into the upper portion of the **vagina** through the **external os.** The wall of the cervix consists of three layers: *endometrium* (mucosa), *myometrium,* and *perimetrium.*

The **perimetrium** is a loose connective tissue at the **vesical surface,** and serosa (not shown in this figure) at the lateral and the **posterior surfaces** of the cervix. The **myometrium** is composed of a limited number of **smooth muscle fibers** arranged in spiral bundles, and a large amount of **dense collagenous connective tissue** containing numerous elastic fibers. This layer is also rich in **blood vessels.**

The **endometrium (mucosa),** also called **endocervix,** is composed of **simple columnar epithelium** and the underlying dense connective tissue of the **lamina propria,** which is devoid of spiral arteries. The epithelium invaginates into the lamina propria to form the **cervical glands,** which secrete mucus. The mucosa displays numerous longitudinal and transverse folds, the **plicae palmate.** The portion of the cervix that projects into the vagina is called the **portio vaginalis,** and is covered by **nonkeratinized stratified squamous epithelium.** The **border** between the stratified squamous and the simple columnar epithelium is an abrupt change, usually occurring just inside the cervical canal. The epithelium and the lamina propria of the portio vaginalis constitute the **exocervix,** which possesses no glands in its lamina propria. The endocervix does not slough off during menstruation, although the glands undergo some change in secretory activity.

Fig. 13-17.

Uterine Cervix: Cervical Glands

The **cervical glands** are extensively branched, and are covered by columnar mucous epithelial cells, which are continuous with the **simple columnar epithelium** on the surface of the **cervical canal.** The cells of the glands have a pale-stained cytoplasm filled with secretory material, and a small ovoid nucleus located basally. Between the glands is the **dense connective tissue** of the **lamina propria,** which contains numerous **collagen fibers** and **fibroblasts** as well as **blood vessels.**

The glands secrete mucus, which contains the enzyme *lysozyme,* capable of cleaving the proteoglycans of bacterial cell walls. During the menstrual cycle, the cervical glands undergo cyclical changes in their secretory activity. Around the time of ovulation, the mucous secretion is thin and watery, which allows the passage of the sperm into the cavity of the uterus. In the luteal phase and in pregnancy, under the influence of progesterone, the secretion becomes highly viscous, forming a plug that prevents the entry of microorganisms, as well as sperm, into the uterine cavity from the vagina.

Figure 13-16. Uterine Cervix

Human • Sagittal section • H.E. stain • Very low magnification

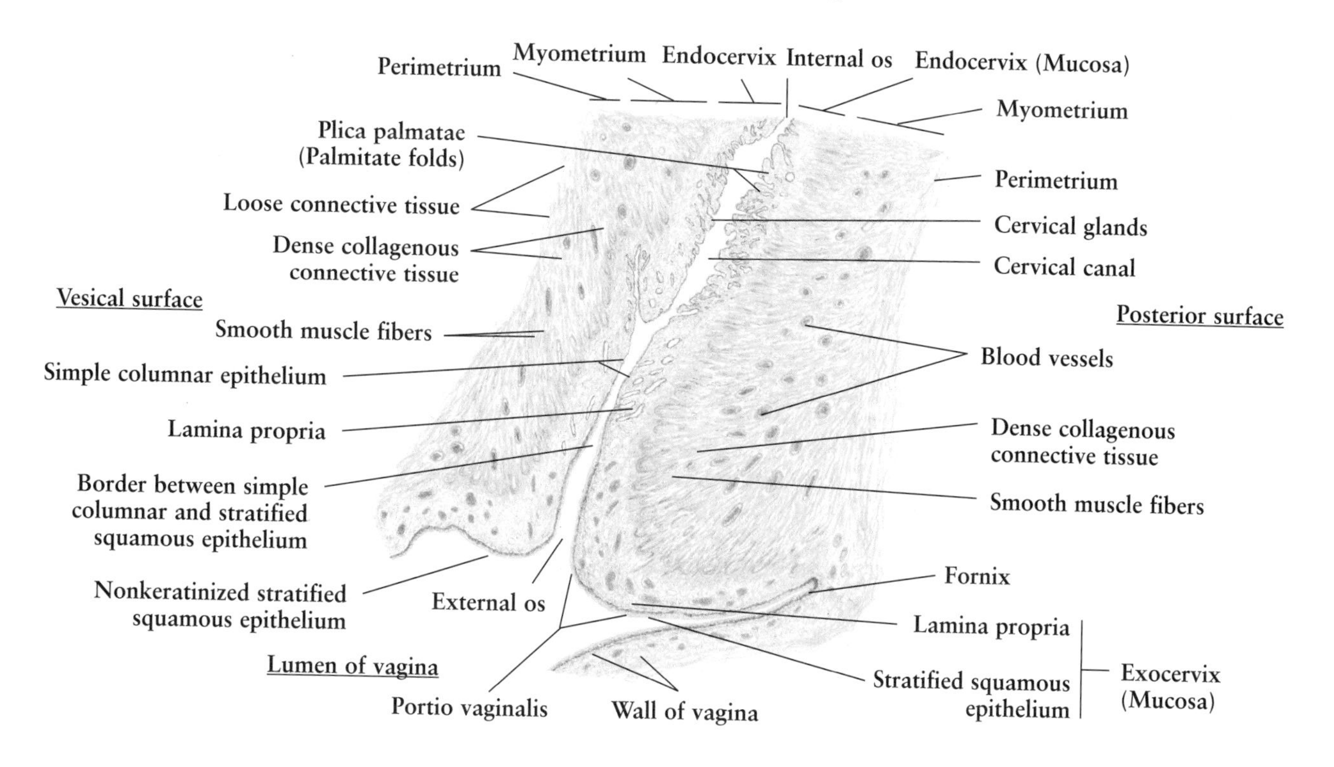

Figure 13-17. Uterine Cervix: Cervical Glands

Human • H.E. stain • Medium magnification

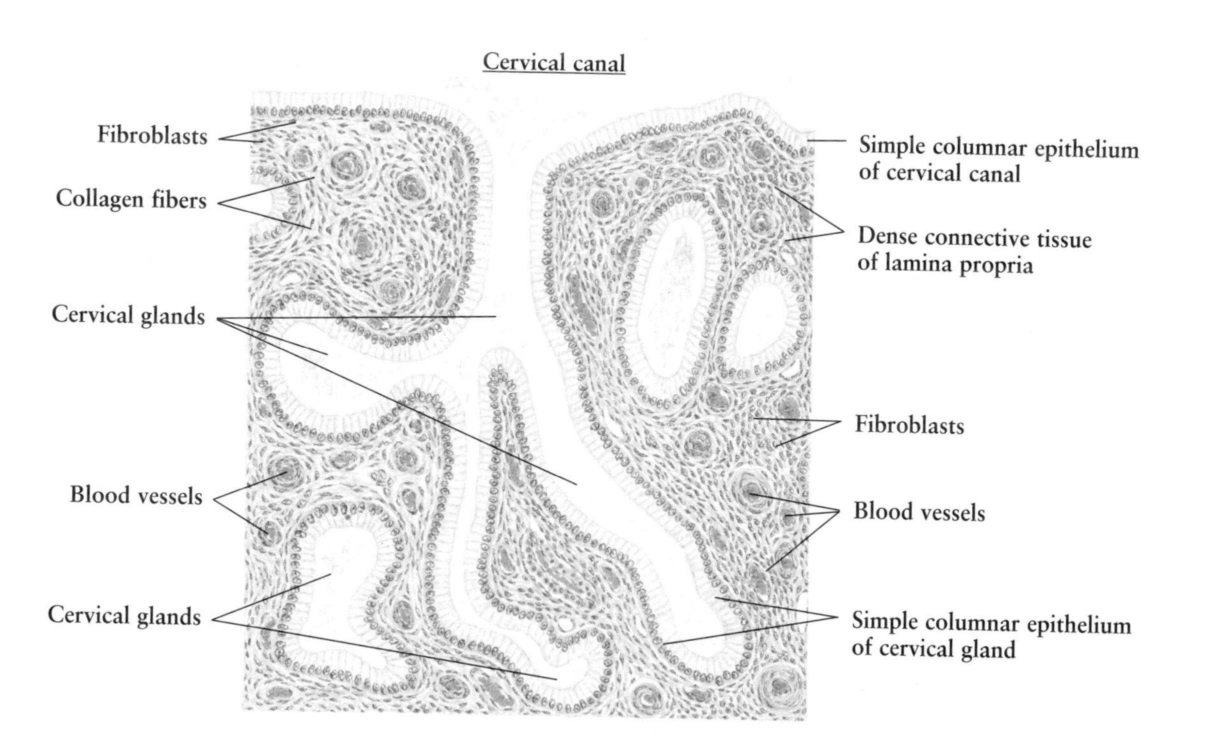

Fig. 13-18. — **Vagina**

The **vagina** is a fibromuscular, tubular organ situated between the uterine cervix and vaginal vestibule. The wall of the vagina consists of three layers: *mucosa, muscularis,* and *adventitia.*

The **mucosa** is lined by a thick, **nonkeratinized stratified squamous epithelium;** the superficial cells undergo continuous desquamation during the menstrual cycle, and may be studied by the smear method for diagnosis. The **lamina propria** is a dense connective tissue containing numerous elastic fibers in addition to the collagen fibers and fibroblasts, and a **vaginal plexus,** the **plexus venosus vaginalis.** The connective tissue of the lamina propria forms many **papillae,** penetrating the epithelium.

The **muscularis** is composed of smooth muscle fibers arranged in a thin **inner circular layer** and a thick **outer longitudinal layer. Blood vessels** and **nerve fibers** are found between the smooth muscle bundles. At the introitus there is a sphincter of skeletal muscle.

The **adventitia** is a thin layer of loose connective tissue that binds the vagina with the surrounding organs. It may contain large **blood vessels** and **adipose cells.**

From the effect of numerous elastic fibers, the vagina is collapsed under normal conditions, and the mucosa is thrown into transverse folds. The mucosa is devoid of glands, and is lubricated by mucus from the cervical glands.

Fig. 13-19. — **Vagina: Mucosa**

This figure shows the lining of the vaginal mucosa as seen under high magnification. The lining is a typical **nonkeratinized stratified squamous epithelium** with a well-developed papillary layer in the underlying connective tissue. The **basal cells** with round nuclei are stained darkly. The **intermediate cells** enlarge and begin to accumulate glycogen, which gives rise to the pale-stained appearance of the cells. The flattened **superficial cells,** rich in glycogen, are continuously desquamated during the menstrual cycle. The liberated *glycogen* is anaerobically metabolized by commensal *lactobacilli vaginales* in the vagina to form lactic acid, which produces an acid environment, inhibiting the growth of pathogenic microorganisms.

The **connective tissue** of the **lamina propria** contains an abundant amount of elastic fibers and is characterized by the presence of a **vaginal plexus.** The connective tissue penetrates the epithelium as **papillae** with **capillaries.** In addition to **fibroblasts, macrophages** or lymphocytes can also be found in the lamina propria.

Figure 13-18. Vagina
Human • Cross section • H.E. stain • Low magnification

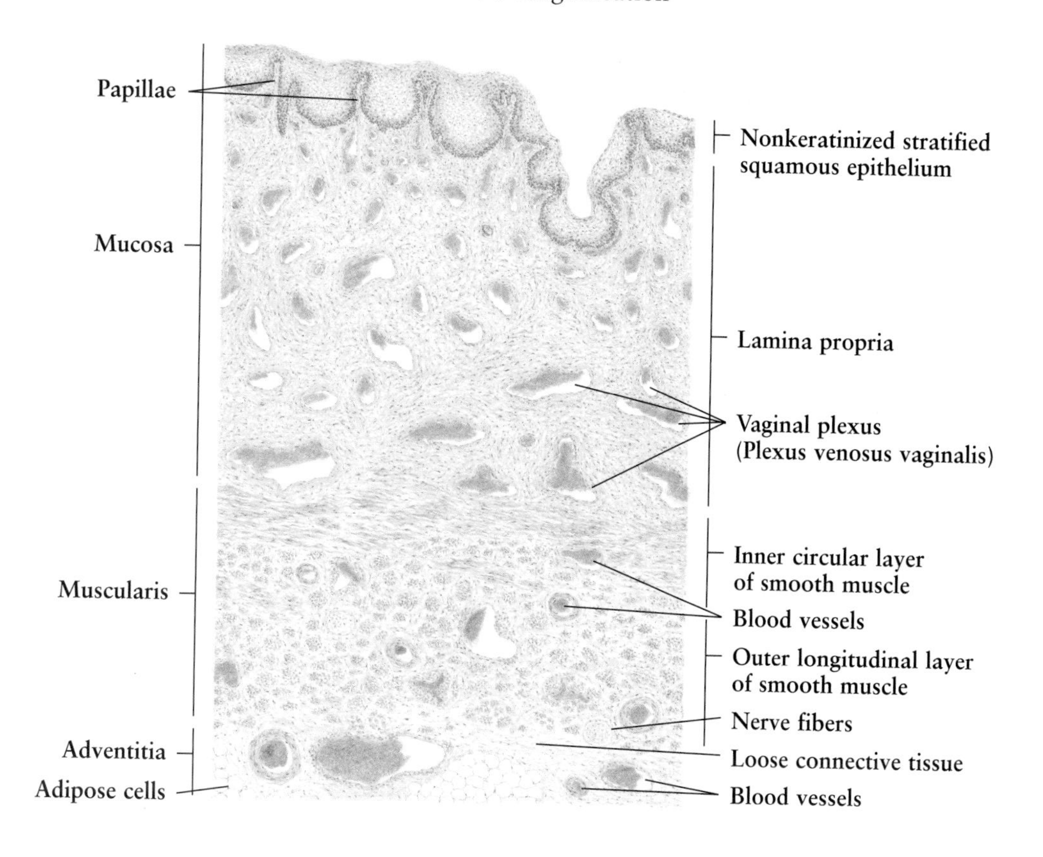

Figure 13-19. Vagina: Mucosa
Human • H.E. stain • High magnification

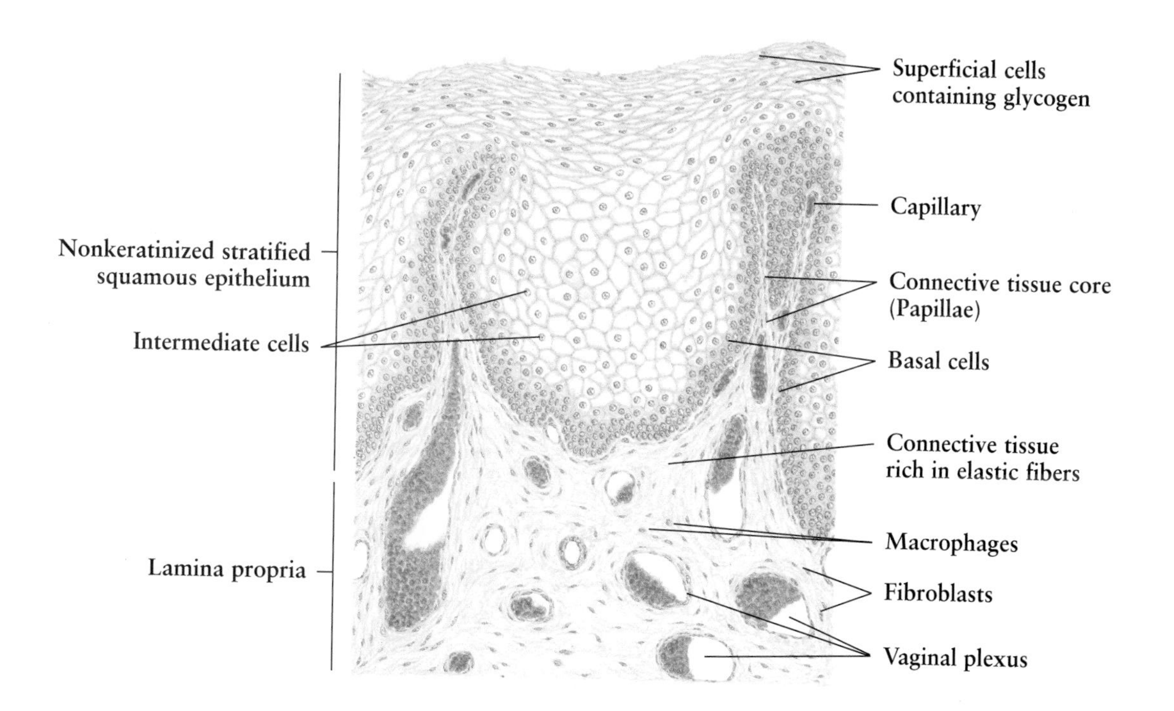

Fig. 13-20.

Labia Minora

The **labia minora** are cutaneous folds of the external female genitalia, consisting of an epithelium and a core of connective tissue, the lamina propria. Both sides of the labia minora are covered by **stratified squamous epithelium** with a thin keratinized layer at the surface, and some pigment in the deeper layers. The **lamina propria** is a spongy **connective tissue** rich in **blood vessels, lymphatic vessels,** and **nerve fibers,** permeated by networks of fine elastic fibers. The connective tissue penetrates the epithelium to form papillae. The labia minora possess neither adipose cells nor hair follicles, but contains numerous large **sebaceous glands** associated with the epithelium on both sides.

Fig. 13-21.

Labia Minora

Figure 13-21 illustrates the details of the epithelium and the underlying connective tissue of the labia minora. The **epithelium** is typical **stratified squamous epithelium** with minimal **keratinization** at the surface. There is some pigment in the deeper layers. The **connective papillae** or cores with **capillaries** are found under the epithelium. **Sebaceous glands** are associated with the epithelium, and their secretion lubricates the surface of the labia minora. The spongy **loose connective tissue,** rich in elastic and collagen fibers, contains numerous **fibroblasts, blood vessels, lymphatic vessels,** and **nerve fibers.**

Figure 13-20. **Labia Minora**

Human • Horizontal section • H.E. stain • Low magnification

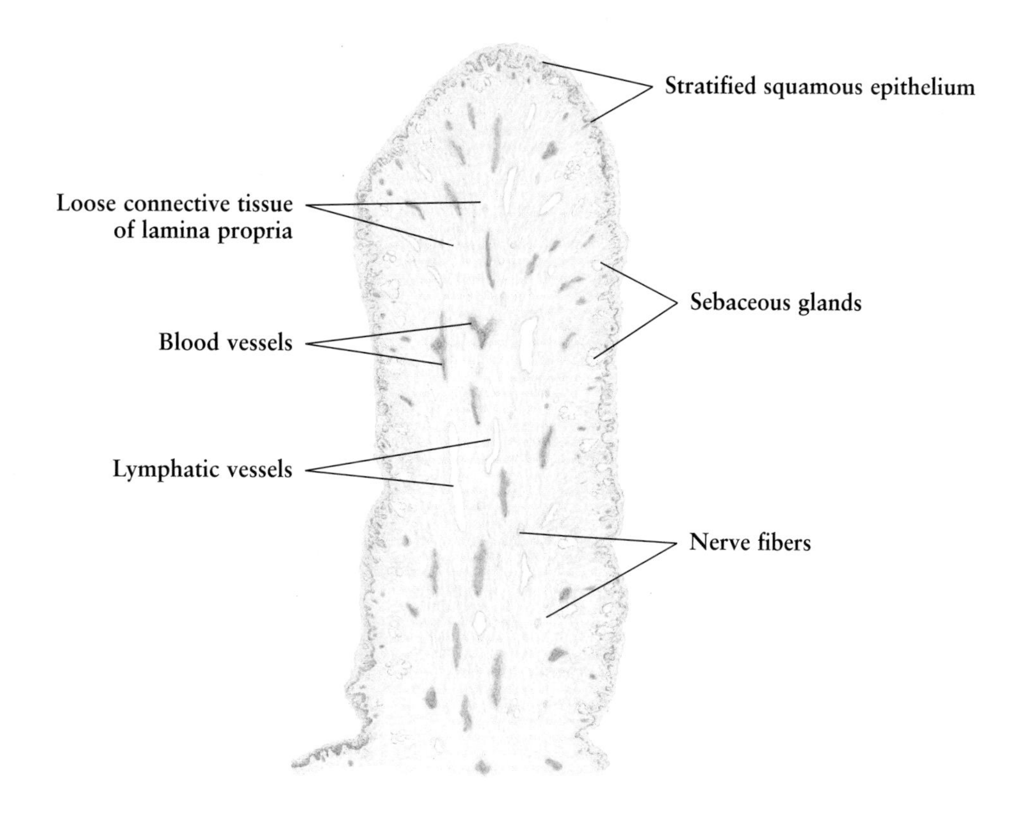

Figure 13-21. **Labia Minora**

Human • H.E. stain • High magnification

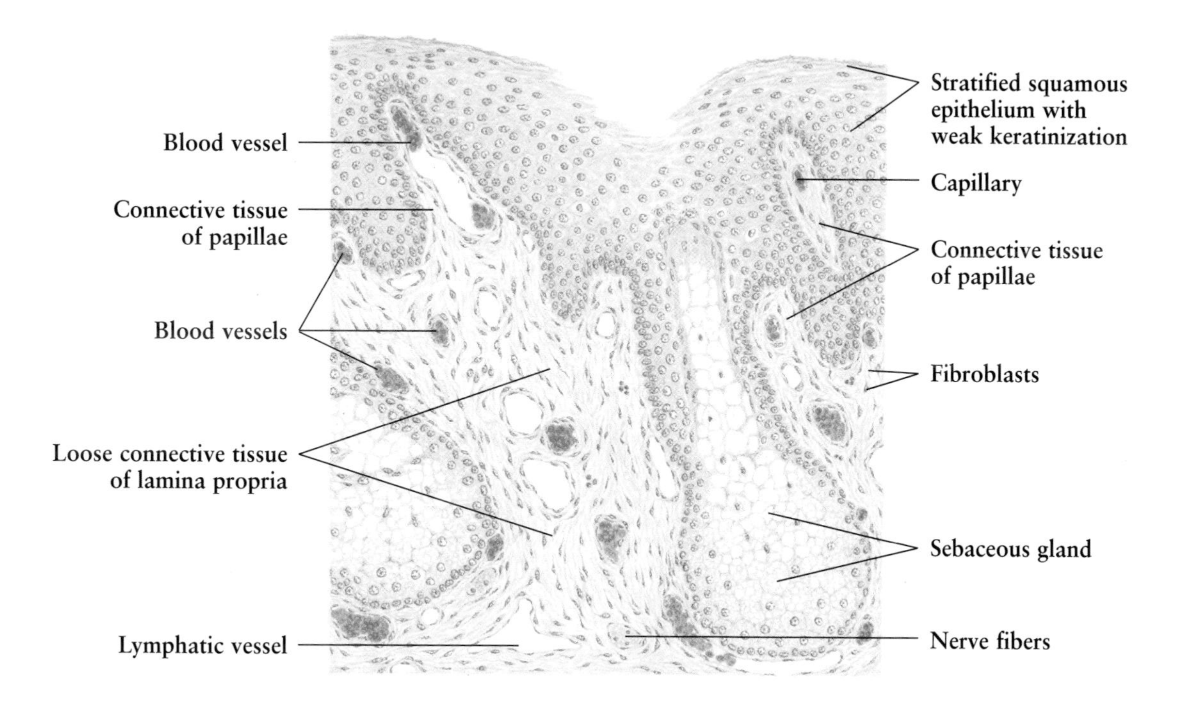

Fig. 13-22. — Placenta

The **placenta** is disc-shaped, with a diameter of about 20 cm, a thickness of 3 cm, and a weight of about 500 g. It is a transient organ serving to mediate physiological exchange between the mother and the fetus. The placenta is histologically composed of a *fetal part* and a *maternal part.*

The **fetal part** consists of the **chorionic plate,** which forms a part of the wall of the **amniotic cavity,** and the **chorionic villi,** which arise from the plate. The **stem villi,** which are directly continuous with the plate, are thick and branch to form numerous secondary **villi.** Each stem villus and its branches are called a fetal cotyledon, which has a separate fetal blood supply. All villi contain **fetal blood vessels,** which are continuous with those in chorionic plate and the umbilical cord. The **intervillous space** between villi is filled with freely circulating **maternal blood** from which nutrients are taken into the fetal blood within the villi and then transported to the fetus. The villi that do not reach the maternal part of the placenta, but float in the maternal blood, are termed **floating** or **free villi;** the villi which attach to the maternal part are referred to as **anchoring villi.** In tissue sections, many **trophoblastic islands** are seen in the intervillous space.

The **maternal part** of the placenta is formed of **decidua basalis.** The portion of the decidua that is eroded more deeply than elsewhere forms a space which is occupied by the villi and intervillous space filled with maternal blood. The portion of the decidua that has not eroded deeply leaves projections of decidual tissue toward the chorionic plate between two stem villi. These projections are known as **placental septa.** As shown in this illustration from a placenta during the 8th week of pregnancy, the **uterine glands** in the decidua basalis become extremely dilated, and fill with secretion. The border between fetal cells and maternal cells is filled with **fibrinoid,** which appears eosinophilic in H.E.-stained preparations. The spiral arteries convey the blood to the **maternal blood sinus,** then to the intervillous space. Fetal and maternal blood do not mix.

The placenta also serves as a temporary endocrine organ that secretes hormones such as ***estrogen, progesterone, relaxin, renin, chorionic gonadotropin,*** and ***chorionic corticotropin.*** All these hormones are believed to be synthesized by the syncytial trophoblast.

Figure 13-22. Placenta

Early human placenta • H.E. stain • Low magnification

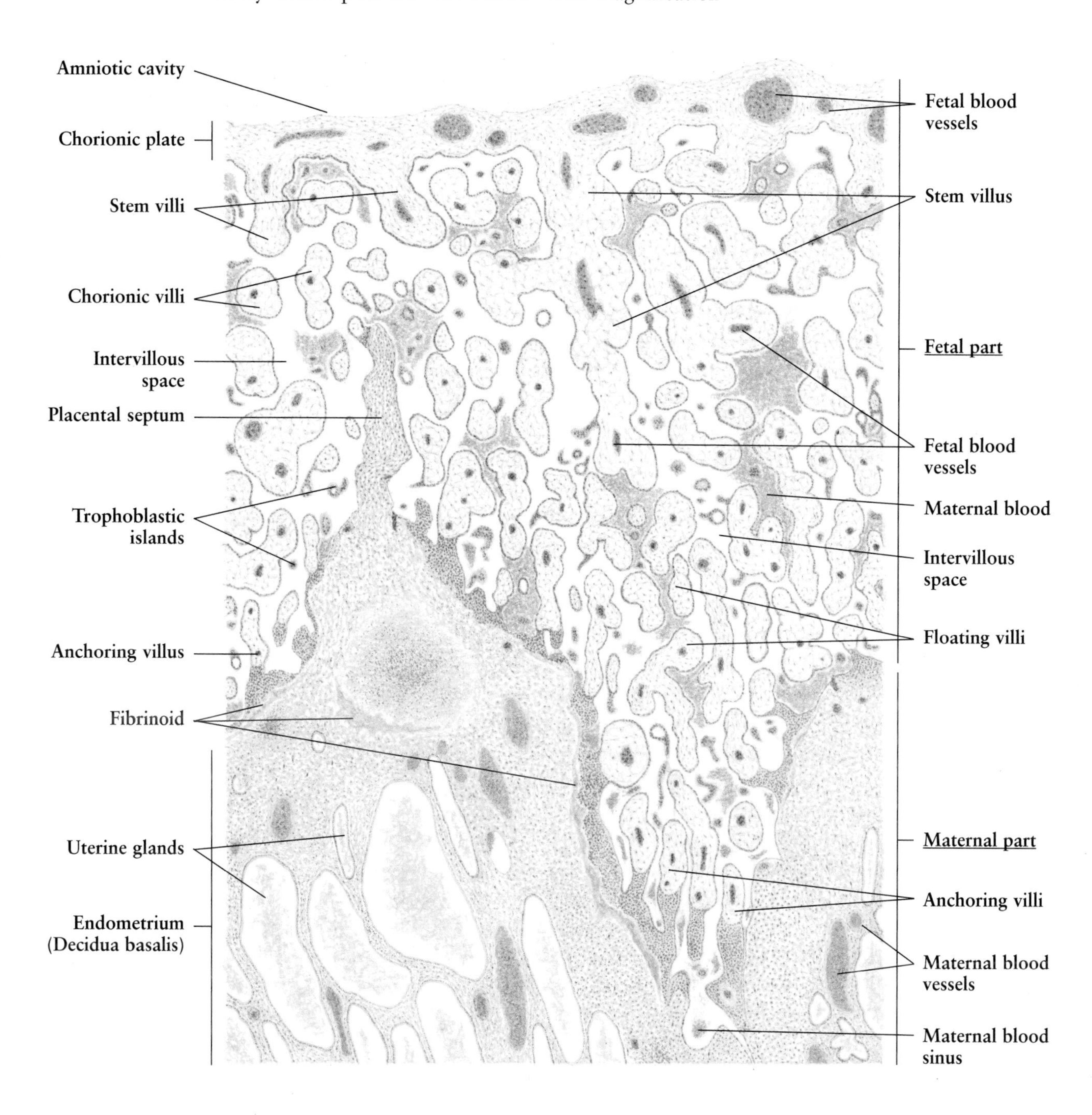

Fig. 13-23. Placenta: Chorionic Villi

The **chorionic villi** consist of a connective tissue core and the surrounding trophoblast layer.

The **trophoblast** is composed of two layers: the inner cytotrophoblast, and outer syncytiotrophoblast. The **cytotrophoblast** is made up of a layer of large, ovoid, discrete cells with a pale-stained cytoplasm and spherical nucleus. The cytotrophoblast is separated from the underlying core connective tissue by a thin **basement membrane.** From the 10th week of pregnancy, the cytotrophoblast begins to disappear until at parturition only some isolated cells remain. The **syncytiotrophoblast** is a dark layer of cytoplasm without any distinct intercellular boundary. It contains numerous small, ovoid nuclei. The cytoplasm is basophilic because of abundant granular endoplasmic reticulum as well as numerous lysosomes. There are numerous microvilli at the free surface of the syncytiotrophoblast, which greatly increase the absorptive surface of the placenta.

The **core** of the chorionic villus is a **loose connective tissue** derived from the extraembryonic mesenchyme. It possesses some collagen fibers and **mesenchymal cells,** the fibroblasts that connect with each other through their processes. The **Hofbauer cells,** a kind of macrophage, are also present in the core connective tissue. These cells are large, and have a big, round nucleus and phagocytic vacuoles within the cytoplasm. The **fetal capillaries** in the core are lined with typical endothelium. Early in pregnancy, **fetal erythrocytes** within the capillaries are still nucleated, as shown in this figure.

In the intervillous space of a early placenta, many **trophoblastic islands** can be found. They are composed entirely of syncytiotrophoblastic cells, and may represent the growing villi.

Fig. 13-24. Placenta: Chorionic Plate

The **chorionic plate** consists of **chorionic mesoderm,** which contains collagen fibers and densely arranged **fibroblasts.** The chorionic mesoderm is continuous with Wharton's jelly in the umbilical cord. The plate is covered by a simple squamous epithelium, the **amniotic epithelium,** on the surface of side of the **amniotic cavity.** On the other side, it is lined by trophoblast composed of a **cytotrophoblastic layer** and a **syncytiotrophoblastic layer,** continuous with that lining the villi. This figure also shows a **stem villus** arising from the chorionic plate. The stem villus structurally resembles the chorionic plate. The chorionic plate contains large **blood vessels** that branch and penetrate the stem villi. Both plate and stem villi may contain macrophages.

Figure 13-23. Placenta: Chorionic Villi
Early human placenta • H.E. stain • High magnification

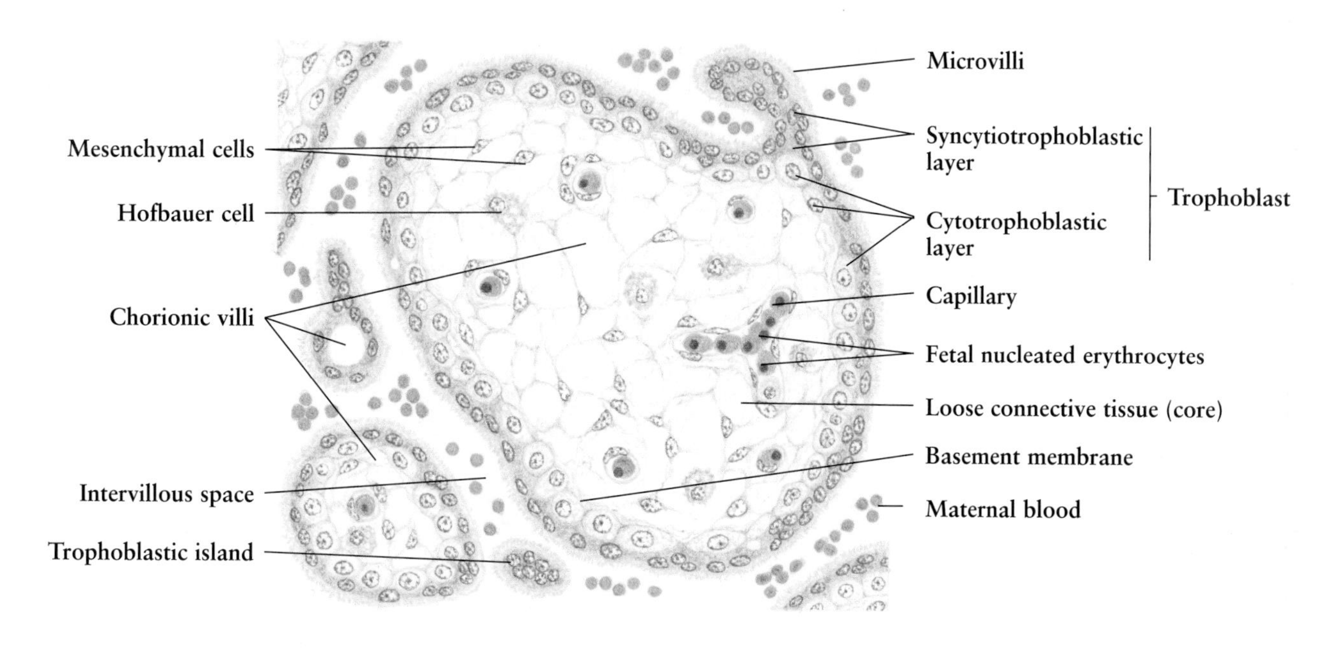

Figure 13-24. Placenta: Chorionic Plate
Early human placenta • H.E. stain • High magnification

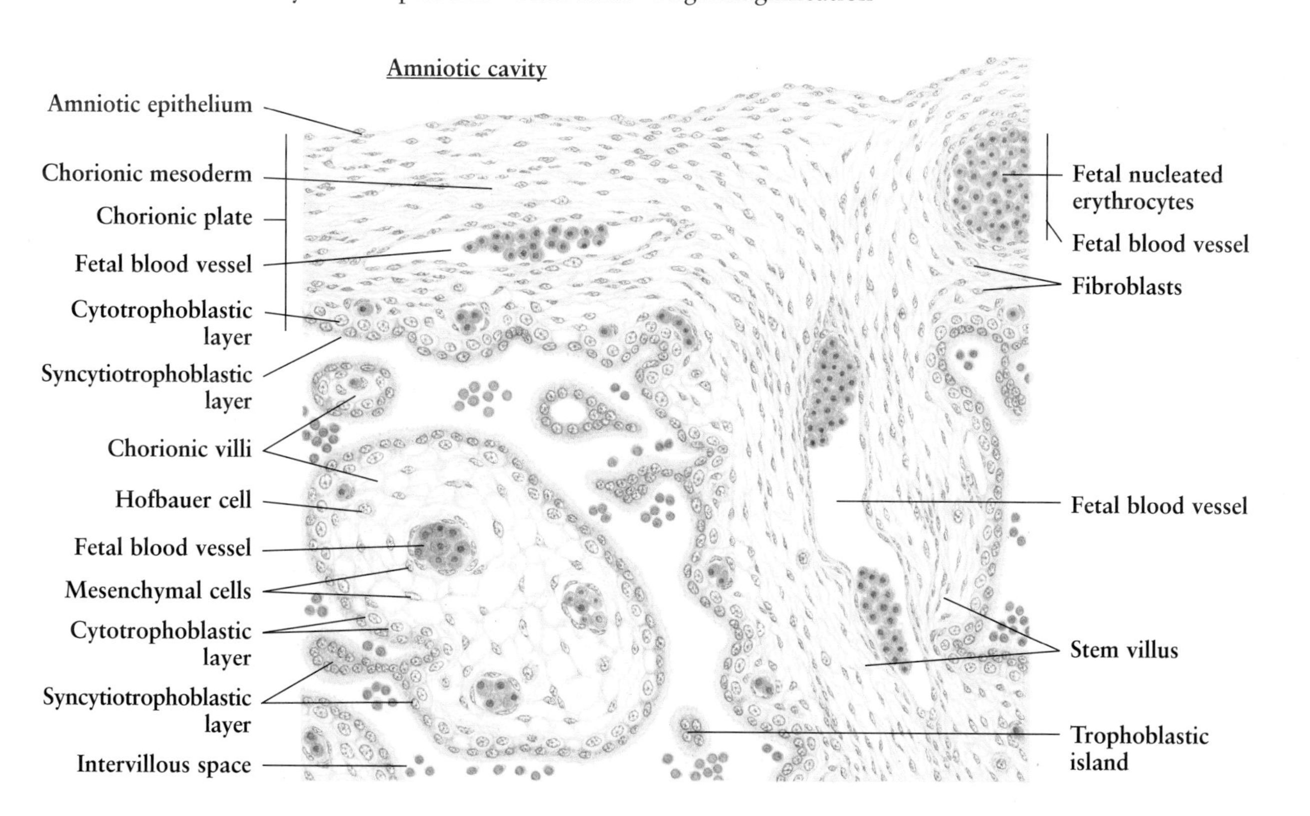

Fig. 13-25. Placenta: Anchoring Villi

The attaching site of the **anchoring villus** (**fetal part** of the placenta) to the **decidua basalis** (**maternal part** of the placenta) is a unique example of direct contact between fetal cells and maternal cells. Although they possess different genotypes, there is no immunological rejection reaction.

At the terminal end of the anchoring villus, cells of the **cytotrophoblastic layer** undergo rapid proliferation and form a **cytotrophoblastic column** that is covered by a layer of **syncytiotrophoblast.** These **fetal cytotrophoblastic cells** break through the superficial **syncytiotrophoblastic layer** and directly contact the **decidual cells** of the endometrial stroma, establishing the attachment of an anchoring villus. On the other hand, the cytotrophoblastic cells, on reaching the decidua basalis, spread laterally along the surface of the decidua basalis, uniting with similar outgrowths from neighboring villi to form a continuous **trophoblastic shell** that covers the decidua basalis. Active **mitosis** can be found within the trophoblastic shell as well as the cytotrophoblastic column. The trophoblastic shell may be interrupted by the communication of the maternal blood vessels with the **intervillous space.** All villi, trophoblastic shells, and **maternal sinuses** are lined by a cuboidal or flattened layer of syncytiotrophoblast. The **syncytic knot** present in the trophoblastic shell has a structure similar to the trophoblastic island in the intervillous space, composed of a group of nuclei with some cytoplasm. Some of these small structures may enter the maternal blood circulation as emboli that become lodged in small vessels of the maternal lungs.

At the **deciduotrophoblastic junctional zone** between trophoblastic cells and decidual cells, an eosinophilic fiber-like structure may be found, called the **fibrinoid,** in which some small dead cells are embedded. Their origin and significance remain unknown. The **decidual cells** are large, polygonal or ovoid cells that have a pale-staining cytoplasm and a large, round or oval nucleus. Between large decidual cells are scattered **fibroblasts** that are fusiform and have a dark-pink cytoplasm. At the bottom left of this illustration, the **uterine gland** and **uterine vein** are shown. Numerous small dying cells are also found, which may be lymphocytes or other white blood cells.

Figure 13-25. Placenta: Anchoring Villi

Early human placenta • H.E. stain • High magnification

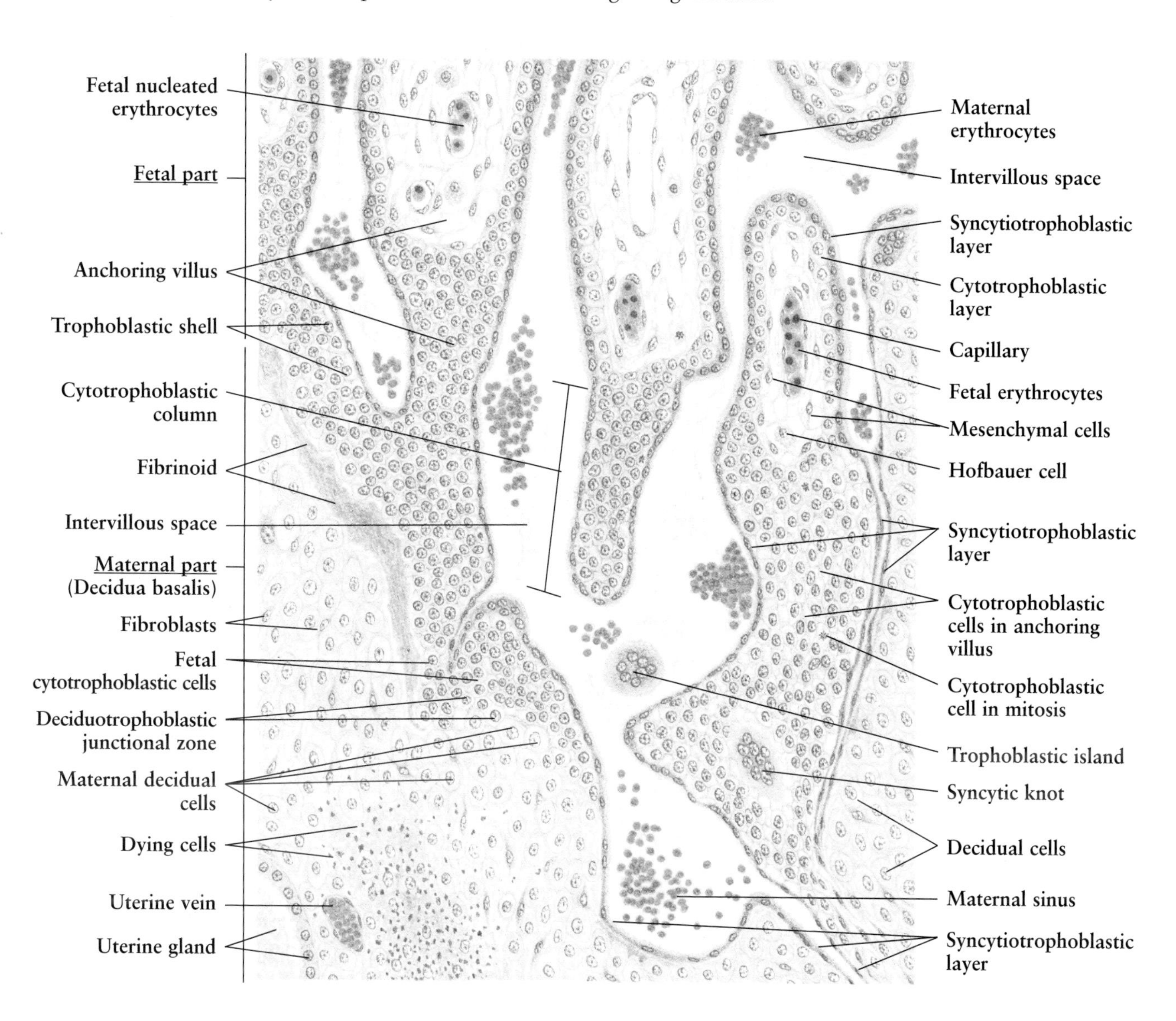

Fig. 13-26.

Placenta: Chorionic Villi

In the late human placenta, the most prominent change in the chorionic villi is the disappearance of the cytotrophoblastic layer. As a result, some **capillaries** attach directly to the syncytiotrophoblastic layer. Some **cytotrophoblastic cells** in the villi, however, persist throughout gestation as isolated ovoid cells. The superficial **syncytiotrophoblastic layer** becomes unevenly distributed. In some areas it becomes thin, while in other areas the **nuclei** aggregate with cytoplasm to form a thick layer or mass. In the connective tissue core of the villus, the **fibrinoid** occurs adjacent to the syncytiotrophoblastic layer; except for the **mesenchymal cells,** other cell types are seldom observed. Note that the erythrocytes within the fetal blood vessels are enucleated.

Fig. 13-27.

Placenta: Decidual Cells

The **decidual cells** (see **Fig. 13-25**) are derived from the stromal cells of the endometrium. At a late stage of pregnancy, the decidual cells enlarge and are separated by the **intercellular matrix.** These cells are ovoid, and have a large nucleus with a prominent nucleolus. The cytoplasm appears basophilic, because it contains a great amount of rough endoplasmic reticulum, with an appearance similar to the Nissl bodies of neurons. Among the huge decidual cells are scattered **fibroblasts** that are fusiform and have a eosinophilic cytoplasm and an ovoid nucleus. The decidual cells are believed to produce proteins.

At the late stage of pregnancy, the trophoblastic shell (see **Fig. 13-25**) has disappeared. The decidual cells are separated from the very thin syncytiotrophoblastic layer by a thick layer of **fibrinoid** within which some decidual cells and fibroblasts may be found.

Figure 13-26. Placenta: Chorionic Villi
Late human placenta • H.E. stain • High magnification

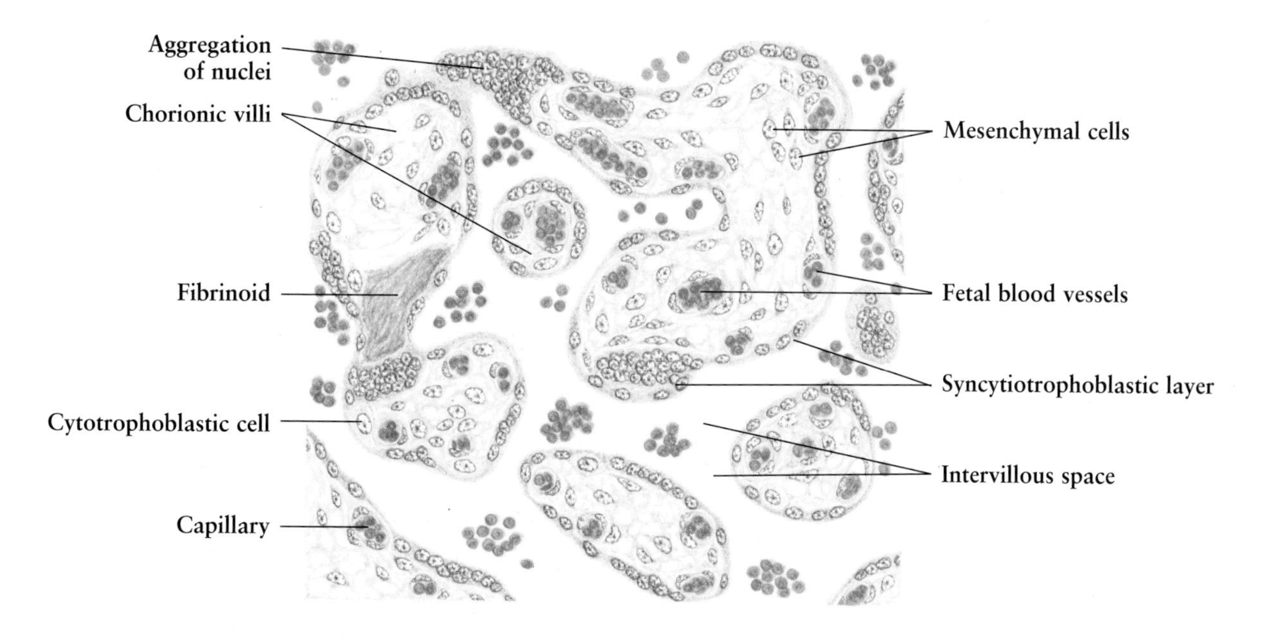

Figure 13-27. Placenta: Decidual Cells
Late human placenta • H.E. stain • High magnification

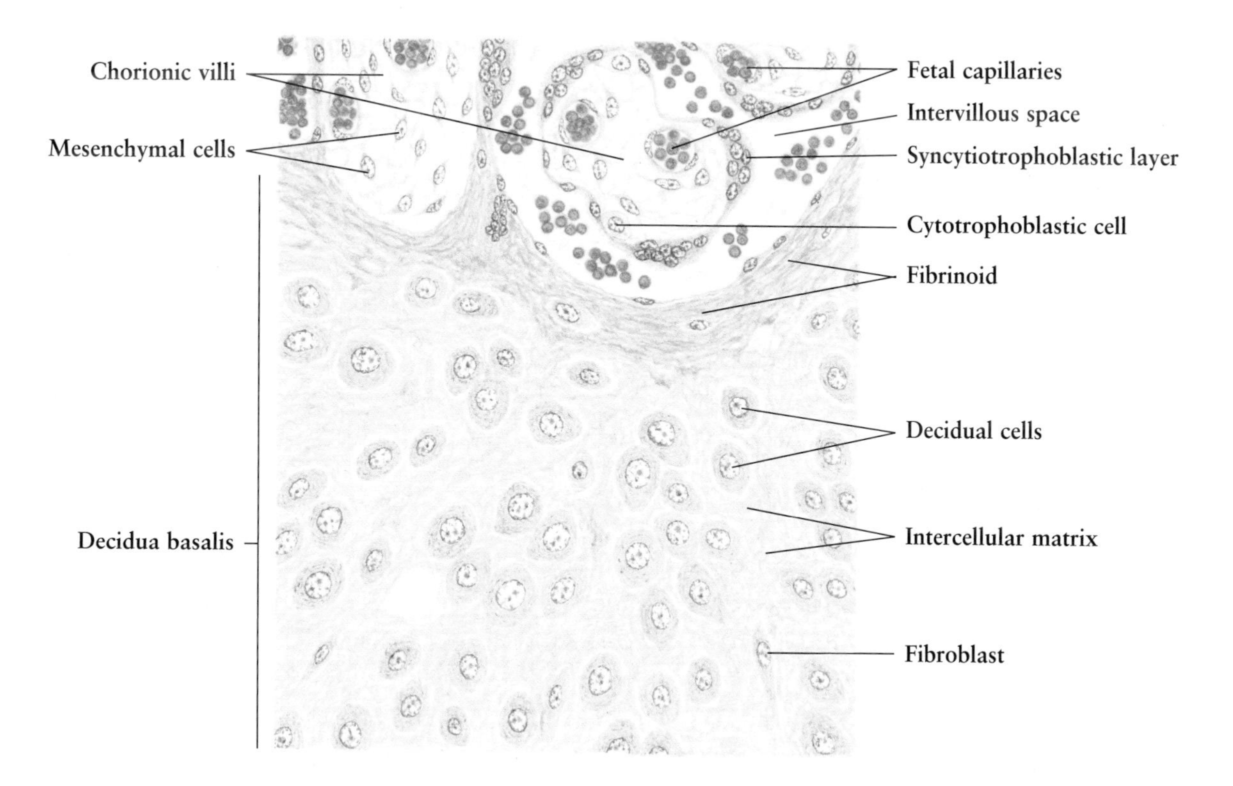

Fig. 13-28.

Umbilical Cord

The **umbilical cord,** 50–60 cm in length, is a white strand, connecting the umbilicus of the fetus to the placenta at its center. The umbilical cord is composed of a **mucous connective tissue,** known as **Wharton's jelly,** covered by the **amniotic epithelium,** a simple cuboidal to stratified squamous epithelium continuous with the epidermis of abdominal skin. The umbilical cord becomes the conduit for the umbilical vessels (two **umbilical arteries** with irregular lumina, and one **umbilical vein**), which traverse its length between the fetus and placenta, and are embedded in the Wharton's jelly. There are no nerve fibers or lymphatic vessels in the umbilical cord, but remnants of the **allantoic duct** and **vitelline duct** may be found near the root at the placenta end.

Fig. 13-29.

Umbilical Cord: Umbilical Artery

This figure illustrates the structure of the wall of the **umbilical artery.** The artery is lined by the **endothelium,** a thin layer of simple flattened epithelium. There is no underlying subendothelial layer or internal elastic lamina, so the endothelium is adjacent to the muscularis. The **muscularis** is composed of smooth muscle fibers arranged in an **inner longitudinal layer** and an **outer circular layer.** In cross sections, the inner longitudinal muscle is always found thrown into folds that give rise to an irregular lumen (see **Fig. 13-28**) filled with blood cells. There is no definite adventitia nor vasa vasorum. The outer circular muscle is continuous with **Wharton's jelly** (mucous connective tissue). The umbilical vein has a structure similar to that of the umbilical arteries.

Figure 13-28. Umbilical Cord
Human • Cross section • H.E. stain • Low magnification

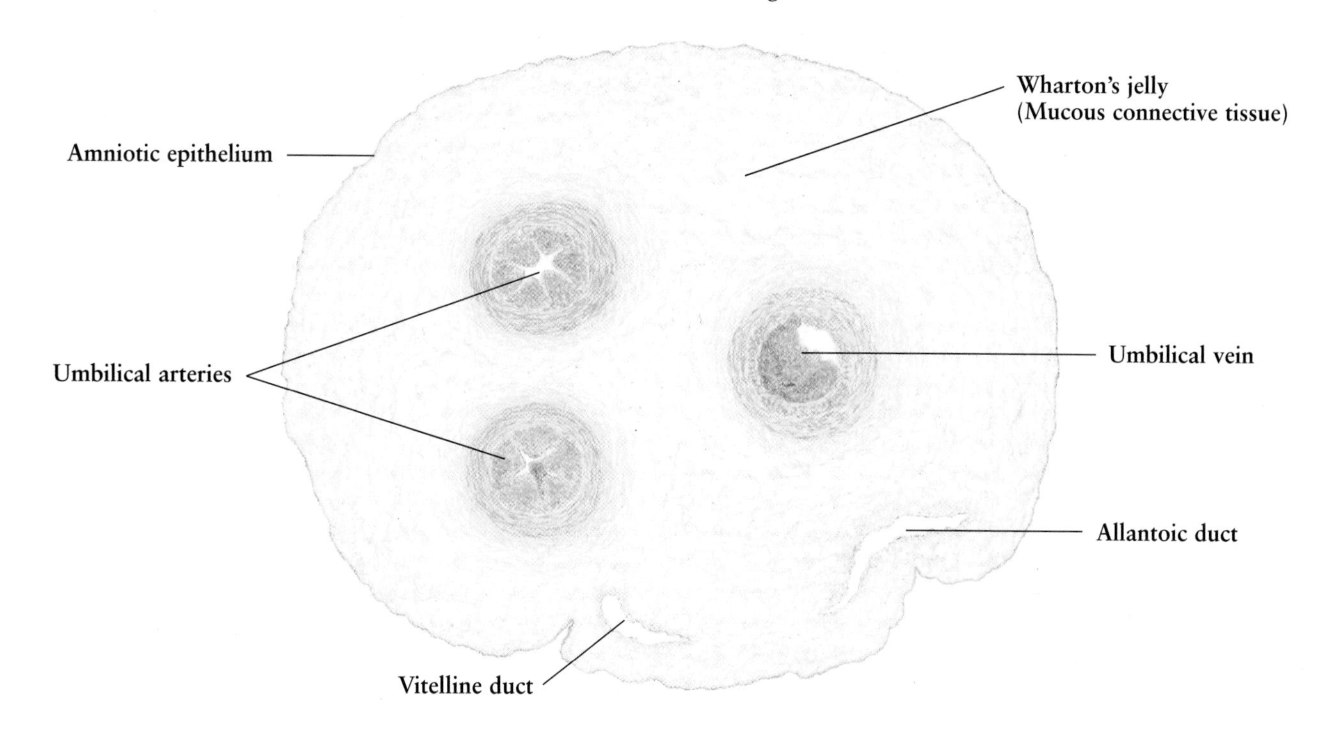

Figure 13-29. Umbilical Cord: Umbillical Artery
Human • Cross section • H.E. stain • Medium magnification

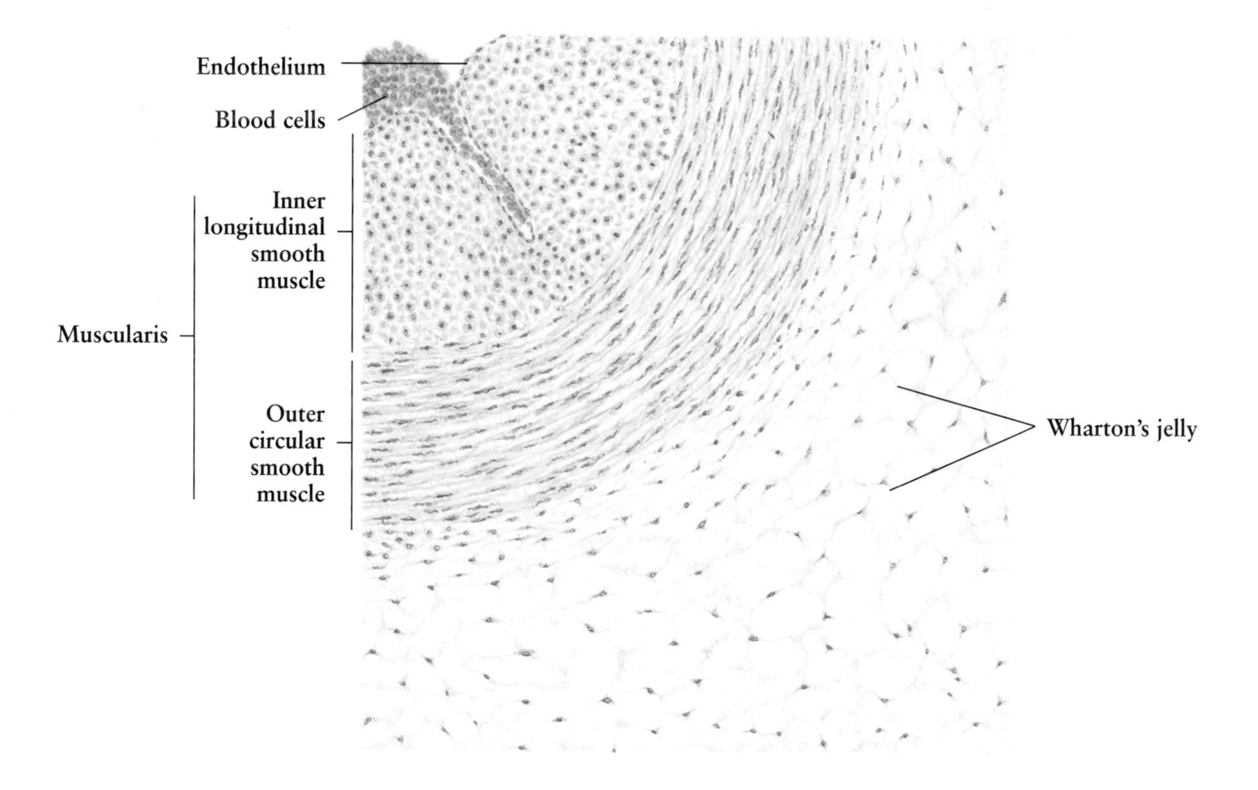

Fig. 13-30.

Umbilical Cord

The **amniotic epithelium** covering the umbilical cord is simple cuboidal or stratified squamous in type, with a very thin layer of keratinization on the surface. **Wharton's jelly** consists of delicate **collagen fibers, ground substance,** and stellate **fibroblasts.** Also shown in this figure is the remnant of the **allantoic duct,** which is lined by a simple squamous epithelium, the **allantoic epithelium.**

Fig. 13-31.

Mammary Gland

The **mammary glands,** which form the major component of the two female breasts, are compound tubuloacinar glands, consisting of 15–25 lobes. The *lobes* are separated from one another by dense connective tissue and **adipose tissue.** Each lobe is divided into a variable number of **lobules** by **dense** (fibrocollagenous) **connective tissue.** The lobules contain the secreting acini and their ducts, enclosed in specialized, cellular, intralobular connective tissue. A system of ducts drains the glandular tissue. The **intralobular ducts,** which drain the acini, join to form the **interlobular ducts** that in turn merge to form 15–25 **lactiferous ducts** which open on the tip of the **nipple.** The intralobular ducts are formed by simple cuboidal epithelium, and the interlobular and lactiferous ducts are lined by stratified (mostly double-layered) cuboidal epithelium. Just before opening on the surface of the nipple, the lactiferous ducts dilate to form the lactiferous sinuses, which are lined by stratified squamous epithelium continuous with the epidermis of the nipple.

The **nipple** and the **areola,** the skin around the nipple, are covered by **keratinized stratified squamous epithelium,** light or dark brown in color because of heavy local pigmentation, which increases during pregnancy. The **dermis** of the nipple and areola is characterized by the presence of numerous **smooth muscle fibers** arranged circularly. The **sebaceous glands,** and the **eccrine** and **apocrine sweat glands,** also can be found in the dermis. In addition, the nipple is rich in sensory nerve endings.

With cessation of lactation, most glandular acini undergo degeneration and the gland returns to a resting state. After menopause the gland undergoes involution, which includes the atrophy of the secretory portion and duct system, and the densification of the connective tissue.

Figure 13-31 shows a resting mammary gland, in which the glandular tissues are not developed. The lobules are composed of only ducts, dense connective tissue, and adipose tissue.

Figure 13-30. Umbilical Cord

Human • H.E. stain • High magnification

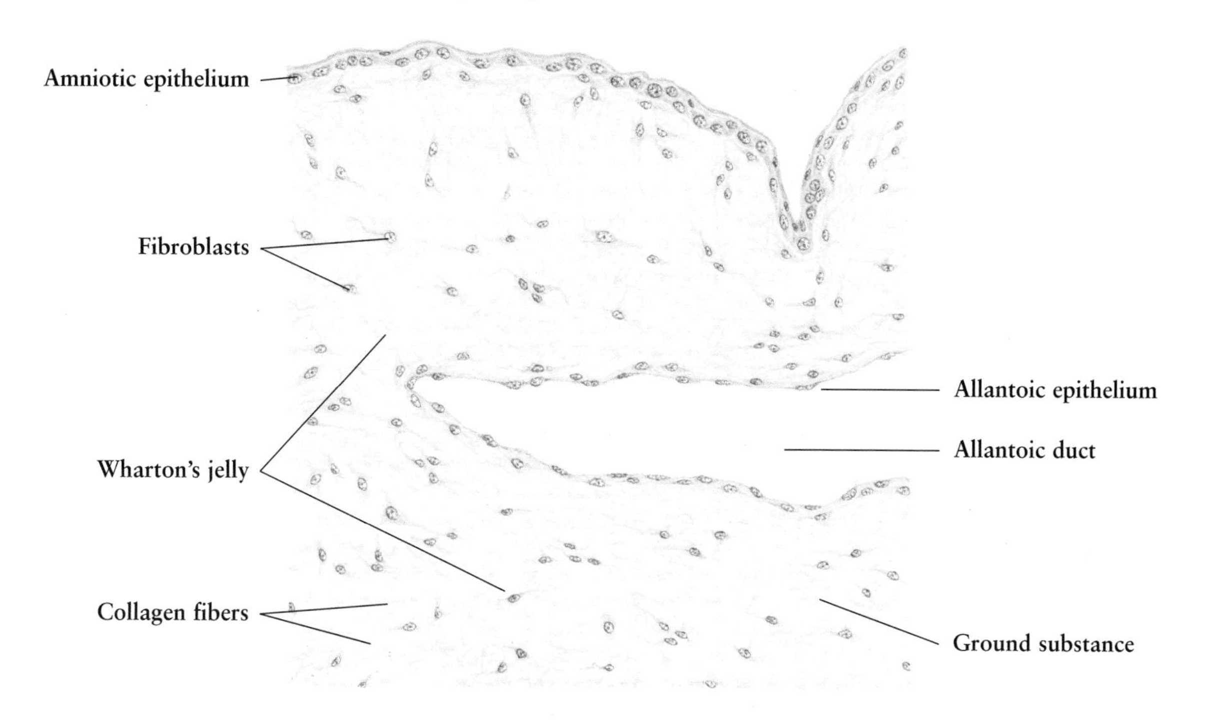

Figure 13-31. Mammary Gland

Human • H.E. stain • Low magnification

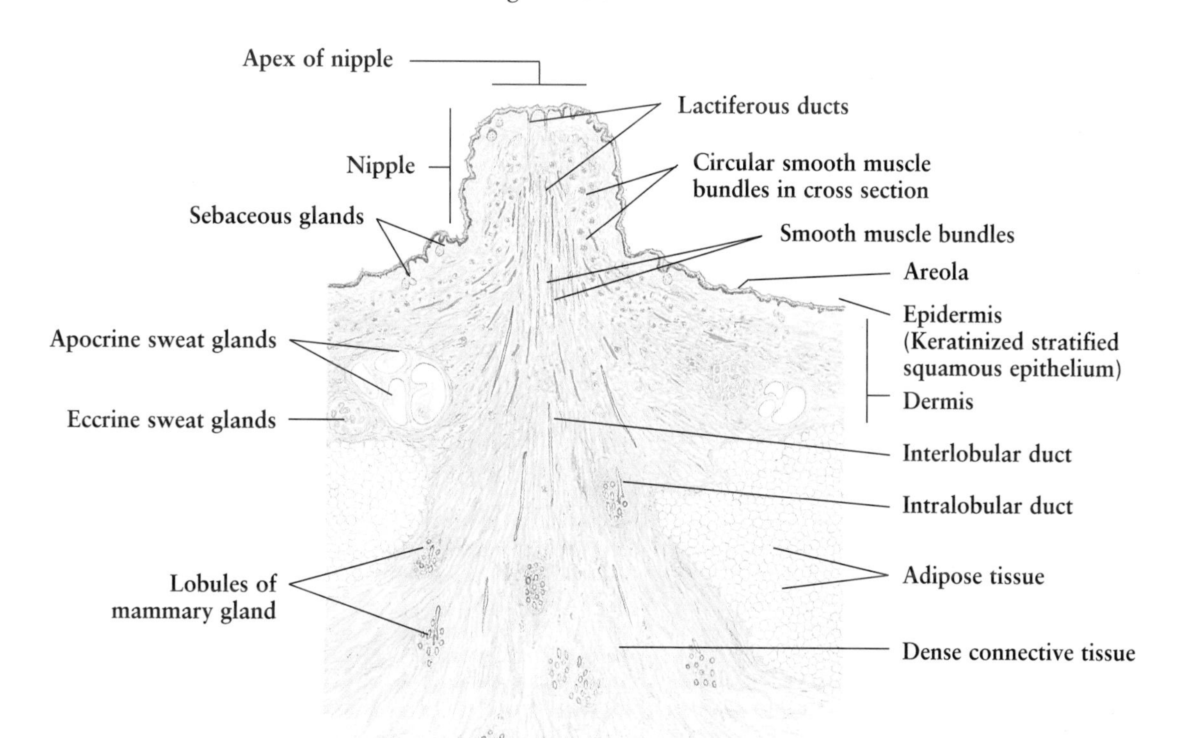

Fig. 13-32.

Resting Mammary Gland

In the **resting mammary gland** the glandular tissue is not active, as it would be during pregnancy or lactation. The **lobules** of the mammary gland appear as islands of glandular tissue surrounded by an extensive mass of **dense connective tissue** and **adipose tissue.** The lobules are composed of **intralobular ducts** and **terminal ductules** (acinar ducts) without real secretory acini. These ducts or ductules are lined by a simple cuboidal epithelium supported by a basement membrane. Between the epithelial cells and the basement membrane there is a discontinuous layer of **myoepithelial cells.** During pregnancy numerous secretory acini develop from the terminal ductules. The intralobular connective tissue is loose and cellular in type, containing a number of **lymphocytes, plasma cells,** and **macrophages.**

Between lobules **interlobular ducts** can be found, lined by a double-layered cuboidal epithelium resting on a basement membrane. The surrounding tissue also contains some lymphocytes and plasma cells.

Fig. 13-33.

Lactating Mammary Gland

During pregnancy, the **mammary glands** undergo extensive histological changes in preparation for lactation. These changes are primarily manifested in two aspects: the *formation* of secretory acini and the relative *decrease* of connective tissue and adipose tissue.

Under the influence of several hormones, such as *estrogens, progesterone, prolactin,* and *human placental lactogen,* the epithelium of the terminal ductules proliferates to form numerous secretory glandular acini. These are spherical in shape and composed of a layer of cuboidal or columnar epithelial cells resting on a basement membrane. Stellate myoepithelial cells are found between the acinar epithelial cells and the basement membrane.

Owing to the expansion of lobules, **interlobular connective tissue** and **adipose tissue** decrease in amount and are replaced gradually by glandular tissue. **Intralobular connective tissue** also diminishes in its amount and becomes infiltrated with lymphocytes.

At the end of pregnancy and after parturition, the glandular tissues reach their highest development and begin to secrete milk. **Figure 13-33** shows a **lactating mammary gland** with its glandular tissue at the state of maximum development. The **glandular acini** are lined by a **cuboidal epithelium,** and some epithelial cells have **secretory vacuoles** within their cytoplasm. The **myoepithelial cells** are located between epithelial cells and basement membrane, and their contraction pushes milk toward the excretory ducts. The acinar lumina are dilated and filled with **secretion** of milk.

The **glandular lobules,** composed of many **acini, intralobular ducts,** and a little **intralobular connective tissue,** are surrounded by a thin layer of **interlobular connective tissue.** The **interlobular duct** with a large lumen is lined by a double layer of cuboidal epithelium. Its lumen is also filled with milk. The interlobular duct is embedded in the interlobular connective tissue, which contains a number of **blood vessels** and **adipose cells.**

The *milk* is rich in nutrients composed of protein, carbohydrate, lipids, ions, vitamins, and antibodies, as well as water. The first milk to appear after birth is referred to as *colostrum.* Colostrum contains more protein and less lipid than the regular milk, and also contains antibodies, principally IgA, that provide a certain degree of temporary, passive immunity to the newborn.

Figure 13-32. Resting Mammary Gland
Human • H.E. stain • Medium magnification

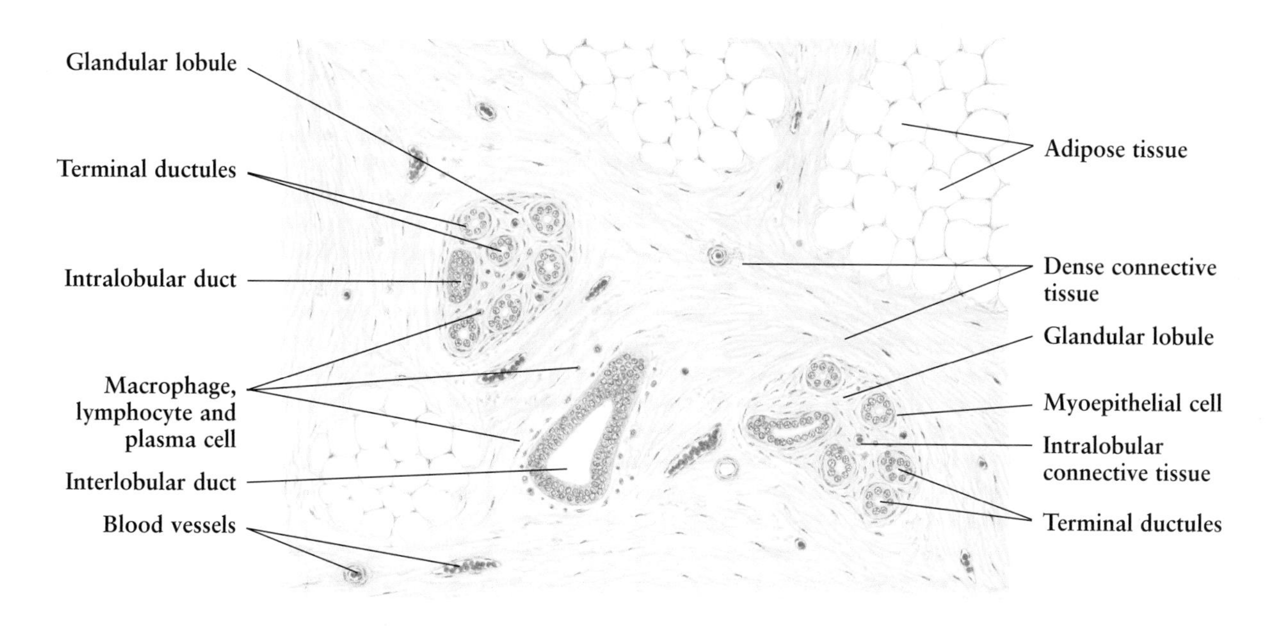

Figure 13-33. Lactating Mammary Gland
Human • H.E. stain • Medium magnification

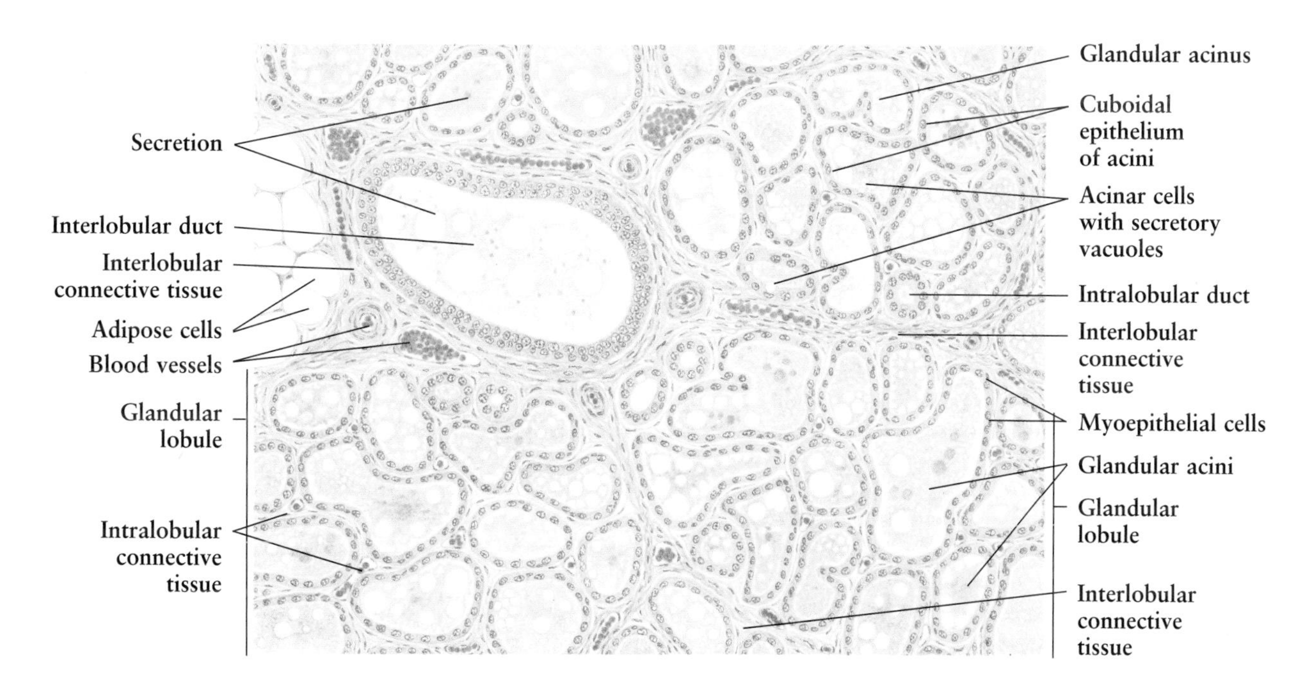

14. ENDOCRINE ORGANS

Secretory cells or **glandular epithelium** (see **Figs. 1-14 and 1-15**) may form two organ types: *exocrine glands* that release their secretory products into a body or organ cavity via a duct (such as the salivary glands), or *endocrine glands* which release their secretory products directly into the blood or lymph circulation (such as the hypophysis).

Endocrine glands lack ducts, and their secretory products, known as *hormones,* are disseminated throughout the body by the blood circulation. Hormones differ greatly in their chemical composition; they may be *biogenic amines, steroids, polypeptides, proteins,* or *glycoproteins.* Each produces an effect upon a particular tissue or organ, known as a *target organ* or a *receptor.* Some hormones have an effect upon the nervous system, and certain endocrine glands are regulated or controlled by neural mechanisms.

In general, endocrine glands have a very simple microscopic structure. The endocrine cells form **sheets, cords, plates,** or irregular **clumps** supported by delicate **connective tissue.** The glands are characterized by a very rich vascular supply of **fenestrated capillaries** or **sinusoids.** The secretory products are released into the perivascular spaces and then are rapidly absorbed into the capillary bed.

Like exocrine glands, endocrine glands arise from epithelial linings, but they eventually lose connection with the surface and become isolated islands of epithelium embedded in a connective tissue matrix. Endocrine cells vary in their embryological derivation; they may arise from **endoderm** (such as the thyroid, the parathyroid, and islets of Langerhans), from **mesoderm** (such as the adrenal cortex, testes, and ovaries), and from **ectoderm** (such as the hypophysis and the adrenal medulla).

Some endocrine glands occur in the form of separate organs, such as the **hypophysis,** the **pineal body,** the **thyroid,** the **parathyroids,** and the **adrenal glands.** Other endocrine tissues are found scattered within exocrine glands or complex organs, such as **islets of Langerhans** in the pancreas, and **Leydig cells** in the testes. This chapter shows all the individual endocrine organs as well as the islets of Langerhans.

Fig. 14-1. Hypophysis (Pituitary Gland)

The **hypophysis,** also known as the **pituitary gland,** is a small but complex endocrine organ, 10 mm in length, 13 mm in width, 5 mm in height, and about 0.5 g in weight. It is located in a bony fossa of the sphenoid bone, the *sella turcica,* and is covered by a dense connective tissue **capsule,** derived from the dura mater.

Histologically, the hypophysis consists of two different tissues: *adenohypophysis* and *neurohypophysis.* The adenohypophysis (glandular portion) develops from the ectoderm at the roof of the oral cavity of the embryo. These cells migrate dorsally and form *Rathke's pouch.* The neurohypophysis (nervous portion) is derived from an outgrowth of the floor of the diencephalon (forebrain).

The **adenohypophysis** is divided by the residual lumen of Rathke's pouch into two unequal parts. The **pars distalis** is anterior to the lumen, and the **pars intermedia** is posterior to the lumen. The **pars tuberalis** is an extension of the pars distalis, enveloping the neural stalk (infundibulum). The **neurohypophysis** is also composed of three portions: pars nervosa, infundibulum, and median eminence. The **pars nervosa** is the major portion; it lies immediately behind the pars intermedia. The pars nervosa is continuous, above, with the **infundibulum** and median eminence (not shown in the drawing). The pars tuberalis and infundibulum form the **hypophyseal stalk.**

The term **anterior lobe** refers to the pars distalis and the pars tuberalis, and the **posterior lobe** refers to the pars nervosa and the pars intermedia. **Figure 14-1** is a sagittal section of the human hypophysis, clearly showing the different parts of the organ.

Figure 14-1. **Hypophysis (Pituitary Gland)**
Human • Sagittal section • H.E. stain • Low magnification

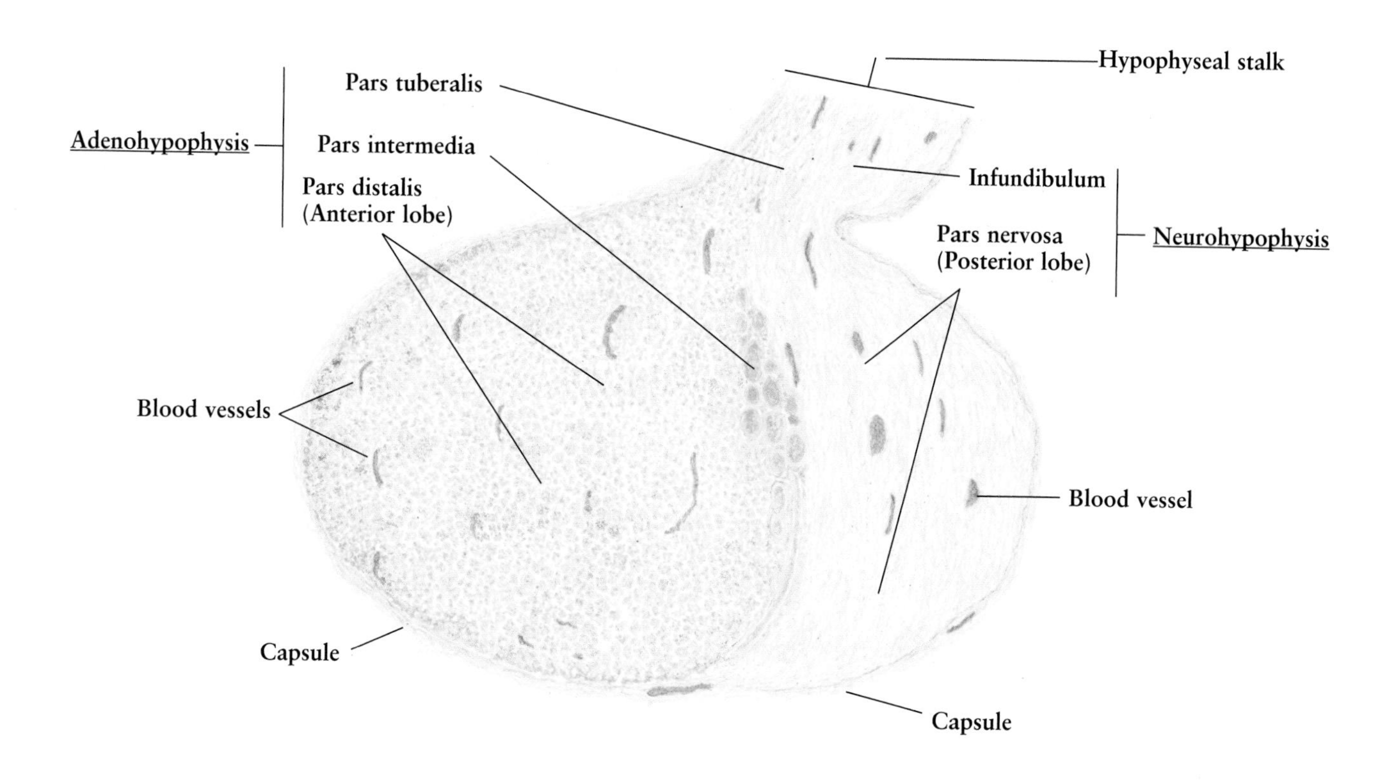

Fig. 14-2. Hypophysis: Pars Distalis

The **pars distalis** is composed of glandular cells that are arranged in irregular clusters or cords, surrounded by a thin network of **connective tissue** rich in reticular fibers. Between the clusters or cords are **sinusoids,** in which the endothelial lining is characteristically fenestrated.

The glandular cells of the pars distalis are classified into two main groups, *chromophils* and *chromophobes,* on the basis of their affinities for routine histological dyes. The **chromophils** may be subdivided into two categories, *acidophils* and *basophils,* according to their staining reaction following the use of hematoxylin and eosin. In general, the acidophils appear centrally and posteriorly, while the basophils are located along the midline and anterior margins of the hypophysis, and the chromophobes often occur in groups in the center of the clusters or cords throughout the adenohypophysis.

The **acidophils,** accounting for 35% of the glandular cells in the pars distalis, are large and round or ovoid in shape. The cytoplasm is packed with small pink or red specific granules, the secretory granules, which are readily stained by acid dyes such as eosin, acid fuchsin, orange G, and azocarmine. The acidophils are composed of two cell types: *somatotrophs* and *mammotrophs,* which can be distinguished by specific immunohistochemical techniques. The somatotrophs produce ***growth hormone,*** which stimulates general body growth, particularly the growth of the epiphyses of long bones. The mammotrophs synthesize ***prolactin,*** which promotes the secretion of milk during lactation.

The **basophils,** representing about 15% of the cell population of the adenohypophysis, are slightly larger in size than the acidophils. They have large, spherical nuclei that are often eccentrically located. The cytoplasm is crowded with small bluish secretory granules when stained with basic dyes (hematoxylin, aniline blue, methylene blue, resorcin fuchsin). These secretory granules also give a PAS-positive reaction. According to immunohistochemical staining or electron microscopy, three kinds of basophils may be classified: the *corticotrophs,* involved in the formation of ***adrenocortico-tropic hormone*** (ACTH), which promotes secretion of glucocorticoids in the cortex of adrenal gland; the *thyrotrophs,* responsible for the secretion of ***thyrotropic hormone (thyroid-stimulating hormone,*** TSH), stimulating the synthesis, storage, and liberation of thyroid hormone; and the *gonadotrophs,* which secrete ***follicle-stimulating hormone*** (FSH) and ***luteinizing hormone*** (LH). FSH stimulates the development of ovarian follicles and the secretion of estrogen in the female, and promotes spermatogenesis in the male. LH ensures the maturation of the follicle, the formation of the corpus luteum, and the secretion of progesterone in the female. The LH, also known as ***interstitial cell-stimulating hormone*** (ICSH) in the male, stimulates the Leydig cells of the testes to secrete an androgen.

The **chromophobes,** about 50% of the cells in the adenohypophysis, are small, rounded or polygonal cells with spherical nuclei and relatively little, pale-stained cytoplasm that gives rise to obscure cell boundaries. In general, these cells show no specific granules when examined by light microscopy; most of them may be degranulated chromophils and some are considered reserve or nonsecretory cells.

Figure 14-2. **Hypophysis: Pars Distalis**
Human • H.E. stain • Higher magnification

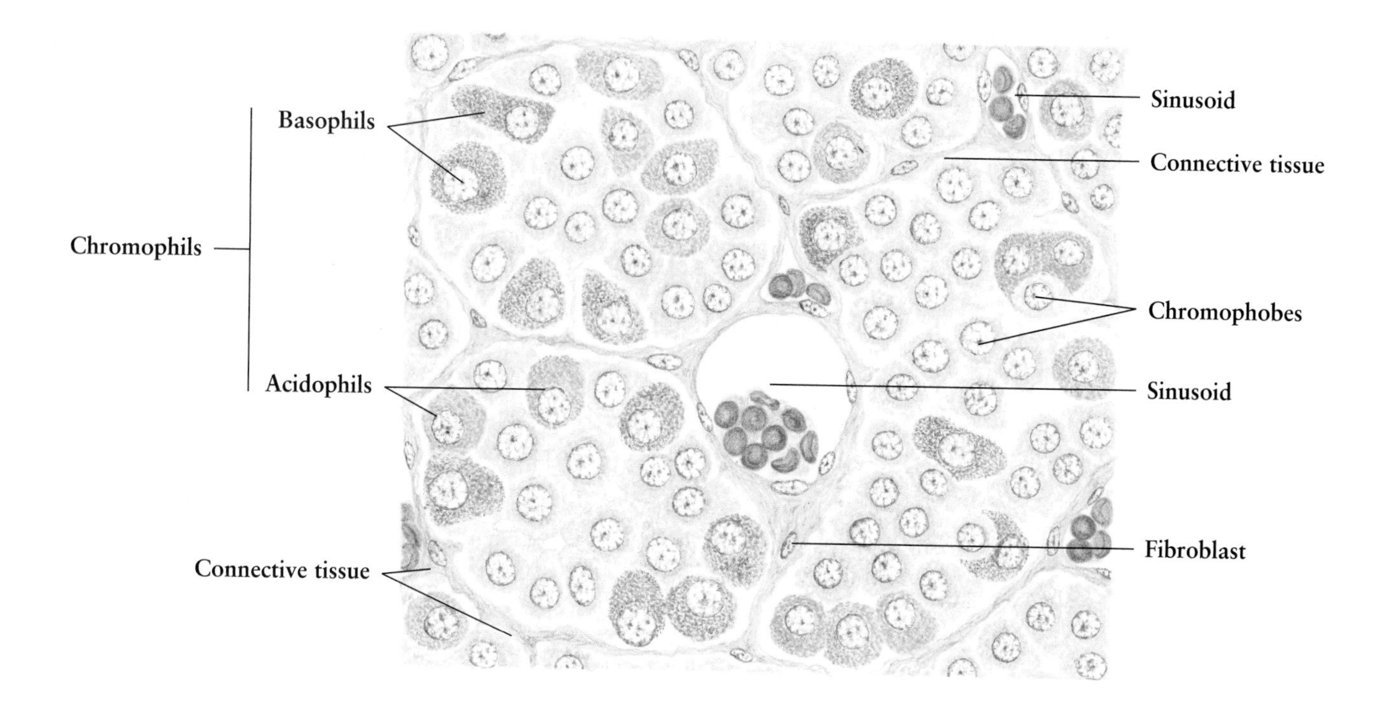

Fig. 14-3.

Hypophysis: Pars Nervosa

The **pars nervosa** is mainly composed of some 100,000 unmyelinated nerve fibers supported by numerous pituicytes. In addition, there is a network of fine, fenestrated **capillaries** and some **fibroblasts.**

The **unmyelinated nerve fibers** are the axons of secretory neurons, whose cell bodies are located in the supraoptic and paraventricular nuclei of the hypothalamus (see **Figs. 6-15** and **6-16**). The neurosecretory granules are synthesized in the neuronal bodies and anterogradely transported through the nerve fibers to the pars nervosa. **Herring bodies** are huge accumulations of granules along the nerve fibers.

The **pituicytes** are small cells with rounded or ovoid nuclei; they are characterized by their cytoplasmic **pigment.** The pituicytes are similar to the neuroglia cells elsewhere in the central nervous system, with short branching processes embracing nerve fibers and Herring bodies.

Two hormones released by the pars nervosa are *oxytocin* and *vasopressin* (*antidiuretic hormone,* ADH), which are synthesized primarily by nerve cell bodies of the paraventricular and supraoptic nuclei, respectively (see **Figs. 6-15** and **6-16**). These hormones are transported down the hypothalamohypophyseal tract to the pars nervosa in association with carrier proteins known as neurophysins. ***Oxytocin*** stimulates contraction of uterine smooth muscle during the final stages of pregnancy, and also induces contraction of the myoepithelial cells of the mammary glands. ***Vasopressin*** promotes the absorption of water by cells of the distal tubules and collecting tubules in the kidney, resulting in a concentration of urine.

Fig. 14-4.

Hypophysis: Pars Intermedia

The **pars intermedia,** which is rudimentary in humans, forms a narrow band between the **pars distalis** and the **pars nervosa.** This part of the hypophysis is sharply demarcated from the pars nervosa, but blends with the pars distalis. Most of the cells constituting the pars intermedia are clear. Some are basophils, but occasionally acidophils can also be found. These cells are often arranged around **vesicles** filled with eosinophilic **colloid,** which contains no iodine. Around the vesicles is a thin loose connective tissue that contains **blood vessels** and **fibroblasts.**

The pars intermedia may be responsible for the secretion of ***melanocyte-stimulating hormone*** (MSH), which promotes the synthesis of melanin by skin melanocytes, resulting in an increase in skin pigmentation.

Figure 14-3. **Hypophysis: Pars Nervosa**
Human • H.E. stain • High magnification

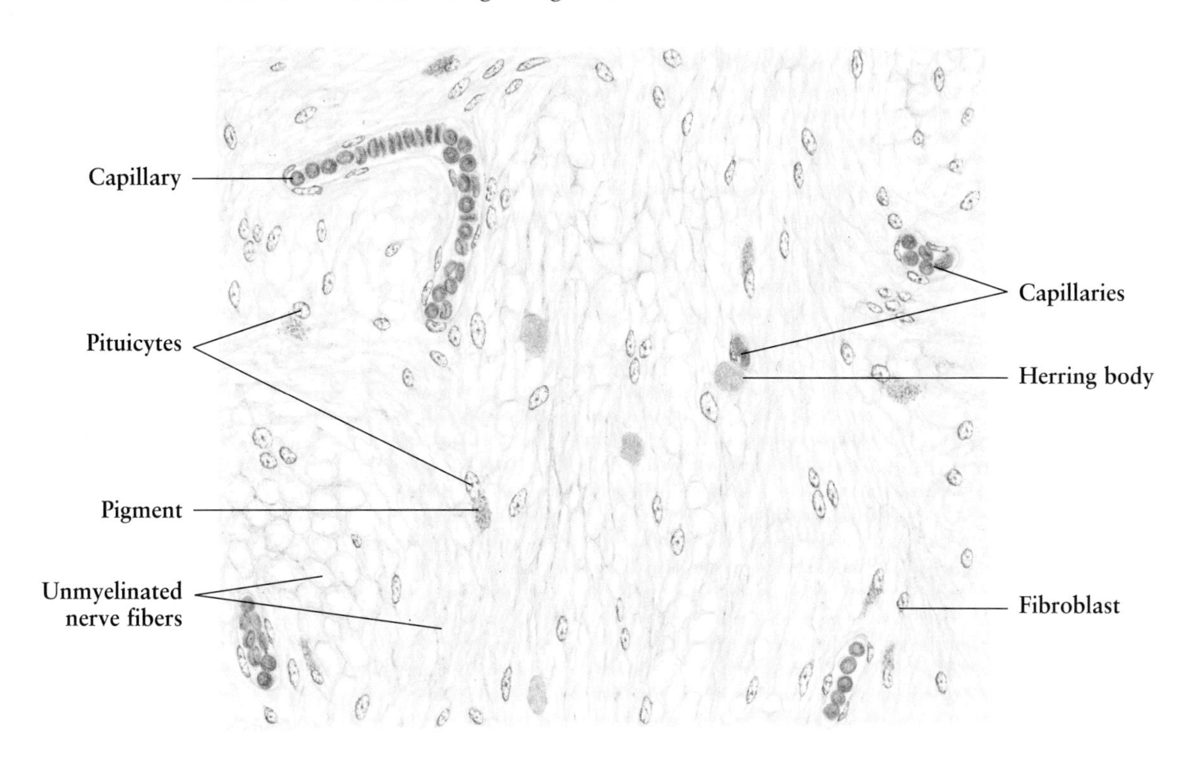

Figure 14-4. **Hypophysis: Pars Intermedia**
Human • H.E. stain • High magnification

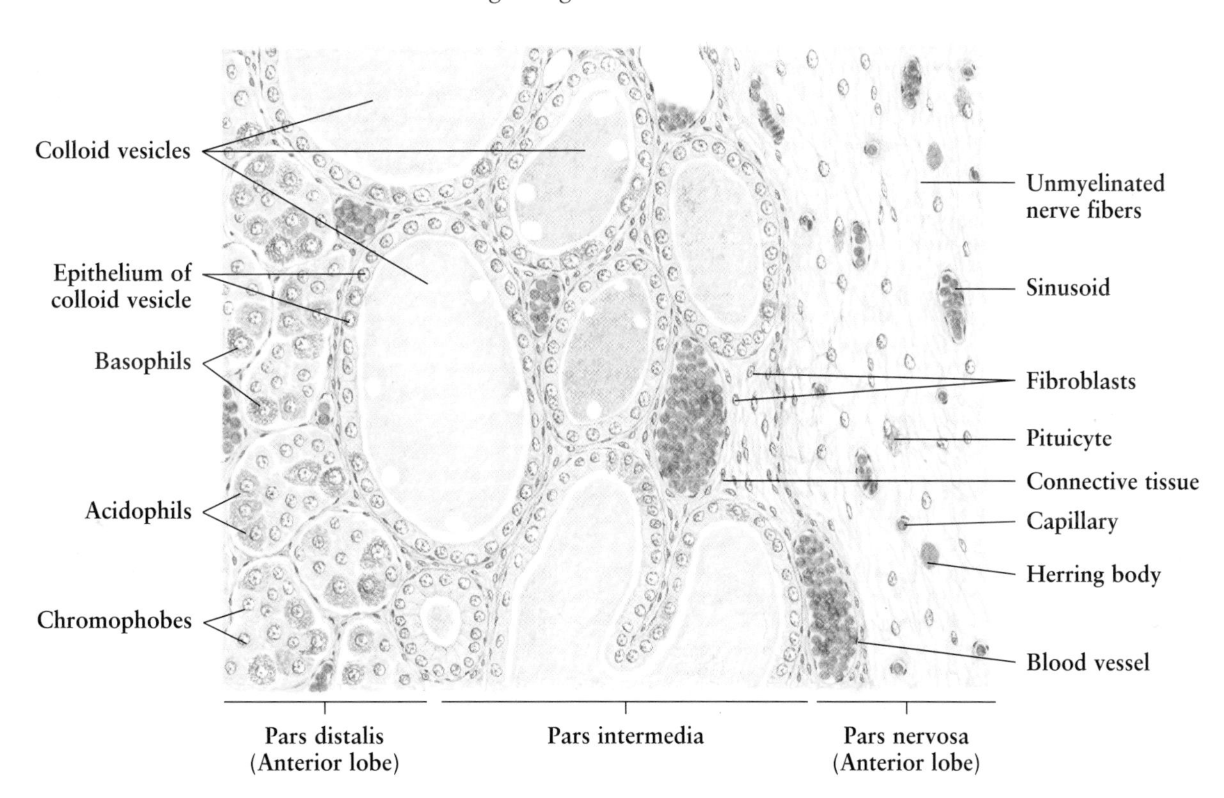

Fig. 14-5.

Adrenal Gland

The **adrenal glands,** overlying the cranial pole of each kidney, are paired, small, triangular endocrine glands, with a definite hilum on the anterior surface. They are embedded in the retroperitoneal adipose tissue. Each gland is invested by a tough connective tissue **capsule,** from which delicate connective tissue fibers pass radially into the parenchyma, supporting the blood vessels that penetrate the gland from the capsule.

The gland consists of an outer **cortex** and a thin inner **medulla;** each has a different function and arises from different sources. The cortex is derived from the mesothelium of mesodermal tissue lining the primitive body cavity (coelom), while the medulla develops from neural crest, the same group of cells as those that form the sympathetic ganglia.

Figure 14-5 is a very low magnification drawing, showing the main structure of the gland, including the capsule, cortex, and medulla. Note the large blood vessels, the **central veins,** in the medulla.

Fig. 14-6.

Adrenal Gland: Medulla

The **medulla** shown in this figure is from a monkey adrenal gland fixed with potassium bichromate and stained with methylene green. This illustration demonstrates the medullary secretory cells, the **chromaffin cells.** All the medullary cells appear brown because the catecholamine granules within the cytoplasm are oxidized by the chromium salts. The glandular cells are arranged in irregular cords, which form an anastomosing network surrounded by **sinusoid** and thin **connective tissue** containing **fibroblasts** and **collagen fibers,** which are stained blue in this illustration. Between the secretory cells are a few **ganglion cells** supported by **satellite cells.** In this figure the medulla is sharply marked off from the adjacent **zona reticularis** of the adrenal cortex, and all red blood cells (**erythrocytes**) are stained yellow.

The chromaffin cells synthesize the catecholamines, *epinephrine* and *norepinephrine.* ***Epinephrine*** causes increases in oxygen consumption, glycogenolysis, and cardiac output. ***Norepinephrine,*** a precursor of epinephrine, regulates blood pressure of the heart and blood vessels via adrenergic nerves.

Figure 14-5. Adrenal Gland
Human • H.E. stain • Very low magnification

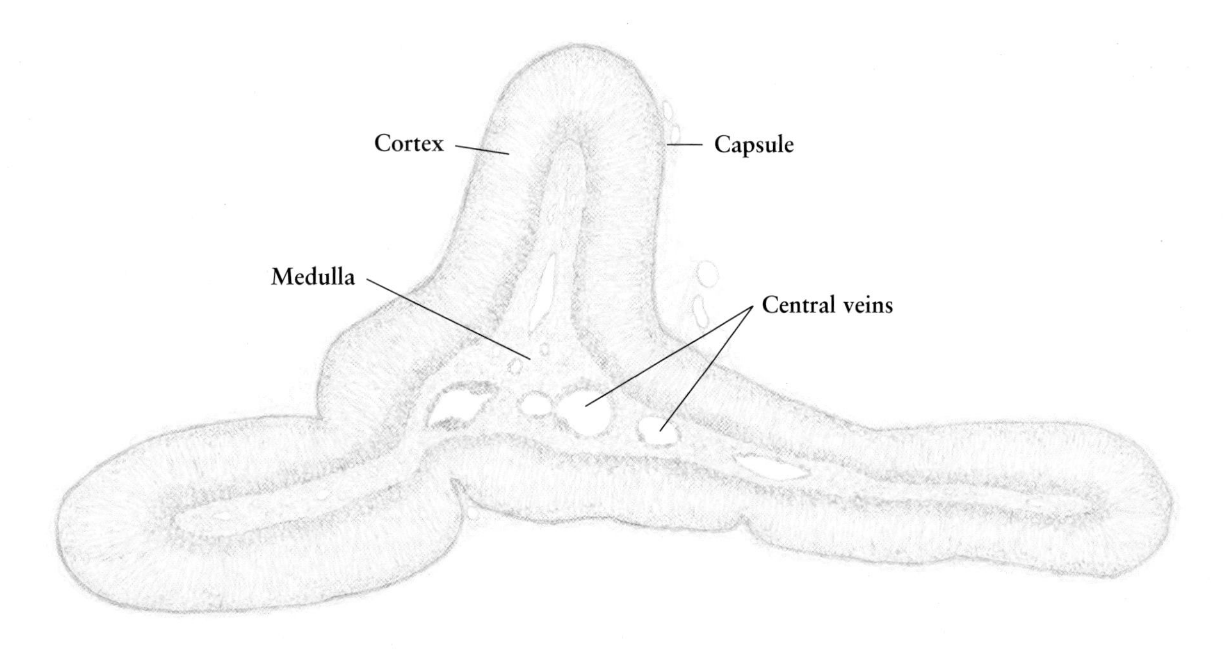

Figure 14-6. Adrenal Gland: Medulla
Monkey • Potassium bichromate fixation • M.G. stain • High magnification

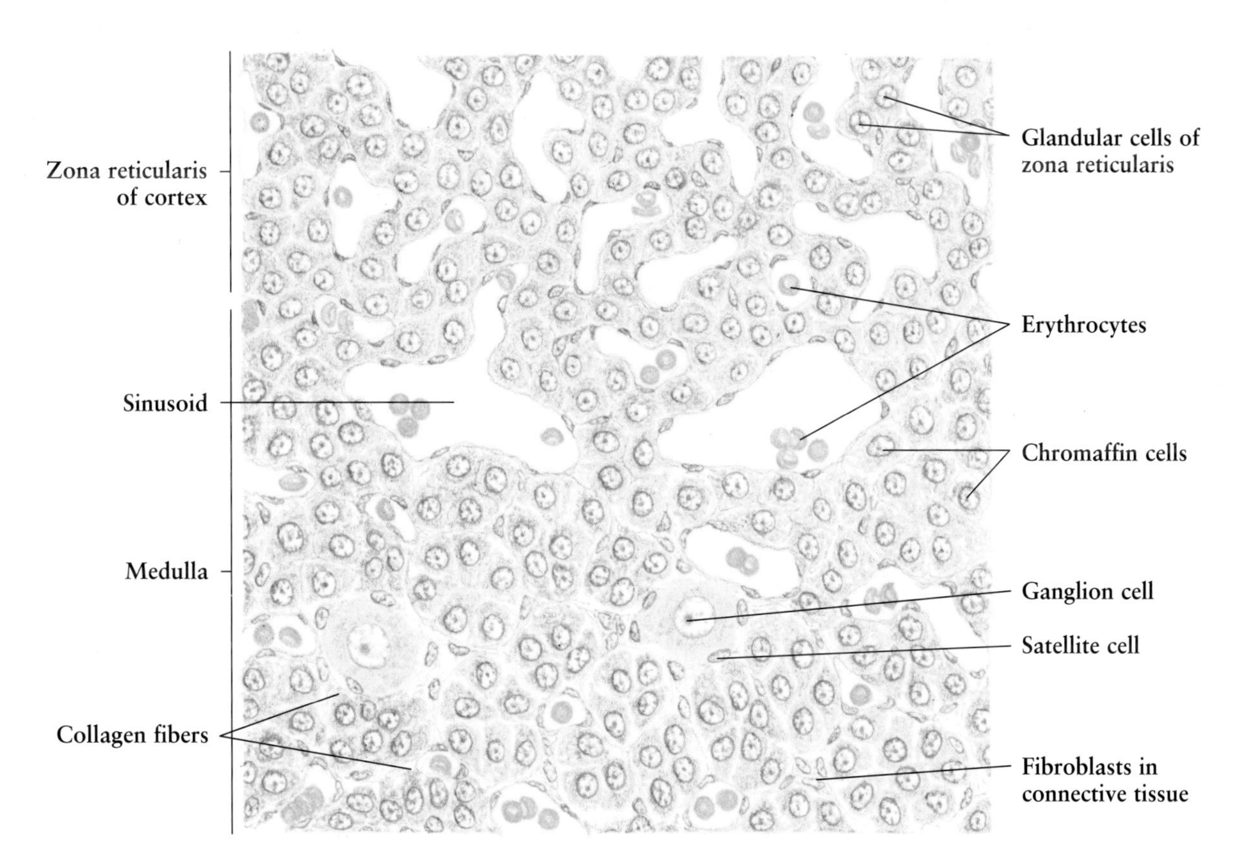

Fig. 14-7.

Adrenal Gland

The **capsule** is a **dense connective tissue** made up of numerous collagen and elastic fibers and **fibroblasts.** It contains **arterioles, venules, capillaries, lymphatic vessels,** and **nerve fibers.** Occasionally ganglia can be seen in the connective tissue of the capsule. Delicate collagen and reticular fibers from the capsule penetrate the parenchyma, supporting the **capillaries** and **sinusoid** in the cortex and medulla.

The **cortex,** which forms the major part of the gland, consists of three layers: *zona glomerulosa, zona fasciculata,* and *zona reticularis.* These three zones secrete different hormones.

The **zona glomerulosa,** located immediately beneath the capsule, is composed of cuboidal, columnar, or pyramidal cells arranged in irregular, ovoid clumps, surrounded by delicate connective tissue containing **capillaries.** Some clumps show central lumina, similar to the acini in other kinds of glands. The cells have darkly stained spherical nuclei and slightly basophilic cytoplasm. The zona glomerulosa produces ***mineralocorticoids*** (principally ***aldosterone***), which control water and electrolyte balance by the regulation of the sodium and potassium ion level. The secretion of aldosterone is independent of ACTH.

The **zona fasciculata** is the intermediate and thickest layer of the three zones of the cortex. It is composed of large, polyhedral cells arranged in long, radial cords usually two cells wide. These cords are perpendicular to the surface of the gland, and separated from one another by parallel **capillaries.** The secretory cells have round, centrally located nuclei and pale-stained cytoplasm filled with lipid droplets that give rise to the characteristic foamy appearance of the cells. The zona fasciculata secretes ***glucocorticoids (hydrocortisone*** and ***cortisone).*** Secretion of this class of hormones is controlled by the hypothalamus via the anterior pituitary hormone ACTH. Glucocorticoids have wide-ranging effects, such as those on the metabolism of carbohydrates, proteins, and lipids, as well as on suppressing the immune response.

In the **zona reticularis,** the innermost zone of the adrenal cortex, the cell cords anastomose to form an irregular network separated by numerous wide capillary **sinusoids.** The glandular cells are somewhat smaller and darker than those of the adjacent zona fasciculata, and the cytoplasm has accumulations of lipofuscin pigment granules. The zona reticularis is responsible for the secretion of small quantities of steroid sex hormones, including both ***female sex hormones (estrogen*** and ***progesterone)*** and several ***androgenic hormones*** that are not of physiological significance.

The **medulla** is composed of large, polyhedral secretory cells with spherical nuclei and slightly basophilic cytoplasm. The cells form irregular cords or clumps separated by **capillaries** and very thin connective tissue containing **fibroblasts.** The medulla produces the catecholamine hormones ***epinephrine*** and ***norepinephrine*** under the direct control of the sympathetic nervous system.

In the medulla, a few sympathetic **ganglion cells** may be found surrounded by **satellite cells,** usually occurring singly or in a small group. The **central vein** is one of the characteristic structures of the medulla. It has a wide lumen, and the wall is surrounded by a thick layer of **smooth muscle fibers** arranged longitudinally. The central vein collects blood from capillaries and venous channels in the cortex.

Figure 14-7. Adrenal Gland
Human • H.E. stain • High magnification

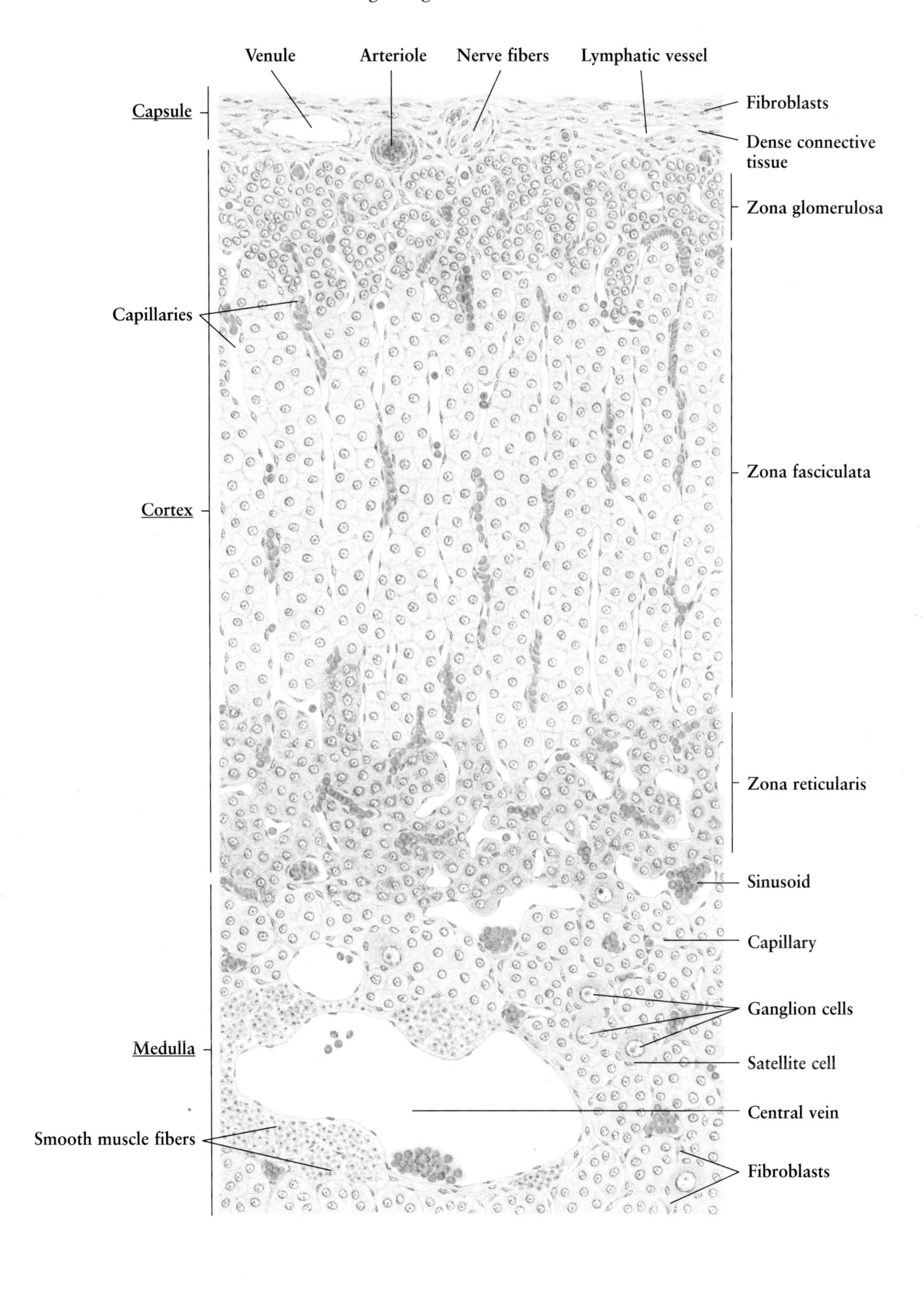

Fig. 14-8.

Thyroid Gland

The **thyroid gland** consists of two lateral lobes joined by a connecting narrow isthmus. It is located in the neck in front of the upper part of the trachea. Embryologically, the thyroid gland develops as a median endodermal downgrowth from the base of the tongue, and later becomes disconnected from the oral cavity. The gland is invested by an outer connective tissue sheath derived from the cervical fascia, and by an inner capsule of fibroelastic connective tissue. From the inner capsule, the trabeculae penetrate the gland, dividing it into lobules and providing a pathway for the vascular and nerve supply of the gland.

The functional units of the thyroid gland are the thyroid *follicles,* which are spherical structures of various sizes, covered by a layer of cuboidal epithelium. **Figure 14-8** shows that the **lobules** are composed of numerous **follicles** filled with **colloid secretion** in the lumina, and are surrounded by **interlobular connective tissue** containing **blood vessels** and **lymphatic vessels.**

Fig. 14-9.

Thyroid Gland: Follicles

The **follicles,** the functional units of the thyroid gland, are composed of a single layer of epithelium resting on a **basement membrane.** The **follicular epithelial cells** are generally cuboidal, but vary with different functional states of the thyroid. They may become columnar in the hyperactive state, or flattened in the hypoactive state. Surrounding the follicles is a rich network of fenestrated **capillaries,** which are embedded in a thin connective tissue containing **fibroblasts,** and collagen and reticular fibers.

The lumina of the follicles are usually filled with **colloid,** which is secreted and stored by the follicular cells. In H.E.-stained preparations, the colloid appears pink and structureless, often showing **spaces** between colloid and epithelium, and **vacuoles** within the colloid; these are artifacts of fixation. The thyroid hormones ***triiodothyronine*** (T_3) and ***tetraiodothyronine*** (T_4) are bound to the thyroglobulin, which is stored in the colloid. When stimulated by *thyroid-releasing hormone* (TRH) from the anterior pituitary, thyroglobulin is pinocytosed into the cytoplasm of the follicular epithelial cells. There it is broken down into small molecules, T_3 and T_4, which are discharged into the underlying capillaries. The most striking effect of these hormone is to accelerate body metabolic processes by increasing oxygen consumption and heat production.

In addition to the follicular epithelial cells, the thyroid contains a small population of **parafollicular cells,** also known as **C** (clear or light) **cells.** These cells are seen either in small groups in the connective tissue between follicles, or singly between follicular cells. C cells that are part of follicles sit on the basement membrane, but are separated from the lumen of the follicle by slender cell processes of the neighboring follicular cells. Usually they are large and have a centrally located nucleus and a light cytoplasm. Parafollicular cells secrete ***calcitonin,*** a polypeptide hormone, which reduces blood calcium levels by stimulating the deposition of calcium in bone.

Figure 14-8. **Thyroid Gland**
Human • H.E. stain • Low magnification

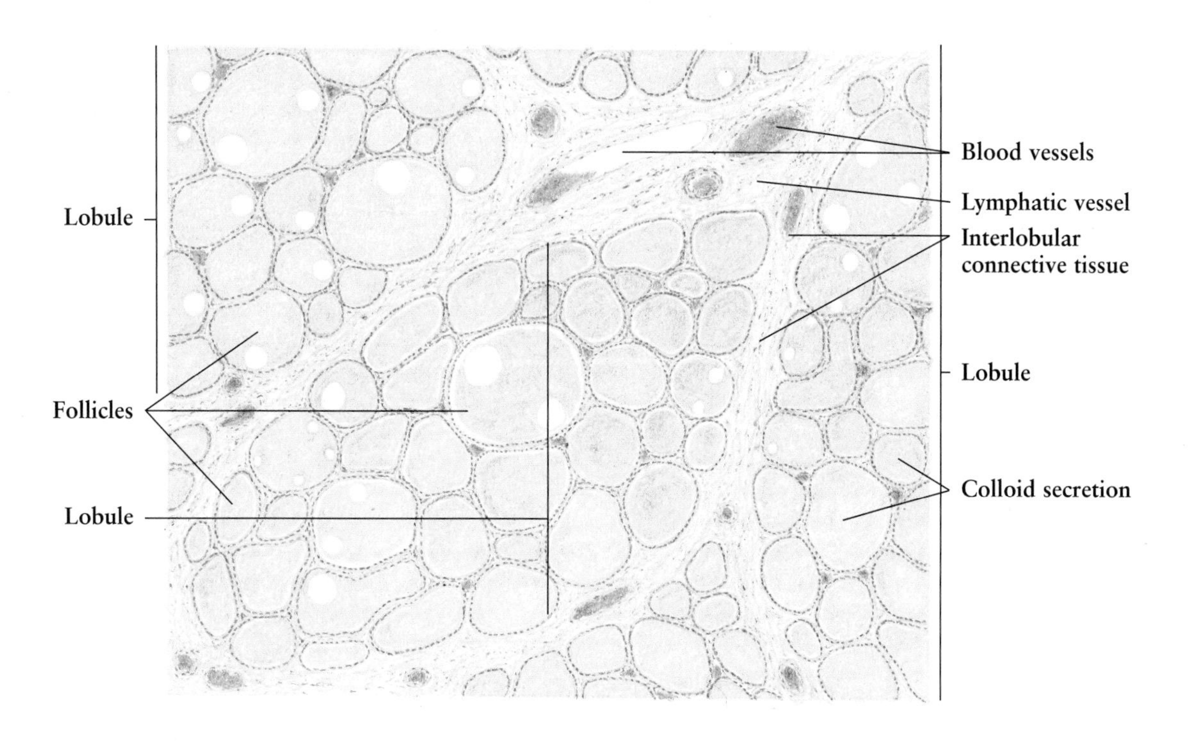

Figure 14-9. **Thyroid Gland: Follicles**
Human • H.E. stain • High magnification

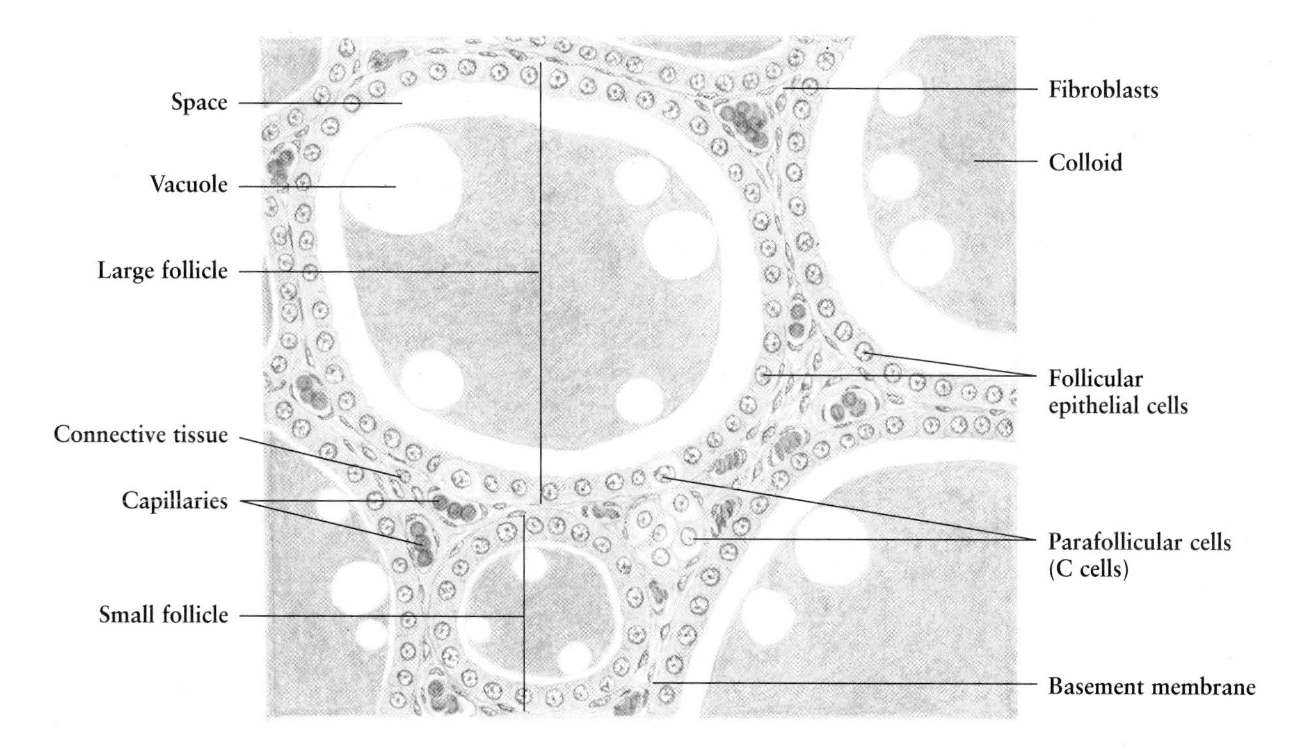

Fig. 14-10.

Parathyroid Gland

The **parathyroid glands** consist of two pairs of small, oval bodies. One pair is situated on the upper posterior surface of the thyroid and the other at the lower pole on each side. The superior parathyroids develop from the endoderm of the fourth pharyngeal pouch, while the inferior parathyroids develop from the third pouch.

Each parathyroid gland is invested by a connective tissue **capsule** that separates it from the **thyroid gland.** From the capsule, delicate **connective tissue** strands pass into the parenchyma carrying **blood vessels** and nerve fibers. The glandular cells do not form definite lobules, but form **cords** that are supported by delicate reticular connective tissue. **Figure 14-10** shows the relationship between parathyroid and thyroid. Outside the parathyroid there is a rich **adipose tissue.** Within the parathyroid, however, only a few **adipose cells** can be found, which increase in number and replace the glandular cells with increasing age.

Fig. 14-11.

Parathyroid Gland: Glandular Cells

The parenchyma of the **parathyroid gland** is arranged in **cords** of epithelial cells supported by delicate **connective tissue** containing **blood vessels.** It is composed of two kinds of glandular cells, *chief cells* and *oxyphil cells.*

Chief cells, also called **principal cells,** form most of the population of glandular cells. They are polyhedral in shape, and have a spherical nucleus and a pale-stained, light cytoplasm. The chief cells are responsible for the secretion of ***parathyroid hormone,*** a polypeptide in the active form. The hormone increases the concentration of calcium in the blood by stimulating resorption of bone matrix by osteoclasts, absorption of calcium from the small intestine, and resorption of calcium from the renal tubules. Parathyroid hormone also decreases the concentration of phosphate ions in the blood.

Oxyphil cells are also polyhedral in shape but larger than chief cells. They are scattered among the chief cells as single cells or small groups. The nucleus is round and centrally located, and the cytoplasm is filled with acidophilic granules. The oxyphil cells do not appear in humans until the age of 5–7 years and thereafter increase in number, especially after puberty. Their function remains unclear.

Figure 14-10. Parathyroid Gland
Human • H.E. stain • Low magnification

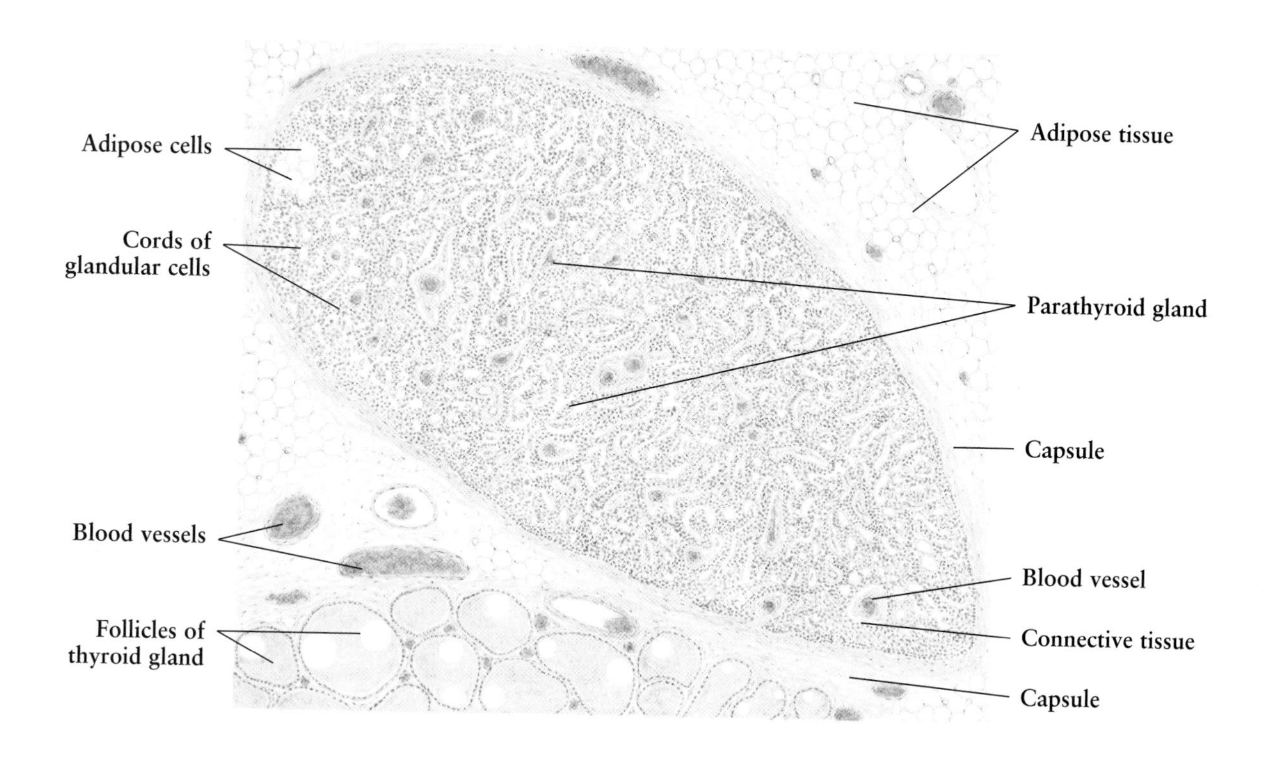

Figure 14-11. Parathyroid Gland: Glandular Cells
Human • H.E. stain • High magnification

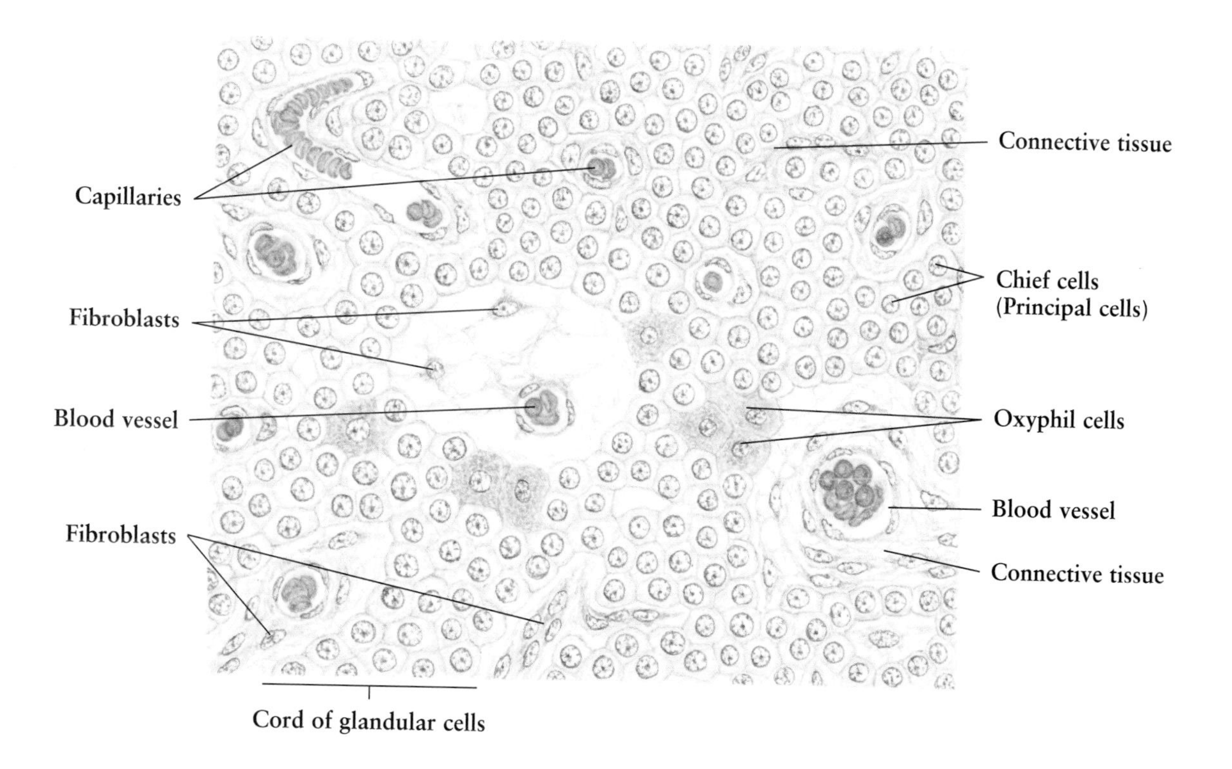

Fig. 14-12.

Pineal Body

The **pineal body,** also known as epiphysis cerebri, is a small, cone-shaped organ, 6–8 mm in length and 3–5 mm in width. It is an endocrine gland of neuroectodermal origin, and is attached to the roof of the **third ventricle** by a short **stalk** containing nerve fibers, many of which communicate with the hypothalamus.

The pineal body is surrounded by a connective tissue **capsule** composed of pia mater. From the capsule **septa** pass into the organ, and divide it into incomplete **lobules.** The **blood vessels** enter the organ from the capsule along with the septa. The parenchyma of the pineal body is composed of *pinealocytes* and *neuroglial cells.* One of the characteristic features of this organ is the presence of **brain sand,** which is composed of calcified concretions in the connective tissue.

Fig. 14-13.

Pineal Body: Pinealocytes

The parenchyma of the pineal body is composed of two cell types: *pinealocytes* and *neuroglia.* **Pinealocytes** are highly modified neurons with many branching cytoplasmic processes, some of which terminate as bulbous endings around **blood vessels.** The cytoplasm is slightly basophilic. The nucleus is large and spherical, with a great amount of peripherally placed dense heterochromatin and a prominent nucleolus. **Neuroglia** are identified as astrocytes by electron microscopy by reason of the presence of fibrillar elements in the cytoplasm. The neuroglia are fusiform in shape and have a dark, ovoid nucleus. Between these cells is a network of fenestrated **capillaries.**

Brain sand, found in the pineal body, are small, mulberry-shaped calcifications consisting of a concentric organization of successive layers of hydroxyapatite, calcium carbonate apatite, and organic matrix. These lamellated bodies gradually increase in number and size with age. The cause of their formation and their significance are not known.

The pineal body secretes ***melatonin,*** which is synthesized in the dark. The formation of melatonin is interrupted when light stimulates the retinae of the eyes. Melatonin inhibits growth and maturation of the gonads until puberty; after that, the pineal body gradually undergoes involution.

Figure 14-12. Pineal Body
Human • Sagittal section • H.E. stain • Low magnification

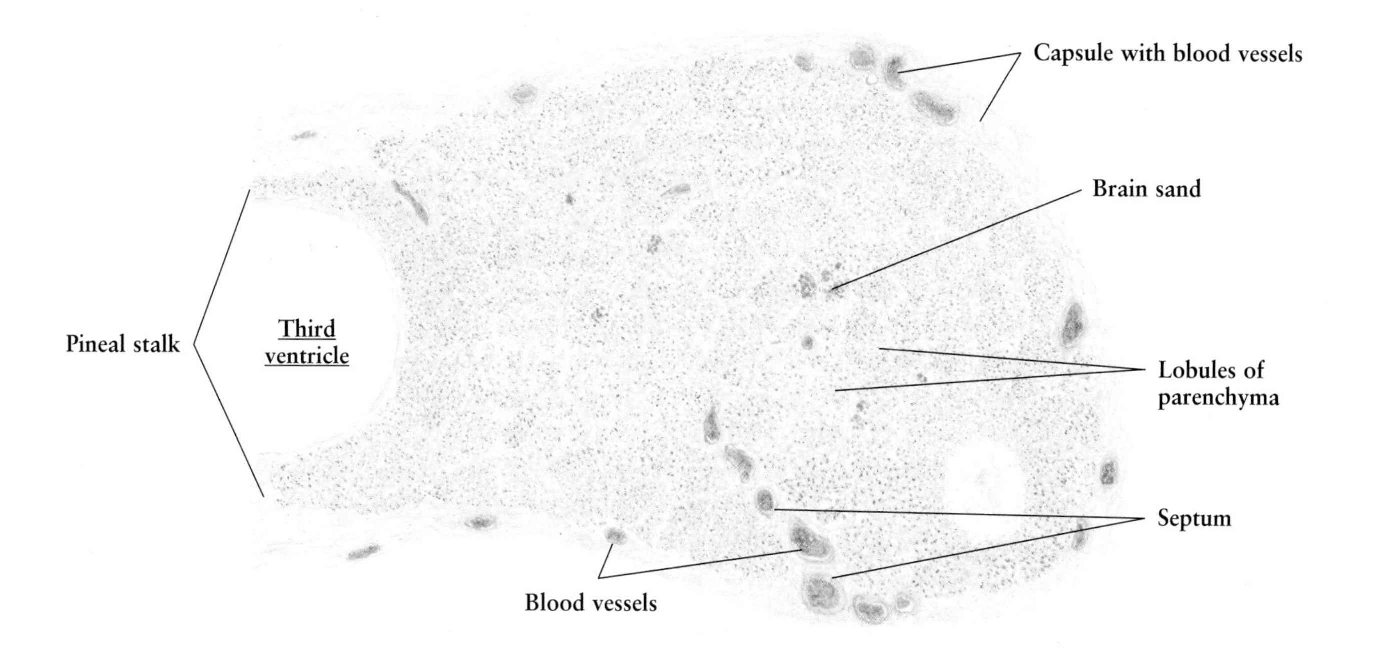

Figure 14-13. Pineal Body: Pinealocytes
Human • H.E. stain • High magnification

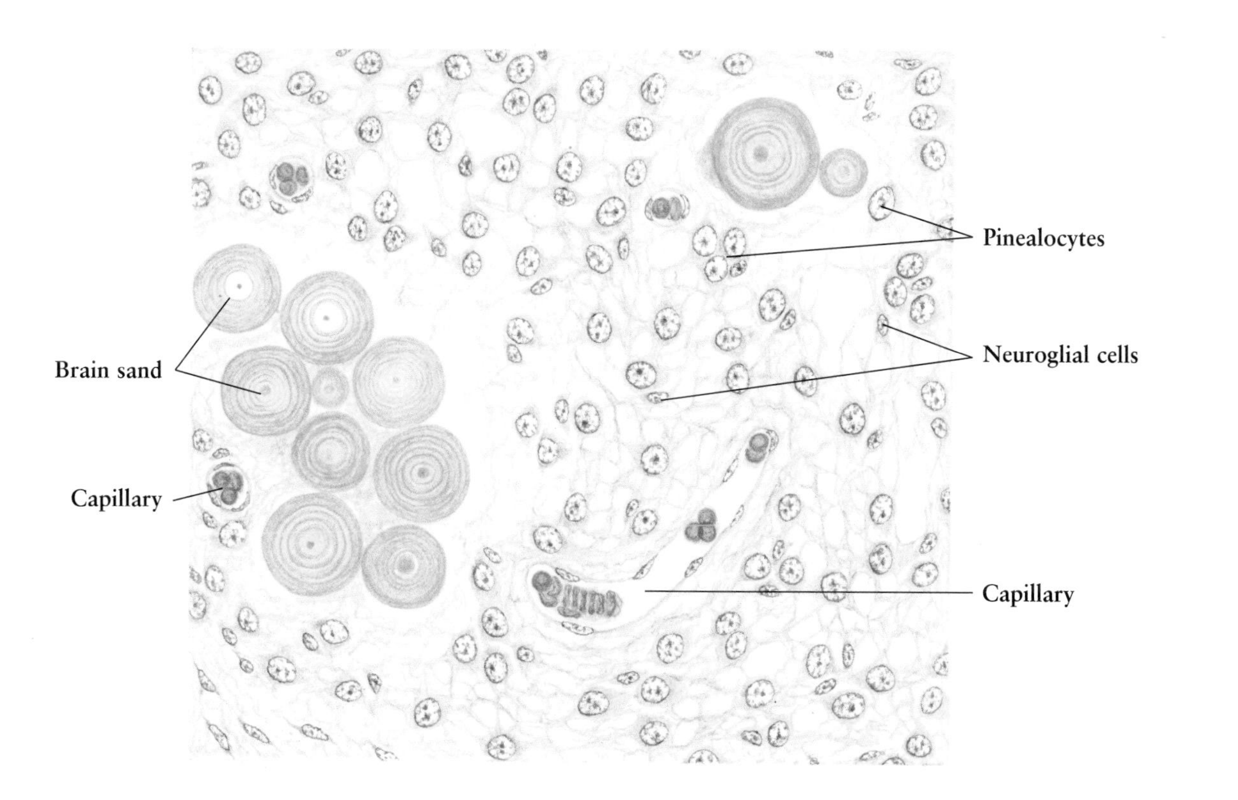

Fig. 14-14. —

Islet of Langerhans

The **islets of Langerhans,** the endocrine portion of the pancreas, are scattered throughout the pancreas and surrounded by the **acini** of the exocrine portion (also see **Figs. 10-54,** and **10-55**). They are readily distinguished from the surrounding exocrine portion by their pale-staining appearance, which contrasts with the dark staining of the exocrine acini. The islets of Langerhans are masses of endocrine cells arranged in irregular cords or spheroidal forms. They are invested by a delicate **connective tissue** that penetrates the islets, surrounding the cords with a rich network of fenestrated **capillaries.**

The islets of Langerhans are composed of secretory cells of three types: *alpha, beta,* and *delta cells.* They are irregular and polygonal in shape with spherical, centrally located nuclei and pale-staining cytoplasm. These three cell types produce ***glucagon, insulin,*** and ***somatostatin,*** respectively. Although they cannot be differentiated from one another in H.E.-stained preparations such as the section shown in this figure, they can be distinguished by special staining methods or immunohistochemical techniques (see **Fig. 14-15**).

Fig. 14-15. —

Islet of Langerhans

Figure 14-15 is a human **islet of Langerhans** stained by the VPL (Victoria blue, phroxin, and light green) method, highlighting cells of three types: *alpha, beta,* and *delta cells,* which possess 20%, 75%, and 5% of the islet cell population, respectively. All the nuclei are stained pink. Surrounding the islet of Langerhans are the **acini** of the exocrine pancreas. The delicate **collagen fibers,** which appear blue, **fibroblasts,** and **capillaries** are present in both endocrine and exocrine portions of the pancreas.

The **alpha cells** are stained pink and are usually located in the periphery of the islet. They secrete ***glucagon,*** which raises the plasma glucose concentration by stimulating the release of glucose from the liver via glycogenolysis.

The **beta cells** stained with this method are filled with blue secretory granules in their cytoplasm. They are responsible for the secretion of ***insulin,*** which promotes glucose uptake from the blood and the conversion of glucose to glycogen by cells, especially hepatocytes and skeletal myocytes. The effect of insulin is thus to lower plasma glucose levels.

The **delta cells** in this illustration have a pale-staining appearance, without any granules. They produce ***somatostatin,*** which may inhibit the secretion of both glucagon and insulin, and diminish the mobility of the stomach, small intestine, and gallbladder.

Figure 14-14. Islet of Langerhans
Human • H.E. stain • High magnification

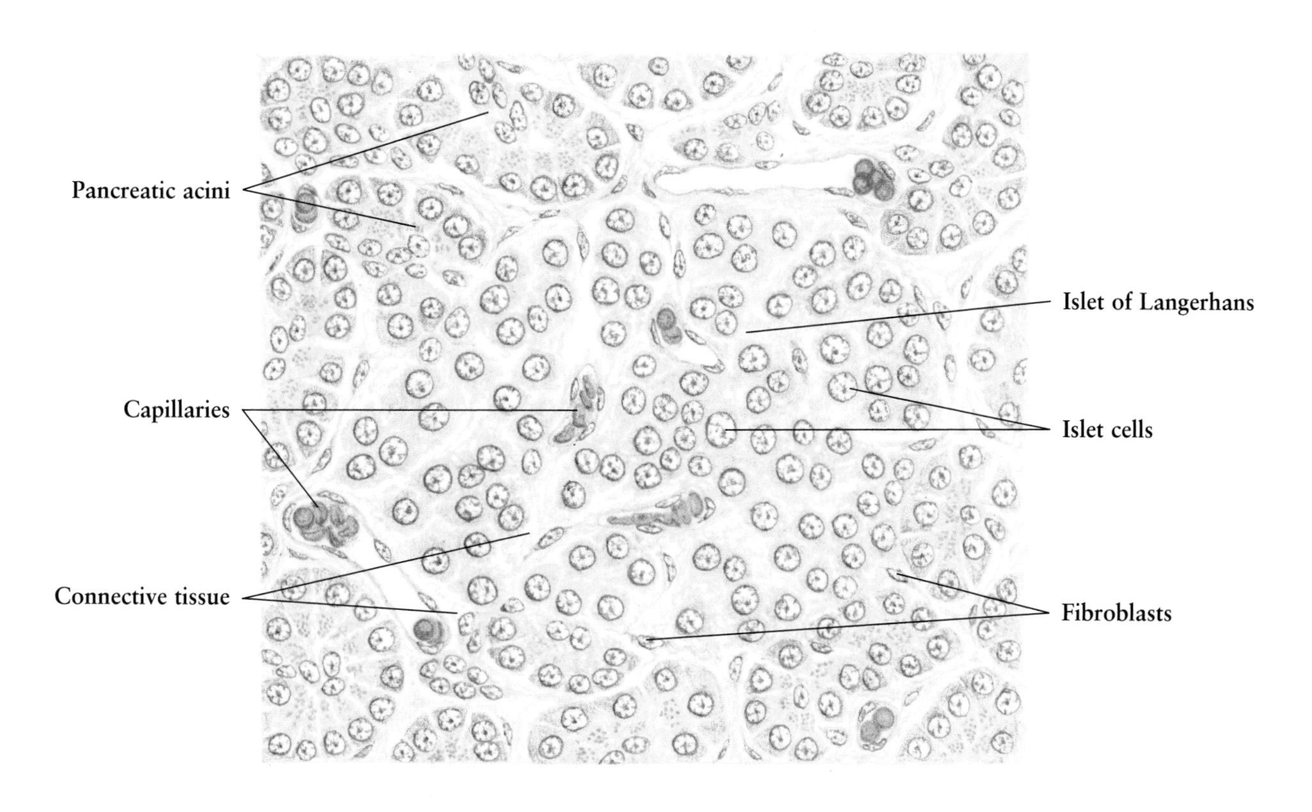

Figure 14-15. Islet of Langerhans
Human • VPL stain • High magnification

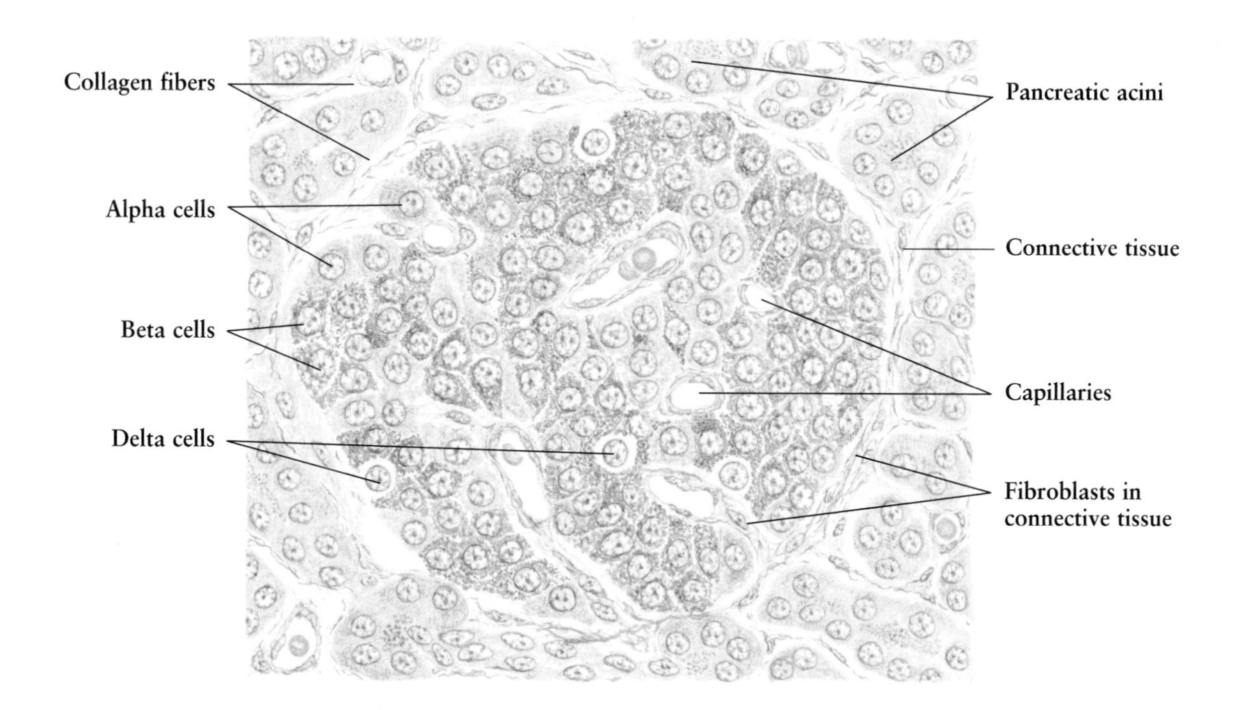

15. The Integument

The **integument,** consisting of *skin* and its *appendages,* is the heaviest organ of the body. It accounts for about 16% of the body weight. The **skin** that covers the surface of the body, composed of *epidermis* and *dermis,* lies directly on the *hypodermis,* the subdermal connective tissue.

The **epidermis** is a **keratinized stratified squamous epithelium,** derived from the ectoderm. The **dermis** is a layer of **dense connective tissue** of mesodermal origin. The interface between the epidermis and the dermis is highly irregular. It contains projections of the dermis, called **dermal papillae,** and downgrowths of the epidermis, known as **interpapillary pegs.** The dermal papillae and epidermal interpapillary pegs fit each other very well to form a firm junction.

The **hypodermis,** also called subcutaneous tissue, is loose connective tissue that generally contains various amounts of **adipose tissue.** The hypodermis, corresponding to the superficial fascia of gross anatomy, is not considered to be part of the skin. It connects the dermis loosely with the subjacent tissue, allowing great mobility of skin over most regions of the body. The exceptions are in the palm and sole, where skin mobility is limited.

The **appendages** of the skin are specialized derivatives of the epithelium. They include *hair, nails, sweat glands,* and *sebaceous glands.* The details of these structures are shown in different figures of this chapter.

The skin has several main *functions.* It protects the body against mechanical, physical, chemical, and biological injuries such as collision, ultraviolet light, corrosive action, and microorganisms. The skin is the largest sensory organ of the body for the reception of touch, pressure, pain, and temperature stimuli. It acts as an excretory organ, excreting water and various waste products of metabolism by sweating. The skin is also able to regulate body temperature via the functions of the sweat glands, blood vessels, hair, and adipose tissue.

Fig. 15-1.

Thick Skin: Fingertip

Based on the thickness of its epidermis, **skin** is classified as *thick* and *thin.* **Thick skin** covers the palms and soles. **Thin skin** is present over the rest of the body. **Figure 15-1** is a low magnification of thick skin of a **fingertip,** demonstrating the *epidermis, dermis,* and *hypodermis.*

The **epidermis** is composed of **stratified squamous epithelium** with an extremely thick keratinized surface layer, known as the **stratum corneum.** The downgrowths of the epithelium into the underlying dermis form numerous **interpapillary pegs.**

The **dermis,** located just beneath the epidermis, is composed of irregular dense connective tissue (see **Fig. 2-3**). It is characterized by thick bundles of **collagen fibers** in various orientations, which constitute a strong three-dimensional fibrous network capable of resisting stress or traction from all directions. At the epidermal–dermal interface, the connective tissue of the dermis forms numerous projections toward the surface of the skin, known as **derminal papillae.** These mesh with interpapillary pegs, forming a very strong junction between the epidermis and the dermis. The dermis contains large **arteries** and **veins, blood capillaries, lymphatic capillaries,** and **bundles of nerve fibers.** The **glomus,** an apparatus of an arteriovenous anastomosis (see **Fig. 7-17**), is found in the connective tissue of the deep dermis.

The **hypodermis,** also called the superficial fascia in gross anatomy, consists mainly of loose connective tissue and adipose tissue. The **adipose tissue** is divided into **lobules** by the **interlobular septa** of connective tissue, which may contain **blood vessels, bundles of nerve fibers,** and **Vater-Pacinian corpuscles** (see **Fig. 6-27**).

The **sweat glands** (see **Fig. 15-10**) are located in the deep portion of the dermis and in the hypodermis, opening onto the surface of the skin by way of the **duct,** which passes through the dermis and epidermis.

Figure 15-1. Thick Skin: Fingertip
Human • Vertical section • H.E. stain • Low magnification

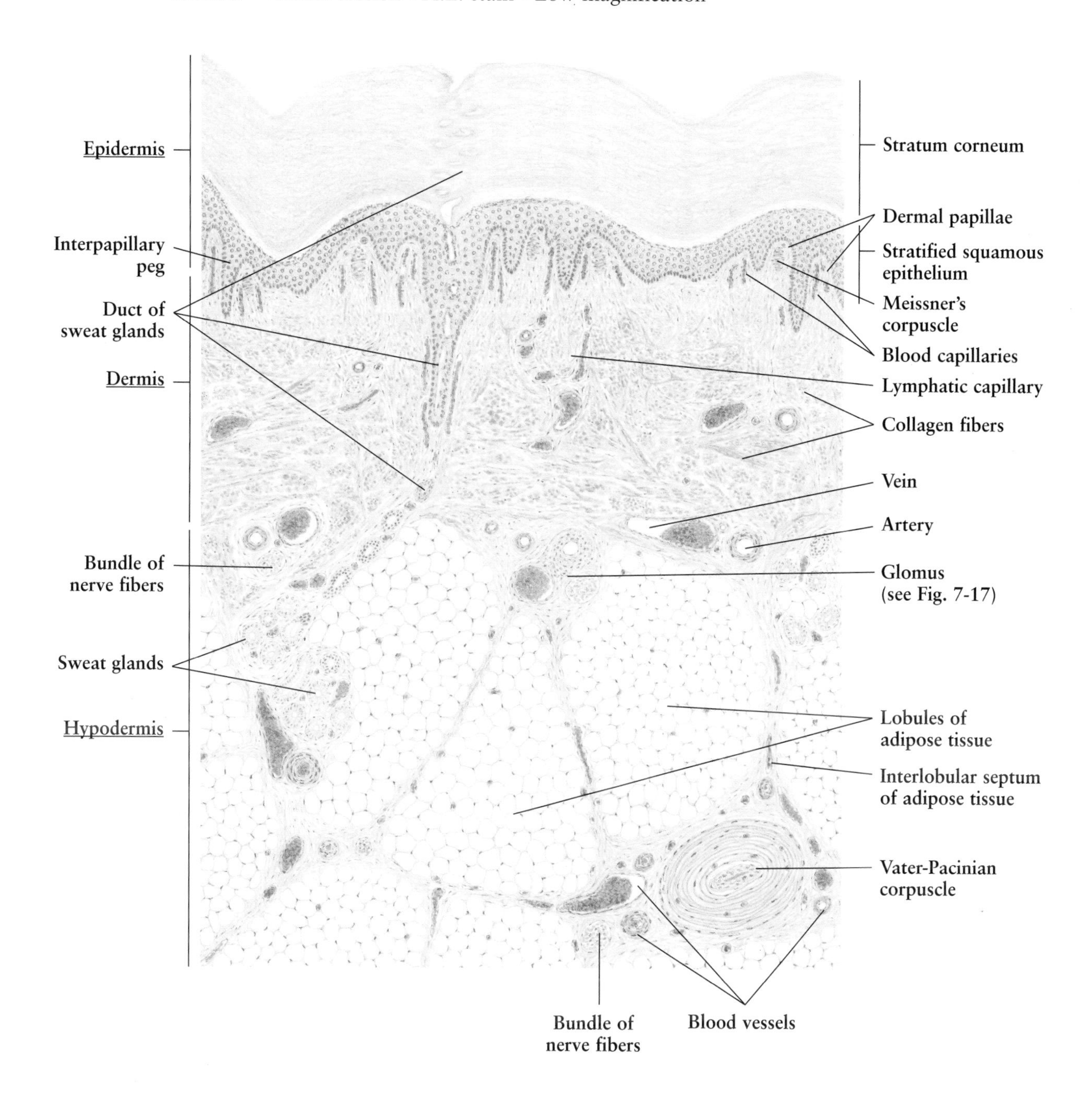

Fig. 15-2.

Thick Skin: Fingertip

The human **fingertip** is covered by typical thick skin, composed of *epidermis* and *dermis*. The **epidermis** consists of five layers: from deep to superficial, they are *stratum basale, stratum spinosum, stratum granulosum, stratum lucidum,* and *stratum corneum*. The **dermis** is divided into a *papillary layer* and a *reticular layer.*

Stratum basale, the deepest layer of the epidermis, is composed of a single layer of columnar or cuboidal cells that have ovoid nuclei and basophilic cytoplasm with tonofibrils. This layer is separated from the underlying connective tissue of the dermis by a **basement membrane.** Active cell mitosis in the stratum basale is responsible for the continual renewal of the epithelium.

Stratum spinosum, several cells thick, is formed of irregular, polyhedral cells with large, spherical nuclei and basophilic cytoplasm exhibiting more tonofibrils. The epithelial cells in this layer are characterized by the presence of "intercellular bridges"(a shrinkage artifact of fixation) formed by protoplasmic processes from adjacent cells. At the level of electron microscopy, the processes of the "intercellular bridges" are found to be connected by desmosomes. Usually, the stratum spinosum together with the stratum basale is called the *stratum germinativum* or *stratum Malpighii.*

Stratum granulosum consists of two to five layers of flattened cells. Their principal distinguishing feature is the presence within the cytoplasm of keratohyalin granules, which stain intensely with basic dyes. These granules contribute to the process of keratinization. In the stratum granulosum, epithelial cells begin to die, and the intercellular space is sealed by the lipid-rich secretory product of the cells, forming the epidermal permeability barrier.

Stratum lucidum is a hyaline, refractile band between the stratum granulosum and the stratum corneum. It consists of several layers of closely packed flattened cells that appear homogeneous. Nuclei are no longer recognizable, and the cytoplasm is filled with semifluid keratohyalin and tonofibrils arranged parallel to the surface of the skin. The stratum lucidum is only present in the thick skin of the palms and soles.

Stratum corneum, the outermost and thickest layer of the epidermis, is composed of layers of dead, clear, and scalelike keratinized cells. The cytoplasm is replaced with keratin, a birefringent filamentous scleroprotein derived from the tonofibrils. The nuclei and organelles have disappeared, and the cytoplasmic membranes become thicker. The keratin filaments are embedded in an opaque, structureless material derived from the keratohyalin granules. In contrast to the hard keratin of nails and hair cortex, the keratin of the stratum corneum is of the soft type, which contains less sulphur and is more elastic. The thickened cytoplasmic membranes and the lipid intercellular substance form a waterproof barrier. Toward the surface, the keratinized cells become more flattened, fused, and detached. In humans, the time required for a basal cell to develop into a mature keratinized cell is 30–90 days.

In addition to the keratinocytes, three other cell types are present in the epidermis: *melanocyte, Langerhans' cell,* and *Merkel cell.* The **melanocytes,** with pale cytoplasm and a round or ovoid nucleus, are seen scattered in the stratum basale. They produce *melanin* and transport it through their processes to the surrounding keratinocytes. Melanocytes themselves contain little melanin, and often appear clear under the light microscope. The **Langerhans' cells,** found in the stratum spinosum, have a kidney-shaped nucleus and a clear cytoplasm. They are macrophages, originating in bone marrow, and are able to detect foreign antigens. The Merkel cells are not seen in routine histological preparations, but studies with electron microscopy have shown that they lie on the stratum basale and are associated with sensory nerve endings. They may function as mechanoreceptors and form part of a diffuse neuroendocrine system.

Figure 15-2. Thick Skin: Fingertip

Human • Vertical section • H.E. stain • High magnification

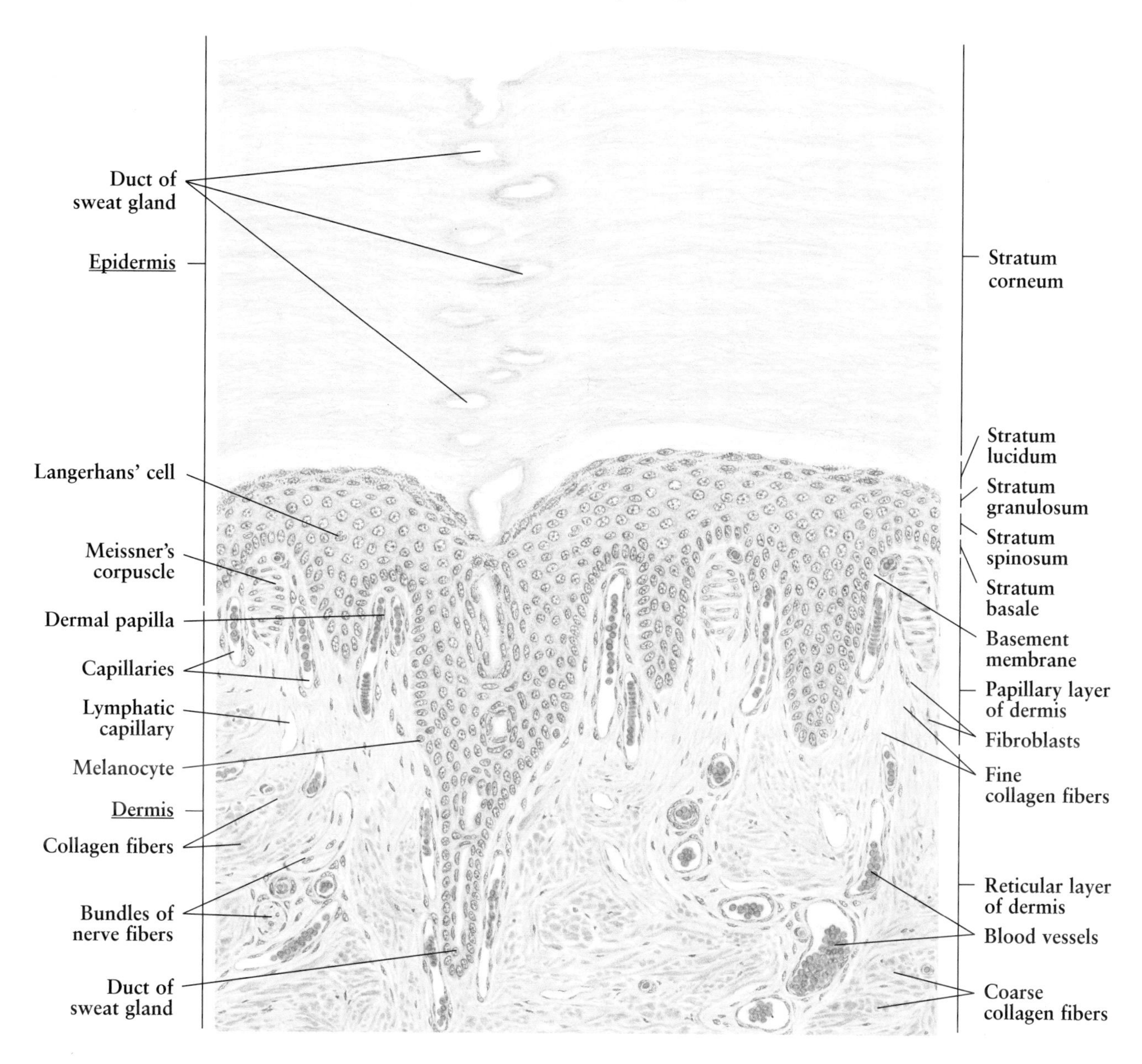

The **papillary layer** of the dermis sits immediately below the epidermis and projects into it as **dermal papillae,** which contain **blood capillaries** or **Meissner's corpuscles.** The papillary layer is formed of loose connective tissue, containing **fibroblasts, fine collagen fibers,** and blood and **lymphatic capillaries.**

The **reticular layer** is thicker, and composed of irregular dense connective tissue (see **Fig. 2-3**). It is characterized by bundles of **coarse collagen fibers** that form a network. Most of the collagen fibers are arranged parallel to the surface of the skin. Interspersed among these fibers are **blood vessels** and **nerve fibers.**

Note that in this figure a **duct of the sweat gland** passes through the dermis and the layers of the epidermis, opening on the surface of the skin. Reaching the epidermis, the duct loses its own wall, forming a zigzag channel, especially in the stratum corneum.

Fig. 15-3.

Thin Skin: Abdomen

This figure illustrates **thin skin** from the human abdomen at high magnification, showing the slender **epidermis** and part of the **dermis.**

The **stratum basale** is composed of a single layer of low columnar cells with **melanin granules,** resting on the **basement membrane.** Scattered among these cells are **melanocytes** with a light cytoplasm and a round nucleus. The **stratum spinosum** consists of a few layers of polyhedral cells, which attach to each other by their **intercellular bridges. Langerhans' cells,** which have a clear cytoplasm and a kidney-shaped nucleus, are found in this layer. The **stratum granulosum** contains only one or two layers of flattened cells with keratohyalin granules. The **stratum corneum** is very slender, and the stratum lucidum is absent.

The **dermis** immediately below the epidermis is a well-vascularized loose connective tissue. It contains a rich network of **capillaries,** fine **collagen fibers,** and many **fibroblasts.**

Fig. 15-4.

Scalp

Figure 15-4 demonstrates the structural features of the human **scalp.** The skin of the scalp is of the thin type, consisting of a thin **epidermis** and a thick **dermis** that contains a great number of **collagen fibers.** The **hypodermis,** usually called superficial fascia, is filled with **adipose tissue.** Beneath the adipose tissue is the **galea aponeurotica,** a flat tendon that attaches forward to the *frontalis* and backward to the *occipitalis.* The dermis and the galea aponeurotica are connected by a number of dense connective tissue septa, the **retinacula cutis,** which limit the movement between the skin and galea aponeurotica. In fact, sliding of the scalp over the cranium is possible because of the presence of a loose connective tissue, the *subaponeurotic fascia,* between the galea aponeurotica and the pericranium.

The scalp is characterized by numerous closely packed **hairs** and **hair follicles.** Hairs are highly modified elastic keratinized threads derived from invaginations of the epidermal epithelium. Each hair, which slopes at an obtuse angle to the skin surface, consists of a free **shaft** and a **root,** which is enclosed by a tubular follicle composed of an epithelial **external root sheath** and a **connective tissue sheath.** The lower end of the follicle expands into a **hair bulb,** which is indented at the basal end by a connective tissue **papilla.**

Associated with the hair follicle are one or more **sebaceous glands.** The secretions of these glands are delivered through a short **duct** to the space between the hair shaft and the invaginating epidermis. On the obtuse angle side between the hair root and the skin surface there is a bundle of smooth muscle fibers, known as **arrector pili,** attached at one end to the middle of the connective tissue sheath of the follicle, and at the other to the papillary layer of the dermis. Its contraction causes erection of the hair accompanied by the formation of "gooseflesh", and simultaneously aids in the expulsion of sebum.

In this figure, **eccrine sweat glands** are found in the deeper portion of the dermis. In addition, part of a **quiescent follicle** is shown, formed by matrix cells that have ceased to proliferate. A quiescent hair can be easily plucked from the follicle.

Figure 15-3. **Thin Skin: Abdomen**
Human • H.E. stain • High magnification

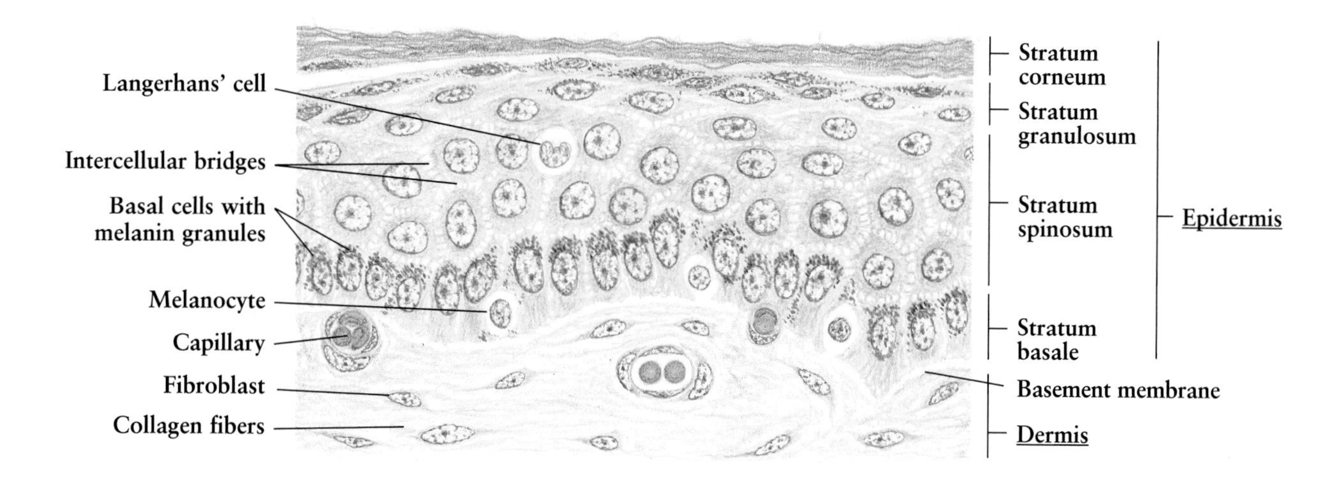

Figure 15-4. **Scalp**
Human • Longitudinal section • H.E. stain • Low magnification

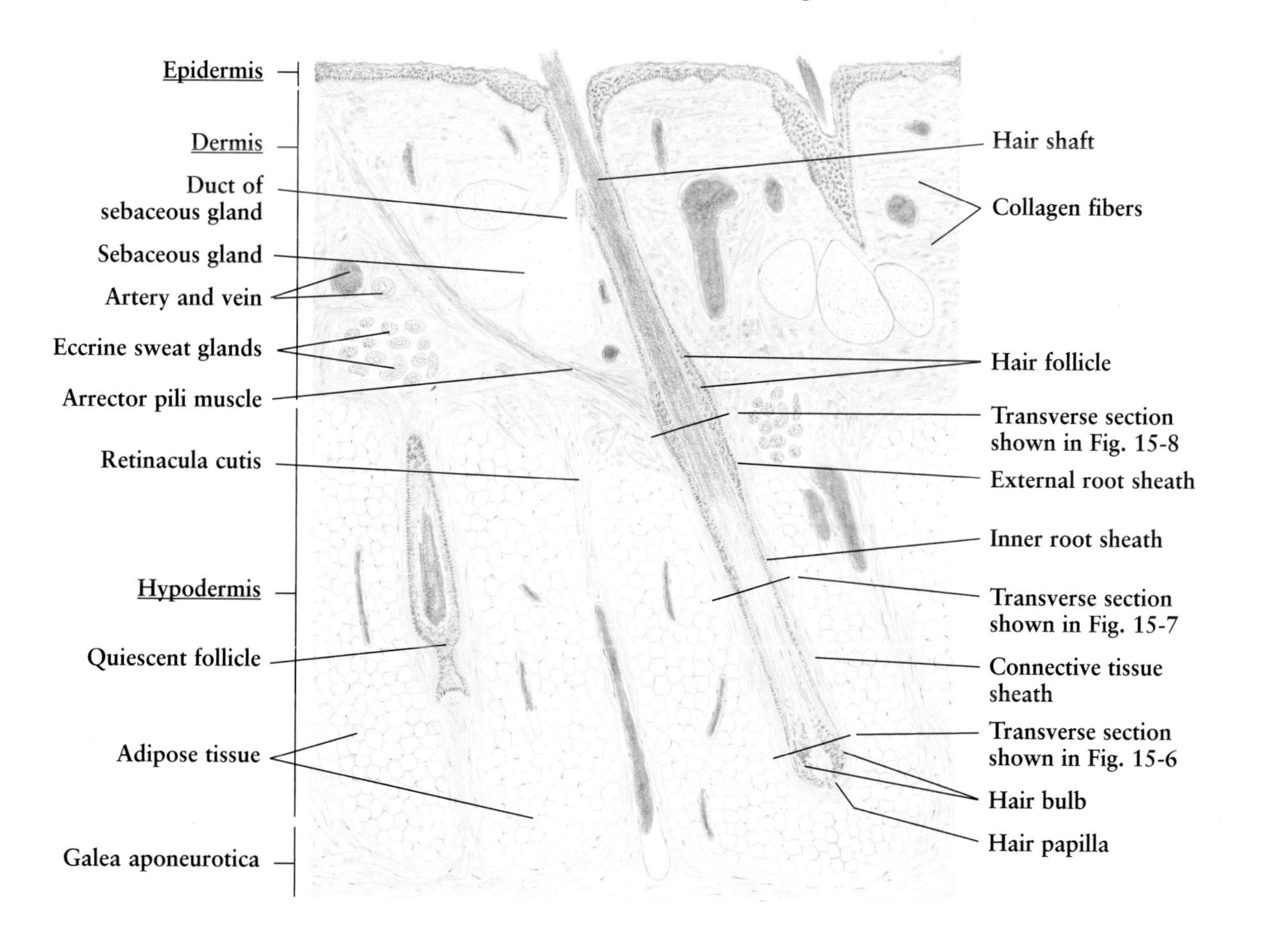

Fig. 15-5.

Hair and Hair Follicle

Figure 15-5 is a longitudinal section of a follicle, giving details of the structure of **hair** and **hair follicle.**

Hair. The hair is composed of keratinized epithelial cells arranged in three concentric layers: the *medulla,* the *cortex,* and the *cuticle.*

The **medulla** consists of two or three layers of keratinized cuboidal cells. These cells are characterized by a clear cytoplasm, a round nucleus, and cytoplasmic keratohyalin granules. The keratin of the medullary cells is of the "soft" type.

The **cortex,** which makes up most of a hair's thickness, is composed of several layers of long fusiform or flattened keratinized cells arranged longitudinally. Toward the upper portion of a hair, the nucleus shrinks and becomes smaller; the pigment granules are found in the cells, and the keratin is of the "hard" type.

The **cuticle of hair** is composed of a single layer of thin eosinophilic cells. These are keratinized cells that overlap each other. Their free edges project upward, resembling shingles on a roof.

Hair follicle. The hair follicle is a tubular sheath composed of an inner *epithelial root sheath,* continuous with the epidermis, and an outer *connective tissue sheath.*

The **epithelial root sheath** consists of two parts, *internal* and *external root sheaths.* The **internal root sheath** is composed of *cuticles, Huxley's layer,* and *Henle's layer.* The **cuticle** of the internal root sheath is a single layer of small clear cells with their free edges directed downward and interlocking with the upward free edges of the cells of the hair cuticle. Immediately external to the cuticle of the internal root sheath is **Huxley's layer.** It is composed of one or two layers of elongated cells that contain trichohyalin granules and tonofibrils. Huxley's layer is surrounded by **Henle's layer,** a single layer of flattened eosinophilic cells that contain soft keratin. The **external root sheath** consists of an inner stratum of polygonal cells that correspond to the cells of the stratum spinosum of the epidermis, and an single layer of columnar cells which rest on the glassy membrane and are continuous with the stratum basale of the epidermis.

The **connective tissue sheath** is made up of three layers. The **inner layer** is a narrow homogeneous band, the **glassy membrane,** continuous with the basement membrane of the epidermis. The **middle layer** is composed of connective tissue containing numerous circular **collagen fibers.** The **capillaries** in this layer are also circularly arranged. The **outer layer** is composed of coarse bundles of longitudinally aligned **collagen fibers** accompanied by longitudinal **capillaries.**

At the lower end, the root and follicle expand to form the **hair bulb.** It is invaginated at the base by a conical projection of the connective tissue, known as a **hair papilla.** The hair papilla is formed of connective tissue containing numerous **fibroblasts** and **collagen fibers.** It is also rich in **blood vessels,** which provide nourishment for the growing and differentiating cells of the hair bulb. The cells of the hair bulb are not arranged in layers but rather form a continuous matrix of growing cells, known as **matrix cells.** Matrix cells immediately above the apex of the papilla transform into the medulla; those on the slope and sides differentiate into cortex and cuticle of the hair, respectively; those lateral to the papilla develop into the inner root sheath and those at the bottom of the follicle form the outer root sheath. Scattered among matrix cells are **melanocytes.** They produce melanin and transfer their melanin granules to the cells of the hair matrix and cortex in a manner similar to that occurring in the epidermis.

Figure 15-5. Hair and Hair Follicle

Human • Longitudinal section • H.E. stain • High magnification

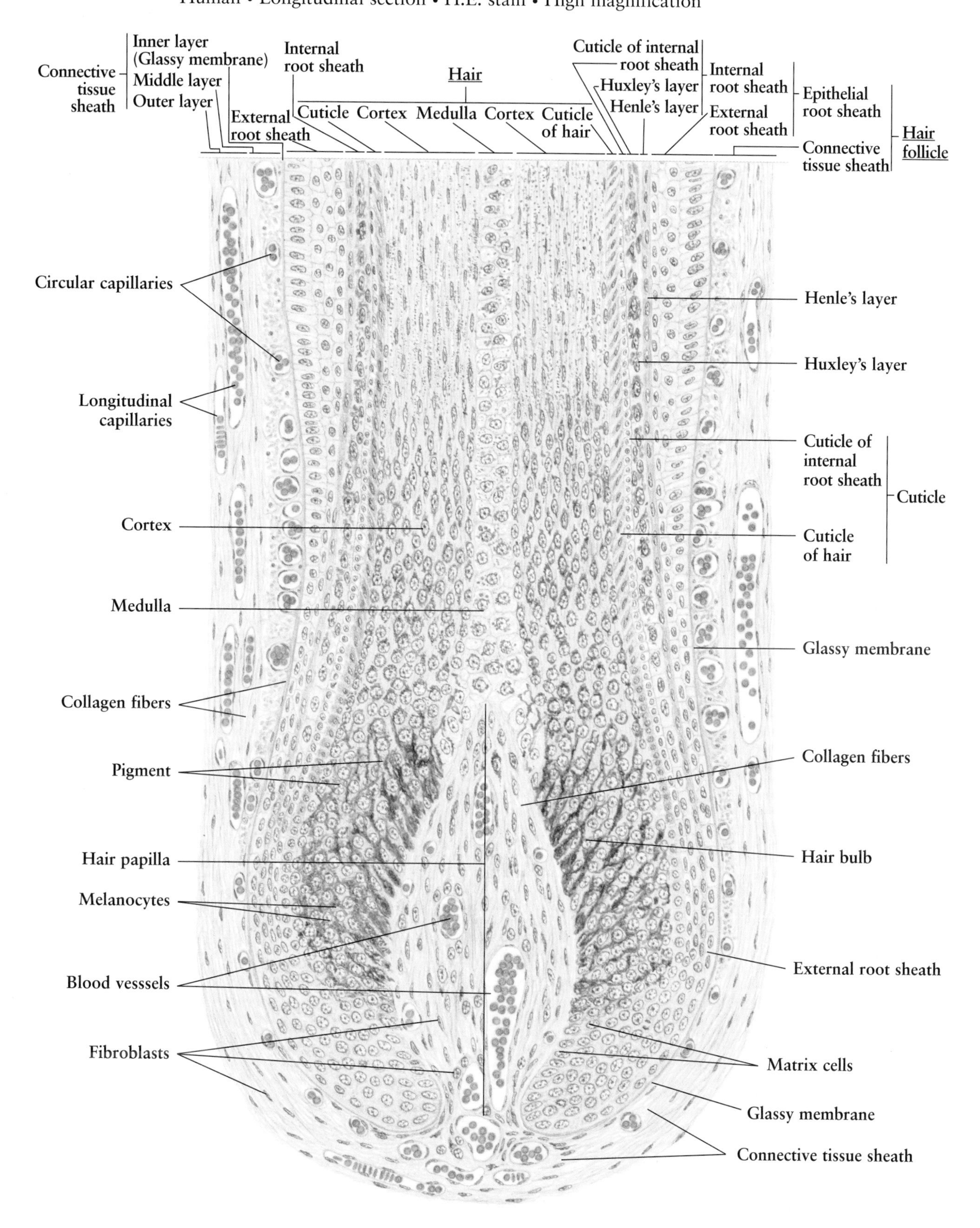

Fig. 15-6.

Hair and Hair Follicle

This figure shows a transverse section of a follicle at the level of the hair papilla illustrated in **Fig. 15-4.** The center of this illustration is a part of the **hair papilla** containing **fibroblasts** and **capillaries.** Surrounding the hair papilla is the **matrix.** Matrix cells in the inner layer are columnar and rest on a basement membrane. All these cells have round or ovoid nuclei and slightly basophilic cytoplasm with pigment. Immediately adjacent to the matrix is the **cuticle of the hair,** a single layer of flattened cells with darkly stained cytoplasm.

At this level, the **cuticle of the internal root sheath** and **Huxley's layer** are both single layers of flattened clear cells with small, round nuclei. **Henle's layer** is a single layer of flattened darkly stained cells. The **external root sheath** consists of two layers of clear cuboidal and flattened cells, which are supported by the **glassy membrane.** The **connective tissue sheath** contains **circular collagen fibers** in the inner layer and **longitudinal collagen fibers** in the outer layer. **Capillaries** are found in both inner and outer layers of the connective tissue sheath.

Fig. 15-7.

Hair and Hair Follicle

This figure demonstrates the structure of a transverse section of a follicle at the level of its lower one third, as shown in **Fig. 15-4.**

In the **medulla,** several large keratinized cells can be found. They are polygonal and clear, with round nuclei and keratohyalin granules in their cytoplasm. The cells in the **cortex** are keratinized, with shrunken nuclei. The cytoplasm, which becomes obscure, is filled with pigment. The cells of the **cuticle of the hair** become more flattened, and the nuclei become smaller and darker.

The **cuticle of the internal root sheath** is composed of a single layer of clear flattened cells. **Huxley's layer** has two rows of keratinized cells, surrounded by **Henle's layer,** which consists of a layer of keratinized, darkly stained flattened cells. The **external root sheath** increases in the number of cells and become three or four cells thick. The **glassy membrane** and the **connective tissue sheath** may also be discerned.

Figure 15-6. Hair and Hair Follicle
Human • Transverse section • H.E. stain • High magnification

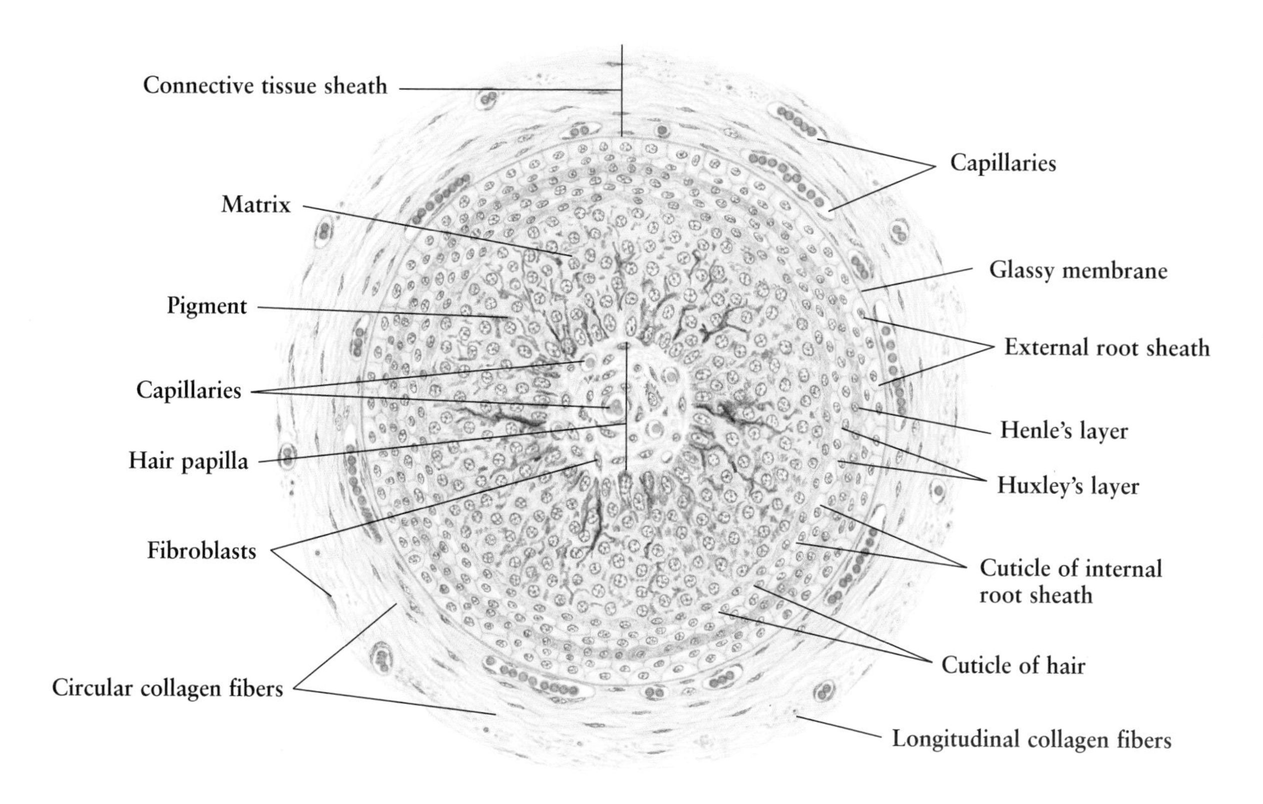

Figure 15-7. Hair and Hair Follicle
Human • Transverse section • H.E. stain • High magnification

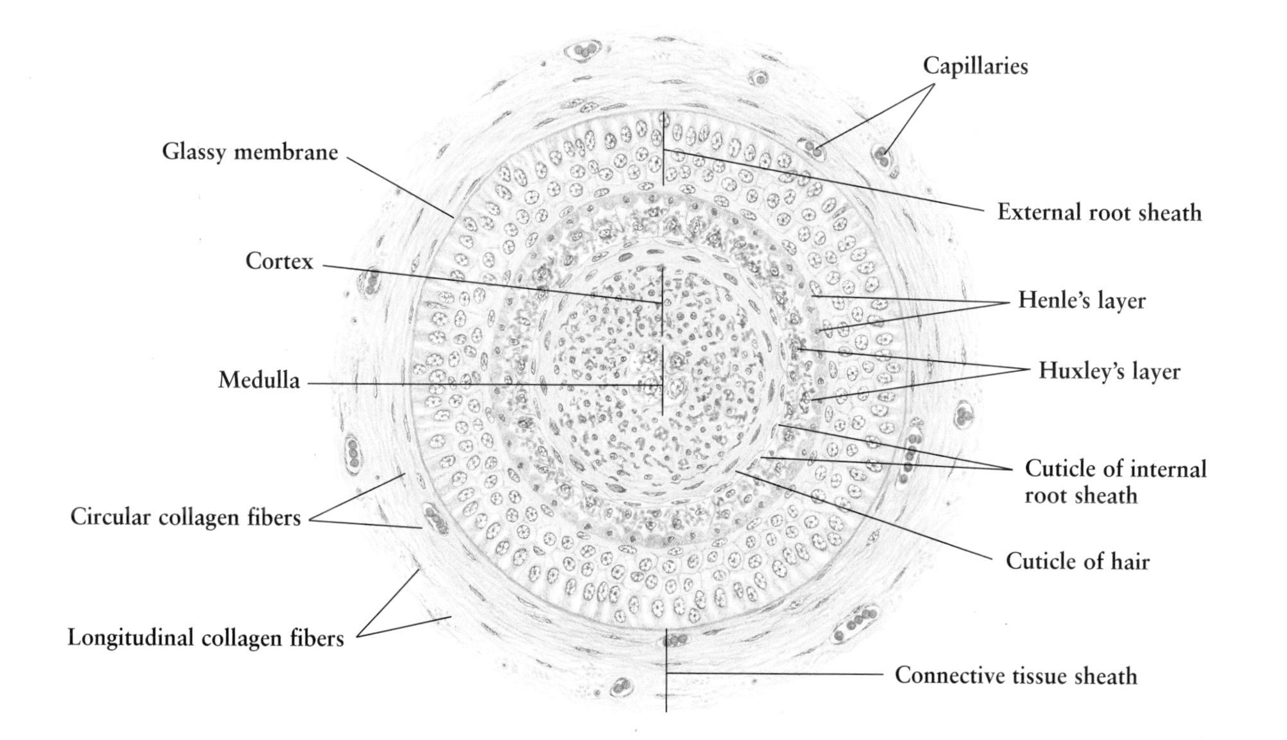

Fig. 15-8.

Hair and Hair Follicle

Figure 15-8 is a transverse section of a follicle at the level of the middle of the hair follicle, as shown in **Fig. 15-4.** At this level, the **cortex** constitutes the highly keratinized components of the hair, but the medulla has disappeared. The **cuticle of the hair** covering the cortex becomes a very thin layer.

The three layers of the **internal root sheath,** that is, the **cuticle of the internal root sheath, Huxley's layer,** and **Henle's layer,** fuse to form a thick keratinized layer. The cuticle of the internal root sheath can be recognized as a very thin, darkly stained layer. A **space** occurs between the cuticle of the hair and the cuticle of the internal root sheath. The **external root sheath** consists of four or five layers of clear cells; its outermost layer of cells rests on the **glassy membrane.**

Similar to that in other regions of the hair follicle, the **connective tissue sheath** at this level also contains circularly and longitudinally arranged **collagen fibers, fibroblasts,** and **capillaries.**

Fig. 15-9.

Sebaceous Gland

The **sebaceous glands** (also see **Figs. 6-28, 6-29,** and **15-4**) are distributed throughout the skin, except in the palms and the soles. Generally their ducts are associated with hair follicles, but those of the eyelid tarsal glands, lip, glans penis, inner surface of the prepuce, and labia minora (see **Fig. 13-21**) open directly onto the skin surface.

Sebaceous glands are located in the dermis, surrounded by a thin layer of **connective tissue.** These are saccular glands, with a short **duct** that opens into the **space,** which lies between the **external root sheath of the hair follicle** and the **cortex of the hair.** The secretory acini are supported by a delicate **basement membrane.** The **basal cells,** which rest on the basement membrane, are small cuboidal cells and form a single layer continuous with the basal cells of the dermis. Toward the center of the acinus, they become larger **secretory cells,** which possess clear foamlike cytoplasm filled with lipid droplets, and the nuclei gradually shrink and then disappear. Near the duct, secretory cells degenerate and break down into a fatty mass and cellular debris; thus, the entire cell becomes secretion, *sebum.* This process is known as *holocrine secretion.* The sebum is an oily secretion, containing lipids, triglycerides, squalene, wax esters, and sterol. It lubricates skin and hairs. Cells lost in the secretory process are replaced by division of the basal cells of the acini. Discharge of secretion is aided by contraction of the **arrector pili muscle.** It is attached at one end to the middle of the connective tissue sheath of the follicle and at the other end to the papillary layers of the dermis (see **Fig. 15-4**).

The top-right portion of this figure shows a part of the **hair follicle** and **cortex of the hair.** Surrounding the sebaceous glands is **connective tissue** containing **blood** and **lymphatic vessels** and **nerve fibers.**

Figure 15-8. Hair and Hair Follicle
Human • Transverse section • H.E. stain • High magnification

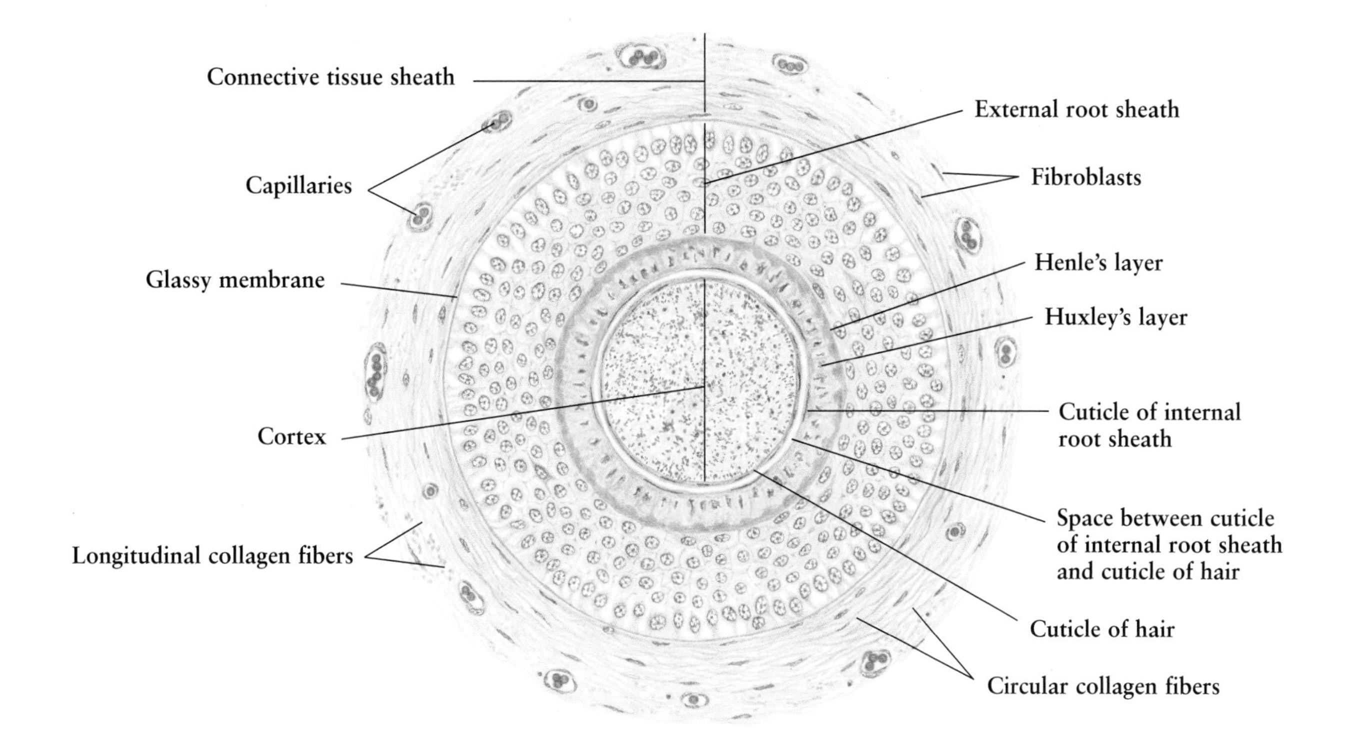

Figure 15-9. Sebaceous Gland
Human • H.E. stain • High magnification

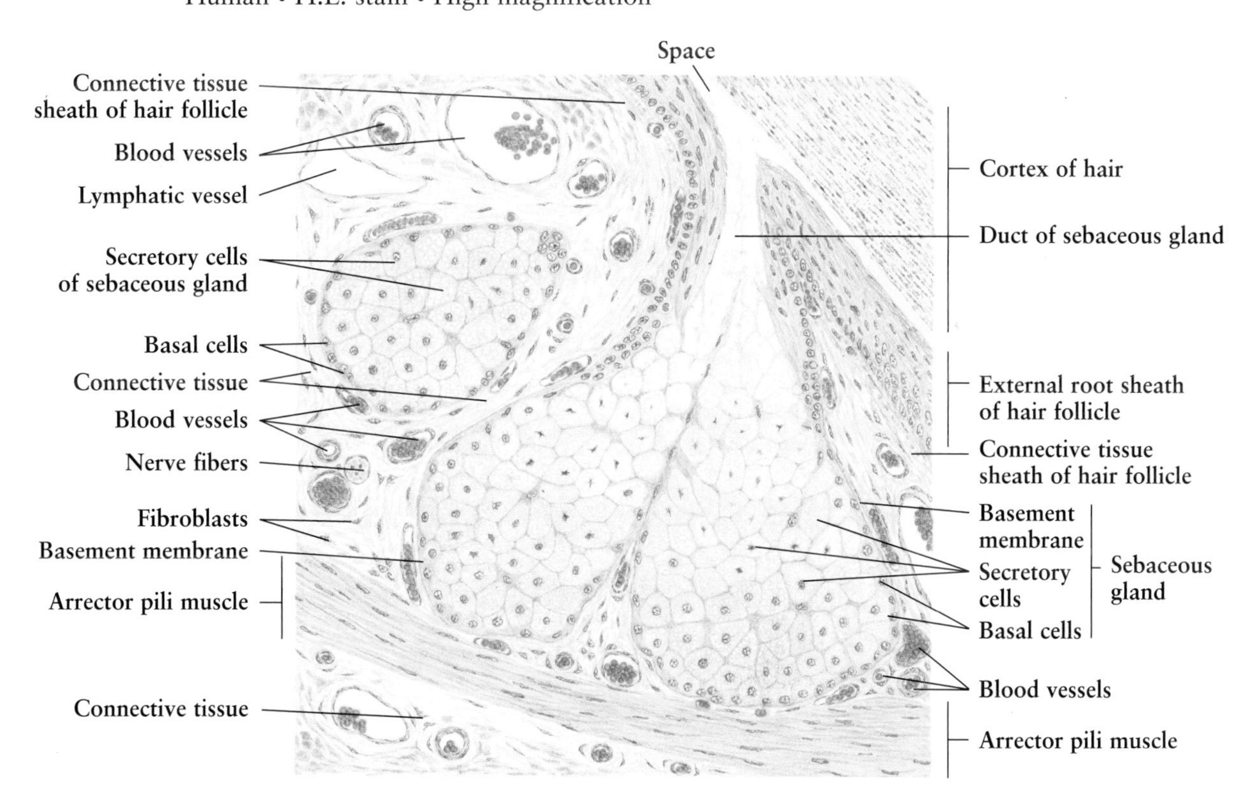

Fig. 15-10.

Eccrine Sweat Gland

The **eccrine sweat glands** are simple tubular glands. They occur in most areas of the skin, but are not found in the lip margins, nail bed, glans penis, inner surface of the prepuce, clitoris, and labia minora. Each gland consists of a *secretory portion* and an *excretory duct.*

The **secretory portion** is coiled into a glomerulus, located deep in the dermis and hypodermis. It is composed of two cell types, *clear cells* and *dark cells.* The **clear cells,** sitting on a **basement membrane,** are cuboidal or pyramidal, with clear cytoplasm and round nuclei. Between them are **intercellular canaliculi.** The clear cells are believed to secrete a watery product containing sodium chloride, potassium chloride, urea, uric acid, and ammonia. The **dark cells,** with dark-stained cytoplasm, are broad and cover the clear cells at the luminal surface. These cells produce a small amount of glycoprotein. Interposed between clear cells and the basement membrane are **myoepithelial cells.** They form a discontinuous layer, aiding in the discharge of secretion. Surrounding the secretory portion and excretory ducts are a rich network of **capillaries.**

The **excretory duct** is lined by two layers of cuboidal cells supported by a **basement membrane.** Myoepithelial cells are absent. The excretory duct passes through the dermis to reach the epidermis, opening on the skin surface. As it transverses through epidermis, the secretory duct loses its own wall, becoming a convoluted and zigzag channel (see **Figs. 15-1** and **15-2**).

Fig. 15-11.

Apocrine Sweat Gland

The **apocrine sweat glands,** also known as large sweat glands, are coiled, tubular glands, mainly distributed in the axilla, areola of the nipple, labia majora, and circumanal region. They often open into the upper portions of hair follicles through their excretory ducts. The apocrine sweat glands do not function until puberty, and produce a thicker secretion that becomes odoriferous after decomposition by skin bacteria.

The **lumen** of the **secretory portion** of the apocrine sweat gland is much wider than that of the eccrine sweat gland. The **epithelial cells** are columnar or cuboidal, depending on the functional situation. The nuclei are round or ovoid, and the cytoplasm is eosinophilic and often contains secretory product at the **cellular apices.** The **myoepithelial cells** are larger and form a more complete layer between the epithelial cells and the **basement membrane.** The **excretory duct,** which empties into a hair follicle, is lined by two layers of epithelial cells. Surrounding the secretory portions and the excretory ducts is loose connective tissue in which **blood vessels, fibroblasts, nerve fibers,** and **adipose cells** can be recognized. The secretion of the apocrine sweat glands, like that of eccrine sweat glands, is merocrine in type, involving no loss of cellular structure. However, traditionally they are called apocrine glands.

The ceruminous (wax) glands of the external auditory meatus (see **Fig. 17-2**) and the glands of Moll in the margins of the eyelid (see **Fig. 16-8**) also belong to this group of apocrine sweat glands.

Figure 15-10. **Eccrine Sweat Gland**
Human • H.E. stain • High magnification

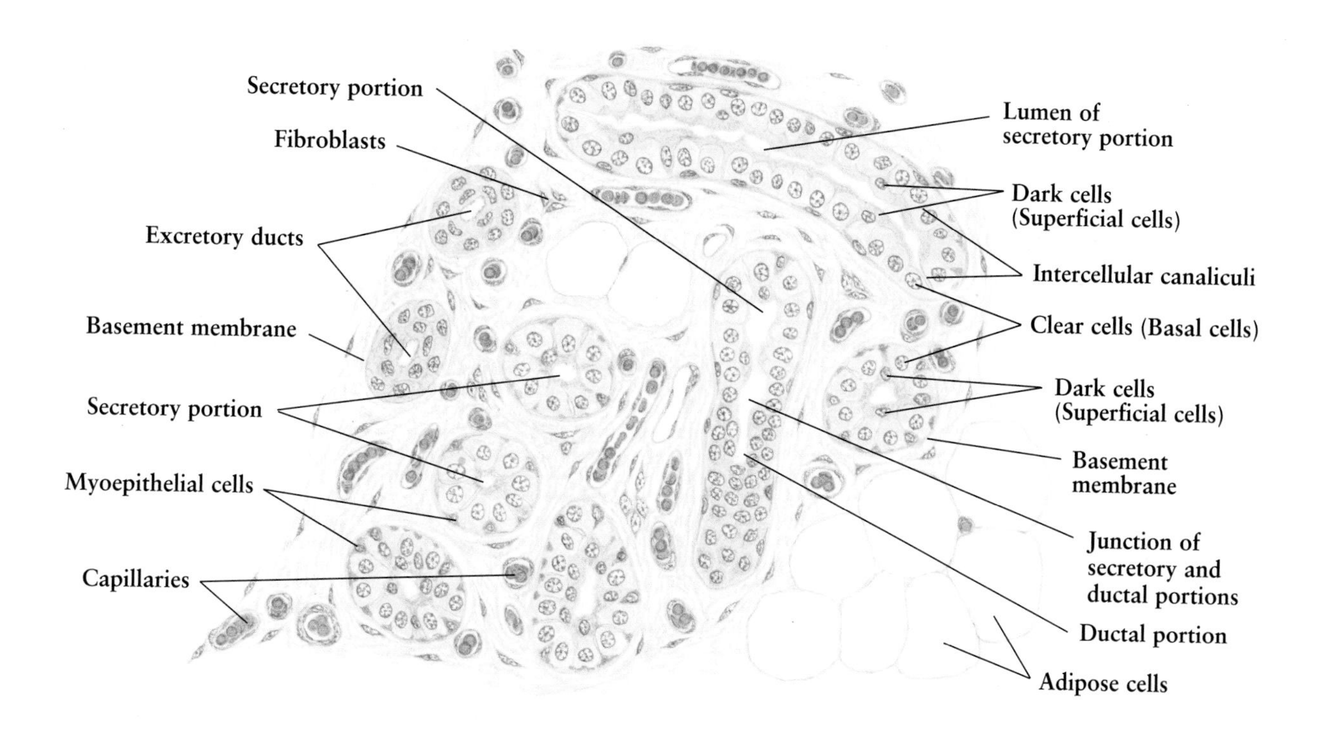

Figure 15-11. **Apocrine Sweat Gland**
Human • H.E. stain • High magnification

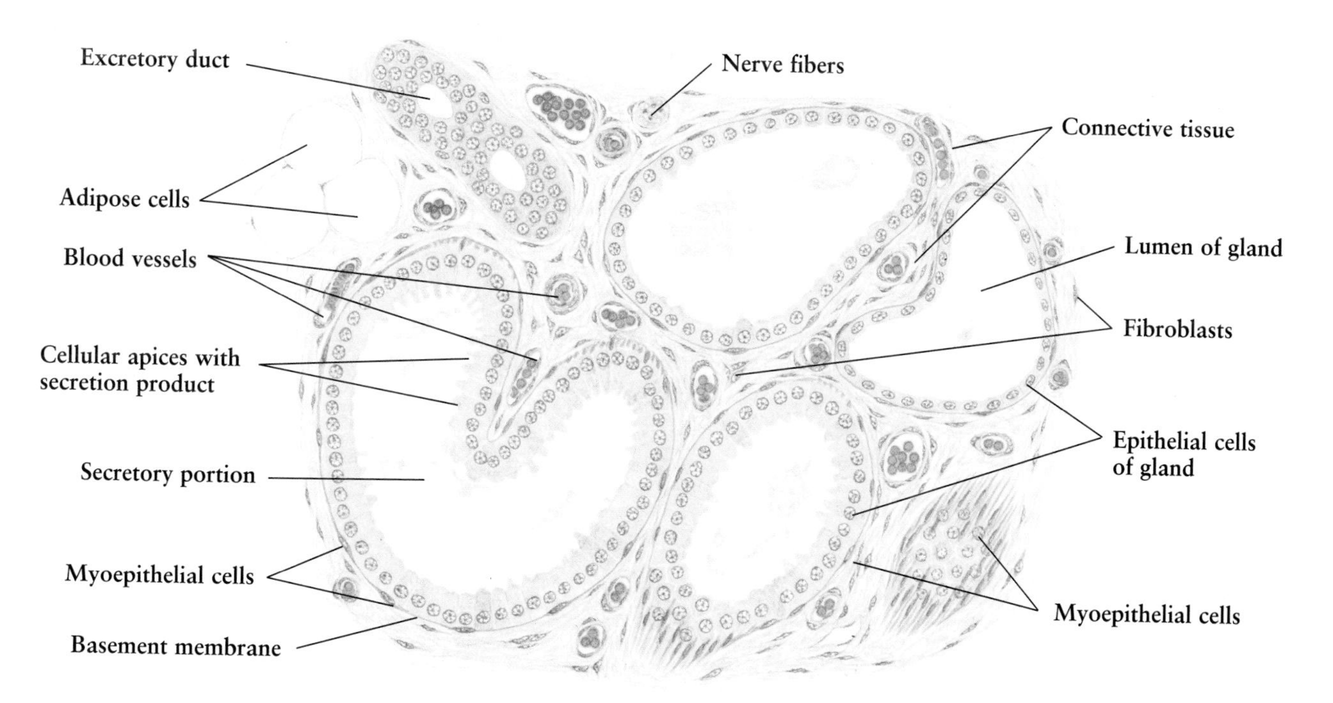

Fig. 15-12. — **Nail**

The **nail** is a highly keratinized, hard structure that forms a protective plate covering the dorsal surface of the tips of fingers and toes. Each nail consists of a **nail plate** that constitutes the bulk of the nail, and a proximal part, the **nail root,** which is embedded in a groove formed by the **nail fold** of the skin.

The nail rests on a stratified squamous epithelium known as the **nail bed.** The nail bed becomes thicker under the root, forming the **ventral matrix.** The nail matrix also covers the nail root on the back, and constitutes the **dorsal matrix.** The keratinized free edge of the nail fold is called the **eponychium,** and the keratinized epidermis attached to the underside of the nail at its free end is known as the **hyponychium.**

The **dermis** of the nail bed is connected to the **periosteum** of the **distal phalanx** by the **reticular cutis,** made up of firm collagen and elastic fibers. The dermis forms numerous longitudinal papillae (not shown in this figure).

Fig. 15-13. — **Nail: Root**

The **nail plate** is composed of many layers of flattened, fully keratinized cells that contain shrunken, **degenerated nuclei** and are filled with hard keratin. These keratinized cells are firmly held together and do not undergo desquamation. The nail is clear and translucent; at its root is an opaque crescentic area, the lunula (not shown in this figure).

The **nail bed** is formed of many layers of epidermal cells beneath the nail plate. These cells correspond to the **stratum germinativum** of the **epidermis,** but they do not participate in the formation of the nail plate.

The **nail matrix** is similar to and continuous with the nail bed. It is however thicker and possesses **dermal papillae.** The basal cells are columnar and frequently undergo mitosis to replace cells lost in the process of nail formation. The cells in the upper layers of the matrix become keratinized and add to the nail root, resulting in a slow movement of the nail plate over the nail bed. Both the **ventral and dorsal matrix** have the same function.

Note that the nail bed and the nail matrix show no pigmentation. A network of **blood vessels** in the **dermis** under the nail bed and nail matrix gives nails their pink appearance.

Figure 15-12. Nail
Human finger • Longitudinal section • H.E. stain • Very low magnification

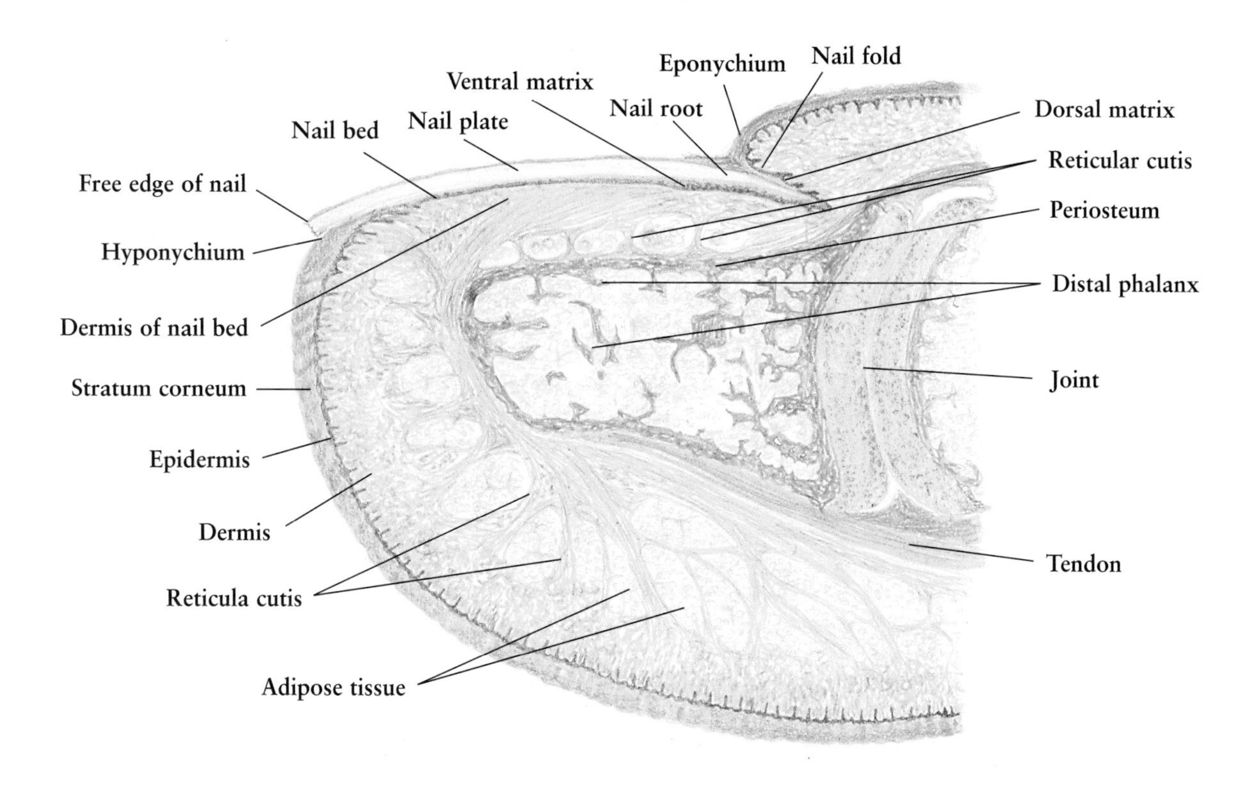

Figure 15-13. Nail: Root
Human finger • Longitudinal section • H.E. stain • Low magnification

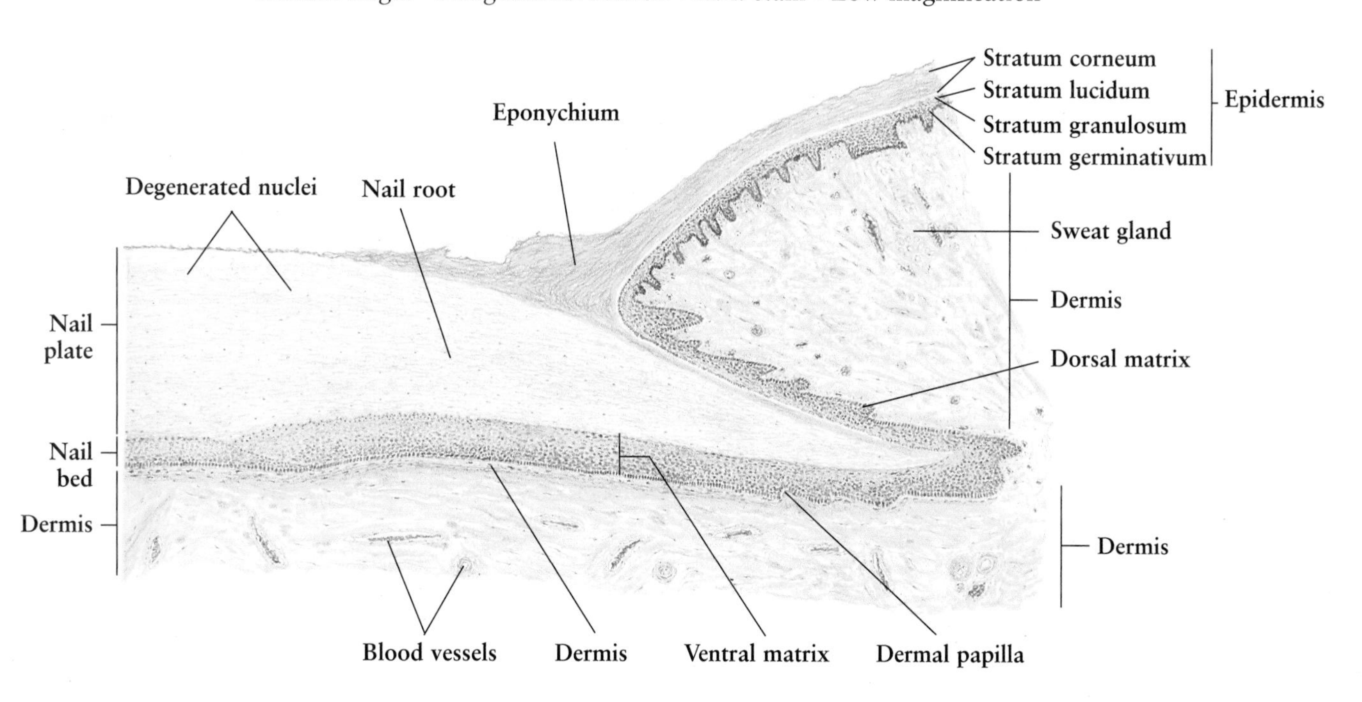

16. The Eye

Eyes are highly specialized organs of vision, which transmit and focus light on the retina and convey the image formed on the retina to the visual area of the cerebral cortex. Each eye is located within the bony orbit of the skull, cushioned from behind with adipose tissue and protected anteriorly by the eyelids and eyelashes. The eye consists of the *eyeball* and *accessory organs.*

The **eyeball** has a hollow globular structure with a thick, fairly elastic wall. The content of the eyeball is formed of an **anterior chamber,** a small **posterior chamber,** a **lens,** and a large **vitreous cavity** occupied by a globular transparent gelatinous mass known as the **vitreous body.** The wall of the eyeball is composed of three basic layers: the outer *corneoscleral layer,* the intermediate *uveal layer,* and the inner *retinal layer.*

The **corneoscleral layer** consists of cornea and sclera. The **cornea** is located in the anterior one-sixth of the wall. It is a transparent structure and acts as one of the principal refracting media. The **sclera,** composed of dense connective tissue, forms the posterior five-sixths of the wall, supporting and protecting the eyeball.

The **uveal layer** is a highly vascularized layer that forms three components: choroid, ciliary body, and iris. The **choroid** is a loose connective tissue characterized by a rich supply of blood vessels and pigmentation that absorbs the light reaching the choroid. The **ciliary body** contains smooth muscle fibers surrounding the lens and controls the focusing of the lens. The **iris** is located between the anterior and posterior chambers. The central **pupil** of the iris regulates the amount of light entering the eyeball.

The **retinal layer** contains two types of photoreceptors: rod cells and cone cells. The **rod cells** are responsible for the perception of light at dusk and for black-and-white vision. The **cone cells** are responsible for color perception and visual acuity. The nerve fibers in the retina merge into the **optic nerve,** projecting to the visual area of the cerebral cortex.

The accessory organs, mainly the **eyelid** and **lacrimal gland,** are also shown in this chapter.

Fig. 16-1. Eyeball

Figure 16-1 is a low-magnification illustration of an eyeball in meridional section. It demonstrates the structures of the content and wall of the eyeball. The wall of the **eyeball** is composed of three basic layers: the outer *corneoscleral layer,* the intermediate *uveal layer,* and the inner *retinal layer.*

The **corneoscleral layer,** corresponding to **dura mater,** forms a tough, fibrous coat that supports the eye. The posterior five-sixths of the eyeball, the **sclera,** is opaque; the anterior one-sixth, the **cornea,** is clear and transparent. The anterior corneal epithelium is continuous with the **conjunctiva.** The cornea has a smaller radius of curvature than the sclera, and is one of the principal refracting media. The corneoscleral junction is referred as to the **limbus,** where the circular **canal of Schlemm** is located (see **Figs. 16-4** and **16-5**).

The **uveal layer** is a highly vascularized structure that consists of three components: the *choroid,* the *ciliary body,* and the *iris.* The **choroid,** corresponding to the pia mater of the brain, lies between the sclera and the retina in the posterior five-sixths of the eye. The pigmentation of the choroid is responsible for the absorption of light that has passed through the retina. It is fundamentally a layer of loose connective tissue rich in blood vessels and melanocytes. The **ciliary body** is formed surrounding the coronal equator of the lens and contains smooth muscle, the **ciliary muscle.** The **lens** is a biconvex transparent structure and attached to the ciliary body by the **ciliary zonule.** The tone of the ciliary muscle controls the shape of the lens. The **iris** is an anterior extension of the ciliary body, and acts as a highly pigmented diaphragm that regulates the amount of light falling on the retina. The central aperture of the iris is known as the **pupil.** The iris divides the anterior compartment of the eyeball into the **anterior chamber** and the **posterior chamber.** These two chambers are filled with a watery aqueous humor, which is secreted into the posterior chamber by the ciliary body and circulated through the pupil to the anterior chamber, where it is absorbed by the canal of Schlemm.

The **retinal layer** is the innermost layer of the eyeball. It is divided by a scalloped line, the **ora serrata,** into an anterior portion and a posterior portion. The anterior portion is composed of a nonphotosensitive epithelial layer covering the ciliary body and the posterior surface of the iris. The posterior portion is a layer of photosensitive epithelium, the **retina,** which lines the rest of the eyeball. The **fovea centralis** is the deepest central zone of the **macula lutea** in the posterior pole of the eyeball. It is the point through which the visual axis of the eye passes and thus is the area of greatest visual acuity. The afferent nerves from the retina converge in the **optic papilla** at the nasal side to form the **optic nerve,** which leaves the eye through the **lamina cribrosa,** a part of the sclera.

The large posterior compartment behind the lens and ciliary body is known as the **vitreous cavity.** It is filled with a viscous, transparent gel called the **vitreous body,** which provides an optical medium.

Figure 16-1. **Eyeball**
Human • Right eyeball • Meridional section • H.E. stain • Low magnification

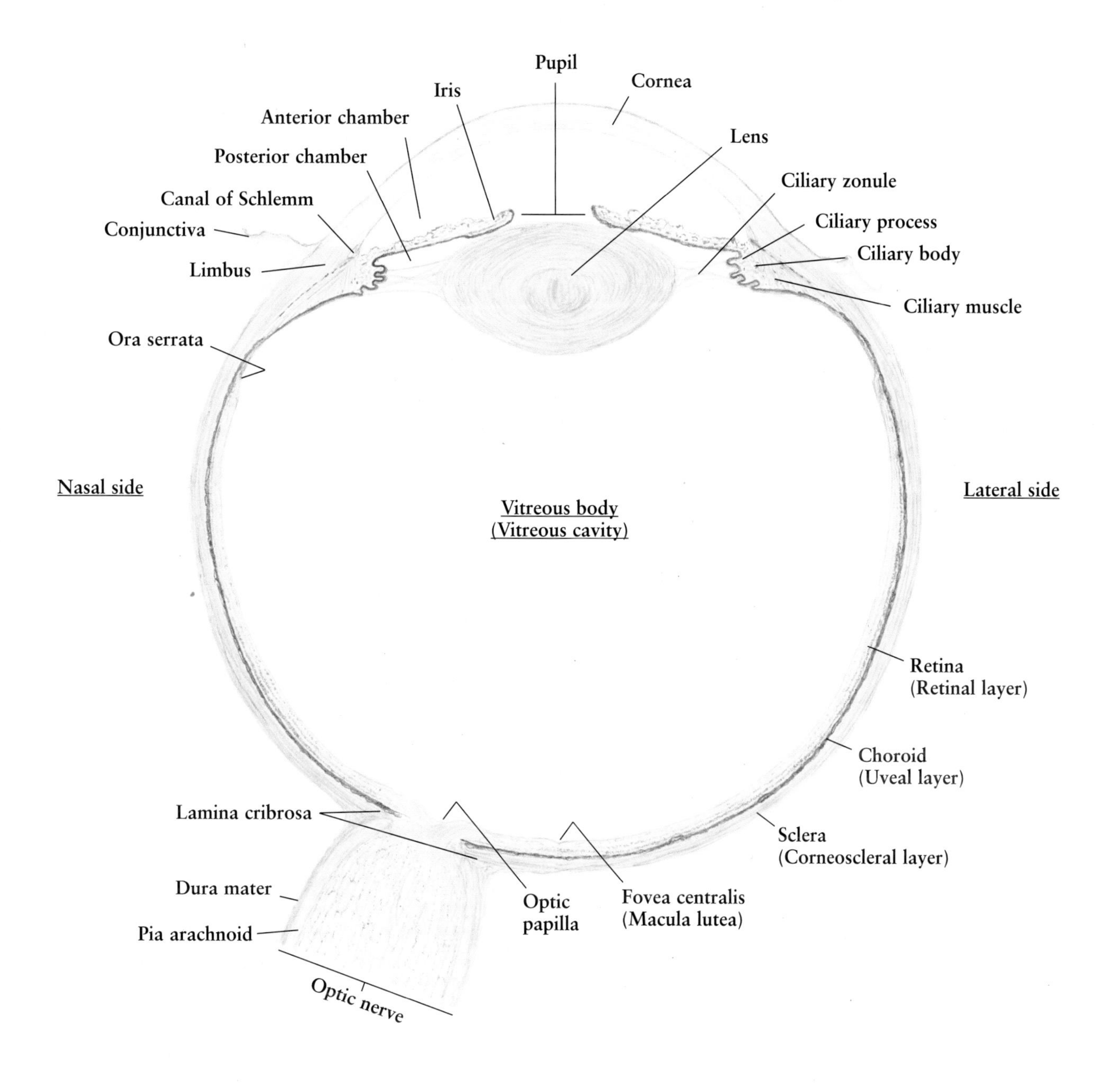

Fig. 16-2. Cornea

The **cornea** is a transparent, avascular structure with a smooth surface. It consists of five layers: *corneal epithelium, anterior limiting membrane, substantia propria, posterior limiting membrane,* and *corneal endothelium.*

The **corneal epithelium** is a nonkeratinized stratified squamous epithelium, with four to six layers of cells. The basal layer is a row of columnar cells. Overlying the basal layer are two or three layers of polyhedral cells. The superficial cells are flattened. The epithelium is rich in sensitive nerve endings and has strong regenerative powers.

The **anterior limiting membrane,** also known as **Bowman's membrane,** is a thick, specialized basement membrane that ends abruptly at the limbus.

The **substantia propria,** which forms the bulk of the cornea, is composed of a highly regular form of dense collagenous connective tissue. The collagen fibers are arranged in thin parallel **lamellae** with different angles. The **fibroblasts,** also called **keratocytes,** are stellate and extremely flattened, scattered in the ground substance between lamellae.

The **posterior limiting membrane,** or **Descemet's membrane,** is a thin basement membrane that supports the corneal endothelium.

The **corneal endothelium** is composed of a single layer of flattened or cuboidal cells that is continuous with the endothelium of the trabecular meshwork at the iridocorneal angle.

Fig. 16-3. Choroid and Sclera

The **choroid,** located between the **retina** internally and the **sclera** externally, is a layer of highly vascular loose connective tissue with dense pigmentation. The **choroidocapillary layer** (see **Fig. 16-6**), a layer of capillaries, is the innermost layer of the choroid. It is separated from the **pigmented epithelium** of the **retina** by a thin, clear glassy membrane (see **Fig. 16-6**). The main function of the choroidocapillary layer is to provide nutrient to the outer layer of the retina. The choroid propria contains many **arterioles** and **venules,** branches of the ciliary arteries and veins. The choroid is characterized by the presence of numerous **melanocytes.** These cells synthesize the pigment melanin, responsible for the absorption of light rays passing through the retina.

The **sclera** forms the tough, opaque fibrous wall of the posterior five-sixths of the eyeball. It is composed of dense fibroelastic connective tissue, whose **collagen fibers** are arranged in bundles parallel to the surface but oriented in various directions. The sclera contains **blood vessels, nerve fibers,** and a few elongated, flattened **fibroblasts** between the collagen bundles. Some **melanocytes** are also found scattered in the deep portion. The sclera maintains the shape and size of the eye, protects its interior, and provides the site of attachment for the extrinsic muscles of the eye.

Figure 16-2. Cornea
Human • H.E. stain • High magnification

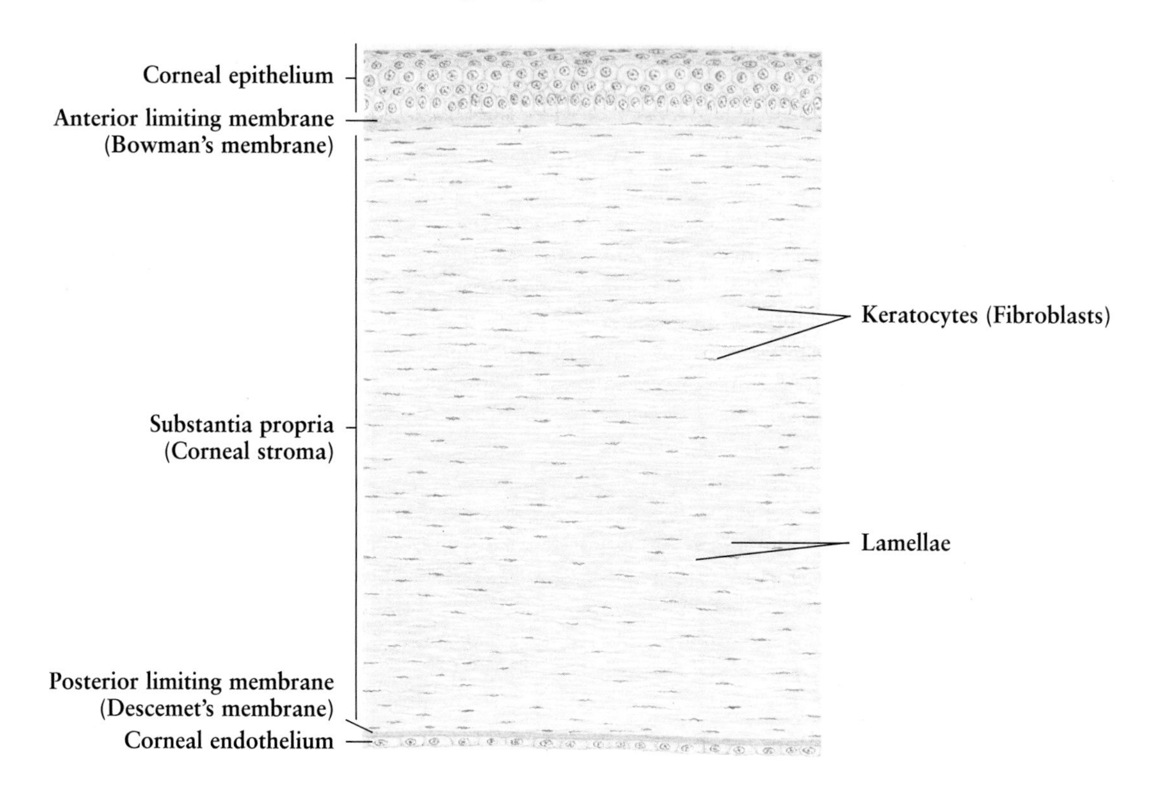

Figure 16-3. Choroid and Sclera
Human • H.E. stain • Low magnification

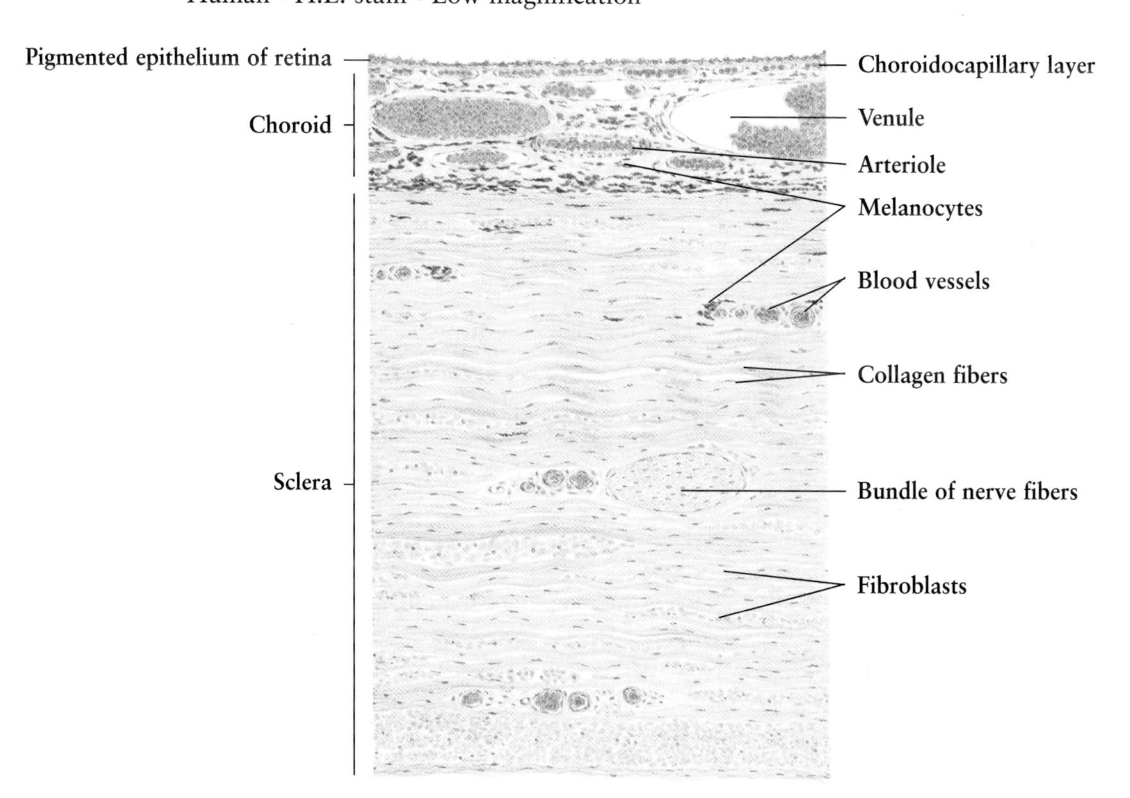

Fig. 16-4.

Eyeball: Internal Structures

This figure demonstrates the internal structure of the eyeball, including the *iris, ciliary body,* and *lens.*

The **iris** is a pigmented disc-like structure with a central circular aperture called the **pupil.** The iris attaches at the periphery to the ciliary body, dividing the anterior compartment of the eye into **anterior** and **posterior chambers.** The iris consists of highly vascular, loose connective tissue containing numerous **melanocytes,** fibroblasts, and **blood vessels** in its **stroma.** The **anterior border layer** of the iris is irregular. It lacks a covering epithelium, which is present in the fetus but disappears during early childhood. The posterior surface of the iris is smooth and lined by the **posterior epithelium,** composed of double layers of pigmented cells. The **anterior layer** of the posterior epithelium is made up of myoepithelial cells that form the radially arranged **dilator pupillae,** and is innervated by the sympathetic nervous system. The **posterior layer** of the posterior epithelium is a layer of heavily pigmented epithelial cells. The **sphincter pupillae** is a circumferentially arranged band of smooth muscle, located anterior to the dilator pupillae, and surrounding the pupil. It is innervated by the parasympathetic nervous system.

The **ciliary body** is a circular structure between the ora serrata and the outer edge of the iris. In meridional section, it appears triangular with its apex attached to the **scleral spur.** The **ciliary muscle,** which forms the bulk of the ciliary body, is arranged in three layers: **meridional, radial,** and **circular.** The meridional and radial fibers have a common origin from the **scleral spur** and **trabecular meshwork** (see **Fig. 16-5**). The ciliary muscle, innervated by the parasympathetic nerve fibers, controls the shape of the lens and therefore its focal power. The ciliary body is covered by a double layer of cuboidal epithelium. The outer layer, resting on a **basement membrane,** is highly **pigmented** and continuous with the pigmented epithelium of the retina; the inner layer is **nonpigmented** and is a nonphotosensitive layer of the anterior portion of the retina. **Melanocytes** are scattered throughout the ciliary body. In the front part of the ciliary body, its inner surface is formed by a number of radially arranged, branching epithelial folds called **ciliary processes,** which contain a highly vascularized connective tissue core. The ciliary processes function to produce a continuous supply of aqueous humor. Aqueous humor, which flows from the posterior chamber into the anterior chamber via the pupil, provides nutrients for the lens.

The **lens** is an elastic, transparent biconvex structure located immediately behind the pupil, between the iris and vitreous body. It is suspended in place by numerous **ciliary zonules,** which attach to the ciliary body. The lens consists of a *lens capsule, lens epithelium,* and *lens substance.* The **lens capsule** envelops the entire lens. It is structurally an elastic basement membrane that supports the lens epithelium. The **lens epithelium** is a single layer of cuboidal cells, located immediately beneath the capsule. It only covers the anterior surface of the lens. The posterior surface lacks covering epithelium. Toward the equator of the lens, the lens epithelial cells increase in height and transform into **lens fibers,** which form the bulk of the **lens substance.** The lens fibers have the shape of extremely elongated, hexagonal prisms, most of them concentrically arranged and parallel to the lens surface. The lens grows throughout life by addition of new fibers to the periphery of the lens substance.

Attached to the lens capsule are the ciliary zonules. These are small bundles of fine filaments, which terminate at the **basement membrane** of the ciliary body.

Figure 16-4. **Eyeball: Internal Structures**

Human • Meridional section • H.E. stain • Medium magnification

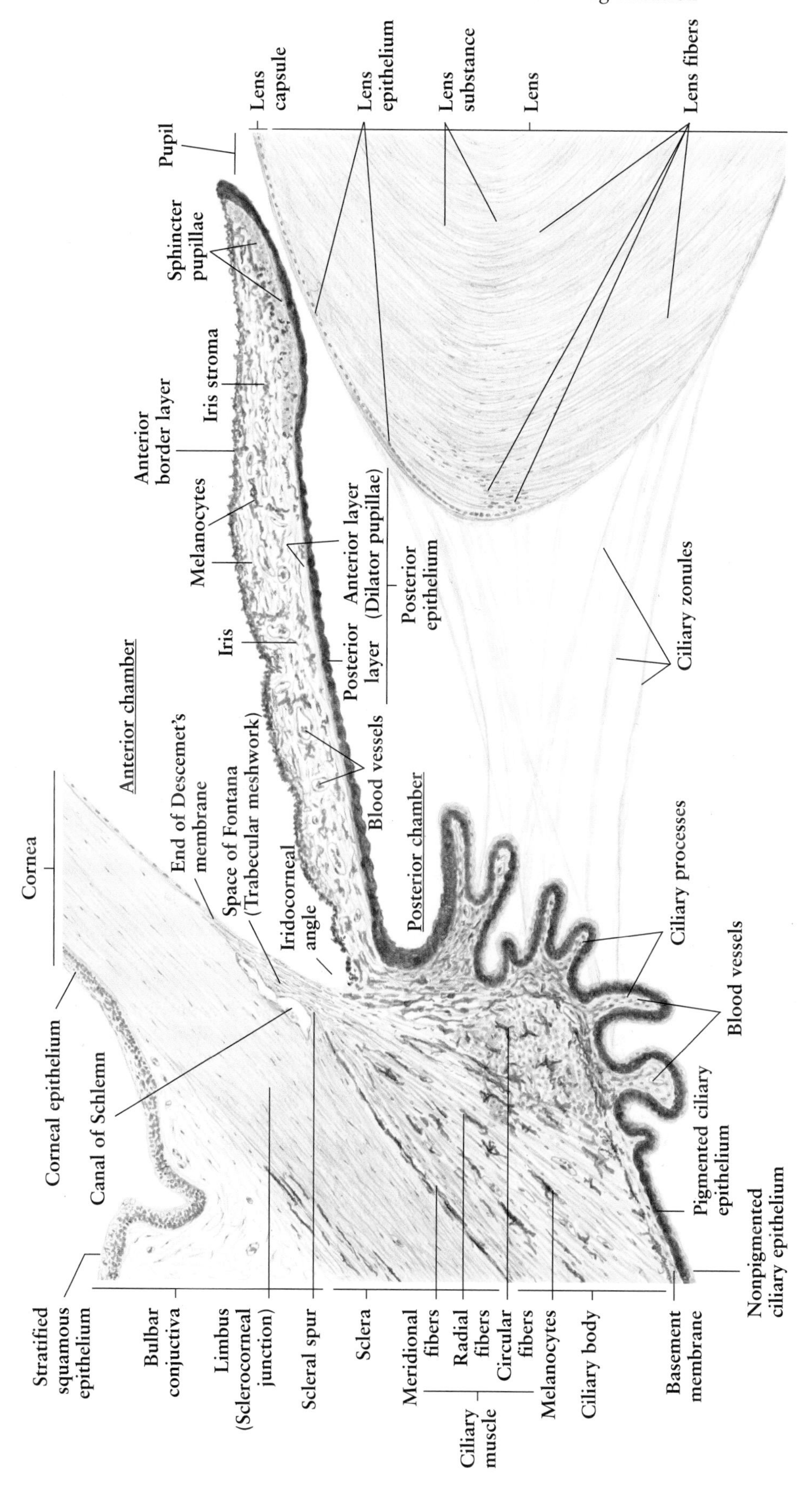

Fig. 16-5.

Canal of Schlemm

The **scleral spur** is a small ridge of the **sclera** projecting from the inner side of the limbus. The **canal of Schlemm,** lined by a thin layer of **endothelium,** is an annular vessel that encircles the limbus just anterior and external to the scleral spur. It usually has a single lumen, but sometimes has two. Between the canal of Schlemm and the **anterior chamber** is a spongelike tissue called **trabecular meshwork,** lined by a layer of **endothelial cells.** The trabeculae are composed of fine connective tissue containing collagen fibrils. Between the trabeculae are **intertrabecular spaces** through which aqueous humor is reabsorbed by the canal of Schlemm from the anterior chamber. The canal of Schlemm drains into the episcleral venous system via minute channels through the sclera. Obstruction to the drainage of aqueous humor leads to increased intraocular pressure, as in the disease known as *glaucoma.*

Note that in this figure **Descemet's membrane** and the **corneal endothelium** of the **cornea** end at the trabecular meshwork; the **ciliary muscle** is shown attaching to the scleral spur and the trabecular meshwork; and there is no epithelium covering the **anterior border layer** of the **iris,** which is rich in **melanocytes** and **blood vessels.**

Figure 16-5. **Canal of Schlemm**
Human • H.E. stain • High magnification

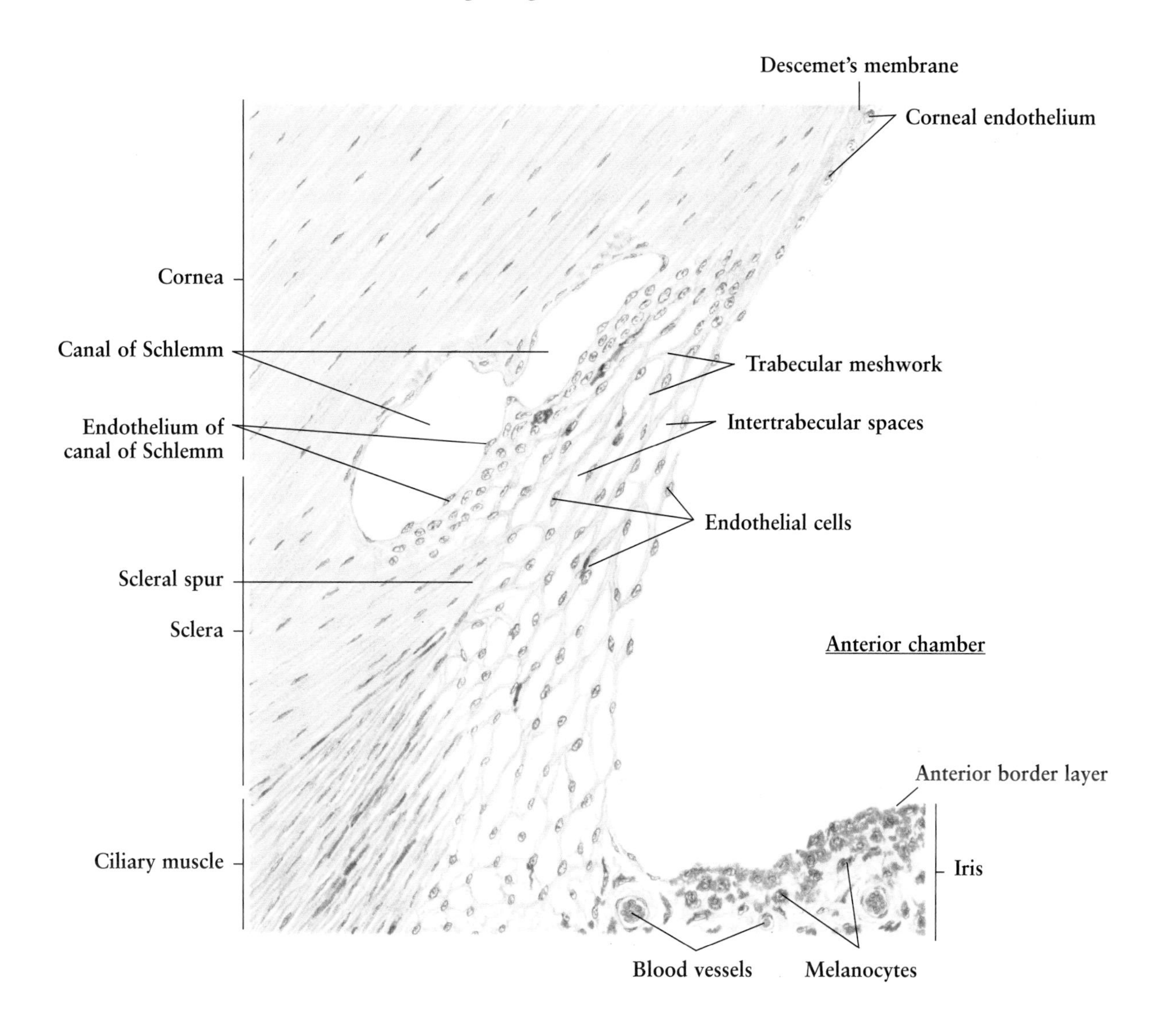

Fig. 16-6. Retina

The **retina** contains three basic cell types: *neurons, neuron supporting cells,* and *pigmented epithelial cells,* whose nuclei are arranged in 4 different layers (layers 1, 4, 6 and 8). Histologically, the retina is traditionally divided into 10 distinct layers. As shown in this figure, these layers are, from outer layer to inner layer:

1. Pigmented epithelium
2. Rod and cone layer
3. Outer limiting membrane
4. Outer nuclear layer
5. Outer plexiform layer
6. Inner nuclear layer
7. Inner plexiform layer
8. Ganglion cell layer
9. Nerve fiber layer
10. Inner limiting membrane

The (1) **pigmented epithelium** is the outermost layer. It is composed of a single layer of cuboidal pigmented epithelial cells resting on the **glassy membrane,** which separates the retina from the **choroid.** The (2) **rod and cone layer** is made up of rod and cone processes of the photoreceptor cells. The (3) **outer limiting membrane** is a thin eosinophilic structure and is formed by the processes of **Müller's cells,** giant, long supporting cells. Their nuclei lie in the inner nuclear layer, and their cytoplasmic processes extend into the inner and outer limiting membranes. The (4) **outer nuclear layer** is composed of densely packed nuclei of the photoreceptor cells, the **rod and cone cells.** In general, the cell bodies of the rod cells are located in the inner part of this layer while those of the cone cells are in the outer part. The (5) **outer plexiform layer** marks the synaptic connections between the photoreceptor cells and integrating neurons. The (6) **inner nuclear layer** contains cell bodies of the integrating neurons including the **bipolar cells, horizontal cells,** and **amacrine cells,** and those of the supporting cells, **Müller's cells.** The (7) **inner plexiform layer** is the site of synapsis between integrating cells and ganglion cells. The (8) **ganglion cell layer** contains large **ganglion cells** and small **neuroglia.** The (9) **nerve fiber layer** is formed of **axons of the ganglion cells,** the unmyelinated nerve fibers that converge to form the optic nerve at the optic papilla. In this layer, longitudinal **Müller's fibers** also can be seen. The (10) **inner limiting membrane** is the innermost aspect, separating the retina from the vitreous body, and is made up of the feet of Müller's cells and their basement membrane. **Blood vessels** are distributed from the inner nuclear layer to the inner limiting membrane.

In this illustration, the structure of the **choroid** is also shown in detail. Note the **glassy membrane, choroidocapillary layer, melanocytes,** and the **venules.**

Fig. 16-7. Macula Lutea: Fovea Centralis

The **macula lutea** is an area with a diameter of 1.5 mm, located about 4 mm temporally from the optic papilla (see **Fig. 16-1**), and appearing yellow in a fresh specimen. The center of the macula lutea, about 0.5 mm in diameter, is a depression called the **fovea centralis.** At the fovea centralis, the photoreceptors consist exclusively of **cone cells** with elongated and closely packed cones. All integrating cells in the **inner nuclear layer** and the **ganglion cells** are obliquely placed around the margins of the fovea centralis. Each cone cell synapses with only one bipolar cell; therefore, the fovea centralis is the area of great visual acuity. There are no blood vessels in the fovea centralis.

Figure 16-6. **Retina**

Human • H.E. stain • High magnification

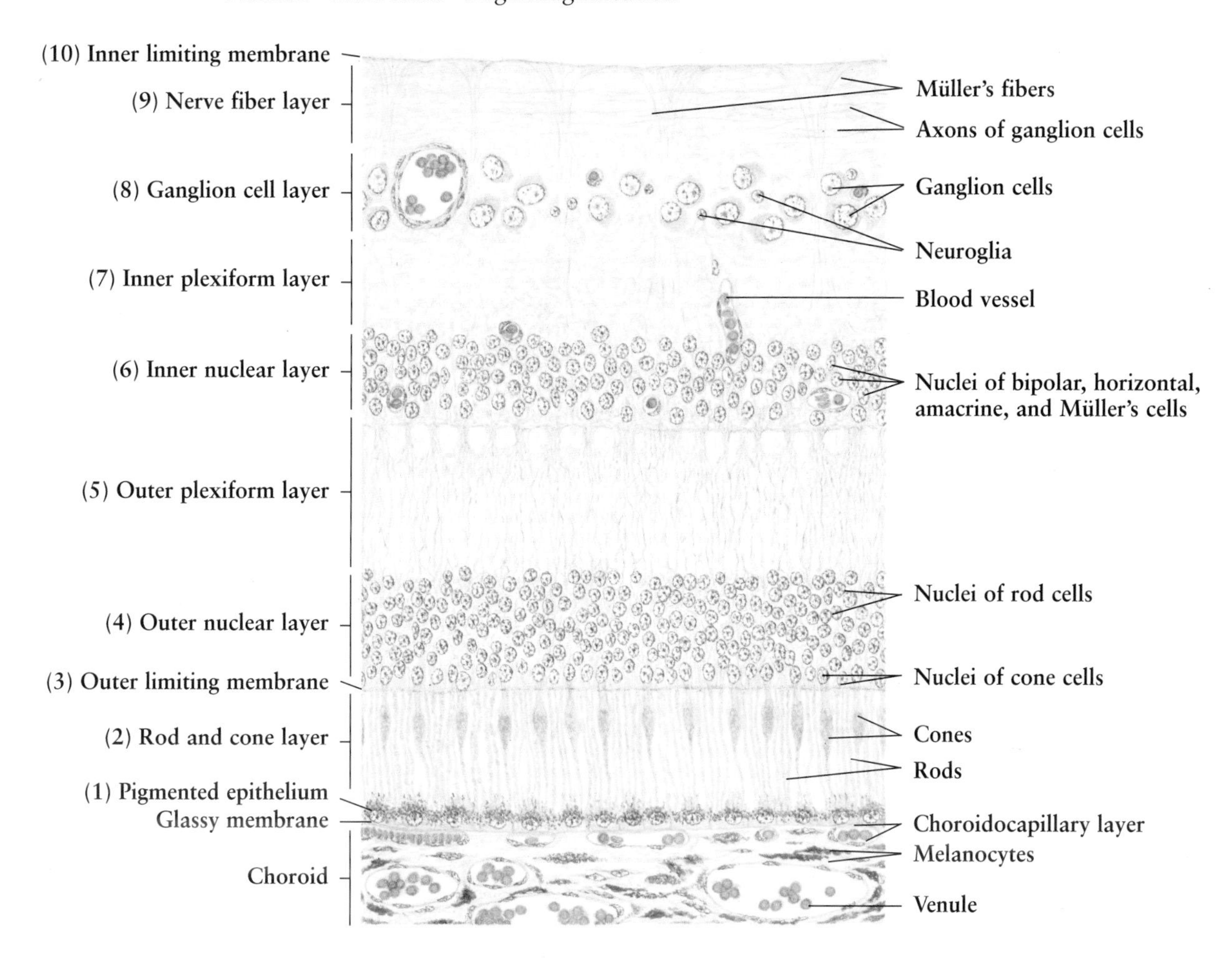

Figure 16-7. **Macula Lutea: Fovea Centralis**

Human retina • H.E. stain • High magnification

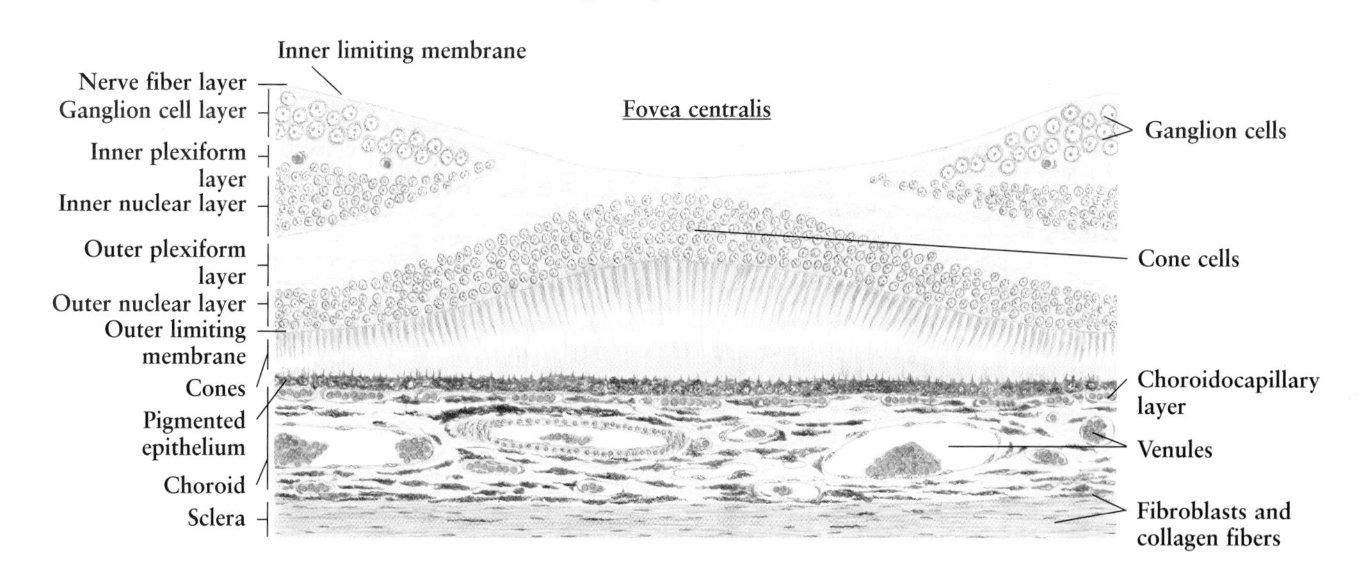

Fig. 16-8. Eyelid

The **eyelid,** covering the anterior surface of the eyeball, basically consists of *skeletal muscle* covered by *skin* externally and a *mucosa* internally.

The skeletal muscle is the palpebral portion of the **orbicularis oculi muscle,** which forms the bulk of the core of the eyelid.

The **skin** consists of a keratinized **epidermis** and a very loose and elastic **dermis** containing **hair follicles, sebaceous glands,** and **sweat glands.** Between the dermis and the orbicularis muscle is a thin **subcutaneous layer,** devoid of adipose tissue. At the **lid margin,** the dermis is denser and contains two or three rows of long, stiff hairs, the **eyelashes.** Small sebaceous glands, the **glands of Zeis,** are associated with these eyelashes. In this area, apocrine sweat glands, the **glands of Moll** and the **ciliary muscle of Riolan** are present.

The **mucosa** in the inner surface of the eyelid is a layer of **palpebral conjunctiva.** It is lined by a **stratified columnar epithelium.** Between the palpebral conjunctiva and the orbicularis oculi muscle lies the **tarsal plate,** a sheet of dense connective tissue. The **tarsal glands (of Meibom),** embedded in the tarsal plate, are modified sebaceous glands. They are oriented vertically and have a long **central duct** opening at the lid margin. The secretion of these glands forms an oily layer that covers the tear layer on the surface of the eyeball, preventing evaporation of the tears.

Above the tarsal plate are located the **superior tarsal muscle (of Müller),** a smooth muscle, and the **tendon of the levator palpebrae superioris muscle.** In addition, the **accessory lacrimal glands (of Wolfring** or **of Krause)** are sometimes present.

Fig. 16-9. Lacrimal Gland

The **lacrimal gland,** with the size of an almond, is located in the superolateral corner of the orbit. It is formed by several separate glandular lobes, with 6–12 excretory ducts opening at the superior conjunctival fornix.

As shown in this figure, the glandular parenchyma is divided into **lobules** by interlobular connective tissue. The **glandular acini** are composed of typical serous epithelial cells, which are columnar with a round nucleus and basophilic basal cytoplasm. Numerous **myoepithelial cells** lie between the bases of the glandular epithelial cells and the basement membrane. The **intralobular duct** is lined by a single layer of cuboidal cells. **Intralobular connective tissue** contains a network of **capillaries,** some **adipose cells,** and **nerve fibers.** The **interlobular connective tissue,** derived from the capsule, is loose and elastic, and is rich in **blood vessels** and **nerve fibers.** It contains a large **interlobular duct,** which is formed by a double layer of epithelium. **Macrophages, plasma cells,** lymphocytes, and **fibroblasts** are found in both the interlobular and intralobular connective tissues.

The lacrimal gland is responsible for the production of the *tears,* a watery fluid, which lubricates and flushes the anterior surface of the eyeball and the interior surface of the eyelid. The tears contain the antibacterial enzyme *lysozyme* and electrolytes of similar concentration to that of plasma. Small accessory lacrimal glands are observed in the inner surface of the eyelid (see **Fig. 16-8**).

Figure 16-8. Eyelid

Human • Sagittal section • H.E. stain • Low magnification

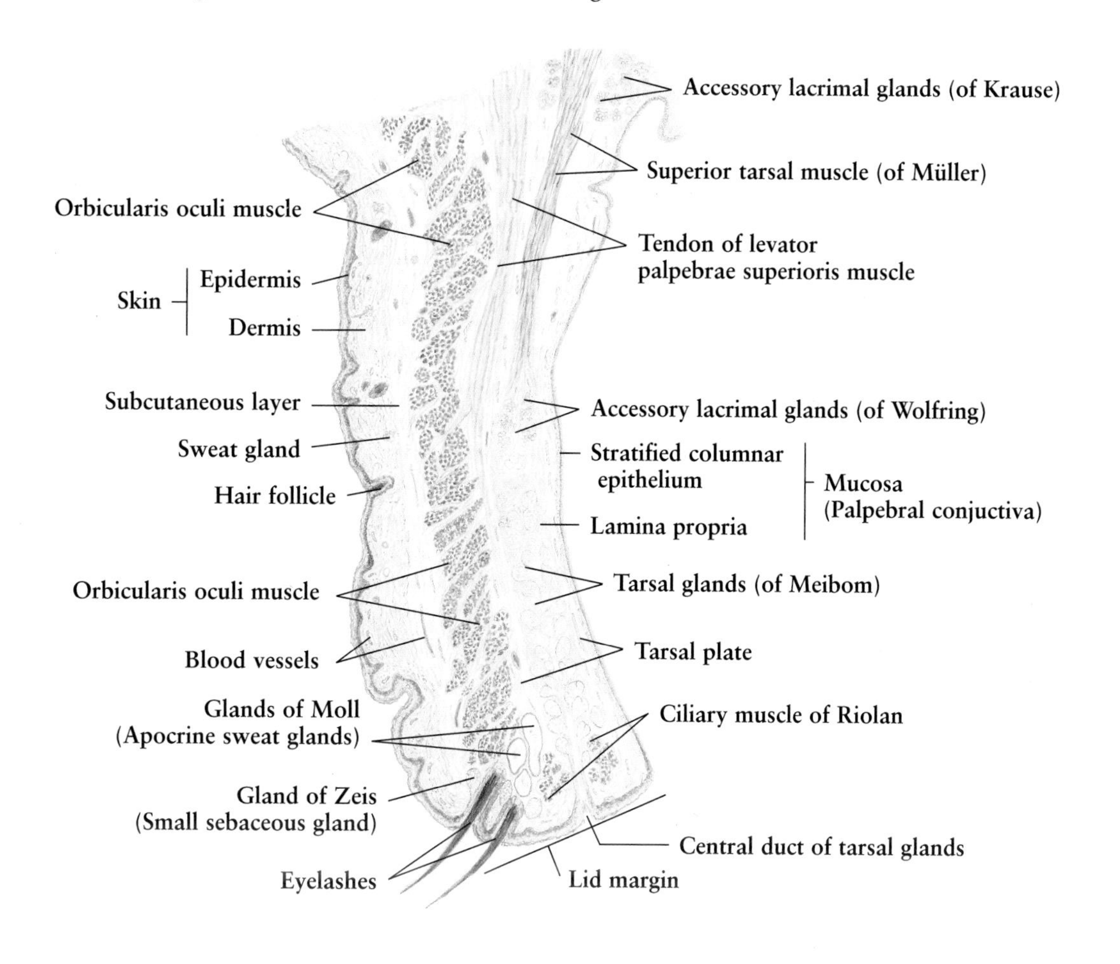

Figure 16-9. Lacrimal Gland

Human • H.E. stain • High magnification

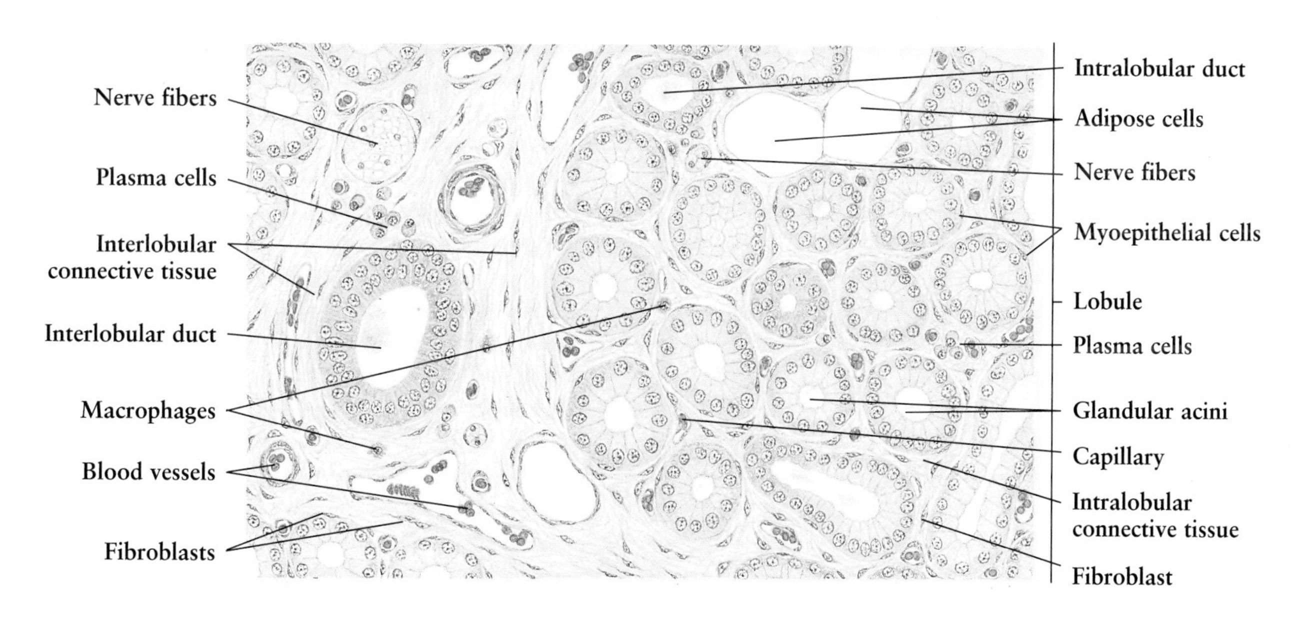

17. THE EAR

The functions of the **ear** (vestibulocochlear apparatus) are associated with *hearing* and *equilibrium.* The organ may be divided into three parts: *the external ear, the middle ear,* and *the inner ear.*

The **external ear** consists of the auricle, the external auditory meatus, and the tympanic membrane or eardrum. The **auricle** is an irregular elastic plate with elastic cartilage as its core, covered by a thin layer of skin. The **external auditory meatus** is a short canal lined by thin skin with **hair follicles, sebaceous glands,** and **ceruminous glands.** The **tympanic membrane** is composed of two layers of collagen fibers (the outer layer is radial and the inner layer is circular in arrangement), covered externally by very thin skin and internally by a low layer of cuboidal epithelium. The external ear picks up the sound waves, which enter the external auditory meatus and reach the tympanic membrane.

The **middle ear,** also called the **tympanic cavity,** is an irregular, air-containing space in the petrous part of the temporal bone. It communicates anteriorly with the nasopharynx by the **pharyngotympanic** (Eustachian or auditory) **tube.** The middle ear is characterized by a chain of three auditory ossicles: the **malleus, incus,** and **stapes.** The malleus contacts the tympanic membrane; the base of the stapes closes the oval window of the inner ear; between them lies the incus. The vibrations of the tympanic membrane caused by sound waves are transmitted by the chain of auditory ossicles to the fluid-filled chambers of the inner ear.

The **inner ear** is formed of an irregular system of canals, the **membranous labyrinth,** walled in by the **bony labyrinth,** which is composed of the **vestibule,** the **semicircular canals,** and the **cochlea.** The vibrations from the middle ear are transduced into specific nerve impulses in the cochlea, and conducted by the acoustic nerve to the hearing area of the cerebral cortex. Specialized structures in the vestibule and in the semicircular canals are responsible for maintaining the body's equilibrium.

In this chapter, the auricle, external auditory meatus, bony labyrinth, and membranous labyrinth are shown in detail.

Fig. 17-1.

Auricle

The **auricle** of the external ear is an irregular plate of **elastic cartilage,** covered on all surfaces by thin **skin.**

The **elastic cartilage** is enveloped by a thick **perichondrium** with a high content of elastic fibers. The **dermis** of the skin is elastic and dense, and contains **hair follicles, sebaceous glands,** and **sweat glands.** These dermal components are in general poorly developed. On the **anterior surface,** there is no hypodermis, and the dermis of the skin directly blends with the perichondrium. On the **posterior surface, hypodermis** is present. It contains **adipose tissue** and **skeletal muscle fibers,** in addition to **blood vessels** and **nerve fibers.** The skeletal muscle fibers are vestigial in humans, but they are well developed in lower animals, capable of more pronounced ear movements.

Fig. 17-2.

External Auditory Meatus

The **external auditory meatus,** about 2.5 cm long, is a S-shaped canal leading from the auricle to the tympanic membrane.

The wall of the outer third is formed by **elastic cartilage** continuous with the auricular cartilage, whereas the inner two-thirds is formed by the temporal bone. The canal is lined by keratinized, thin skin, which has hairs and associated sebaceous glands but is lacking the usual dermal papillae. The **dermis** consists of elastic and dense collagenous connective tissue, and is firmly attached to the **perichondrium** or periosteum. The wall of the canal is characterized by the presence of the **ceruminous glands,** modified apocrine sweat glands. Their **ducts** open directly onto the surface or, together with adjacent sebaceous glands, into hair follicles. The ceruminous glands secrete waxy cerumen, which functions to protect the skin of the canal from moisture and infection.

Also shown in this figure are **blood vessels, nerve fibers,** and some **adipose cells.**

Figure 17-1. **Auricle**
Human • H.E. stain • Low magnification

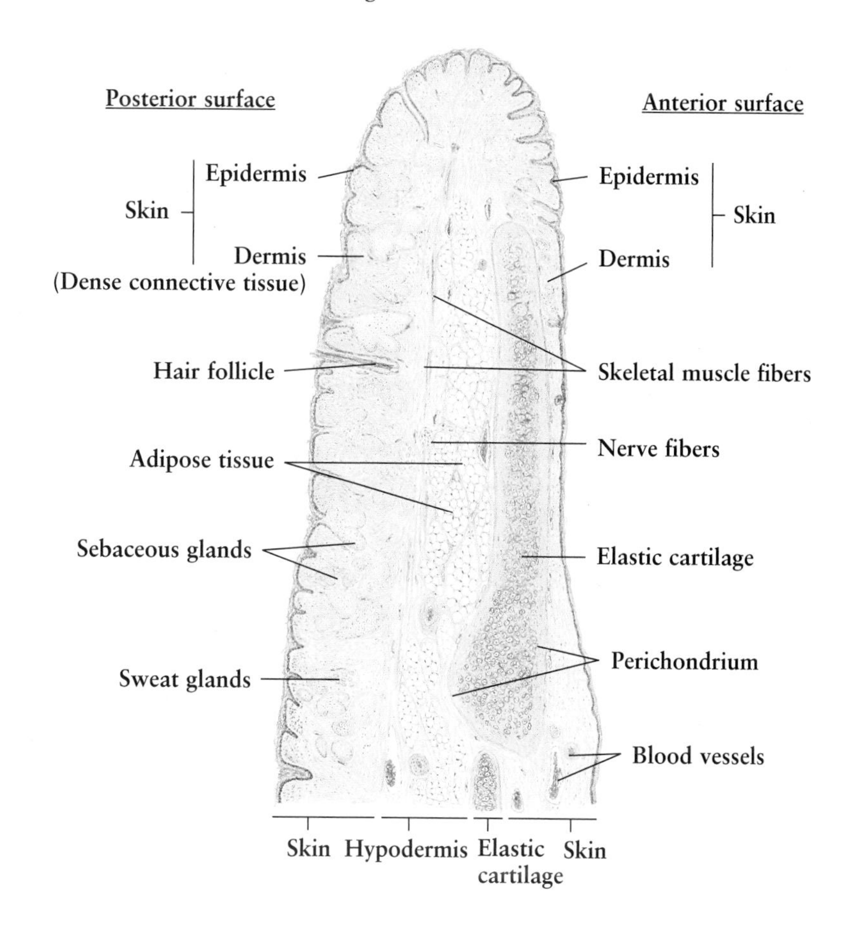

Figure 17-2. **External Auditory Meatus**
Human • H.E. stain • Low magnification

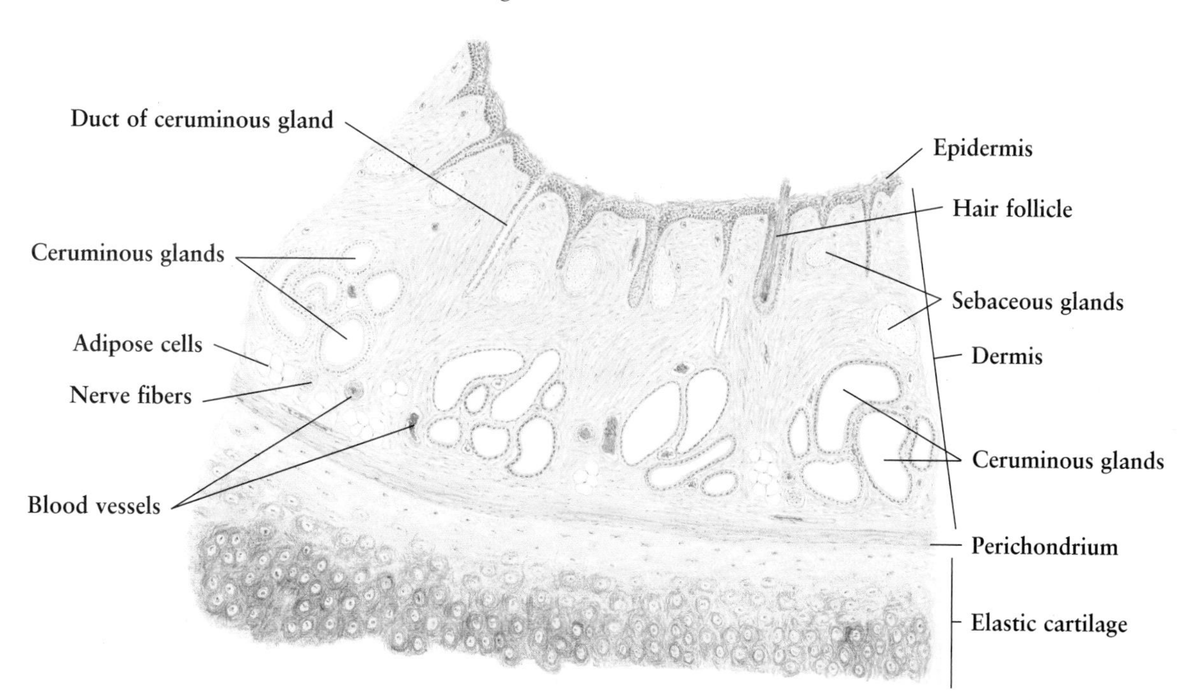

Fig. 17-3.

Bony Labyrinth

The **bony labyrinth,** which contains the membranous labyrinth, is a system formed by a series of irregular bony spaces of the inner ear located in the temporal bone. It consists of a *vestibular portion, semicircular canals,* and a cone-shaped *cochlear part.*

The **vestibule** is the centrally located irregularly rounded cavity, composed of two recesses: the **spherical recess** and **elliptical recess.** It is separated from the tympanic cavity by the **oval (vestibular) window.** The vestibule is continuous posteriorly with the semicircular canals and anteriorly with the cavity of the cochlea.

There are three **semicircular canals** named according to their position. The **anterior** and **posterior semicircular canals** are set vertically and perpendicular to each other, whereas the **lateral semicircular canal** lies in a horizontal plane. Each canal has a dilatation known as an **ampulla** at one end. The **anterior ampulla** and **lateral ampulla** lie close together above the oval window, and the **posterior ampulla** opens into the posterior part of the vestibule. The other nonampullated ends of the anterior and posterior semicircular canals fuse to form the **common crus,** opening into the medial part of the vestibule. The nonampullated end of the lateral semicircular canal opens separately into the upper part of the vestibule.

The **cochlea** has a conical shape resembling that of a snail shell, with a canal following a spiral course from the vestibule to the apex of the cone. From the **basic cochlea** to the **cochlear cupula** it surrounds an axial bony stem, the *modiolus* (see **Fig. 17-7**) with two three-quarter turns. The cochlea has its apex pointing forward and laterally. Near the vestibule, the cochlea communicates with the tympanic cavity through the **round (cochlear) window.**

Fig. 17-4.

Membranous Labyrinth

The **membranous labyrinth,** which lies within the bony labyrinth, has a form resembling that of the bony labyrinth. It consists of a system of interconnected parts lined by epithelium and containing endolymph. Occupying the space between the membranous labyrinth and the bony labyrinth is **perilymph.**

The vestibule is occupied by two chambers: utricle and saccule. The **utricle** is located in the posterior end of the vestibule; it has an epithelial structure, the **macula of the utricle.** The utricle communicates via five orifices with the three membranous **semicircular ducts.** Each membranous semicircular duct has a **crista ampullaris** at the ampulla. In the anterior end of the vestibule is the spherical **saccule** with **macula of the saccule.** The utricle and saccule are connected through a Y-shaped duct that forms the **endolymphatic duct** leading to a blind sac, the **endolymphatic sac.** On the other hand, the saccule is connected via the **ductus reuniens** with the **cochlear duct** along which the **organ of Corti** is located.

The cristae and maculae are responsible for the sense of equilibrium, whereas the organ of Corti is the organ of hearing.

Figure 17-3. Bony Labyrinth
Human • Right side • Frontolateral aspect • Very low magnification

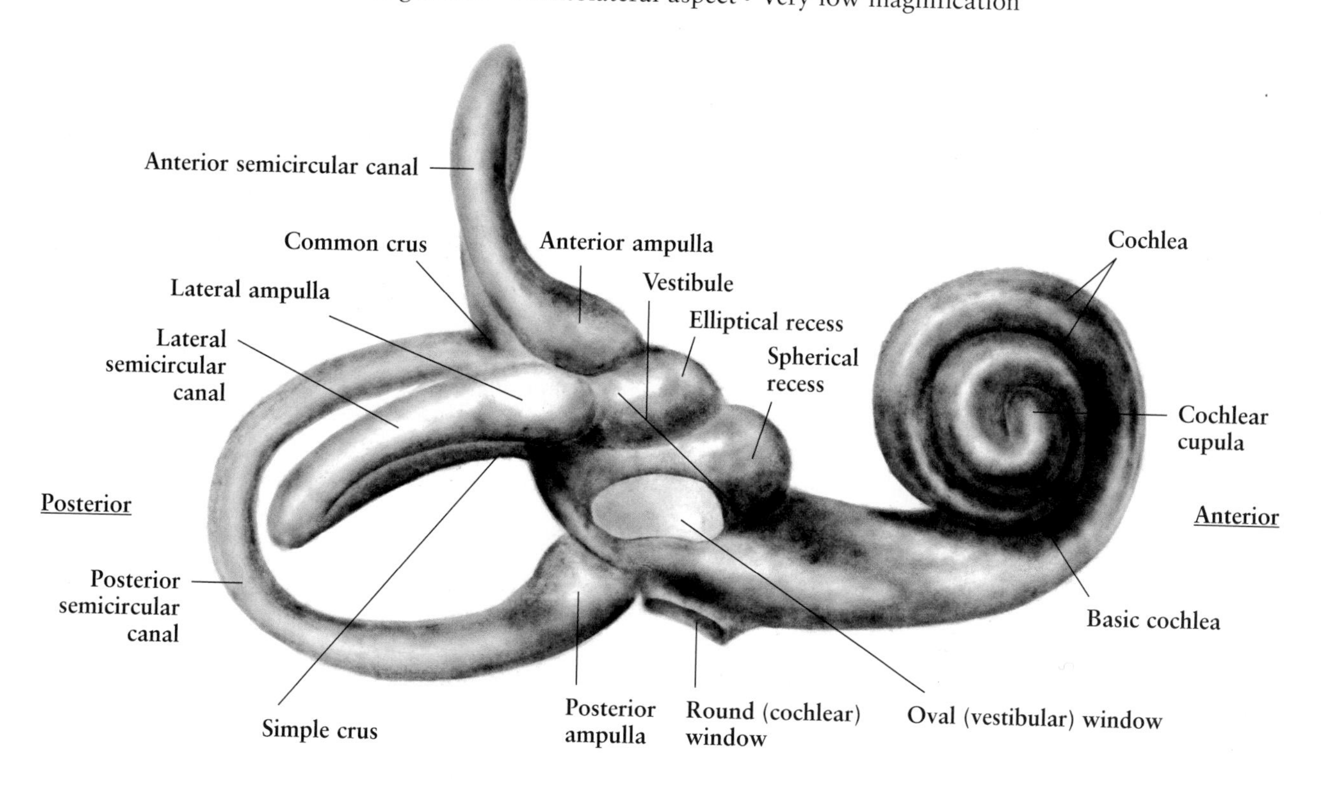

Figure 17-4. Membranous Labyrinth
Human • Right side diagram • Frontolateral aspect • Very low magnification

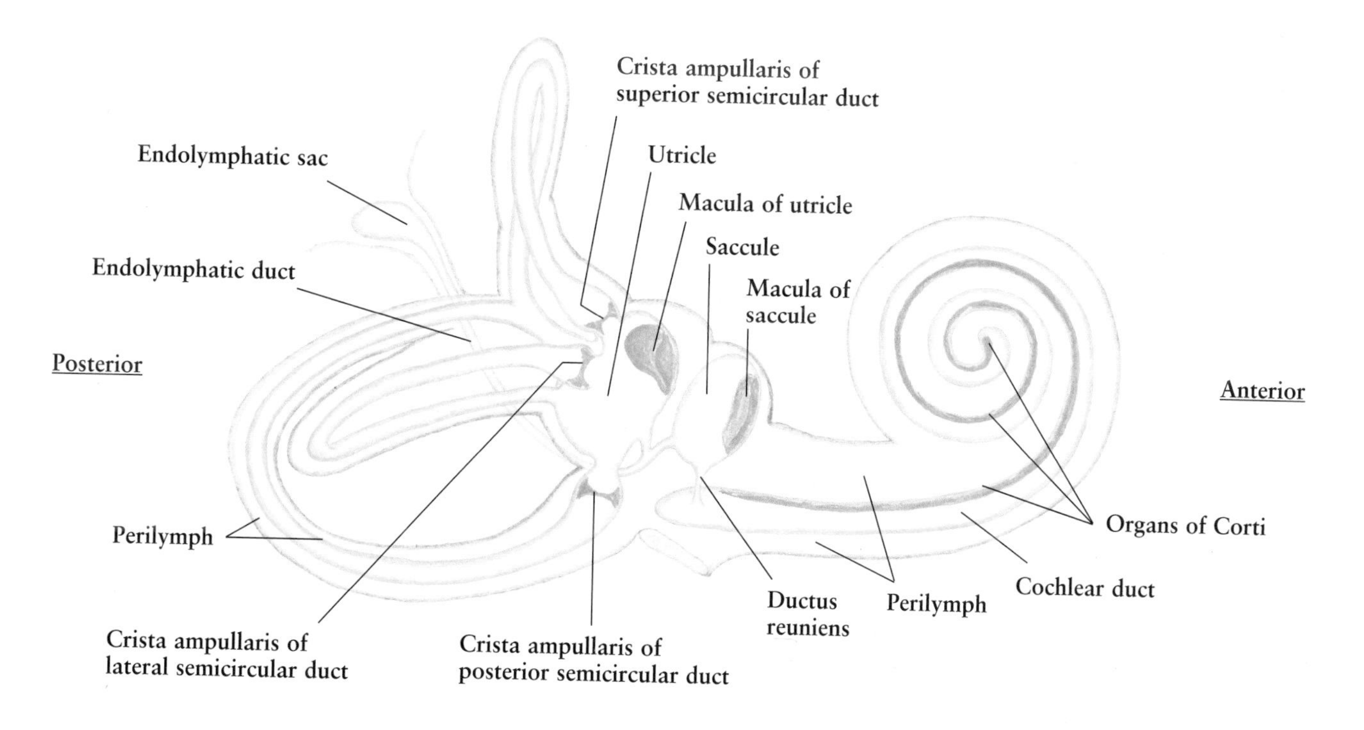

Fig. 17-5.

Maculae

As shown in **Fig. 17-4,** the **maculae** are located in both utricle and saccule. The maculae are spotlike neuroepithelial structures, composed of sensory *hair cells* and *supporting cells,* and covered over by a layer of *otolithic membrane.*

There are two types of **hair cells:** type I and type II. The **type I hair cell** is flask-shaped with a nucleus-containing rounded base and a short neck. A large part of the cell is enveloped by a net of afferent nerve endings with some inhibitory efferent nerve fibers. The **type II hair cell** is columnar with a round and predominantly apical nucleus. It is surrounded by numerous separate afferent and efferent nerve endings at its base. Type I and II hair cells are characterized by the presence, on the free surface, of a sensory **hair bundle** composed of 30–100 stereocilia and one cilium. The **supporting cells** are tall columnar cells, rest upon a **basement membrane,** and extend to the free surface where apical microvilli are present.

The microvilli of the supporting cells, and the stereocilia and cilia of the hair cells, are embedded in the **otolithic membrane,** which is a thick, gelatinous glycoprotein membrane, secreted by the supporting cells. The membrane contains numerous small **otoliths** composed of calcium carbonate and protein. Beneath the neuroepithelium is the **lamina propria** of **connective tissue** containing numerous **fibroblasts, capillaries,** and **nerve fibers.** Note the **regular epithelial cells** of the utricle or saccule at the left edge of this figure.

Changes in the position of the head result in the stimulation of the hair cells as a result of the influence of gravity on the otoliths. The stimulus is detected by nerve endings that contact the hair cells. The two maculae are set at right angles to each other, and are responsible for the sensation of linear acceleration and deceleration.

Fig. 17-6.

Crista Ampullaris

The **crista ampullaris** occurs in the ampulla of each membranous semicircular duct (see **Fig. 17-4**). It is a small, ridgelike sensory organ, perpendicular to the axis of the semicircular duct, and extends across the ampulla, almost touching its opposite wall.

Structurally similar to the macula, the cristae are also composed of **supporting cells, type I hair cells,** and **type II hair cells** (see **Fig. 17-5**). Their microvilli and **hair bundle** including stereocilia and cilia are embedded in a **cupula ampullaris,** a very thick gelatinous mass, similar to the otolithic membrane but lacking otoliths. The underlying lamina propria is a well-vascularized **connective tissue** containing a large number of **fibroblasts, nerve fibers,** and **capillaries.**

During the movements of rotational acceleration or deceleration of the head, the hair cells are stimulated by the passive movement of the endolymph, generating an action potential that is received by surrounding nerve endings.

Figure 17-5. **Maculae**
Human • H.E. stain • High magnification

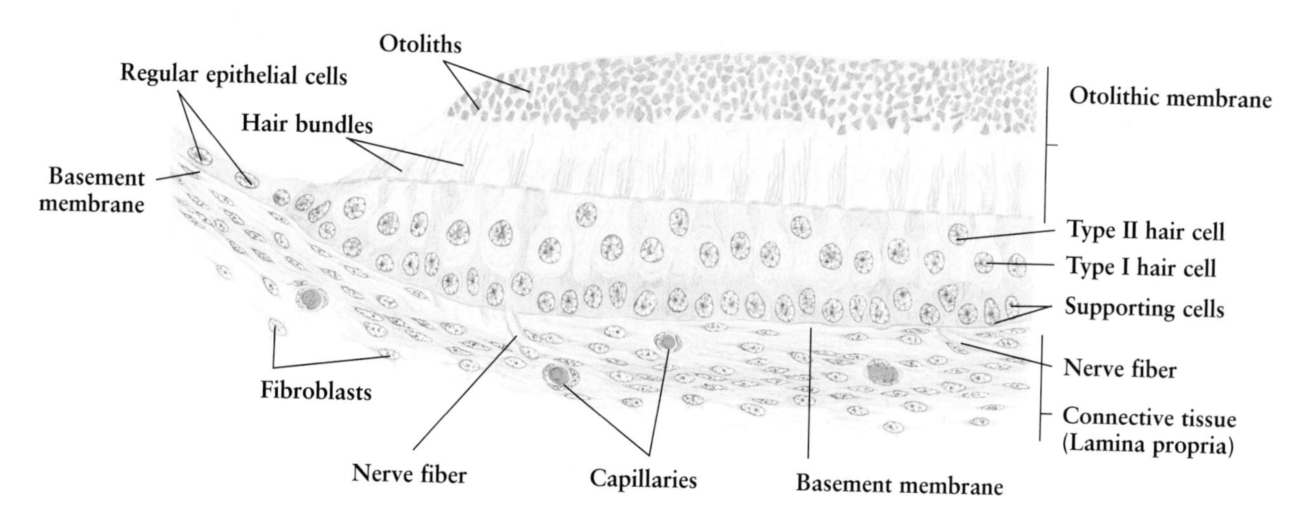

Figure 17-6. **Crista Ampullaris**
Human • H.E. stain • High magnification

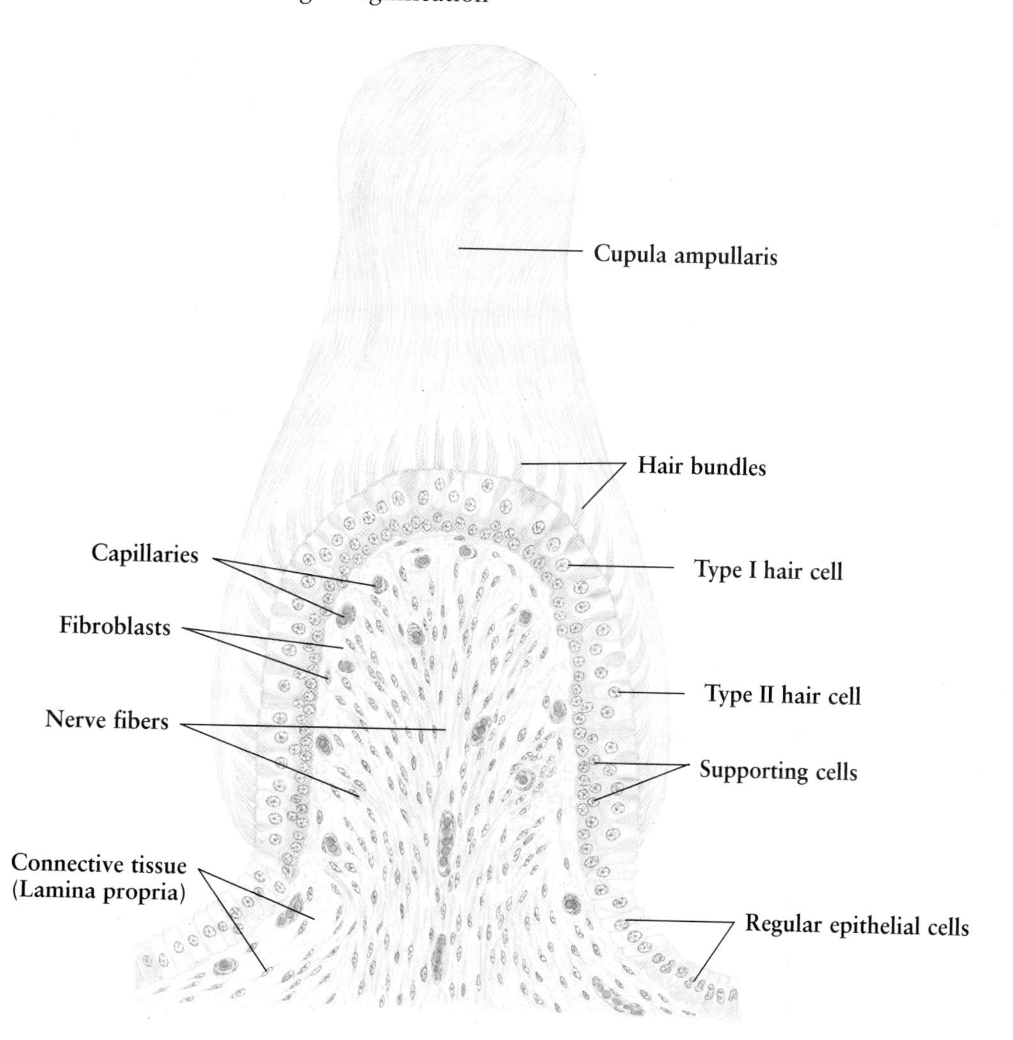

Fig. 17-7. Cochlea

The **cochlea** is the portion of the bony labyrinth that contains the auditory sensory organ. It is a spiral bony canal about 35 mm in length, which spirals for two three-quarter turns around a conical pillar of spongy bone, the **modiolus,** forming a conical, spiral-shaped contour.

In vertical section as shown in this figure, five separate **cross sections of the cochlea** can be found. The modiolus forms the central axis of the cochlea, containing **spiral ganglia** and the **cochlear nerve.** The **osseous spiral lamina** projects from the modiolus into the cochlear canal. From the edge of the osseous spiral lamina, the **basilar membrane,** a sheet of connective tissue that supports the **organ of Corti,** extends across the canal to its opposite wall and joins the **spiral ligament,** a thickening of the periosteum. Therefore, the basilar membrane, the spiral ligament, and the **vestibular membrane** (which arises from the upper surface of the osseous spiral lamina) together divide the cochlear canal into three parts: above, the **vestibular duct;** middle, the **cochlear duct;** and below, the **tympanic duct.** The vestibular duct communicates with the tympanic duct through a small opening at the apex of the cochlea, the **helicotrema.**

Fig. 17-8. Cochlear Canal

Figure 17-8 is an enlargement of one cross section of the cochlea in **Fig. 17-7,** showing the structure of the **cochlear canal.** It consists of three parts: *scala vestibuli, cochlear duct,* and *scala tympani.*

Both **scala vestibuli** and **scala tympani** are lined by a simple squamous epithelium supported by the connective tissue blending with the periosteum. These are perilymphatic spaces filled with perilymph. The two scalae are in communication through the helicotrema at the apex of the cochlea (see **Fig. 17-7**). The perilymph of the scala vestibuli is continuous with that in the perilymphatic space of the vestibule and thus reaches the inner surface of the oval window (see **Fig. 17-3**), which is occluded by the base plate of the stapes. The perilymph within the scala tympani is separated from the tympanic cavity by an elastic membrane (the secondary tympanic membrane), which closes the round window.

The **cochlear duct** is triangular in cross section with its apex at the **spiral limbus.** Its roof is formed by the **vestibular membrane,** and its floor consists of an **osseous spiral lamina** and a **basilar membrane.** Its lateral wall is the **stria vascularis** and **spiral prominence,** both of which are supported by the **spiral ligament,** a triangular thickening of the periosteum. The **organ of Corti,** the auditory sensory organ, is supported on the basilar membrane, while the **tectorial membrane** is formed on the surface of the spiral limbus.

Bundles of afferent **nerve fibers** arise from the base of the organ of Corti, run through the canal of the osseous spiral lamina, and converge toward the **spiral ganglion,** which is composed of bipolar neurons located in the bone of the **modiolus.** Their central processes (myelinated axons) run together to form the **cochlear nerve,** which is the auditory component of the VIIIth cranial (acoustic) nerve.

Figure 17-7. **Cochlea**

Human • Vertical section • H.E. stain • Very low magnification

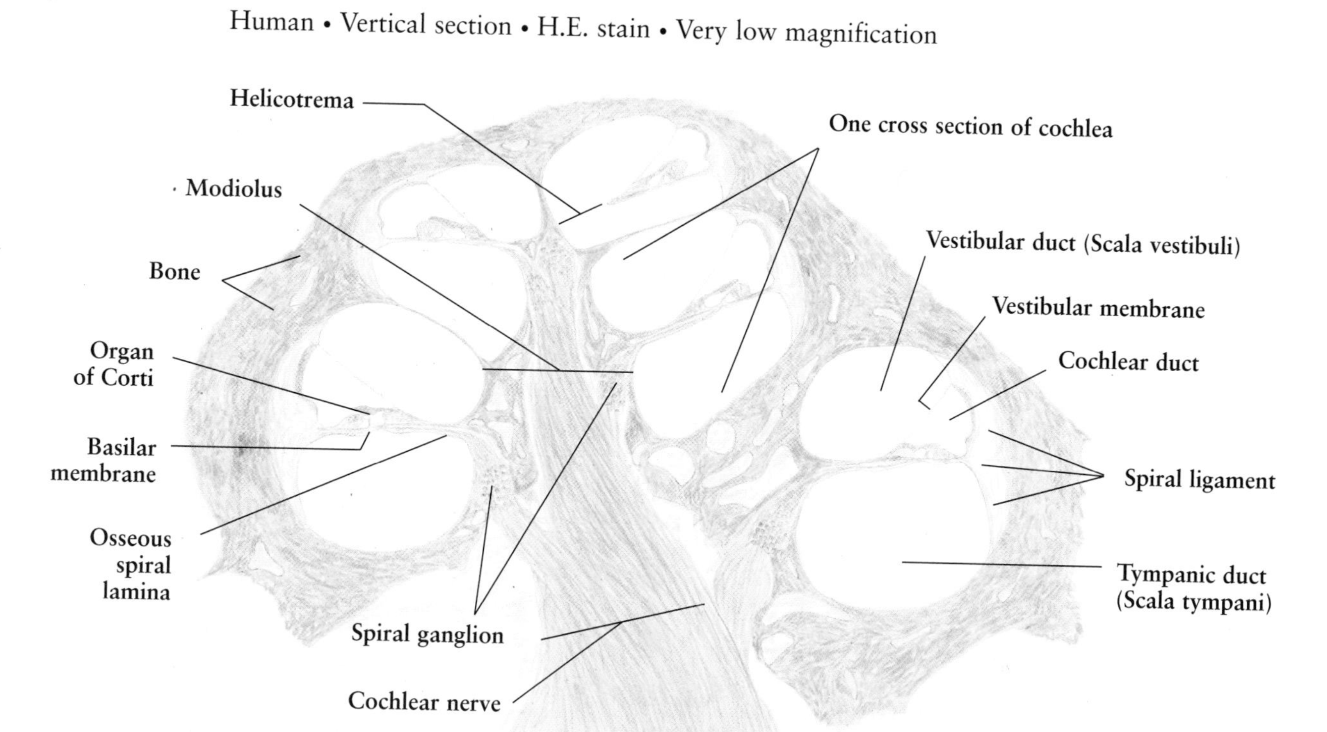

Figure 17-8. **Cochlear Canal**

Human • Vertical section • H.E. stain • Low magnification

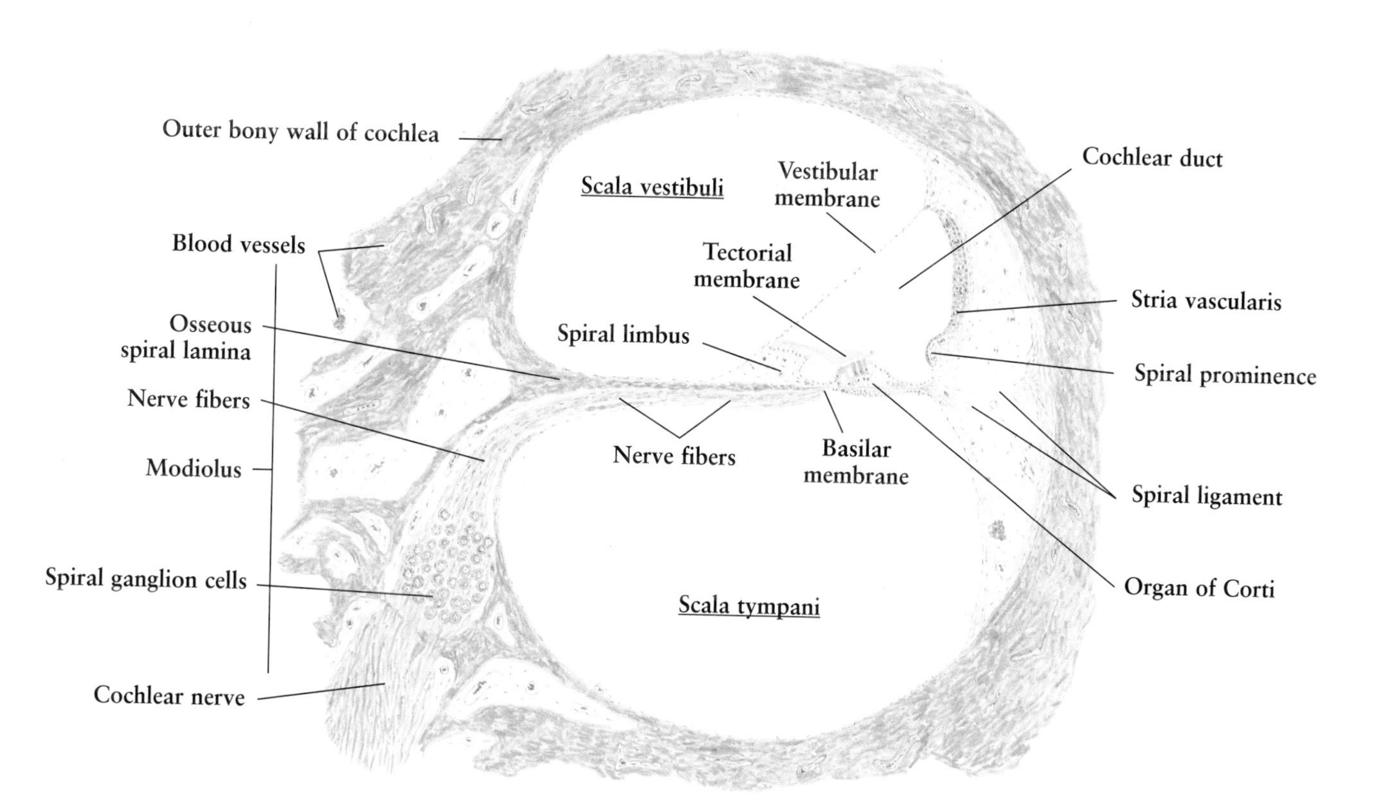

Fig. 17-9.

Organ of Corti and Stria Vascularis

The **organ of Corti** is an epithelial auditory receptor composed of *supporting* and *hair cells,* resting on the **basilar membrane** of the **cochlear duct.**

Two rows of pillar cells, the **inner** and **outer pillar cells,** form a triangular canal, the **inner tunnel,** which is an important landmark of the organ of Corti. The pillar cells are tall columnar cells with a broad base containing a spherical nucleus lying against the basilar membrane. Each pillar cell contains a dense bundle of microtubules that extends from the base of the cell to its apex. At their apices, the inner and outer pillar cells converge to form a thin, hoodlike structure. One row of **inner phalangeal cells** and three (to five) rows of **outer phalangeal cells** lie on the basilar membrane adjacent to the inner and outer pillar cells, respectively. Each phalangeal cell supports one hair cell. Thus there is one row of **inner hair cells** and three (to five) rows of **outer hair cells.** The hair cells have a structure similar to that of type I hair cells of the macula or ampulla. The **nerve fibers,** composed of the dendrites of the spiral ganglion neurons, contact the hair cells, running in the canal within the **osseous spiral lamina.**

At the inner side of the organ of Corti, neighboring the inner hair cell and inner phalangeal cell are the **border cells,** which vary from columnar to cuboidal to squamous in shape, surrounding the **internal spiral tunnel** over the **tympanic lip.** Between the outer pillar cells and the innermost outer hair cell and the outer phalangeal cells is a space called **Nuel's space.** Another small space, the **outer tunnel,** occurs between the outermost outer hair cells and **cells of Hensen,** which are very tall columnar cells with a round nucleus at the upper portion of the cell. In the basal turn of the cochlea, two layers of cells known as the **cells of Claudius** and the **cells of Boettcher** are adjacent to the cells of Hensen. The **outer spiral tunnel** is surrounded by the cells of Hensen and the cells of Claudius.

The **spiral limbus** is a periosteal connective tissue of the osseous spiral lamina protruding into the cochlear duct. On its surface occurs the flaplike **tectorial membrane,** which overlies the hair cells of the organ of Corti. The tectorial membrane is formed by the fibrillar material and the mass of glycosaminoglycans produced by the **interdental cells,** located in a row on the upper surface of the spiral limbus. The interdental cells are flasklike in appearance, with a basal nucleus and narrow neck. Their thin apical cytoplasmic sheath covers the surface of the spiral limbus and anchors the root of the tectorial membrane.

The **vestibular membrane** rises obliquely from the inner edge of the spiral limbus and joins the spiral ligament, separating the cochlear duct from the scala vestibuli. It is a very thin membrane, composed of extremely delicate fibrous tissue lined on both sides by an apposed layer of squamous epithelial cells.

The **basilar membrane,** supporting the organ of Corti, is stretched between the osseous spiral lamina and spiral ligament. This membrane is about 30 mm in length and 0.16–0.52 mm in width, varying from the basal to apical turn. It contains radially arranged **auditory strings,** which are composed of compact bundles of fine collagen fibrils embedded in an amorphous substance. At the inner one-third of the basilar membrane, the auditory strings appear to be a single layer, but at the outer two-thirds they form two layers. Beneath the inner tunnel is a blood capillary called the **spiral vessel.** The undersurface of the basilar membrane, exposed to the perilymph of the **scala tympani,** is lined by a mesothelium that is composed of cuboidal **epithelial cells** and has an irregular free surface. Vibrations of the auditory strings are transmitted to hair cells by the displacement of the tectorial membrane over the hair cells and then converted into adequate bioelectrical impulses, which result in auditory sensations.

Figure 17-9. Organ of Corti and Stria Vascularis
Human • H.E. stain • High magnification

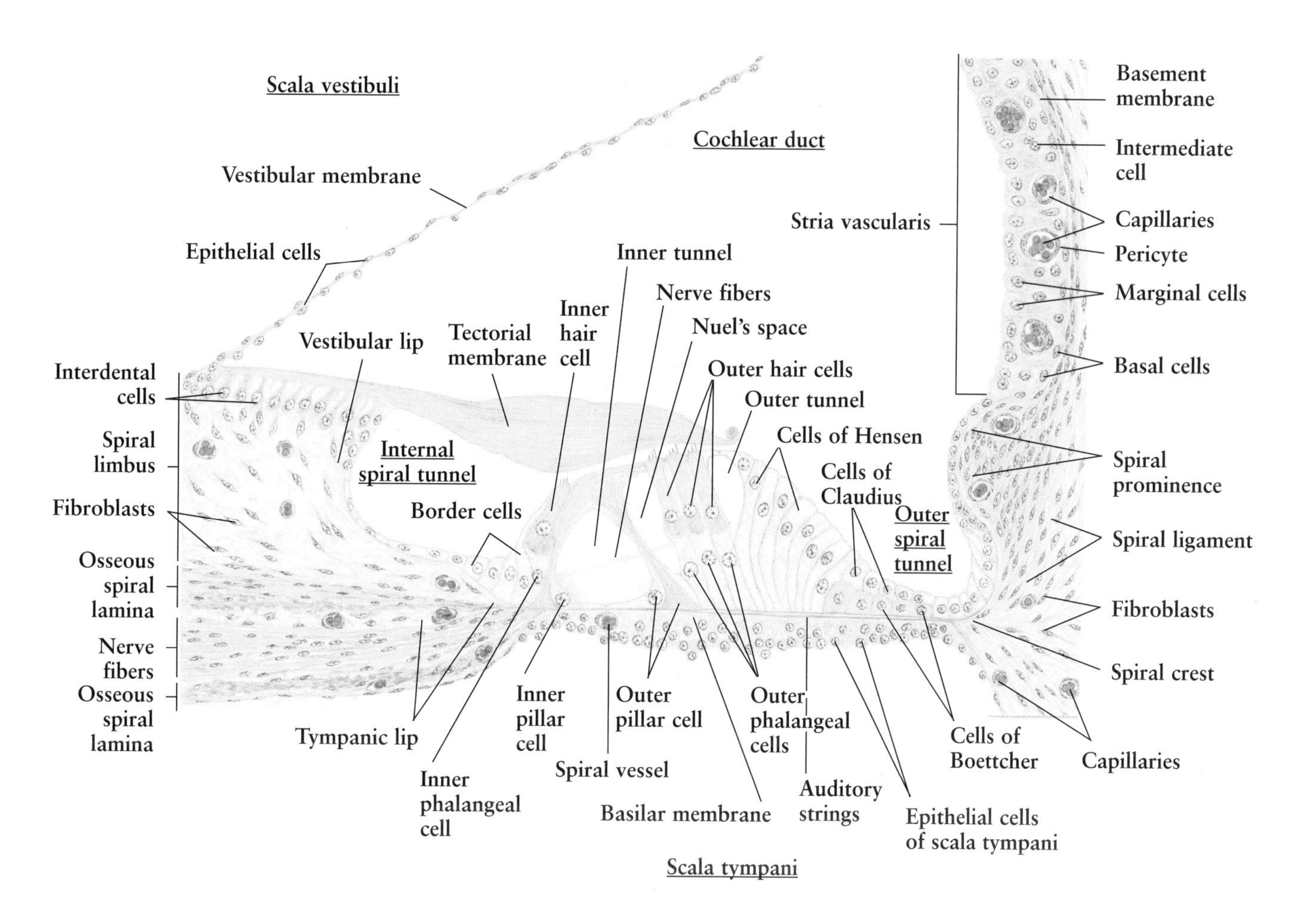

At the lateral wall of the cochlear duct is the **stria vascularis,** a band of stratified columnar epithelium with an intraepithelial plexus of capillaries. The stria vascularis is composed of marginal cells, intermediate cells, and basal cells. The **marginal cells** are columnar, with a round nucleus and eosinophilic cytoplasm. Their luminal surface is covered with microvilli, and the basal cytoplasm contains numerous vesicles and mitochondria that give the appearance of striations under the light microscope. The **intermediate cells** are irregular, and do not contact the luminal surface. The **basal cells** are flattened or pyramidal in shape, resting on the **basement membrane.** The characteristic feature of the stria vascularis is the presence of the intraepithelial **capillaries,** which are enveloped by **pericytes.** The stria vascularis is responsible for producing endolymph and maintaining its normal ionic composition.

The **spiral prominence** inferior to the stria vascularis is a highly vascularized protrusion of the spiral ligament. It follows a spiral course along its entire length. It is covered by a simple squamous or cuboidal epithelium, continuous above with the stria vascularis and below with the cells of Claudius, which surround the outer spiral tunnel.

References

1. Amenta PS: *Elias-Pauly's Histology and Human Microanatomy* (5th ed.). John Wiley, New York, 1987.
2. Bacon RL, Niles NR: *Medical Histology, A Text-Atlas with Introductory Pathology.* Springer-Verlag, New York, 1983.
3. Berman I : *Color Atlas of Basic Histology.* Appleton & Lange, Stamford, CT 1993.
4. Bevelander G, Ramaley JA: *Essentials of Histology.* C.V. Mosby, St. Louis, 1979.
5. Borysencho M, Beringer T: *Functional Histology* (2nd ed.). Little, Brown, Boston, 1984.
6. Bradbury S: *Hewer's Textbook of Histology for Medical Students* (9th ed.). Year Book Medical Publishers, Chicago, 1973.
7. Burkitt HG, Young B, Heath JW: *Wheater's Functional Histology, A Text and Colour Atlas* (3rd ed.). Churchill Livingstone, Edinburgh, 1993.
8. Carlson BM: *Human Embryology and Developmental Biology.* C.V. Mosby, St. Louis, 1994.
9. Carpenter MB: *Core Text of Neuroanatomy* (4th ed.). Williams & Wilkins, Baltimore, 1991.
10. Cormack DH: *Ham's Histology* (9th ed.). J.B. Lippincott, Philadelphia, 1987.
11. Cormack DH: *Introduction to Histology.* J.B. Lippincott, Philadelphia, 1984.
12. Eroschenko VP: *di Fiore's Atlas of Histology with Functional Correlations* (8th ed.). Williams & Wilkins, Baltimore, 1996.
13. Fawcett DW: *Bloom and Fawcett: Concise Histology.* Chapman & Hall, New York, 1997.
14. Fawcett DW: *Bloom and Fawcett: A Textbook of Histology* (12th ed.). Chapman & Hall, New York, 1994.
15. Fujita T, Tanaka K, Tokunaga J: *SEM Atlas of Cells and Tissues.* Igaku-Shoin, Tokyo, 1981.
16. Gartner LP, Hiatt JL: *Color Atlas of Histology* (2nd ed.). Williams & Wilkins, Baltimore, 1994.
17. Gartner LP, Hiatt JL: *Color Textbook of Histology.* W.B. Saunders, Philadelphia, 1997.
18. Geneser F: *Color Atlas of Histology.* Munksgaard, Copenhagen, 1985.
19. Geneser F: *Textbook of Histology.* Lea & Febiger, Philadelphia, 1986.
20. Greep RO, Weiss L: *Histology* (3rd ed.). McGraw-Hill, New York, 1973.
21. Hammersen F (ed.): *Sobotta/Hammersen Histology, Color Atlas of Microscopic Anatomy* (3rd ed.). Urban & Schwarzenberg, Baltimore, 1985.
22. Han SS, Holmstedt JOV: *Human Microscopic Anatomy.* McGraw-Hill, New York, 1981.
23. Heimer L: *The Human Brain and Spinal Cord, Functional Neuroanatomy and Dissection Guide* (2nd ed.). Springer-Verlag, New York, 1995.
24. Junqueira LC, Carneriro J, Kelley RO: *Basic Histology* (8th ed.). Appleton & Lange, Stamford, CT 1995.
25. Kelly DE, Wood RL, Enders AC: *Bailey's Textbook of Microscopic Anatomy* (18th ed.). Williams & Wilkins, Baltimore, 1984.
26. Krause WJ: *Essentials of Human Histology* (2nd ed.). Little, Brown, Boston, 1996.
27. Krause WJ, Cutts JH: *Concise Text of Histology* (2nd ed.). Williams & Wilkins, Baltimore, 1986.

28. Krstic RV: *Human Microscopic Anatomy, An Atlas for Students of Medicine and Biology.* Springer-Verlag, Berlin, 1991.
29. Krstic RV: *Illustrated Encyclopedia of Human Histology.* Springer-Verlag, Berlin, 1984.
30. Larsen WJ: *Human Embryology* (2nd ed.). Churchill Livingstone, New York, 1997.
31. Leeson CR, Leeson TS, Paparo AA: *Textbook of Histology* (5th ed.). W. B. Saunders, Philadelphia, 1985.
32. Mathews JL, Martin JH: *Atlas of Human Histology and Ultrastructure.* Lea & Febiger, Philadelphia, 1971.
33. Reith EJ, Ross MH: *Atlas of Descriptive Histology* (3rd ed.). Harper & Row, New York, 1977.
34. Rhodin JAG: *An Atlas of Ultrastructure.* W.B. Saunders, Philadelphia, 1963.
35. Rhodin JAG: *Histology, A Text and Atlas.* Oxford University Press, New York, 1974.
36. Rogers AW: *Cells and Tissues, An Introduction to Histology and Cell Biology.* Academic Press, London, 1983.
37. Ross MH, Romrell LJ, Kaye GI: *Histology, A Text and Atlas* (3rd ed.). Williams & Wilkins, Baltimore 1995.
38. Snell RS: *Clinical and Functional Histology for Medical Students.* Little, Brown, Boston, 1984.
39. Stevens A, Lowe J: *Human Histology* (2nd ed.). C.V. Mosby, St. Louis, 1997.
40. Weiss L (ed.): *Histology, Cell and Tissue Biology* (5th ed.). Elsevier Science, New York, 1983.

Index

A

B

C

D

E

F

K

L

M

P

Q

R

T

U